The stamen, and its fertile pollen-bearing part, the anther, have received relatively little scientific attention, in spite of their fundamental role in the reproductive cycle of flowering plants, and their importance in interpreting plant evolution. To help begin to redress this shortcoming the contributions in this volume give an indication of the kinds of studies now being undertaken with a view to stimulating further work on this neglected plant organ. Summaries of traditional and current concepts of stamen construction, function, and terminology are accompanied by new evidence, drawn from the fossil record, for the evolution of the anther and of anther-bearing plants. Current studies of growth and structure, including detailed studies of several families, help explain the dynamics of stamen evolution. New and striking illustrations of many plant structures aid our understanding of these concepts. An exhaustive bibliography and details of techniques for investigating the anther are also included. Emerging agreement can be seen on the nature of structures and functions of anthers, but the scope of some definitions and some features of the stamen and its evolution await consensus.

THE ANTHER

THE ANTHER

Form, function and phylogeny

Edited by

WILLIAM G. D'ARCY
Missouri Botanical Garden, St Louis

and

RICHARD C. KEATING
Missouri Botanical Garden, St Louis
Southern Illinois University, Carbondale and Edwardsville

CAMBRIDGE
UNIVERSITY PRESS

CAMBRIDGE UNIVERSITY PRESS
Cambridge, New York, Melbourne, Madrid, Cape Town, Singapore, São Paulo, Delhi

Cambridge University Press
The Edinburgh Building, Cambridge CB2 8RU, UK

Published in the United States of America by Cambridge University Press, New York

www.cambridge.org
Information on this title: www.cambridge.org/9780521120036

First published 1996
This digitally printed version 2009

A catalogue record for this publication is available from the British Library

Library of Congress Cataloguing in Publication data

The anther : form, function, and phylogeny / edited by William G. D'Arcy and
Richard C. Keating.
    p.   cm.
Papers presented at the 1993 International Botanical Congress in Yokohama, Japan.
Includes index.
ISBN 0 521 48063–9
1. Anther – Congresses. 2. Stamen – Congresses.   I. D'Arcy, William G.,
1931–   . II. Keating, Richard C.   III. International Botanical Congress (1993 :
Yokohama-shi, Japan)
QK658.A57   1996
582. 13′04463–dc20   95–20222   CIP

ISBN 978-0-521-48063-5 hardback
ISBN 978-0-521-12003-6 paperback

Contents

List of contributors   ix

Preface   xi

1 **Anthers and stamens and what they do**   1
WILLIAM G. D'ARCY

2 **The fossil history of stamens**   25
WILLIAM L. CREPET AND KEVIN C. NIXON

3 **The origin and early evolution of angiosperm stamens**   58
LARRY HUFFORD

4 **Diversity and evolutionary trends in angiosperm anthers**   92
PETER K. ENDRESS

5 **Are stamens and carpels homologous?**   111
WILLIAM BURGER

6 **Heterochrony in the anther**   118
JEFFREY P. HILL

7 **Diversity of endothecial patterns in the angiosperms**   136
JOHN C. MANNING

8 **The calcium oxalate package or so-called resorption tissue in some angiosperm anthers**   159
WILLIAM G. D'ARCY, RICHARD C. KEATING AND STEPHEN L. BUCHMANN

9 **Anther adaptations in animal pollination**   192
PETER BERNHARDT

10 **Anther differentiation in the Asclepiadaceae – Asclepiadeae: form and function**   221
SIGRID LIEDE

11 Stamen structure and development in legumes with   236
    emphasis on poricidal stamens of caesalpinioid tribe
    Cassieae
    SHIRLEY C. TUCKER

12 Anther investigations: a review of methods   255
    RICHARD C. KEATING

13 A bibliography of stamen morphology and anatomy   272
    ANNA H. LYNCH AND MARY GREGORY

    Index   337

Contributors

PETER BERNHARDT, Department of Biology, St Louis University, 3507 Laclede Boulevard, St Louis, MO 63103, USA.

STEPHEN L. BUCHMANN, Carl Hayden Bee Research Center, 2000 East Allen Road, Tucson, AR 85719, USA.

WILLIAM BURGER, Botany Department, Field Museum of Natural History, Roosevelt Road and Lake Shore Drive, Chicago, IL 60605, USA.

WILLIAM L. CREPET, L.H. Bailey Hortorium, 462B Mann Library, Cornell University, Ithaca, NY 14853, USA.

WILLIAM G. D'ARCY, Missouri Botanical Garden, P.O. Box 299, St Louis, MO 63166, USA.

PETER K. ENDRESS, Institut für Systematische Botanik der Universität Zurich, Zollikerstrasse 107, CH-8008 Zurich, Switzerland.

MARY GREGORY, Jodrell Laboratory, Royal Botanic Gardens, Kew, Richmond, Surrey TW9 3AB, UK.

JEFFREY P. HILL, Department of Biological Sciences, Campus Box 8007, Idaho State University, Pocatello, ID 83209, USA.

LARRY HUFFORD, Department of Botany, Washington State University, Pullman, WA 90164, USA.

RICHARD C. KEATING, Missouri Botanical Garden, P.O. Box 299, St Louis, MO 63166, USA.

SIGRID LIEDE, Abteilung Spezielle Botanik (Biologie V) Ulm, Albert–Einstein Allee 11, W-7901 Ulm, Germany.

ANNA H. LYNCH, Jodrell Laboratory, Royal Botanic Gardens, Kew, Richmond, Surrey TW9 3AB, UK.

JOHN C. MANNING, Compton Herbarium, National Botanic Institute, Kirstenbosch, Private Bag X7, Claremont 7735, Republic of South Africa.

KEVIN C. NIXON, L.H. Bailey Hortorium, 462B Mann Library, Cornell University, Ithaca, NY 14853, USA.

SHIRLEY C. TUCKER, Department of Botany and Plant Pathology, Louisiana State University, Baton Rouge, LA 70803, USA.

Preface

The small attention given to anthers and stamens greatly understates their importance. Because they are fundamental to the sex cycle of seed plants, they underly most of the plant life we depend upon. These structures are the analogy in plants to male sex organs in birds and mammals. They are mostly inconspicuous and unseen, and for birds and mammals as well as plants, most human focus is on other parts. In addition to contributing genes for the next generation, much of the organisms' vital efforts are centered around achieving success in the mating processes that are key to their continued existence.

For botanists, stamens are important in classifying plants. Many families and genera are arranged largely or entirely on characters found in the stamens. And as the content of this volume denotes, the anther is of great interest in interpreting the evolution of both fossil and living plants. The anther is a rich source of information independent of most other kinds of study.

We know of few traditional uses for stamens, but in non-traditional ways, anthers have become economically important. As a source of haploid plants, they can be used in plant breeding to derive pure breeding lines. Such haploid plants are also used as intermediates in producing altered plants by genetic engineering. The anther is also a focus of attempts to breed male sterile plants for producing hybrid seed. Its product, pollen, is a basic foodstuff of bees and the reason for honey in beehives and on human mealtables. It also causes misery to many during hayfever seasons. But the study of pollen is another discipline, which has more adherents than the anther itself.

This book is really about the stamen. But because the spore-bearing region is the fundamental unit, our focus is on the anther itself rather than the stalk, and hence the book gets its name. One result of our approach is a series of somewhat overlapping and partly parallel treatments, where workers describe stamens and concepts in their own terms. This seeming redundancy strengthens the overall presentation, adding insight into the diversity of viewpoints from which the subject can be seen. Emerging agreement can be seen on the nature of structures and the functions of anthers, but the scope of some definitions and some features of the stamen are still awaiting consensus.

As this volume shows, upgrading of old-time methods is good for getting new information from the stamen. However, promising data is being derived from new fields of study such as gene analysis and DNA studies, gene probes, and many other methodologies and such concepts as cladistics that were unavailable only a few years ago. There is now greater integration of information from the various ways of examining floral structures, which is aided by electronic tabulation of literature sources and content. In coming years we will have a much better understanding of how anthers grow, what they do and what they mean. We hope this book will help the process.

To draw attention to the need for study of stamens, we organized a symposium which was held at the 1993 International Botanical Congress in Yokohama, Japan, that had the same title as this volume. The presentation was joined and greatly aided by Hiroshi Tobe, Kyoto University. The program was a sample of the many ways that anthers can be viewed and studied. Here we present the papers given there plus papers by Burger, D'Arcy and Hufford. In addition, methods for studying stamens and a bibliography relating to their study have been included.

The following busy scientists kindly lent their talents to a review of all manuscripts, for which the authors are indebted. Joseph E. Armstrong, Illinois State University, Normal; Herbert G. Baker, University of California, Berkeley; Peter Bernhardt, Saint Louis University; John Bozzola, Southern Illinois University, Carbondale; Sherwin Carlquist, Santa Barbara, California; William C. Dickison, University of North Carolina, Chapel Hill; David L. Dilcher, Florida Museum of Natural History; William Friedman, University of Georgia, Athens; James W. Grimes, The New York Botanical Garden; Patrick S. Herendeen, Field Museum of Natural History, Chicago; John M. Herr, Jr., University of South Carolina; Peter C. Hoch, Missouri Botanical Garden; William A. Jensen, The Ohio State University; Richard J. Jensen, St Mary's College, Indiana; Donald R. Kaplan, University of California, Berkeley; Elizabeth M. Lord, University of California, Riverside; Lazarus. W. Macior, University of Akron, Ohio; Steven R. Manchester, Florida Museum of Natural History; V. Raghavan, The Ohio State University; Jennifer H. Richards, Florida International University, Miami; David W. Roubik, Smithsonian Tropical Research Institute; Scott D. Russell, The University of Oklahoma; Warren D. Stevens, Missouri Botanical Garden; Dennis W. Stevenson, The New York Botanical Garden; Walter J. Sundberg, Southern Illinois University, Carbondale; Leonard B. Thien, Tulane University, New Orleans; Shirley C. Tucker, Louisiana State University, Baton Rouge; Garland R. Upchurch, Southwest Texas State University, San Marcos; Scott L. Wing, Smithsonian Institution.

WILLIAM G. D'ARCY
RICHARD C. KEATING

1 • Anthers and stamens and what they do

WILLIAM G. D'ARCY

INTRODUCTION

Stamens or anthers are short-lived but complex organs that show an overwhelming diversity based on a seemingly simple model. Unfortunately, they have all too often been ignored. This book presents a sample of current studies on a variety of topics relating to the anther and stamen in hope of stimulating further interest. As these chapters show, the stamen merits attention for its own sake and is worthy of inclusion in other kinds of floral study.

This discussion is a primer for readers who may appreciate learning what a stamen is and the terms and concepts used to describe its structure and function. It will serve as a guide to the literature describing stamens and anthers. For plant morphologists, much of it will seem elementary, but new conjectures may stimulate readers of canonical bent into reaction and perhaps reassessments.

Good outlines of stamen structure along with current views of its evolutionary origins and current terminology may be found in many standard texts on plant morphology (Eames 1961; Bierhorst 1971; Bold *et al.* 1987; Weberling 1981, 1989; Hess 1983; Mauseth 1988; Gifford & Foster 1989; Kaussmann & Schiewer 1989) and plant anatomy (Carlquist 1961; Esau 1965; Fahn 1990).

The stamen is the male reproductive organ in flowers, and the anther is the top of the stamen, usually elevated by means of a filament, which contains the pollen. The definitions from Webster's (1983) dictionary describe the subject of this discussion. In simple terms, the anther must do two things: it must make pollen and have it taken to the proximity of an ovule for the making of seed. In expanded terms, the anther must first produce haploid plants (that are reduced to pollen grains). And,

second, it must present this pollen-plantlet for delivery at a distance from where the pollen was made. The pollen must arrive at a stigma where sperm can be delivered to ovules which are held in a megasporangium (carpel or gynostrobilus).

To understand the stamen, we must be aware of how scientific advances are changing views of plant evolution. Cladistics has changed the processes of thinking about phylogenetic relationships. Molecular analyses are adding new information to challenge placements of plant groups based on morphology alone. The century-old discourse about the basic nature of the plant body, the cell and organismal theories, has been revived (Kaplan & Hagemann 1991; Kaplan 1992), which re-examines the concepts of homology and form that underlie most assumptions about floral structure and evolution. A valuable service in Kaplan's (1984) work is a restatement of the tests of homology provided earlier by Remane (1952).

Major recent changes in plant classification also affect the understanding of stamens. Recent analyses at the molecular level (especially of DNA sequences using the chloroplast gene *rbcL*) are rewriting the traditional evolutionary history of plant groups (Chase *et al.* 1993). Although the Magnoliales and Ranales, now Magnoliidae, remain near the base of the evolutionary tree and the Asteraceae, Asteridae, stay near the top, there are many rearrangements within the flowering plant lineage. On the basis of both morphological and *rbcL* DNA analysis, the coontail, *Ceratophyllum*, has been postulated to be the most primitive living flowering plant (Les 1988; Les *et al.* 1991; Chase *et al.* 1993). And the Gnetophytes, *Gnetum*, *Ephedra* and *Welwitschia*, are now thought to be the closest relatives of the flowering plants (angiosperms).

This outline is based largely on traditional under-

1

standing – a body of knowledge that will be reflected in botanical literature for long in the future. Keeping straight what is still valid and how changed facts affect previous thinking is a continuing challenge. Updating of traditional canons with the new tools is one of the interesting tasks for contemporary biologists.

THE STRUCTURE OF STAMENS AND ANTHERS

What to call an anther

The word 'anther' comes from the Greek *anthos*, which means a flower. The *Oxford English Dictionary* (1989) explains how the term was used in medieval times for medicines made from internal parts of flowers. By the 1700s, the word referred to the pollen-bearing, apical part of the stamen.

Pollen-producing organs are sometimes called microsporangia and the structures that bear them are called microstrobili or microsporophylls, depending on how they are thought to relate to stems or leaves. The words 'stamen' and 'anther' are more commonly used, but in English writing these terms are usually reserved for parts of flowering plants. The term 'stamen' comes from the Greek *stamos* and Latin *stamen*, meaning something standing up as a thread in a loom, which has no taxonomic implication. The taxonomic implication of flower in the prefix *anthos-* does not prevent use of the word 'anthers' for parts of non-flowering seed plants (gymnosperms), since the word 'antheridium', from the same Greek stem, is used for sperm-bearing structures of non-seed plants.

Rather, some botanists have thought that the microsporangia of non-flowering seed plants are not homologs of the ones in flowering plants and, therefore, they should not have the same names. However, the corresponding parts in most gymnosperms, such as *Gingko*, *Ephedra* and conifers (Mauseth 1988: 396), resemble those in flowering plants. They function in much the same way as stamens, anthers and filaments, and they may turn out to be homologous in origin. Thus, these words are applied less because of their own nature than because of the kind of female parts their products pair with: if they supply pollen to ovules in carpels (flowering plants), they are anthers and stamens, but if their pollen goes to naked ovules (gymnosperms), they are microsporangia, parts of cones or strobili. Non-English literature sometimes fearlessly refers to gymnosperm microsporangial elements with words translating as 'stamens' and even sometimes as 'anthers' (Kugler 1970: 47; Kunze 1978; Martens 1971; Vidakovic 1991).

What did anthers come from?

When and in which ancestors stamens first arose is unknown. Other distant relatives of the seed plants – bryophytes, *Equisetum*, *Isoetes*, *Lycopodium*, *Psilotum* and ferns – differ in life cycle and structure. None of them bear organs much like stamens nor do they produce pollen grains. They are thought to have diverged from the seed plant line over 350 million years ago (Chase *et al.* 1993: 537). The fossil groups of seed plants that seem most likely to be direct ancestors of the flowering plants (Crepet & Nixon, Hufford, this volume) are found in strata of mid-Cretaceous and earlier ages. Many bore stalked microsynangia that might be interpreted as homologous with or are somewhat similar to modern anthers (Rothwell & Scheckler 1988: 110; Taylor 1988; Burger, Hufford, Crepet, this volume), but the present fossil record is too incomplete to be useful in identifying the male strobili ancestral to those of flowering plants.

Life cycles

Most of the organisms we see around us, plant and animal, follow a life cycle with an overall similarity; the conspicuous phase is diploid, the nucleus of each cell having a double set of chromosomes, one from each parent. In plants this phase is called a sporophyte. The process of meiosis separates the chromosomes and leads to a haploid phase of minuscule proportion. The male haploid cells – pollen grains or sperm – travel to female haploid cells – egg cells – that are retained in their parent structure. These features are analogous to sex in animals and underlie the concept of male and female sexuality in plants. The pairs of haploid cells fuse in an action called fertilization or syngamy. Fertilization results first in a one-celled organism (zygote) with a nucleus that combines the chromosome sets of both haploid parent cells. This is the first cell of the new diploid organism that grows into a new conspicuous sporophyte generation.

Seed plants (spermatophytes) have an important difference from other organisms with this life cycle. In seed plants, the male haploid phase or microgametophyte, called pollen, develops three to six cells before releasing two sperm nuclei. These two sperm then fuse with female haploid cells located in a megagametophyte. The pollen is carried from the parent by various agents to near the ovule where the sperm can be released. Through an opening (aperture) or weakness in the pollen wall, a tube emerges carrying the sperm toward the ovule (siphonogamy). Flowering plants differ from this general seed plant life cycle because both sperm fuse independently with haploid cells in the retained ovule. This creates a new diploid generation and a supporting nutritive endosperm tissue. Double fertilization has also been reported in some gymnosperms (Knox *et al.* 1994), but no endosperm is formed.

Pollen was a major innovation in the grand success of seed plants. The delivery of sperm in pollen grains with durable walls permits their safe aerial transfer within a desiccating atmosphere. Land plants without pollen are restricted in their sexual phase to habitats or atmospheric conditions where the sperm can move independently through at least a film of water to reach the egg cells. The haploid, pollen phase in seed plants is sometimes regarded as being reduced through evolution from unknown ancestors with more complex haploid bodies. The female structures and the female haploid phase (megagametophyte) are quite different in development from the stamens and pollen in the male sex (see Burger's postulation in this volume that stamens and carpels are not homologous).

The seed plant life cycle includes two expensive and risky stages when the plant is wholly dependent on outside agents. These are during delivery of pollen and during the dispersal of seed. At both points in the life cycle, plants usually create large numbers of dispersal units. Each unit has a low individual chance of success, but even this low success rate permits continuation of the species. Pollen must contact and adhere to tiny areas of a stigma of the same species at a time when they are receptive, while seeds usually have a much wider latitude in target area and timing. Seeds also face constraints such as predation and germination needs, which while much broader in general nature, are often severe. Different determinants for success should be expected for these

two dispersal phases. This book discusses arrangements for the launching of pollen from the stamen in ways that guarantee delivery of some of it to appropriate targets.

Spermatophytes are commonly divided into gymnosperms and angiosperms. This reflects a major discontinuity in the evolutionary history of ovule-bearing structures. The discontinuity is signaled by two innovations in flowering plants: the closed carpel and the presence of an endosperm. This evident discontinuity is not mirrored in the history of pollen-bearing structures. Details of microsporangia and cell development in microspores differ in the major groups (Divisions) of seed plants, but all seed plants have male structures much like stamens and anthers. These structures are similar both in plants with flowers and in plants without them (gymnosperms).

Environments of the various plant lineages have changed and reversed many times over geological time, and trends in stamen evolution have changed and reversed as well. This hampers perception of trends in anther structure. Remarkably, however, the historical thread of evolution in stamens can still be traced in considerable detail. Some consequences of this are noted by Endress (this volume).

The classical and other theories of plant design

Botanists have digested or discarded many theories to explain the evolution of plant structures. One central theory that has kept its adherents through controversial periods is the classical theory of plant development and homology. Some botanists question its validity for stamens, but many still apply it to stamens as well as to other parts of the plant.

Classical theory begins with Goethe's 1790 *Essay on the Metamorphosis of Plants*. This work has been translated into English (Arber 1946; Mueller 1952) and other languages. Goethe held that the life of the flowering plant has six stages alternating in size: seed, stem leaves, calyx, corolla, sex organs and fruit. Axillary buds and shoots are seedlings that were not formed sexually. More fundamentally, the plant is made up entirely of leaves that progressively differentiate (metamorphose) through the six stages, with intergrades and reversals sometimes occurring.

In this theory, as later developed (Eames 1931; Can-

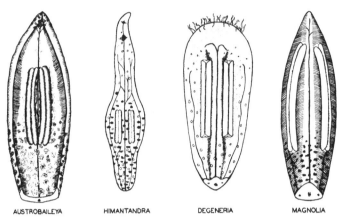

Fig. 1. Examples of laminar stamens. *Austrobaileya* and *Magnolia* are shown looking out from the centre (adaxial views), and *Galbulimima* (*Himantandra*) and *Degeneria* are viewed from the outside of the flower (abaxial views). From Canright (1952).

right 1952) and eloquently explained by Eyde (1975), stamens are homologs of leaves. Hence those resembling foliage leaves are primitive (plesiomorphic). Such stamens tend to be laminar, that is, flat, broad and leaf-like (Fig. 1) They are widespread in many plants that constitute the Magnoliidae. Likewise, stamens whose vasculature suggests that of leaves – with more than one vascular trace – are also primitive. These occur in Magnoliales and some Nymphaeaceae. From such laminar, multiveined stamens, all other stamens can be derived along multiple lines. The formation of anther and filament zones is derivative (Canright 1952: 488).

The long-standing classical theory was the centerpiece of views on plant morphology championed by Eames (1931, 1961) and others during much of this century. It has been discussed in detail by its proponents (Arber 1937; Eyde 1975) and by advocates of conflicting theories. The conflicting telome and gonophyll theories focus on stems or strobili rather than leaves or phyllomes. The classical theory has also been applied to gymnosperms. In conifers, the microsporangium-bearing structure is 'the pollen cone [which] is a simple shoot axis whose lateral appendages are microsporophylls instead of photosynthetic leaves' (Mauseth 1988: 385).

One challenge to the classical theory (Thomas 1932, 1936) held that the reproductive and foliar structures are independent lines and cannot be regarded as homologous.

The telome theory (Wilson 1937, 1942; Zimmermann 1952, 1959; Stewart 1964) proposed that the vascular plant is a system of branches, not leaves alone as in the classical theory. The uppermost parts, telomes, may be sterile or fertile. The stamen is derived from a complex branch system bearing terminal sporangia. Plants with fasciculate or branched stamens are thus basal in the angiosperms. This theory traces the origins of the anther through an ancestral line that includes all known fossil and living plants. It may support the Batyngia–Davis classification of anther types that is discussed and questioned below.

The gonophyll theory (Melville 1960, 1962, 1963) modified the telome theory in an attempt to harmonize it with the classical theory. In this theory, the flower arises from condensation of inflorescences of ancestral gymnosperms. The basic fertile structure is called a gonophyll. It is a leaf-borne fertile branch. This implies that laminar stamens are homologous with branches or branch–leaf complexes rather than with leaves. Because this theory held that the angiosperms and their vascular systems are polyphyletic, the laminar stamen is not necessarily basal in the angiosperms.

Literature dealing with variations of the above theories is indicated by Ronse Decraene & Smets (1993a, b). An iconoclastic review of many issues relating to floral theory was presented by Carlquist (1969). The rebuttals that followed (Kaplan 1971; Schmid 1972) left most of his dicta undiminished. The nature of the flower and terms for its parts are reviewed by Esau (1965) and others.

Stamen number and arrangement

Stamens together form the androecium (Greek *andros* = man, *oicos* = dwelling). They occur singly or in multiples, free standing, or variously fused to one another and to other floral parts. Surveys of androeciums with terminology for their many parts and conditions are found in some taxonomy texts (Lawrence 1951; Cronquist 1968). Good information is also found in morphology texts (Eames 1961; Weberling 1989), in botanical dictionaries (Jackson 1916; Font Quer 1970) and in glossaries to floras. Good discussions of stamen arrangement and number are given by Eames (1961), Cronquist (1968) and Weberling (1989).

Most flowers are perfect, containing both stamens and carpels. In the many dioecious or monoecious plants, the androeciums and gynoeciums are found in different flowers called staminate or pistillate. The perfect or staminate flower commonly includes several stamens, some of which may be reduced to infertile structures called staminodes.

In perfect flowers, the stamens are almost always situated outside (below) the gynoecium, but in the recently described *Lacandonia* (Márquez-Guzmán *et al.* 1989), the positions are reversed, which might be thought to contravene the classical theory.

The classical view holds that stamens in primitive plants are arranged spirally on the floral axis and are of variable number. Androeciums with numerous stamens are also somewhat primitive, while stamens of fixed number in one or more cycles or whorls are more advanced. The arrangement of stamen whorls in relation to one another and to other parts of the flower has its own terminology. Diplostemony, the commoner case, is where stamens alternate with petals or corolla lobes. Obdiplostemony is where stamens are opposite the petals. Such contrasting arrangements and their variants are not obviously explained by the classical theory. Stamens and stamen whorls have divided or multiplied (*dédoublement*: Moquin–Tandon 1826) and reduced or fused in various groups (Kunze 1978; Tucker, this volume). These are common kinds of evolutionary rever-

sal or parallelism in stamens. The subject, included under the terms *dédoublement* and polygenesis, was recently reviewed by Ronse Decraene & Smets (1993a, b). Which floral whorls actually arise from floral primordia may be under the control of few genes (Mandel *et al.* 1992).

In some cases, stamens have modified into petals and perhaps vice versa (*Nymphaea* – Goethe 1790; Cronquist 1968: 90; Crepet & Nixon, this volume). Support for this comes partly from teratological, or aberrant, flower forms where such shifts of the classical whorls can be seen to have taken place. Whether stamens develop centrifugally or centripetally has long been given great systematic weight (Payer 1857: 4) and used to arrange families of dicotyledons into high taxonomic groups (Cronquist 1968). In a continuing series of publications, Ronse Decraene & Smets (1987, 1993a, b) reduced the many arrangements of stamens to two contrasting types. They derive all extant types from positions of primordia that show alternations of whorls as in the classical theory. These authors re-examine classification of dicotyledonous family groups in light of their conclusions, and provide new terminology to describe their concepts.

In species with heterostyly, two or three classes of individuals differ in style length and usually in stamen insertion or morphology, sometimes including different classes of pollen.

Parts of the stamen

Although angiosperm stamens come in many forms and configurations, they are all fashioned on the same groundplan and most retain some features of this foundation. Above the basal portion four parallel sporangia are separated into two pairs, each called a theca (Green 1980). Each theca has two sporangia or locules. The sporangia generally open lengthwise with the aid of a mechanical tissue called an endothecium. Whether or not the stamen differentiates into a filament and anther are modifications of this groundplan.

Some stamens lack distinct filaments. In some cases these have resulted from reduction of normal, slender filaments and consist of sessile anthers. Other cases are thought to antedate the evolution of filaments. These, referred to as laminar, consist of more or less flat, almost leaf-like, elongate structures. The sporangia are often

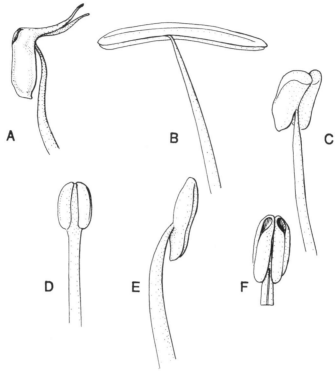

Fig. 2. Anther types and types of fixation on the filament. A. *Enkianthus cernuus*. The apex of the anther points down and the spurs or tails point up. Although resembling a versatile anther, it is actually rigid, fused for a short distance to the filament apex. After Wilson 6895 MO. B. Stamen of *Epilobium speciosum*: long, straight, horizontal, versatile anther. C. Stamen of *Rosa multiflora*: short, curved versatile anther. D. Stamen of *Ranunculus abortivus*: small rigid anther. E. Stamen of *Salvia splendens*. The anther appears to be versatile but is actually rigid, fused for a short distance to the filament apex. After Hess (1983: 298). F. Stamen of *Solanum umbellatum*: rigid, porose anther.

embedded in the surface of the apical part on either inner (adaxial) or outer (abaxial) surfaces, or they may be marginal (Fig. 1) (Eames 1961). Such laminar stamens may have multiple vascular traces (Eames 1931; Bailey & Nast 1943; Canright 1952; Sampson 1987), and they tend to be stiff and unbending. Most kinds of laminar stamens have been described and illustrated in the series of papers by Endress (see the bibliography by Lynch & Gregory, this volume). Laminar stamens have long been regarded as the basal angiosperm type, but this is now being questioned (Burger, Hufford, this volume). In this book, Hufford presents an improved classification and terminology for the arrangement of parts of rigid anthers.

More common than laminar stamens are stamens differentiated into an anther and a basal stalk or filament (Fig. 2). The anther thecas or microsporangia are sometimes referred to as lobes or pollen sacs. In divided anthers this term often means only the basal, separated portions of the thecas, which might perhaps better be called tails or spurs. While most anthers develop four microsporangia or locules, by maturity (anthesis) the paired locules of each theca are usually confluent and then often appear as two pollen chambers rather than four. A few families (Endress & Stumpf 1990) have more sporangia. Eight locules were mistakenly reported in *Bixa* (Davis 1966: 7). In this genus, the anther is bent over (Keating 1972: 293) and serial sections usually cut twice through each theca giving the idea of eight locules when, in fact, there are only four. In other cases the thecas may be chambered, either crosswise or lengthwise (Tobe & Raven 1986a; Mauseth 1988: 397). Stamens, especially in zygomorphic flowers, often have unequal or suppressed anther lobes, and when they are reduced and infertile, they are called staminodes.

The interior of the anther is differentiated into several regions (Fig. 3). Against the epidermis in the arc surrounding the theca is a mechanical tissue called endothecium that is capable of hygroscopic movements. Opening (dehiscence) is usually effected by the endothecium, which also acts as a stiffening framework to keep the sporangium wall rigid as pollen is shed. In many cases, thecal arcs of the two thecas open wide and flatten against the arcs of the opposite thecas to stand rigid in a plane at 90 degrees to the plane of the septums (Richter 1929). Anchoring of the endothecium is often improved by thickening at the inner side of the anther, which in some cases extends inward beyond the thecal arc. An endothecium is absent or reduced in anthers where its mechanical services are not needed, such as in submerged, aquatic flowers, cleistogamous flowers (Lord 1981: 425), and in tubular, poricidal anthers. A thickened endothecium layer is reported to be lacking in some small families in the myrtalean alliance, yet it is present in many monocots with porose anthers (Gerenday & French 1988). In some monocotyledons (*Mayaca*, *Xyris* – Stevenson 1983) a fibrous-thickened epidermis acts to open the anther. Nature is never without exceptions!

Anthers that open lengthwise do so at the stomium (Greek *stoma* = mouth), which is the outer region between the paired locules of each theca (Hufford & Endress 1989). It is usually marked by a stomial groove. Even when the locules of a theca join into one, the endothecium is discontinuous between them at the stomium.

The term connective, in the broadest sense, includes all internal parts of the anther except the sporogenous parts. In a narrower sense, it includes all except the thecal arcs, and in a still narrower sense it includes only the part nearest the central vascular trace and between the two thecas. Because of such varied usage, it is probably best to use the term in a broad sense and use other terms for more specific regions of the anther interior. From outside, much of the connective often appears like

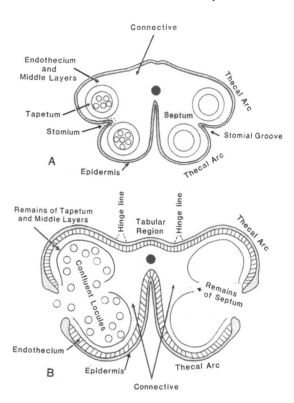

Fig. 3. Anther of *Lilium* showing major features. A. Developing anther. B. Dehisced anther. The endothecium is braced across the dorsal side in a tabular configuration. Redrawn from Haupt (1953: 379).

a continuation of the filament running between the two thecas. It usually includes a continuation of the vascular supply coming from the filament.

The septum is the part of the connective that passes between the two locules of each theca (Matthews & Knox 1926: 240), and the cortex is the part that surrounds the central vascular trace. In some families – many members of the Tubiflorae (Chatin 1870; Hartl 1963; Weberling 1989) and some Centrospermae – expansion of the septum into septal arms or placentoids make the locules 'C'-shaped. Use of the term septum has also been extended to refer to connective tissue between locules in multi-sporangiate anthers (Lersten 1971; Skvarla *et al.* 1975). At the outer end of the septum there is usually an undifferentiated opening tissue that breaks down before confluence of adjacent locules. In a few families, this is replaced by a calcium-oxalate-accumulating tissue (D'Arcy *et al.*, this volume).

Inward of the endothecium in the thecal arc is the outer tapetum. Flanking each septum is an inner tapetum that is continuous with and resembles the outer tapetum. The tapetum surrounds and apparently furnishes material such as sporopollenin to the sporogenous cells. It collapses (shrinks or disintegrates) before pollen release. The tapetum and pollen formation is a major subject in itself beyond the scope of this book.

In some anthers, an extension of the connective forms an appendage or apicule. This distal extension of the anther was once thought (Parkin 1951) to be evidence for the leaf-like origin of the stamen. It was also thought to support the telome theory. In most cases, it is probably derived (Canright 1952; *Skytanthus* – Fallen 1986: 274).

Narrow filaments often contract at their apex into a slender neck, which acts like a hinge allowing the anther to be versatile, that is, to swing around freely. Anthers joined to the filament at their lower end are basifixed, and those joined at the middle are medifixed. Versatile anthers are often attached above the base, while the bottom part of the anther is divided into two lobes or tails. When the anther appears to be attached on the abaxial side it may be termed dorsifixed, on the adaxial side ventrifixed. In such cases, the determination often rests with the appearance of the surface of the anther and not on any histological specialization within it. These conditions are sometimes variants that grade into basifixed or medifixed insertion. Unfortunately, most texts have illustrated versatile anthers by showing medifixed anthers (Fig. 2B). While this is the commonest type, basifixed anthers also are sometimes versatile (*Cecropia, Cyclamen, Physalis, Tropaeolaeum*). Their filaments have a clearly identifiable apical neck. Filaments lacking an apical neck have rigid or fixed rather than versatile anthers. Some examples (*Hamamelis, Ranunculus*) are in groups not much above basal angiosperms in current angiosperm phylogeny (Chase *et al.* 1993). Other examples with slender filaments are stamens dispersing pollen by wind (*Pinus, Salix, Quercus*). These filaments are generally flexuous but lack an apical neck. Some groups with mainly versatile anthers have exceptions with fixed anthers (*Barleria, Salvia*) that may

have arisen recently in their evolution. Eames (1961: 113) employed different terminology for some of these conditions, but his nomenclature has not been adopted. For example, he restricted the term basifixed to slightly modified laminar stamens where anthers are rigidly fixed on distinct filaments.

Filaments sometimes separate from the flower, launching the whole anther rather than just the pollen. In a few cases, filaments separate basally from the flower axis and set the anther free (*Ceratophyllum*). However, in most cases of anther dispersal (orchids, Asclepiadaceae, *Cecropia*) separation occurs at the filament apex. Filaments are sometimes pubescent or glandular and may have appendages at the apex (*Mahonia*) or base (*Viola*) and other modifications (see Bernhardt, this volume).

The simplest kinds of stamens are free from one another and from other parts of the flower. Many flowers have stamens with various kinds of fusion. In simpler kinds of fusion, the basal part of the filament is fused (adnate) to the perianth. In these, parts of the perianth are also fused (connate). In such cases, the flower may be perigynous or epigynous, that is, with the stamens arising around or above the ovary.

Filaments connate into a tube around the style are monadelphous (many Iridaceae, Malvaceae, *Nierembergia*–Solanaceae) or, if united into two groups, diadelphous.

Anthers are connate or coherent into tubes in some groups (Lobeliaceae, Calyceraceae, *Echeandia*–Anthericaceae, *Lycopersicon*), and this is characteristic of one of the largest of plant families, the Asteraceae. In some Apocynaceae, this connation involves sclerified wings (Fallen 1986: 274). In most Asclepiadaceae and orchids, stamens and the style are fused into a single column or gynostegium or gynostemium. In these two groups, pollen is agglutinated within each anther locule into massulae (Arditti 1992: 115) forming a single pollinium.

There is general agreement that stamens with more fusion are more advanced than free stamens. This evolutionary trend does not entirely mirror views about primitive and advanced plant groups, whether traditional or recent classifications are used. Thus in the basal Myristicaceae there is extensive fusion of anther and receptacle (Armstrong & Tucker 1986). Filament fusion in many Malvaceae and some Solanaceae (*Nierembergia*) is quite similar, and anther fusion in some Solanaceae

(*Lycopersicon*) and many Asteraceae is quite similar. These family pairs are usually regarded as representing quite different levels in flowering plant phylogeny.

Tissues of the stamen

Stamen histology and development mostly resemble those of other parts of the plant, notably the leaf. However, some features related to the basic functions of the anther do not appear elsewhere in the plant body. Schmid (1976) provided a useful review of tissues in the stamen, and the subject is surveyed in many standard texts.

The epidermis, which is continuous over both anther and filament, is derived from a single set of initiating cells that divide as the organ expands. The epidermis may have stomata, especially on the connective (Eames 1961: 108; Mauseth 1988: 395), and it may bear trichomes. In a few cases it emits scents (osmophoric) (*Cyphomandra* – Sazima *et al.* 1993). Anthers seldom or never have nectaries (Bernhardt, this volume), but in a few cases they exude oils as a pollinator reward (*Mouriri* – Buchmann & Buchmann 1981). At times, the epidermis over the thecal arc has been referred to as an exothecium (Mohl 1830; Purkinje 1930; Weberling 1989), which sometimes includes the endothecium (Staedtler 1923). Gymosperms are sometimes considered to have an exothecium and angiosperms an endothecium, but this concept is unclear. In *Mayaca* and *Xyris* (Stevenson 1983 and pers. comm.) a fibrous-thickened epidermis has been called an exothecium.

Just beneath the epidermis, other parts of the anther are produced from a subapical, tunica portion of the apical meristem. Different regions of the anther have various patterns of differentiation. Most study has been devoted to developments in the thecal arc. In this region, the layers underlying the epidermis are derived from the same initials as the sporogenous tissue (archesporial cells). The first layer inwards divides periclinally into a primary parietal layer and the inward sporogenous layer. The primary parietal layer divides further into two secondary parietal layers, which divide further in various ways. Hill (this volume) discusses tissue origins of these layers.

Division of the secondary parietal layers was outlined for the Ericaceae and Gramineae by Batygina

et al. (1963). Soon after, Davis (1966: 9) extended these generalities into a general scheme. She named four types – Basic, Dicotyledonous, Monocotyledonous and Reduced – according to how cells in this layer divide. Later workers (e.g., Gupta & Nanda 1978: 397; Palser 1975; Siddiqui & Khan 1988) have often followed this classification. However, the scheme has been of limited utility, and Hill (this volume) questions its validity. The types do not sort into taxonomic alignments at ordinal or class levels. In at least six families (Davis 1966), and some orders (Tobe & Raven 1983), more than one type has been reported. Phylogeny imputed to this four-variant scheme rests partly on an assumption that these cells are not subject to much selective pressure, and evolution in them must be extremely conservative. Therefore, the layers must reflect homologies and a great degree of homoplasy. However, as noted later, the thecal arc region functions differently in different angiosperms. Reversals may have been frequent, and the elegant postulated *a priori* phylesis classification of Batygina–Davis is suspect. It exceeds probability that details of these cell divisions in all the angiosperms and predecessor plants should be homologous, except in the general sense that non-sporogenous tissues of the microsporangium may be homologous.

A conspicuous product of the secondary parietal layers in most anthers is an endothecium with thickened, patterned cell walls (Manning, this volume). This tissue frequently extends into other parts of the anther. Many writers (Hufford & Endress 1989) restrict the term endothecium to the thickened cells in the thecal arc, and others (Tobe & Raven 1984, 1986b) use it to mean the hypodermal layer in the thecal arc, whether or not it is thickened. Restriction of the term to cells in the thecal arc ignores functionality. And it expresses the idea that cells in the thecal arc are somehow different from – not homologous with – cells in other parts of the anther. It assumes that the primary parietal layer, which spawns the sporogenous layer, is not homologous with layers in other parts of the anther, and it assumes an unwarranted determinism for the hypodermal layer. It is just as tenable to regard the parietal layers simply as parenchyma forming part of the connective tissue of the anther (in the broadest sense). From this, specialized cells, such as the sporogenous layer, the endothecium, septum or cortex, may be derived. Such cells are all initiated from the tunica layer beneath the epidermis. The homology of the fibrous layers in the thecal arc with those sometimes present in other parts of the anther meets all three of Remane's (1952) criteria for homology: equivalent position, equivalent quality, and connection to one another by intermediate forms. To paraphrase Carlquist (1969: 354), when we look at structures as biological phenomena rather than primarily as histological ones, we discover some interesting features not hitherto appreciated.

In seed plants, the term endothecium has from its inception (Purkinje 1830) been used mainly for the mechanical tissue in the anther. An endothecium is found in angiosperms and some gymnosperms (*Gingko* – Jeffrey & Torrey 1916). The mechanical tissue in *Zamia* (Jeffrey & Torrey 1916) and *Gnetum* (Maheshwari & Vasil 1961: 62) is external around the anther. In bryophytes the term refers to the entire sporangial area (Limpricht 1890; Blomquist & Robertson 1941: 571; Kreulen 1972: 68).

The seed plant endothecium is often one cell layer thick and confined to the thecal arc. In many anthers, however, extra anchoring for its mechanical operation is provided by a thickening to several layers, sometimes with extension inward from the thecal arc. Patterns of endothecium thickening usually follow one of two dissimilar-appearing configurations. In many anthers the thickened mechanical tissue extends across the connective, often joining to form a pair of tabular plates (conceived three-dimensionally), one dorsal and one ventral, which hinge at the inner edge of the thecal arcs (Fig. 3B). In the other configuration, the mechanical tissue is thickened inward and curls around the inner side of the locules, sometimes running along part of the septum. In three-dimensional view, this endothecium is canaliculate, resembling a trough or gutter (Figs. 3, 4). The tabular extension of endothecium into the connective region is perhaps more common in large anthers such as in *Magnolia* (Canright 1952). Many examples in the Hamamelidae are illustrated by Hufford & Endress (1989). In the divided basal lobes of many medifixed anthers, the tabular configuration is not possible: there is no tissue between the lobes. In *Schizanthus* and probably other taxa, the endothecium extends plate-like between the thecas in the apical part of the anther. Lower, where the anther is divided, it curves around the inner side of the lobes, or tails (Fig. 4B).

The extended canaliculate endothecium seldom joins two locules of a theca, but in a few cases (*Cuscuta* – Batygina *et al.* 1987: 202) it is continuous on the inner side.

Often termed spirally thickened or fibrous in earlier literature, endothecial cells actually have U-shaped rather than spiral thickenings. They are described in detail in this book by Manning, who evaluates their systematic importance.

The remaining cells in the thecal arc, the secondary parietal layer or its derivatives, are usually crushed (or perhaps resorbed) against the rigid endothecium and are usually not evident in the mature anther. Some exceptions have been reported (*Ranunculus, Lilium* – Mauseth 1988: 396).

In the central connective region, cell layers immediately beneath the epidermis derive from peripheral layers of the connective. They either specialize as spirally thickened cells forming part of the endothecial structure, or they remain as undifferentiated, parenchymatous ground tissue. Often parts of the connective which outwardly appear quite distinct from other parts of the anther show no differentiation at all within when examined by serial sections. Similarly, parts of the connective which are scent producing (osmophoric) may have no evident internal differentiation, although lipid stains may provide evidence in the epidermal layer.

In the stomium, the hypodermal and immediately subhypodermal cells usually disintegrate leading to confluence of adjacent locules. In some taxa, they develop an oxalate package of crystals (D'Arcy *et al.*, this volume).

The tapetum is continuous around the locules but is reported to be of two distinct origins (Periasamy & Swamy 1966; Gupta & Nanda 1978; Nanda & Gupta 1978). The outer tapetum, found in the thecal arc, derives from the division of archesporial cell parietal layers; the inner tapetum is derived from parenchymatous ground tissue flanking the septum. In anthers with C-shaped locules, the outer tapetum shows little change during pollen development, but the inner tapetum may conspicuously expand and contract again. As the inner tapetum undergoes contraction, the septal arms often retract as well. In small anthers, they may completely disappear, but in anthers of larger diameter they often persist into maturity of the anther (D'Arcy *et al.*, this volume). The tapetum is one of the few usually poly-

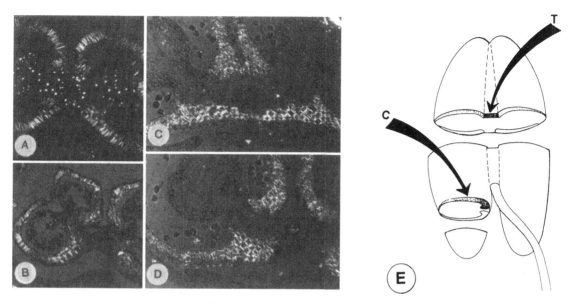

Fig. 4. Cross-sections of anthers with endothecium in canaliculate and tabular configurations. The dorsal (abaxial) side of the anther follows the bottom edge of each photograph. Light is partially polarized, and bright areas are endothecium. A. *Petunia hybrida*. Cross-section from near the anther apex. In most of the thecal arc endothecium is hypodermal and one layer thick. Towards the center of the anther, as it nears the connective, the endothecium thickens to more than one cell, and then it reverses direction to run into the septum. When viewed in three dimensions, this structure is trough-like, or canaliculate. In this species, the endothecium configuration is canaliculate from base to apex. B. *Dunalia lycioides*. Details are as in A (*Petunia hybrida*). C.D. *Schizanthus pinnatus*. Cross-sections of one anther at different vertical levels (heights). C. Cross-section near the apical (distal) end of the anther. The endothecium on the dorsal side of the anther is tabular across the anther. On the ventral side, endothecium from the two thecae unite but do not form a tabular plate. D. Cross-section about halfway down the anther shown in C. The lobes are separated by the narrow filament (center right). At this level and below the endothecium is canaliculate. The endothecium is greatly thickened near the connective and extends into the area of the septum. E. Endothecium configuration from *Schizanthus:* canaliculate (C), tabular (T).

nucleate seed plant tissues (Mechelke 1952; Carniel 1963; Periasamy & Swamy 1966; Mauseth 1988: 386). Other polynucleate tissues are the nucellus of some gymnosperms and the endosperm in many groups of flowering plants. After pollen is developed, the tapetum undergoes necrosis. In longitudinally dehiscent anthers, a break in the stomium permits confluence of the adjacent locules and egress of the pollen (Keijzer 1987a). Within the tapetum boundary, sporogenous cells develop into pollen, and the tapetum acts to supply the developing pollen in a series of steps outlined by Keijzer (1987b).

An analogy with the derivation of the endothecium in the thecal arc and elsewhere can be found in the origin of the inner and outer tapetums. They are mostly similar in appearance and perhaps in function. Gene-tracing in the tapetum by Koltunow *et al.* (1990) did not disclose differences between inner and outer tapetums. There have been no proposals to restrict the term tapetum to the outer tapetum, which alone derives directly from the primary parietal layer. Although only outer tapetum derives from the secondary parietal layer, the inner tapetum does derive from initials antecedent to the secondary parietal layer, so views about non-homology of the tapetum parts may be specious.

Most filaments have one vascular trace (Wilson 1942; Schmid 1976), but sometimes there are more. Laminar stamens may have multiple vascular traces, and three-veined stamens have been postulated as primitive (Bailey & Nast 1943; Eames 1961; Sampson 1987). The vascular trace supplying the filament usually ascends unbranched to near the apex of the connective, but sometimes one or even two branches curve out of the cortex region. Branched vasculature in stamens was used by Wilson (1942) as evidence to support the telome theory. Hufford (1980) found that the extent of branching of stamen bundles varies with the form of the stamen. In most of the Ericales (Matthews & Knox 1926) and some members of the Melastomataceae (Wilson 1950), the vasculature bends within the cortex region, reversing with the orientation of the anther (Fig. 5). Thus, the morphological apex becomes basal and the base and its lobes become distal. Esau (1965: 549) discussed parenchyma associated with filament cells. In the vascular bundles, phloem surrounds the xylem (amphicribral – Schmid 1976).

Beyond the above generalized statement about stamens, much could be said about exceptional cases and abnormalities. One such subject is cleistogamous flowers. 'A common denominator that characterizes flowers in all cleistogamous species is a reduction in the corolla and the androecium, usually of anther size and/or number in the latter' (Lord 1981: 441).

Fig. 5. Diagram of anther of *Kalmia latifolia* showing the bend in the vascular trace typical of the ericalian alliance of plants. After Matthews & Knox (1926: 248)

THE FUNCTIONING OF ANTHERS AND STAMENS

Stamens present pollen in different ways

Recognition of sexuality in plants and the need for pollination, reviewed by Baker (1983), is widely attributed to a letter published in 1694 by Rudolf Jakob Camerarius (Faegri & van der Pijl 1979: 1). Cross-pollination and the role of insects were noted by several workers in the second half of the eighteenth century. This information became common currency as Charles Darwin in England and Frederico Delpino in Italy published many observations on pollination in the second half of the next century. Students of pollination then became more numerous, and in 1898–1905, Knuth published a monumental three volumes of pollination observations that is still useful. Now many workers study pollination from diverse viewpoints. Efforts to improve seed set of crops and honey production have focused on activities of the honeybee, *Apis mellifera*. Plant systematic studies now examine pollination relationships for taxonomic clues and to explain evolutionary shifts. Stamen function is usually considered under the subject of pollination, for which there are a number of texts (Percival 1965; Faegri & van der Pijl 1979; Leppik 1977; Jones & Little 1983; Wyatt 1992).

The evolution of flowers in response to pollen vectors has received considerable attention. But how stamens have evolved in this theater has hardly been addressed. The following overview looks at salient features of the process. Much of the story is missing, and some speculations are added where support for them seems likely to be found later. This presentation is more a *pro forma* outline of what should be studied and documented than a completed analysis.

In the epochs since pollen evolved as a vehicle for delivering sperm for fertilization, plants have evolved many systems of pollen delivery. Because pollination is vital, it is not surprising that many plants use more than one mode of pollen presentation. Plants often retain earlier evolved pollination capabilities for a long time. By ingenuity or by hazard, many insects visit and pollinate flowers for which they are poorly adapted pollinators, thus blurring specialization relationships with better-adapted major pollinators. Stamen structure reflects a complex history of shifts and reversals. There is overwhelming diversity in modifications of the pollination equation – the match of stamen structure and pollinator. Yet, the great majority of stamens are of relatively few types.

Direct and indirect pollen delivery

Direct pollen delivery, where air, water or animals collect pollen from anthers and deliver it to stigmas, is usual. Less commonly, pollen delivery is indirect, where pollen is placed outside the anther on other, often specialized, surfaces for uptake by pollinators. Indirect pollen delivery is discussed by Bernhardt (this volume) and by Yeo (1993). Yeo noted two major kinds of indirect delivery. In one kind, all the pollen is delivered at once, often explosively, usually following an insect visit. More often, pollen is exposed on other parts of the flower and taken up in several doses. Movement of pollen by the style occurs in many flowers that present pollen indirectly. Only a few families (25) were said by Yeo to have indirect pollination, but some of these families are large, e.g., Asteraceae, Leguminosae, Rubiaceae, with many species using these systems.

Inanimate delivery

In flowers with air transport of pollen (anemophily), most stamens have a well-identified filament and anther, and the filaments are long, slender and flexible. Pollen quantity is large, and pollen tends to be dry. There may be an ideal size range for pollen success (Paw & Hotten 1989). Anthers are usually in elevated or pendant structures (cones and catkins) where they meet air currents. Delivery distances are often of kilometer magnitudes, and development of pollen is often seasonal. Dioecy is common in anemophilous species. Most gymnosperms (conifers, *Ephedra, Gingko*) are wind pollinated, but even here, anemophily is regarded (Robertson 1904; Vogel 1981) as deriving from other pollination modes. Hence its structural attributes are modifications of forms previously adapted for other kinds of delivery. Niklas (1987) was puzzled by the imprecision of aerodynamic adaptation in many kinds of pollen he examined in studying reception of air-transported pollen. Anemophily dates from near the beginning of the fossil record for stamens

(Chloranthaceae, see Crepet & Nixon, this volume). Both indirect (*Artemesia*) and direct (*Plantago*) delivery systems have yielded to anemophily more than once.

Water delivery of pollen (hydrophily) is relatively rare and is confined to plants that grow in the water. Flowers presenting pollen for water delivery often have few stamens and sessile, bisporangiate anthers which lack endothecium. Pollen may be delivered below the water surface (Cymodoceaceae – Ducker *et al.* 1978; *Cerato-phyllum* – Jones 1931; *Callitriche* – Philbrick 1993), or it may be buoyant and carried on the surface (*Ranunculus* – Philbrick & Anderson 1987). The dynamics of pollen dispersal by water were examined by Cox & Knox (1986) and by Cox & Humphries (1993) in discussions of pollination in the seagrass alliance. *Ceratophyllum* and *Acorus*, which are currently regarded as basal to angiosperms or monocots respectively, are both water dwellers, arguing that aquatic habitats are ancient for flowering plants and that pollen delivery by water is a long-established system.

Animate delivery

Pollination by animal vectors, especially those that fly, is one of the great innovations that led to such grand diversity of flowering plants. Most angiosperms and some gymnosperms, for example *Welwitschia*, some species of *Gnetum*, and some cycads (Norstog 1987), are pollinated by insects. Taylor (1982: 24) suggested that at least some Carboniferous plants were insect pollinated, and Crepet & Nixon (this volume) note ample insect pollination in the Cretaceous.

Coevolution of flower and pollinator types holds that 'flower types have evolved in response to the sensory abilities of their contemporary pollinators' (Leppik 1977:1). A number of coincident trends were recognized by Leppik in a scheme now regarded as simplistic, if not incorrect. He described levels in floral evolution of insect-pollinated (entomophilous) plants: amorphic, haplomorphic, actinomorphic, pleomorphic, stereomorphic and zygomorphic. Flowers at the base of this regression series are without geometric form, at the next levels flowers progress through a series of geometric forms, and at the highest level they are bilaterally symmetrical. The lowest levels are fertilized by unskilled pollinators (beetles, flies, bugs), the next levels by insects that can visually distinguish forms (bees, moths and butterflies), and the upper levels by creatures with refined sensory perception (bees, birds). Plausible on the surface, limited testing of this scheme (Thien 1980) has shown that laminar stamens are indeed pollinated by beetles, flies, thrips, and sometimes primitive moths. Some beetles are now known to have acute olfactory perception. Leppik's idea that floral form is related to insect abilities may take different shape as field observations and the fossil record provide more detailed information. Bernhardt & Thien (1987) show that primitive angiosperms are pollinated by a variety of insects.

Pollen, one of the most nutritious materials made by plants, is sought by many insects for food. Protection of pollen from all except the vectors that effect pollination is vital for pollen-offering systems. Although not always obvious, this function underlies much of the structural complexity of flowers. Yeo (1993) regarded pollen protection as an important selective 'canalization' leading to indirect pollen presentation. How some arthropods deal with pollen and its sporopollenin is discussed by D'Arcy et al. (this volume).

While the flower–pollinator coevolution regression relates floral form to pollinator patterns, it does not examine how stamens participated in such coevolutionary advancement. Parallel coevolution regressions between stamen form and pollinator activities can be observed. Yeo's analysis of structural shifts in relation to indirect pollination is an example. The following discussion suggests some possible structural changes that have resulted from altered pollinator behaviors. Details of the regression between stamen form and pollinator behavior are a challenge for future work.

Plants attract animals for pollination by offering a variety of signals. Attractants include scents, shapes and colors, which are usually provided by parts of the flower other than the stamen. However, the stamens themselves often act as attractants. Perhaps most striking are flowers of Zingiberaceae, Cannaceae and Maranthaceae where staminodes are modified into conspicuous, petaloid flags, and even fertile stamens have colorful areas of tissue. Stamens are often colored differently from the surrounding perianth, and in many cases they contrast under ultraviolet light, which insects can perceive (reviewed by Biedinger & Barthlott 1993: 10). Anthers or pollen are

often scented (Percival, 1965: 38, 41; Buchmann 1983: 88; D'Arcy et al. 1991; Sazima et al. 1993).

Attraction of visitors is only part of the delivery paradigm. Visitors must keep visiting to move pollen from one flower to another, and for this a reliable clientele of pollinators is needed. Thus, most animal-pollinated flowers provide rewards for visitors (Simpson et al. 1983) or delude them in reward-fulfillment. Stamens offer pollen and scents as rewards. Oils are offered in *Mouriri* (Buchmann & Buchmann 1981). Calcium oxalate may be a stamen-offered reward (D'Arcy et al., this volume). In some cases anthers or parts of them are offered as bait (feeding anthers) to insects that visit flowers to eat them. The insects take up pollen incidentally when they feed (Faegri & van der Pijl 1979: 62). However, most rewards are presented by other parts of the flower.

Animal pollination may be passive or active (Bernhardt, this volume). In passive pollination, the vector (wind, water, animal) removes pollen from the anther and delivers it incidentally to other actions. Systems using mimicry to draw visitors for sex or nonexistent food are also passive. In active pollination, on the other hand, the pollen itself is the reward sought, and its delivery takes place in the course of seeking more pollen. The best known pollen-seeking insects are bees, but beetles, butterflies and flies are also active pollinators.

Stamen form relates to pollinator behavior

Insects and other animals visit and take pollen from flowers while crawling, hovering, gleaning, walking, buzzing, nesting, and even attempting copulation. To understand the structure of an anther, it is more important to know how the visitor behaves than what it is seeking. Although there are many, many modes of pollination and pollen presentation, the vast majority are represented by only a few morphological systems. The major systems are outlined here.

Laminar stamens

Stamens with laminar structure (Fig. 1) are mostly visited by insects that crawl or rub on the anther surfaces. In some cases (*Calycanthus* – Grant 1950; *Victoria* – Prance & Arias 1975), the flower holds the insect for hours or longer, ensuring that enough pollen is rubbed onto it. Only then is the insect released to go to another flower. Flowers of plants at the primitive end of the evolutionary scale, Magnoliidae (Bailey & Nast 1943; Sampson 1987), tend to have rigid stamens that act as sticky pegs that a crawling visitor can brush against. A contrary view says that such seemingly unspecialized visitation may actually be highly specialized and some cases are derived from more advanced forms. Carlquist (1969: 338) suggested that the extra size and tissue of the laminar anther may protect against tissue loss by chewing or clumsy visitors. In any event, pollination of plants with such stamens is often through highly specialized plant–insect relationships rather than by generalist pollinators.

Rigid anthers with filaments

A more common type of stamen is differentiated into filament and anther. This structure is often rigid. In the classical view, the rigid form is more primitive, probably derived directly from the laminar stamen (Parkin 1951), and it does occur early in the fossil record (Crepet & Nixon, this volume). Such may be the case in *Ranunculus*, which is related to other plants with laminar stamens. Flowers in this genus are pollinated by thrips and other crawling insects (Baker & Cruden 1991) as well as by flies and butterflies. Rutaceae, Sapindaceae and Proteaceae are other examples with rigid anthers that may have descended from laminar, or at least fixed, anthers. Visitors include those with both hovering and alighting kinds of behavior (Adema 1991).

Fusion with the perianth

In many flowering plants, the bases of stamens are fused with (adnate to) the base of the perianth. The top of the adnate part of the filament is often called the insertion of the filament or stamen. Above the insertion, such stamens look much like simpler stamens that lack fusion.

The major innovation here is not so much in the stamen as in the perianth parts, which have fused laterally, a step that seems to have preceded the fusion with the stamen base. This advance may be related more to presenting a better display and to protecting the ovary and nectary than to better anther efficiency. The basal, fused portion of the perianth is sometimes called a hypanthium, and sometimes it is fused to the ovary, which is then called inferior.

Stamens with filament adnation to the perianth are found in a large and diverse array of dicotyledonous families. Formerly, most such plants were classified in the large order Sympetalae. Now they form most of the Asteridae and advanced elements of the Rosidae. There are also many monocotyledons with filament bases adnate to the perianth (Dahlgren & Clifford 1982).

Versatile anthers

The versatile anther (Fig. 2B, C) is an important step up in flowering plant evolution (Parkin 1951: 5), and it may be the most widespread of all simple anther types. This anther when touched in one place swings around to brush pollen from its entire length. It is well adapted for insects, birds or bats that fly and hover, seeking a reward from a part of the flower situated behind or below the anther. It may be especially common in flowering trees. For such a generalist role, a horizontal anther seems best for intercepting the probe of a visitor seeking the reward beneath it. Thus, it is unsurprising that a majority of versatile anthers are medifixed – horizontally balanced on a vanishingly slender filament neck. Fixation of the anther in the middle allows a heavier, longer anther to balance on a smaller neck.

Although a more or less straight, horizontal anther (*Hemerocallis*, *Gloriosa*) with median fixation might seem an ideal, many versatile anthers depart from this in various ways. Such straight anthers are often held upright (*Hemerocallis*, *Lilium*, *Fuchsia*, *Gladiolus*, *Aloe*) or in other positions. Versatile medifixed anthers are not always linear. Length may be reduced so that the anther resembles a small knob on the filament (*Rosa*, *Hibiscus*, *Valeriana*, *Sedum*, *Viburnum*) (Fig. 2C). Sometimes the anther is curved with each end directed toward the base of the filament (*Bixa*). Another common modification is shortening of the connective. Here, the anther may comprise mostly downward or outwardly directed lobes or tails and little or no sporangial tissue beyond the point of attachment (*Passiflora*).

A similar kind of anther fixation, more properly called hinged than versatile, is where the anther moves back and forth in only two dimensions rather than in a wide arc as a fully versatile anther. This is found in many Iridaceae, Liliaceae and in *Physalis*. It functions much as a versatile anther for serving flying pollinators, but visitors are likely to be more directed in their angle of approach to the floral reward.

In a medifixed anther, the basal part of the anther is sometimes divided into two separated lobes, or tails, and a connective is present only in the apical portion. A consequence is that the endothecium in the lobes, if expanded inward beyond the thecal arc, must have a canaliculate configuration (described above under anther tissues). Seeming to contradict this, a number of cases of versatile anthers are basifixed (*Cecropia*, *Physalis*, Tropaeolaeum, *Cyclamen*). The filaments of these stamens have a clearly identifiable apical neck, and the anthers have short or no basal lobes with connective running the full length of the anther. The anthers wobble freely when touched.

Versatile anthers are commonly held above other parts of the flower (*Rosa*, *Hibiscus*, *Koeneria*), but they are often included in tubular flowers (*Cyclamen*), either loosely at the corolla mouth (*Geranium*, *Cestrum*, *Viburnum*) or along the interior walls of the tube (*Lantana*, *Bougainvillea*).

Versatile anthers are common in both monocots and dicots. They are found in some basal taxonomic groups as well as many advanced ones. Preparations shown by Crepet & Nixon (this volume) are suggestive of versatile anthers. This may indicate that they appeared early in the angiosperm fossil record, perhaps soon after there were flying pollinators. Baum (1991: 279) showed that in *Adansonia* (Bombacaceae), versatile anthers are found in more advanced species, which are pollinated by hovering bats and hawkmoths. Species basal in the genus have anthers that are rigid on the filaments. They are usually pollinated by terrestrial mammals which press their heads into the flowers seeking nectar. This activity may not select for anther versatility. Seeming to have evolved in response to generalist flying animal pollinators, the versatile anther can shift with little or no modification to anemophily. This appears to have happened in the Gramineae, Cyperaceae, *Salix* and *Quercus*.

Fixed, free anthers on filaments

A medifixed versatile anther with teetering lobes is poorly adapted for active pollen gleaning from the anther itself. In some plants, fixed anthers may have evolved from versatile anthers as bees replaced generalist visitors. Such fixation would provide greater stability for bees manipulating the anther for pollen. Rigid anthers that are derived from versatile ones might have remnant anther lobes or tails. Other evidence for this derivation would sometimes be expected. Three or four possible cases are suggested here.

In *Cleome*, rigidity without fusion is achieved as the anther base clasps and encloses the filament neck. A common way for a versatile anther to become fixed on the filament is for part of it to fuse along the filament. This can be seen in *Salvia* and *Barleria* where what resembles a versatile medifixed anther is actually fixed on the end of the filament with the lobes directed downward (Fig. 2E). Such examples serve flying pollinators that are directed in predictable ways in their approach to contact the anther. They are found in flowers with few stamens, suggesting efficacy and economy.

In *Enkianthus*, which is postulated as the most basal extant member of the Ericaceae (Kron & Chase 1993), the anther lobes are reduced in size, making the fertile, apical portion relatively larger and heavier (Fig. 2A). The reduced lobes, called spurs in this family, are directed upward. A short part of the fertile portion is fused to the filament, presenting a rigid anther for an insect to take pollen from the longitudinal slits. In the whole ericalean line, anthers have a reversal in the central vascular trace (Fig. 5) (Artopoeus 1903; Matthews & Knox 1926) that could be explained by such a fixation. That is, the apex is down with the tails directed upward so that the anther is inverted. Bee pollination is common in this group, and many cases are vibratile (see below).

A second possible case of a versatile anther becoming fixed on the filament is suggested in the Solanaceae. *Lycium* is thought to be a basal member of the Solanoideae (Olmstead & Sweere 1994), and *Schizanthus* is basal in the whole family (Olmstead & Palmer 1992). Anthers in both genera have short connectives and long separate lobes. In more advanced genera, the connective is relatively longer, and in many cases the lobes are rigid on the filament. In all Solanaceae for which it is known except *Schizanthus*, the endothecium is canaliculate throughout, a condition mandated by separate lobes when increased anchoring is required for rigidity in pollen display. This is taken as evidence that anthers in the Solanaceae were derived from medifixed anthers with divided lobes. Active collection of pollen by specialist pollen gleaners seems to have at least partly replaced likely generalist visitors in ancestors of the Solanaceae. Most genera of Solanaceae provide both pollen and nectar as a reward, but some groups offer scents.

Most genera of Solanaceae still provide nectar as a reward, but in genera with vibratile pollination, such as *Solanum*, no nectar is offered. With development of

porose dehiscence, the endothecium disappeared in these anthers. Canaliculate endothecium occurs in other flowering plant groups, as well, e.g. Gesneriaceae (Trapp 1956a: 77), Scrophulariaceae (Trapp 1956b: 288), *Cuscuta* (Batygina *et al.* 1987: 202), *Mussaenda, Macrosphyra* (Batygina *et al.* 1987: 117–119), *Centranthus, Valerianella* (Batygina *et al.* 1987: Valerianaceae plate, figures 9, 11). It may turn out to be quite widespread. Whether it is more efficient and was selected for its anchoring value alone or was derived in ancestors that had divided lobes is an open question. The second choice points to an ancestry of medifixed anthers and seems more likely.

In a third case, the Inuleae, Mutiseae, and some Astereae of the Asteraceae are characterized by auricles or tails on the anther lobes (Figs. 6, 7) (Cabrera 1950: figure 1; D'Arcy 1975: 838, 1036; Zhang & Bremer 1993: 271). These suggest relict anther lobes that have mostly fused with the filaments. In any case, rigid fixation of anthers on the filaments would be necessary before the anthers could unite into a tube, which is characteristic in this family.

In still another case, basal members of the Apocynaceae (Fallen 1986) all have tubular flowers with the plugged corolla mouth arrangement described by D'Arcy (1978) for zygomorphic groups of Solanaceae. Features of this are high insertion of the filaments in the tube and crowded anthers. By plugging the corolla mouth, access to the pollen and nectar is restricted to insects with long sucking mouthparts, and a versatile anther seems inappropriate for doing this. In the six basal tribes of the

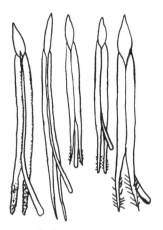

Fig. 6. Diagram of selected species of *Gochnatia* and *Moquinia*, Asteraceae. Redrawn from Cabrera (1950).

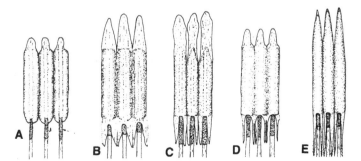

Fig. 7. Diagram of stamens in Asteraceae showing tails and apical appendages. Tails are lacking in *Fleishmannia pratensis*, and the apical appendage is not demarcated in *Onoseris onoserioides*. A. *Fleishmannia pratensis*. B. *Taraxacum officinale*. C. *Pterocaulon virgatum*. D. *Gnaphalium americanum*. E. *Onoseris onoserioides*. After D'Arcy (1975: 838), with permission of the Missouri Botanical Garden.

Apocynaceae, the anthers are mostly dorsifixed slightly above the base. They have short tails that are polleniferous to the base (Endress *et al.* 1990: 158). Is this evidence that their ancestors had versatile anthers?

Porose anthers

Many anthers release their pollen through pores (Fig. 2F). Porose anthers with dry pollen are vibratile. These anthers are grasped by a bee (or the syrphid fly *Volucella*) that buzzes or sonicates the pollen out through the pore (Buchmann 1983). One structural consequence of this pollinator form is that they are rigid on the filaments. Without lengthwise dehiscence, the anther, consisting of four parallel tubes like faschia, maintains its structural rigidity and hence does not need the stiffening and opening assistance provided in other anthers by the endothecial skeleton. Poricidal anthers usually have such endothecial strengthening only in the region of the pore. As a case of active pollination, nectar may not be offered (*Solanum* – D'Arcy 1991). This kind of anther is found in as many as 10% of angiosperm genera (Buchmann, 1983: 4). A few flowers with non-porose anthers also have vibratile pollination (Harder 1990).

Porose anthers with moist or greasy pollen operate differently. In the Araceae, pollen is extruded like toothpaste from a tube. The anther is not sonicated and an endothecium is present (Gerenday & French 1988). Pollination observations are too scant in this group for generalization. Pollen is also extruded in *Rhododendron*, *Zygogynum* and other plants.

Fusion between stamens

Fusion of stamens occurs between the filaments or between the anthers. In members of the Gesneriaceae, Scrophulariaceae and other members of the Asteridae, some stamens are suppressed and the remainder are fused into pairs. This is seen as positioning the anthers to deliver pollen to precise parts of pollinators, and hence is a specialized step.

Another common type of fusion results in formation of a stamen tube. Formation of a filament tube has little consequence for the structure of the anthers, and in groups with connate filaments, anthers are commonly versatile (*Camelia*, *Hibiscus*). In this volume, Tucker describes many variations of filament fusion in the Leguminosae (Papilionaceae).

Anthers united into a tube offer different possibilities for pollination. In *Lycopersicon* and some species of *Lycianthes*, the entire tube is sonicated by bees, and the pollen, ejected through pores or slits, covers their bodies.

In a different kind of tubular structure, where the inside diameter of the tube is smaller, anthers form a column around the style, which resembles a tube and piston (noodle squeezer – *Nudelspritze* – of Yeo 1993). In *Lobelia*, *Campanula*, Calyceraceae and Asteraceae (Juel 1908; Erbar & Leins 1989), the apparatus does act as a tube and piston. Parts of the style or stigma push pollen forth as they grow up within the tube. Pollen delivery is thus indirect.

If the anther tails in the Asteraceae are, in fact, vestiges from medifixed versatile anthers, as suggested above, then this group of families has undergone a series of innovative steps in anther structure. For versatile anthers to fuse into a tube, fixation on the filament would seem a necessary preliminary step. This might well involve a shift from generalist to specialist pollinators. Following fusion of anthers into a tube, the flowers in the Asteraceae and some Calyceraceae became aggregated into heads of floral tubes. The heads act as platforms so that insects can pick up pollen on their feet as they walk on its surface, pollen that has been pushed from the numerous anther columns on the emergent styles. Insects can still gather pollen on slender probosci as they extract nectar. The system has once again become adapted for generalist pollinators. Some members of the Asteraceae have evolved still further, becoming specialized for wind pollination (*Solidago*, *Artemesia*) but with little change in anther morphology. Thus, stamen presentation in this suite of families shows progressive evolution, but the process has included a series of advances and reversals, some of which have affected anther structure and histology.

Fusion of stamens and pistil

In two big plant families, Asclepiadaceae and Orchidaceae, male and female parts are combined into a single structure for pollen delivery. One family is a dicot and the other a monocot, but they have some common features. One feature they share is that pollen is all shed at once. It is stuck together in masses, or pollinia, along with other parts of the anther and parts of the pistil, and the whole unit is called a pollinarium. The pollinia are fastened to visiting insects in ways that bring about exact placement on the stigma of the target flower. In the Asclepiadaceae, part of the connective forms translators that join adjacent pollinia. This mechanism and morphological trends in one group of Asclepiadaceae are described by Liede in this volume. In these, the stamen–pistil device clasps parts of the insect, forcing it to take away the pollinia when it pulls out of the trap. The orchid flower forms stamen–pistil devices which are found in many designs. They were well described and illustrated by Dressler (1993).

In both the Asclepiadaceae and Orchidaceae, the lineage from free to fused flowers is reasonably clear. Both originated from plants that had connate perianth and basally adnate filaments like those noted for the Sympetalae. The Asclepiadaceae and Apocynaceae have long been separated because of differences in their floral fusion. However, only the basal Plumerioideae and Cerberioideae have stamens free from the gynoecium. Since all other genera have some adnation of stamens to pistil, the two customary families are now often put into one, the Apocynaceae. Unity of the perianth in this group has not led to inferior ovaries. In the Orchidaceae, even the most primitive genera (e.g., *Apostasia*) have similar unity of perianth parts and, in addition, the perianth is fused around the ovary making it inferior.

Each of these families is big. The Apocynaceae/Asclepiadaceae have 300–400 genera and 5000 species. The Orchidaceae has 1000 genera and perhaps 30 000 species, the most of all flowering plant families. Each family has many other special features, such as growth form, chemistry, and seed dispersal systems. While all of the evolutionary success of these two groups cannot be attributed to their united flower parts, fusion of the stamens with the pistil was probably at least partly responsible. Pollen in pollinia-like polyads has evolved in some Mimosoideae (Hernández 1989), but there is no fusion of stamens.

One result of releasing pollen in a pollinium is that the anther no longer opens to display the pollen. Thus, there is no need for an endothecium, and in fact an endothecium is lacking in the Asclepiadaceae (Purkinje 1930: 3) and in all but the most primitive orchids (Freudenstein 1991).

Convergence and reversal follow changes in pollinator behaviors

Some rigid anthers may have evolved from versatile anthers. This seems likely in Ericaceae, Apocynaceae, Solanaceae, Asteraceae, and some other advanced families, as suggested above. There are not many alternatives to this seemingly far-fetched idea. One is that ancestors of these stamens were rigid. Rigid anthers are found in laminar stamens of basal angiosperms and in such groups as *Ranunculus*, Rutaceae, Sapindaceae or Proteaceae. But flowers in these groups do not have the united perianths and adnate stamens that are found in the advanced groups with rigid stamens. It is likely that these perianth and stamen unions took place long before the anthers became fixed on the filaments, which increases the likelihood that such flowers had versatile anthers at some time in their ancestry. Although the evidence offered here is scant, it may be enough to spur studies that will strengthen it.

There is a very great number and diversity of flowers with versatile anthers which are suited to generalist flying pollinators. This diversity alone makes for a likelihood that in at least some cases they became rigid in their filaments. They did this to oblige active pollen seekers or to keep pollen and nectar from visitors that do not have long tongues. In large phylads with elaborate fusion of fertile parts – Orchidaceae and Apocynaceae – complex and dedicated systems have replaced earlier generalist morphologies. The success of these and other major changes in ways of delivering pollen shows the strength of selective forces on the anther and its plasticity in coping with shifting challenges. The above speculative scenarios seem all the more likely in view of these examples of success.

This account of anther–pollinator types suggests that convergence and reversals in anther morphology have come about because of shifts between pollinators with different flower-visiting behaviors. Such changes have sometimes been achieved by heterochrony, or altered timing of development stages (Guerrant 1982). Besides the common examples noted here, a multitude of other shifts and reversals have occurred in stamens since the seed plants arose. Structural accommodation of such changes has called for many shifts and reversals in underlying anther histology and has blurred trends in its evolution.

CONCLUSION

Many areas of contention in how scientists view anthers have been reviewed here and are expressed in this volume. A major cause of such differences is insufficient information. This is partly because knowledge about anthers is so scattered in the literature and also because facts about anthers that were available from histological preparations were not recorded. Observations of actual organisms can often be made in the simplest or most complicated settings. Careful study of biological events in both field and laboratory is the foundation for the understanding of stamens and how they work. Derivative activities such as computer analysis and cladistic rearranging are only valid when based on solid primary observations. With the exciting new methods now available for both primary observation and digestion of the findings, attention to the stamen is sure to produce rewarding outcomes. A better framework for considering facts about the stamen may draw more interest in future work. The foregoing tries to suggest some ways that facts and theories about the anther may be tested and realigned. The other papers in this book sample current efforts toward uncovering the unknowns of stamen and anther function and evolution.

LITERATURE CITED:

Adema, F. 1991. *Cupaniopsis* Radlk. (Sapindaceae): a monograph. Lieden Bot. Ser. 15: 1–190.

Arber, A. 1937. The interpretation of the flower: a study of some aspects of morphological thought. Biol. Rev. 12: 157–184.

Arber, A. 1946. Goethe's Botany. Chronica Botanica 10: 67–124.

Arditti, J. 1992. Fundamentals of Orchid Biology. Wiley, New York.

Armstrong, J.E. & S.C. Tucker. 1986. Floral development in *Myristica* (Myristicaceae). Amer. J. Bot. 73: 1131–1143.

Artopoeus, A. 1903. Über den Bau und die Öffungsweise der Antheren und der Entwickelung der Samen der Erikaceen. Flora 92: 309–330.

Bailey, I.W. & C. Nast. 1943. The comparative morphology of the Winteraceae. I. Pollen and stamens. J. Arnold Arbor. 24: 340–346.

Baker, H.G. 1983. An outline of the history of anthecology, or pollination biology. pp. 7–30 in Pollination Biology (edited by L. Real). Academic Press, Orlando.

Baker, J.D. & R.W. Cruden. 1991. Thrips-mediated self-pollination of two facultatively xenogamous wetland species. Amer. J. Bot. 78: 959–963.

Batygina, T.B., E.S. Teriokhin, G.K. Alimova & M.S. Yakeviev. 1963. Genesis of male sporangia in the families Gramineae and Ericaceae. Bot. Zurn 48: 1108–1120.

Batygina, T.B., A.L. Takhtgajan & M.S. Yakovlev, eds. 1987. Comparative Embryology of Plants: Davidiaceae–Asteraceae. Nauka Publishers, Leningrad.

Baum, D.A. 1991. The pollination and floral biology of *Adansonia* (Bombacaceae). Ph.D. Thesis, Washington University, St Louis.

Bernhardt, P. & L.B. Thien. 1987. Self-isolation and insect pollination in the primitive angiosperms: new evaluations of older hypotheses. Pl. Syst. Evol. 156: 159–176.

Biedinger, N. & W. Barthlott. 1993. Untersuchungen zur Ultraviolettreflexion von Angiospermen-Blüten I. Monocotyledoneae. Akad. Wiss. u. Lit. Mainz 86: 1–122.

Bierhorst, D.W. 1971. Morphology of Vascular Plants. Macmillan, New York.

Blomquist, H.I. & L.L. Robertson. 1941. The development of the peristome in *Aulacomnium heterostichum*. Bull. Torrey Bot. Club 68: 569–574.

Bold, H.C., C.J. Alexopoulos & T. Delevoryas. 1987. Morphology of Plants and Fungi. 5th ed. Harper & Row, New York.

Buchmann, S.L. 1983. Buzz pollination in angiosperms. pp. 73–113 in Handbook of Experimental Pollination Biology (edited by C.E. Jones, & R.J. Little). Van Nostrand Reinhold, New York.

Buchmann, S.L. & M.D. Buchmann. 1981. Anthecology of *Mouriri myrtilloides* (Melastomataceae: Memecyleae), an oil flower in Panama. Biotropica. 13 (2, suppl.): 7–24.

Cabrera, A.L. 1950. Observaciones sobre los Géneros *Gochnatia* y *Moquinia*. Notas Museo La Plata, Bot. 15: 37–48.

Canright, J.E. 1952. The comparative morphology and relationships of the Magnoliaceae. I. Trends of specialization in the stamen. Amer. J. Bot. 39: 484–497.

Carlquist, S. 1961. Comparative Plant Anatomy. Holt, Rinehart & Winston, New York.

Carlquist, S. 1969. Toward acceptable evolutionary

interpretations of floral anatomy. Phytomorphology 19: 332–362.

Carniel, K. 1963. Das Antherentapetum. Öst. Bot. Z. 110: 145–176.

Chase, M.W., D.E. Soltis, R.G. Olmstead, D. Morgan, D.H. Les & B.D. Mishler *et al.* 1993. Phylogenetics of seed plants: an analysis of nucleotide sequences from the plastid gene rbcL. Ann. Missouri Bot. Gard. 80: 528–580.

Chatin, A. 1870. De l'anthère. Baillière et fils, Paris.

Cox, P.A. & C.J. Humphries. 1993. Hydrophilous pollination and breeding system evolution in seagrasses: a phylogenetic approach to the evolutionary ecology of the Cymodoceaceae. Bot. J. Linn. Soc. 113: 217–226.

Cox, P.A. & R.B. Knox. 1986. Pollination postulates and two-dimensional pollination. Pp. 48–57 in Pollination '86. School of Botany, University of Melbourne, Melbourne.

Cronquist, A. 1968. The Evolution and Classification of Flowering Plants. Houghton Mifflin, Boston.

Dahlgren, R.M. & H.T. Clifford. 1982. The Monocotyledons: A Comparative Study. Academic Press, London.

D'Arcy, W.G. 1975. Compositae. In Flora of Panama (edited by R.E. Woodson *et al.*). Ann. Missouri Bot. Gard. 62: 835–1306.

D'Arcy, W.G. 1978. A preliminary synopsis of *Salpiglossis* and other Cestreae (Solanaceae). Ann. Missouri Bot. Gard. 65: 698–724.

D'Arcy, W.G. 1991. The Solanaceae since 1976, with a review of its biogeography. pp. 75–138 in Solanaceae III: Taxonomy, Chemistry, Evolution (edited by J.G. Hawkes, R.N. Lester, N. Nee, and N. Estrada). Royal Botanic Gardens, Kew, United Kingdom.

D'Arcy, W.G., N.S. D'Arcy & R.C. Keating. 1991. Scented anthers in the Solanaceae. Rhodora 92: 50–53.

Davis, G.L. 1966. Systematic Embryology of the Angiosperms. Wiley, New York.

Doyle J.A. & M.J. Donoghue. 1986. Seed plant phylogeny and the origin of angiosperms: an experimental cladistic approach. Bot. Rev. 52: 321–431.

Dressler, R.L. 1993. Phylogeny and Classification of the Orchid Family. Dioscorides Press, Portland, Oregon.

Ducker, S.C., J.M. Pettit & R.B. Knox. 1978. Biology of Australian seagrasses: pollen development and submarine pollination in *Amphibolis antarctica* and *Thalassodendron ciliatum* (Cymodoceaceae). Aust. J. Bot. 25: 67–95.

Eames, A.J. 1931. The vascular anatomy of the flower with refutation of the theory of carpel polymorphism. Amer. J. Bot. 18: 147–188.

Eames, A.J. 1961. Morphology of the Angiosperms. McGraw-Hill, New York.

Endress, M.E., M. Hesse, S. Nilsson, A. Guggisberg & J. Zhu. 1990. The systematic position of the Holarrheninae (Apocynaceae). Pl. Syst. Evol. 171: 157–185.

Endress, P.K. & S. Stumpf. 1990. Non-tetrasporangiate stamens in the angiosperms: structure, systematic distribution and evolutionary aspects. Bot. Jb. Syst. 112: 193–240.

Erbar, C. & P. Leins. 1989. On the early floral development and the mechanisms of secondary pollen presentation in *Campanula, Jasione* and *Lobelia*. Bot. Jb. Syst. 111: 29–55.

Esau, K. 1965. Plant Anatomy. 2nd ed. Wiley, New York.

Eyde, R.H. 1975. The foliar theory of the flower. Amer. Sci. 63: 430–437.

Faegri, K. & L. van der Pijl. 1979. The Principles of Pollination Ecology. 3rd ed. Pergamon, Oxford.

Fahn, A. 1990. Plant Anatomy. 4th ed. Pergamon Press, Oxford.

Fallen, M.E. 1986. Floral structure in Apocynaceae: morphological, functional and evolutionary aspects. Bot. Jb. Syst. 106: 245–286.

Font Quer, P. 1970. Diccionario de Botánica. Editorial Labor, Barcelona.

Freudenstein, J.V. 1991. A systematic study of endothecial thickenings in the Orchidaceae. Amer. J. Bot. 78: 766–781.

Gerenday, A. & J.C. French. 1988. Endothecial thickenings in anthers of porate monocotyledons. Amer. J. Bot. 75: 22–25.

Gifford, E.M. & A.S. Foster. 1989. Morphology and Evolution of Vascular Plants. W.H. Freeman, New York.

Goethe, J.W. 1790. Versuch die Metamorphose der Pflanzen zu erklären. [An Attempt to Explain the Metamorphosis of Plants.] Ettinger: Gotha, Germany. [Facsimile reprint Acta Humaniora, Weinheim. 1984.]

Grant, V. 1950. The pollination of *Calycanthus occidentalis*. Amer. J. Bot. 37: 294–297.

Green, J.W. 1980. A revised terminology for the spore-containing parts of anthers. New Phytol. 84: 401–406.

Guerrant, E.O. Jr. 1982. Neotenic evolution of *Delphinium nudicaule* (Ranunculaceae): a hummingbird-pollinated larkspur. Evolution 36: 699–712.

Gupta, S.C. & K. Nanda. 1978. Studies in the Bignoniaceae. 1. Ontogeny of dimorphic anther tapetum in *Pyrostegia*. Amer. J. Bot. 75: 395–399.

Harder, L. D. 1990. Pollen removal by bumble bees and its implications for pollen dispersal. Ecology 71: 1110–1125.

Hartl, D. 1963. Das Placentoid der Pollensacke, ein Merkmal der Tubifloren. Ber. Dtsch Bot. Ges. 76: 70–72.

Hernández, H.M. 1989. Systematics of *Zapoteca* (Leguminosae). Ann. Missouri Bot. Gard. 76: 781–862.

Hess. D. 1983. Die Blüte. Eugen Ulmer, Stuttgart.

Hufford, L.E. 1980. Staminal vascular architecture in five dicotyledonous angiosperms. Proc. Iowa Acad. Sci. 87: 96–102.

Hufford, L.E. & P.K. Endress. 1989. The diversity of anther structures and dehiscence patterns among Hamamelididae. Bot. J. Linn. Soc. 99: 301–346.

Jackson, B.D. 1916. A Glossary of Botanic Terms. Duckworth, London.

Jeffrey, E.C. & R.E. Torrey. 1916. *Ginkgo* and the microsporangial mechanisms of the seed plants. Bot. Gaz. 62: 281–292.

Jones, C.E. & R.J. Little. 1983. Handbook of experimental pollination biology. Van Nostrand Reinhold, New York.

Jones, E.N. 1931. The morphology and biology of *Ceratophyllum demersum*. Univ. Iowa Stud. Nat. Hist. 13: 11–55.

Juel, O. 1908. Om pollinationsapparaten hos familjen Compositae. Svensk. Bot. Tidskr. 2: 350–363.

Kaplan, D.R. 1971. On the value of comparative development in phylogenetic studies – a rejoinder. Phytologia 21: 134–140.

Kaplan, D.R. 1984. The concept of homology and its central role in the elucidation of plant systematic relationships. pp. 51–70 in Cladistics: Perspective on the Reconstruction of Evolutionary History (edited by T. Duncan & T.F. Stuessy). Columbia Univ. Press, New York.

Kaplan, D.R. 1992. The relationship of cells to organisms in plants: problem and implications of an organismal perspective. Int. J. Plant Sci. 153: 828–837.

Kaplan, D.R. & W. Hagemann. 1991. The relationship of cell and organism in vascular plants. Bioscience 41: 693–703.

Kaussmann, B. & U. Schiewer. 1989. Funktionell Morphologie und Anatomie der Pflanzen. Gustav Fischer, Stuttgart.

Keating, R.C. 1972. The comparative morphology of the Cochlospermaceae. III. The flower and pollen. Ann. Missouri Bot. Gard. 59: 282–296.

Keijzer, C.J. 1987a. The processes of anther dehiscence and pollen dispersal. I. The opening mechanism of longitudinally dehiscing anthers. New Phytol. 105: 487–498.

Keijzer, C.J. 1987b. The processes of anther dehiscence and pollen dispersal. II. The formation and the transfer mechanism of pollenkitt, cell-wall development of the loculus tissues and a function of orbicules in pollen dispersal. New Phytol. 105: 499–507.

Knox, R.B., S.Y. Zee, C. Blomstedt & M.B. Singh. 1994. Male gametes and fertilization in angiosperms. New Phytol. 125: 679–694.

Knuth, P. 1898–1905. Handbuch der Blütenbiologie. Engelmann, Leipzig. [English: 1906–1909. Handbook of Flower Pollination. 3 vols. Clarendon Press, Oxford.]

Koltunow, A.M., J. Treuttner, K.H. Cox, M. Wallroth & R.B. Goldberg. 1990. Different temporal and spatial gene expression patterns occur during anther development. Plant Cell 2: 1201–1224.

Kreulen, D.J.W. 1972. Spore output of moss capsules in relation to ontogeny of archesporial tissue. J. Bryol. 7: 61–74.

Kron, K.A. & M.W. Chase. 1993. Systematics of the Ericaceae, Empetraceae, Epacridaceae and related taxa based upon *rbcL* sequence data. Ann. Missouri Bot. Gard. 80: 735–741.

Kugler, H. 1970. Blütenökologie. Gustav Fischer Verlag, Jena.

Kunze, H. 1978. Typologie und Morphogenese des Angiospermen-Staubblattes. Beitr. Biol. Pflanz. 54: 239–304.

Lawrence, G.H.M. 1951. Taxonomy of Vascular Plants. Macmillan, New York.

Leppik, E.M. 1977. Floral Evolution in Relation to Pollination Ecology. Today & Tomorrow's Printers & Publishers, New Delhi.

Lersten, N.R. 1971. A review of septate microsporangia in vascular plants. Iowa State J. Sci. 45: 487–497.

Les, D.H. 1988. The origin and affinities of the Ceratophyllaceae. Taxon 37: 326–345.

Les, D.H., D.K. Garvin & C.F. Wimpee. 1991. Molecular evolutionary history of ancient aquatic angiosperms. Proc. Natl. Acad. Sci. U.S.A. 88: 10119–10123.

Limpricht, K.G. 1890. Die Laubmoose Deutschlands, Oesterreichs und der Schweiz. 1: 44. Kummer, Leipzig.

Lord, E.M. 1981. Cleistogamy: a tool for the study of floral morphogenesis, function and evolution. Bot. Rev. 47: 421–449.

Maheshwari, P. & V. Vasil. 1961. *Gnetum*. Bot. Monograph 1: 1–142. Council of Scientific & Industrial Research, New Delhi.

Mandel, M.A., J.L. Bowman, S.A. Kemplin, H. Ma, E.B. Meyerowitz & M.F. Yanofsky. 1992. Manipulation of flower structure in transgenic tobacco. Cell 71: 133–143.

Martens, P. 1971. Les Gnétophytes. Borntraeger, Berlin.

Márquez-Guzmán, J., M. Engelman, A. Martínez-Mena,

E. Martínez & C. Ramos. 1989. Anatomía reproductiva de *Lacandonia schizmatica* (Lacandoniaceae). Ann. Missouri Bot. Gard. 76: 124–127.

Matthews, J.R. & E.M. Knox. 1926. The comparative morphology of the stamen in the Ericaceae. Trans. Bot. Soc. Edinb. 29: 243–281.

Mauseth, J.B. 1988. Plant Anatomy. Benjamin/Cummings, Menlo Park, California.

Mechelke, F. 1952. Die Entstehung der polyploiden Zellkerne des Antherentapetums bei *Antirrhinum majus* L. Chromosoma 5: 246–295.

Melville, R. 1960. A new theory of the angiosperm flower. Nature 188: 14–18.

Melville, R. 1962. A new theory of the angiosperm flower I. Kew Bull. 16: 1–50.

Melville, R. 1963. A new theory of the angiosperm flower II. Kew Bull. 17: 2–63.

Mohl, H. 1830. Ueber die fibrosen Zellen der Antheren. Flora 13: 697–708, 715–742.

Moquin-Tandon, A. 1826. Essai sur les dédoublements ou multiplications d'organes dans les végétaux. Montpellier.

Mueller, B. 1952. Goethe's Botanical Writings. Ox Bow Press, Woodbridge, Connecticut.

Nanda, K. & S.C. Gupta. 1978. Studies in the Bignoniaceae. 1. Ontogeny of dimorphic anther tapetum in *Tecoma*. Amer. J. Bot. 65: 400–405.

Niklas, K.J. 1987. Aerodynamics of wind pollination. Sci. Amer. July 1987: 90–95.

Norstog, K. 1987. Cycads and origin of insect pollination. Amer. Sci. 75: 270–279.

Olmstead, R.C. & J.D. Palmer. 1992. A chloroplast DNA phylogeny of the Solanaceae: subfamilial relationships and character evolution. Ann. Missouri Bot. Gard. 79: 346–360.

Olmstead, R.G. & J.A. Sweere. 1994. Combining data in phylogenetic systematics: an empirical approach using three molecular data sets in the Solanaceae. Systematic Biol. 43: 467–481.

The Oxford English Dictionary. 2nd ed. 1989. Clarendon Press, Oxford.

Page, C.N. 1972. An interpretation of the morphology and evolution of the cone and shoot of *Equisetum*. Bot. J. Linn. Soc. 65: 359–397.

Palser, B.F. 1975. The bases of angiosperm phylogeny: embryology. Ann. Missouri Bot. Gard. 62: 621–646.

Parkin, J. 1951. The protrusion of the connective beyond the anther and its bearing on the evolution of the stamen. Phytomorphology 1: 1–8.

Paw U, K.T. & C. Hotten. 1989. Optimum pollen and female receptor size for anemophily. Amer. J. Bot. 76: 445–453.

Payer, J.-B. 1857. Traité d'Organogénie Comparée de la Fleur. Librairie Victor Masson, Paris.

Percival, M. 1965. Floral Biology. Pergamon Press, Oxford.

Periasamy, K. & B.G. Swamy. 1966. Morphology of the anther tapetum of angiosperms. Curr. Sci. 35: 427–430.

Philbrick, C.T. 1993. Underwater cross-pollination in *Callitriche hermaphroditica* (Callitrichaceae): evidence from random amplified polymorphic DNA markers. Amer. J. Bot. 80: 391–394.

Philbrick, C.T. & G.J. Anderson. 1987. Implications of pollen/ovule ratios and pollen size for the reproductive biology of *Potamogeton* and autogamy in aquatic angiosperms. Syst. Bot. 12: 98–105.

Prance, G.T. & J.R. Arias. 1975. A study of the floral biology of *Victoria amazonica* (Poepp.) Sowerby (Nymphaeaceae). Acta Amazonica 5: 109–139.

Purkinje, J.E. 1830. De cellulis antherarum fibrosis nec non de granorum pollinarium formis. J.D. Gruesonius, Vratislaviae.

Remane, A. 1952. Die Grundlagen des natürlichen Systems, der vergleichenden Anatomie und der Phylogenetik. Akademische Verlagsgesellschaft, Leipzig.

Richter, S. 1929. Über den Öffnungsmechanismus der Antheren bei einigen vertretern der Angiospermen. Planta 8: 154–184.

Robertson, C. 1904. The structure of the flowers and the mode of pollination of the primitive angiosperms. Bot. Gaz. 37: 294–298.

Ronse Decraene, L.P. & E.F. Smets. 1987. The distribution and the systematic relevance of the androecial characters oligomery and polymery in the Magnoliophytina. Nord. J. Bot. 7: 239–253.

Ronse Decraene, L.P. & E.F. Smets. 1993a. Dédoublement revisited: towards a renewed interpretation of the androecium of the Magnoliophytina. Bot. J. Linn. Soc. 113: 103–124.

Ronse Decraene, L.P. & E.F. Smets. 1993b. The distribution and relevance of the androecial character polymery. Bot. J. Linn. Soc. 113: 1285–1350.

Rothwell, G.W. & S.E. Scheckler. 1988. Biology of ancestral gymnosperms. pp. 85–176 in Origin and Evolution of Gymnosperms (edited by C.B. Beck). Columbia University Press, New York.

Sampson, F.B. 1987. Stamen venation in the Winteraceae. Blumea 32: 79–89.

Sazima, M.S., Vogel, A. Cocucci & G. Hausner. 1993. The perfume flowers of *Cyphomandra* (Solanaceae): pollination by euglossine bees, bellow mechanism, osmophores, and volatiles. Pl. Syst. Evol. 187: 51–88.

Schmid, R. 1972. Floral bundle fusion and vascular conservatism. Taxon 21: 429–446.

Schmid, R. 1976. Filament histology and anther dehiscence. Bot. J. Linn. Soc. 73: 303–315.

Siddiqui, S.A. & F.A. Khan. 1988. Ontogeny and dehiscence of anther in Solanaceae. Bull. Soc. Bot. Fr. 135, Lettres Bot. 2: 101–109.

Simpson, B.B., J.L. Neff & D.S. Seigler. 1983. Floral biology and floral rewards of *Lysimachia* (Primulaceae). Amer. Midl. Nat. 110: 249–256.

Skvarla, J.J., P.H. Raven & J. Praglowski. 1975. The evolution of pollen tetrads in Onagraceae. Amer. J. Bot. 62: 6–35.

Staedtler, G. 1923. Über Reducktionserscheinungen im Bau der Antherenwand von Angiospermen–Blüten. Flora: 116: 85–108.

Stevenson, D.W. 1983. Systematic implications of the floral morphology of the Mayacaceae (Abstr.). Amer. J. Bot. 70(5, part 2): 32.

Stewart, W.N. 1964. An upward outlook in plant morphology. Phytomorphology 14: 120–134.

Taylor, T.N. 1982. Reproductive biology in early seed plants. Bioscience 32: 23–28.

Taylor, T.N. 1988. Pollen and pollen organs of fossil gymnosperms: phylogeny and reproductive biology. pp. 197–217 in Origin and Evolution of Gymnosperms (edited by C.B. Beck). Columbia University Press, New York.

Thien, L.B. 1980. Patterns of pollination in the primitive angiosperms. Biotropica 12: 1–13.

Thomas, H.H. 1932. The old morphology and the new. Proc. Linn. Soc. London. 145: 17–32.

Thomas, H.H. 1936. Palaeobotany and the origin of the angiosperms. Bot. Rev. 2: 397–418.

Tobe, H. & P.H. Raven. 1983. An embryological analysis of Myrtales: its definition and characteristics. Ann. Missouri Bot. Gard. 70: 71–94.

Tobe, H. & P.H. Raven. 1984. The embryology and relationships of *Rhynchocalyx* Oliv. (Rhynchocalycaceae). Ann. Missouri Bot. Gard. 71: 836–843.

Tobe, H. & P.H. Raven. 1986a. Evolution of polysporangiate anthers in Onagraceae. Amer. J. Bot. 73: 475–488.

Tobe, H. & P.H. Raven. 1986b. The embryology and relationships of *Penaeaceae* (Myrtales). Pl. Syst. Evol. 146: 181–195.

Trapp, A. 1956a. Zur Morphologie und Entwicklungsgeschichte der Staubblätter sympetaler Blüten. Bot. Studien 5: 1–93.

Trapp, A. 1956b. Entwicklungsgeschichtliche Untersuchungen über die Antherengestaltung sympetaler Blüten. Beitr. Biol. Pflanzen 32: 279–312.

Vidakovic, M. 1991. Conifers: morphology and variation. Graficki Zavod Hrvatske, Zagreb.

Vogel, S. 1962. Duftdrüsen im Dienste der Bestäubung. Abh. Akad. Wiss. Math. Naturwiss. Kl. (Mainz) 1963: 602–763.

Vogel, S. 1981. Bestäubungskizepte der Monokotylen und ihr Austruk im System. Ber. D. Bot. Ges. 94: 667–675.

Weberling, F. 1981. Morphologie der Blüten und der Blütenstände. Verlag Eugen Ulmer, Stuttgart.

Weberling, F. 1989. Morphology of Flowers and Inflorescences. Cambridge University Press, Cambridge (English translation of Weberling 1981).

Webster's New Universal Unabridged Dictonary, deluxe 2nd ed. 1983. Simon & Schuster, New York.

Wilson, C.L. 1937. The phylogeny of the stamen. Amer. J. Bot. 24: 686–699.

Wilson, C.L. 1942. The telome theory and the origin of the stamen. Amer. J. Bot. 29: 759–764.

Wilson, C.W. 1950. Vascularization of the stamen in the Melastomaceae with some phyletic implications. Amer. J. Bot. 37: 431–444.

Wyatt, R. (ed.) 1992. Ecology and Evolution of Plant Reproduction. Chapman & Hall, New York.

Yeo, P.F. 1993. Secondary pollen presentation, form, function and evolution. Pl. Syst. Evol. Suppl. 6: 1–268.

Zhang, X. & K. Bremer 1993. A cladistic analysis of the tribe Astereae (Asteraceae) with notes on their evolution and subtribal classification. Pl. Syst. Evol. 184: 259–283.

Zimmermann, W. 1952. Main results of the telome theory. Palaeobotanist 1: 456–470.

Zimmermann, W. 1959. Die Phylogenie der Pflanzen. 2nd ed. G. Fischer Verlag, Stuttgart.

2 · The fossil history of stamens

WILLIAM L. CREPET AND KEVIN C. NIXON

INTRODUCTION

Modifications in pollen-bearing organs through time reflect structural and functional changes that often transcend their fundamental role (i.e., production and presentation of pollen) in plant reproduction. Pollen-bearing organs are a potential source of nutrition for insects and, with ovules and pollination droplets (e.g., Crepet 1979; Taylor 1982), were probably critical rewards in the development of the insect–plant relationships that evolved into the mutualism known as insect pollination (Regal 1977; Taylor 1982; Crepet 1983; Crepet & Friis 1987; Friis & Crepet 1987; Crepet et al. 1991). Once established, insect pollination might have resulted in increased propinquity between ovules and pollen organs, and thus insect pollination may have catalyzed the evolution of cosexual (=bisexual) reproductive structures in lineages proximal to angiosperms and, ultimately, in angiosperms themselves (see discussion in Crane 1985; Donoghue & Doyle 1989; Loconte & Stevenson 1990, 1991).

Unfortunately, the record of pollen-bearing organs and other reproductive structures does not provide an unequivocally documented transition from those of other seed plants to angiosperm stamens. For example, Caytoniales, once popular with some authors as possible angiosperm ancestors (e.g., Andrews 1963; Doyle 1978), are relatively distantly placed from angiosperms in recent phylogenetic analyses (Nixon et al. 1994), and we interpret the four-chambered synangiate organs of Caytoniales as a parallelism with the tetrasporangiate anthers of most angiosperm stamens (but see Doyle & Donoghue 1986 for an alternative interpretation of the relationship between angiosperms and Caytoniales). The extinct

Cycadeoidales (Bennettitales) and extant Gnetales are included in a monophyletic group with angiosperms in virtually all phylogenetic analyses of the seed plants that include fossil taxa (Crane 1985; Doyle & Donoghue 1986; Nixon et al. 1994). Cycadeoidales are a diverse assemblage with various types of rather complex pollen-bearing organs. These range from pinnate microsporophylls bearing numerous synangia (*Cycadeoidea*: e.g., Wieland 1906; Delevoryas 1963; Crepet 1974) to wedge-shaped fleshy microsporangia with only four or six synangia (*Williamsoniella*: Thomas 1915). In extant Gnetales, microsporangiate structures in *Gnetum* and *Welwitschia* have terminal microsporangia (two in *Gnetum*, three in *Welwitschia*). In *Welwitschia*, the microsporangia are terminal on pinnae of a tripinnate microsporophyll that occurs in the same taxis as subtending bracts (see Martens 1971; Nixon et al. 1994). Although *Ephedra* is typically scored as having 'terminal' microsporangia in cladistic analyses (e.g., Doyle & Donoghue 1986), published micrographs suggest that the microsporangia are actually abaxial with connective tissue extending between the sacs (see Bierhorst 1971), and the microsporophylls vary from simple to compound (branched) in different species. While it is tempting to draw analogies, and perhaps suggest homologies, between angiospermous stamens and the microsporangiate structures of the aforementioned taxa, we still lack a careful simultaneous analysis of these outgroups that also includes an adequate sample of variation within the angiosperms. Until such analyses are available, it is premature to postulate a primitive type of stamen within angiosperms (Nixon et al. 1994).

The historical pattern of angiospermous stamen variation as evidenced by the fossil record can, in principle, be a useful source of characters for phylogenetic analyses. Fossil data may be included directly in phylogenetic analyses and may provide evidence of the minimum age of lineages and appearances of characters and character complexes. When equally parsimonious hypotheses of phylogeny exist, it may also be possible to use fossil appearance as corroborating evidence for particular phylogenetic hypotheses.

Aside from the phylogenetic perspective, stamen history is of interest because of the importance of stamens in the evolution of flower structure and function. The pattern of stamen evolution is, at some level, a correlate of the evolution of functional floral structure and of the evolution of pollination mechanisms. Perhaps the most obvious correlations involve stamens that have morphological structures with clear functional implications (e.g., porate dehiscence, versatile anthers, elongate filaments) and these, in floral context especially, could provide highly specific insights into the history of various suites of floral characters that are now associated with particular pollination mechanisms. Such information would allow more careful consideration of the possible significance of coevolutionary events with insects in angiosperm history.

The fossil record has already yielded a high number of stamens based on specimens that are either isolated or attached to entire flowers (e.g., Friis et al. 1991 and references therein). In fact, as stamen evolution is a correlate of floral structure, the fossil record of stamens is, to a great extent, a correlate of the burgeoning fossil record of flowers. And, the fossil record of flowers has recently improved significantly in several critical periods in the history of angiosperm radiation (e.g., Friis et al. 1986, 1988, 1994; Crepet & Friis 1987; Friis & Crepet 1987; Crepet et al. 1991). Thus, in addition to providing insights into the pattern and timing of the evolution of stamen types, the fossil record also has the potential of providing insights into the evolution of floral character complexes and, by inference, floral function.

This chapter presents a summary of the fossil record of stamens as it is now understood with the emphasis on the fossil record of Cretaceous stamens. We present the history of stamens in the context of associated floral morphology. Although we include a summary of critical data for earlier and later deposits from the literature, we emphasize and illustrate new findings from Atlantic Coastal Plain deposits of Turonian age (Brenner 1963; Doyle & Robbins 1977; Crepet et al. 1992; Crepet & Nixon 1994; Nixon & Crepet, 1993, 1994; Herendeen et al. 1993, 1994 and unpublished data). Fossils from these localities are so diverse and include so many unique taxa that they dramatically affect our understanding of the history of reproductive structures in Cretaceous angiosperms and have important implications with respect to timing in the evolution of character complexes now associated with several modes of insect pollination. Fossil data are presented in chronological order. The fossils are further organized into general groups according to subclasses of Cronquist (1981), because as yet we see no consensus emerging from recent within-angiosperm cladistic analyses of morphological and molecular data (e.g., Donoghue & Doyle 1989; Loconte & Stevenson 1991; Chase et al. 1993).

The fossil records of dispersed leaves and pollen are *broadly* consistent with the record of stamens and floral remains (Doyle 1969; Muller 1970; Doyle & Hickey 1976; Hickey & Doyle 1977; Hickey 1978; Upchurch 1984; Crabtree 1987; Crepet & Friis 1987; Friis & Crepet 1987; Crane 1989; Lidgard & Crane 1990; Upchurch & Dilcher 1990; Crepet et al. 1991; Friis et al. 1991). There has been more effort expended over a longer period of time in studying the fossil records of leaves and pollen than has been put into the study of fossil stamens and other reproductive remains of angiosperms (e.g., Brenner 1963; Doyle 1969; Dilcher 1974; Hickey & Doyle 1977; Wolfe & Upchurch 1986), and, at this time, there is a more complete record of leaves. Thus, with respect to the appearance of new groups, the leaf record might be expected to 'lead' the flower record at any particular time depending on the level of resolution with respect to identification of new taxa that is available from leaves. Pollen also might be expected to be more complete than the flower/stamen record and to signal accurately the appearance of new groups before megafossils enter the record, *if* the pollen record is reasonably complete and *if* the pollen in question is sufficiently distinctive. Correlation among the leaf, dispersed pollen and stamen fossil records is not the aim of this chapter, but there are times when the increased resolution available from the flower record provides insights into diversity that have not yet been inferred from the pollen or leaf records alone or times when the leaf or pollen records provide insights that do not seem to match the known record of reproductive structures.

APTIAN–ALBIAN

Stamens from Aptian deposits in Portugal (Friis *et al.* 1994) and dispersed molds of anthers with pollen from the Aptian of Argentina (Archangelsky & Taylor 1993) are the earliest megafossil evidence of angiosperm reproductive structures. The stamens from Portugal are fleshy and somewhat flattened with lateral anther sacs and reticulate pollen. While these stamens are similar to those of some modern magnoliids, they do not have a combination of characters that, by itself, restricts possible affinities within the Magnoliidae, and it is difficult to suggest a precise systematic position for these stamens without other floral characters. The Argentine fossils consist of dense masses of *Clavatipollenites*-type pollen without pollen sac walls. This pollen is similar to the pollen of modern *Ascarina* in the Chloranthaceae. Dispersed pollen of this type, *Clavatipollenites hughesii*, is a common Aptian palynomorph and is among the earliest pollen grains assignable to a modern family (Walker & Walker 1984). There is additional diversity of chloranthoid pollen documented in the palynoflora of the Albian with *Asteropollis asteroides* Hedlund and Norris, similar to pollen of modern *Hedyosmum*, and *Stephanocolpites fredericksburgensis* Hedlund and Norris, similar to *Chloranthus* (Walker & Walker 1984; Crane *et al.* 1989). Thus, both commonly recognized major lineages of extant Chloranthaceae (the *Ascarina–Hedyosmum* group and the *Chloranthus–Sarcandra* group) are known as palynomorphs from the Early Cretaceous (Couper 1958; Hughes 1976; Hughes *et al.* 1979; Chapman 1987; Chlonova & Surova 1988).

Dispersed pollen tetrads with distal ulcerate apertures assignable to Winteraceae are also known from the Aptian (Walker *et al.* 1983). Another Aptian palynomorph, *Lethomasites*, is atectate with characters suggesting magnolialean affinities (Ward *et al.* 1989). Doyle & Hotton (1991) point out that similarities between *Lethomasites* and the pollen of Bennettitales might raise questions about the actual taxonomic affinities of these pollen grains. But they note that some of the characters of this palynomorph differentiate it from pollen of some Bennettitales. However, these characters do not differentiate *Lethomasites* from all of the bennettitalean pollen that has been sectioned (see discussion in Crepet & Nixon 1994). In any case, inferences about floral morphology based on pollen morphology unavoidably involve a degree of uncertainty in associating this palynomorph with modern Magnoliales. This is especially true because mosaicism (i.e., an extinct taxon with a suite of characters now in separate taxa or monophyletic groups of taxa) has been observed in many Turonian age Cretaceous flowers (Crepet *et al.* 1992; Nixon & Crepet 1993; Crepet & Nixon 1994), and we see no *a priori* reason to assume mosaicism in floral morphology and pollen characters would have been less common in earlier Cretaceous fossil taxa. Thus, compressed axes with leaves and attached flowers/fruits from Aptian deposits with some piperalean associations of characters (notably the leaves and general fruit morphology: Taylor & Hickey 1990) cannot be (and have not been) assumed to have had stamens similar to those of modern Piperales, without clearly preserved evidence of such stamens.

By the Albian, there is increased diversity of fossil stamens and stamens that are attached to well-preserved flowers. Currently known Albian stamens include fossils assignable to 'lower' Hamamelididae as well as Magnoliidae. Within Magnoliidae, calycanthoid flowers with stamens having four abaxial pollen sacs have recently been discovered in the mid-Albian of Virginia (Crane *et al.* 1994). In Late Albian deposits of Maryland, USA, stamens similar to those of Chloranthaceae have been described by Friis *et al.* (1986) and Crane *et al.* (1989). These are not identical to those of any modern taxon of Chloranthaceae. Instead of being lobed or branched, dorsiventrally flattened structures, like the chloranthoid stamens with which they are most easily compared, each stamen is composed of three terete-in-section units that are fused at their bases. Each of the branches of the stamen is tetrasporangiate, with two opposed pairs of microsporangia. This contrasts with extant species of Chloranthaceae, which have either unbranched tetrasporangiate stamens or three-branched stamens with the central branch tetrasporangiate and each lateral branch bisporangiate (see Herendeen *et al.* 1993, for a discussion of this character). One implication of the number and distribution of pollen sacs on these stamens is the inference that the pollen sac configuration of some (or all) modern chloranthoids with lobed stamens is the result of a historical reduction in the number of microsporangia on the lateral lobes (Crane *et al.* 1989). Micromorphology of *in situ* pollen is typical of Chloranthaceae.

The same Late Albian deposits also include dispersed dithecal, tetrasporangiate stamens with each theca having a single flap for dehiscence. These are similar to the stamens borne by a flower of hamamelidaceous affinity from younger deposits (Santonian–Campanian) in Sweden (Friis *et al.* 1991).

The earliest remains of platanoid reproductive structures are middle Albian staminate inflorescences and dispersed stamens (Crane *et al.* 1986; Friis *et al.* 1988, 1991) from Virginia associated with spherical pistillate inflorescences, inflorescences of uncertain sex (see Friis *et al.* 1988), and *Sapindopsis* foliage (Doyle & Hickey 1976). There is considerable variation among these platanoid staminate structures, with short or long terete filaments, variously expanded connectives, and pollen that is either tricolpate or tricolporate, with exine sculpture varying from open reticulate to almost foveolate. Some of these characters [tricolporate pollen, foveolate sculpture (Bogle & Philbrick 1980); long filaments] suggest rosid affinities for some of the fossils (as pointed out by Friis *et al.* 1988), and it should be noted that some extant Sapindaceae have well-developed connective extensions and vestiges of valvate dehiscence in the anthers (e.g., *Xanthoceras* see Endress & Stumpf, 1991). The mosaicism of the Albian platanoids, particularly as reconstructed with *Sapindopsis* leaves, suggests that these fossils need to be considered in the context of larger phylogenetic analyses before affinities with extant taxa can be firmly established.

Platanoid stamens have also been reported from Late Albian sediments. These stamens are attached to flowers borne in spherical staminate inflorescences. Platanoid remains include various inflorescence and leaf types (*Sapindopsis, Araliaephyllum et al.* Hickey & Doyle 1977), and are relatively common in the Potomac Group of the Atlantic Coastal Plain in the Late Albian through Cenomanian (Friis *et al.* 1988). The associated leaves also have certain rosid characters (Hickey & Doyle 1977). Other platanoids have been reported from later in the fossil record (Santonian–Campanian) of Sweden (Friis *et al.* 1988).

SUMMARY OF THE PRE-CENOMANIAN FOSSIL RECORD OF STAMENS

There is no unequivocal linear sequence to the appearance of taxonomic groups represented by stamens in the pre-Cenomanian. Although the pre-Cenomanian fossil record of stamens is growing, it is, at this time, relatively sparse. The Magnoliidae are present and some components of the tricolpate Hamamelididae have also differentiated before the Cenomanian. Certain rosid characters appear in one type of mid-Albian platanoid stamen. Early anthers generally had large connectives and valvate dehiscence. The enlarged nature of the connectives and their extension beyond the anthers suggest the possibility that they served a protective function or, conversely, were attractive to insect pollinators (Endress 1986, 1987, 1990). Because, from the known examples, flowers tended to lack petals (Endress 1990; Friis & Endress 1990), stamens may have been both attractants and rewards (Friis *et al.* 1991; Crepet *et al.* 1991). Although chloranthoid stamen contents appear early in the record with dispersed chloranthoid pollen (Walker & Walker 1984; Archangelsky & Taylor 1993), the morphology of taxa to which these stamens belonged is unknown and they may have differed from modern Chloranthaceae in floral and vegetative characters. The earliest intact stamens with possible chloranthoid features are Late Albian (Friis *et al.* 1986; Crane *et al.* 1989) and these differ (presumably they are more plesiomorphic) from the stamens of any extant Chloranthaceae or other angiosperm family. At the same time, there is no convincing megafossil evidence of the type of stamen that characterizes Magnoliaceae or other Magnoliales (i.e., obviously laminar stamens). Aptian dorsiventrally flattened stamens with reticulate monosulcate pollen and latrorse pollen sacs described by Friis, Crane and Pedersen (1994) might be allied with several different orders within the Magnoliidae. With the exception of Winteraceae, early palynological evidence of Magnoliales is uncertain and, as noted earlier, the occurrence of mosaic fossil floral evidence in other Cretaceous taxa suggests that inferences from the dispersed pollen record must be regarded cautiously.

Assumptions about modes of pollination in Early Cretaceous angiosperms are difficult due to the paucity of fossil stamen and floral data, uncertainty about the behaviors of Cretaceous relatives of modern insect taxa, and the relatively small literature on pollination in extant Magnoliidae. There is a documented diverse fauna of pre-Cenomanian insects whose modern relatives are significant pollinators. They include thrips, Coleoptera, Micropterigidae, Diptera, and Hymenoptera (e.g., Whalley, 1977; Jarzembowski 1984; Michener & Grimaldi 1988a, b; Carpenter & Raznitzen 1990; Carpenter 1992). The modes of pollination in modern Magnoliidae that are related to early angiosperms usually involve flies and beetles, but include, on occasion, thrips and micropterigid moths (Gottsberger 1977, 1988; Gottsberger et al. 1980; Pellmyr et al. 1990; Thien 1980; Thien et al. 1985). Given the potential pollinator fauna and early angiosperm flora, one can extrapolate from the suggestions of Bernhardt & Thien (1987), Gottsberger (1988), and Lloyd & Wells (1992) regarding pollination in angiosperm ancestors and conclude that insect pollination in early angiosperms involved unspecialized syndromes with pollination by a 'fluctuating spectrum of opportunistic pollinators' (Lloyd & Wells 1992). None of the anthophilous insects present in the Early Cretaceous can be ruled out as participants in early angiosperm pollination based on the fossil record of floral structures.

CENOMANIAN

The Early Cenomanian provides the first unequivocal megafossil evidence of reproductive remains of the Magnoliales. This is a fragment of one laminar stamen with valvate dehiscence, an extended connective and monosulcate reticulate pollen (Friis et al. 1991). Another magnoliid family that first appears in the Cenomanian is Lauraceae. Drinnan et al. (1990), have described flowers of Lauraceae with three whorls of stamens. Stamens of one whorl have lateral sterile appendages, a character typical of extant Lauraceae (Drinnan et al. 1990). There is no preserved pollen in these lauracean flowers. Finally, fruits (Couperites mauldinensis) that are unilocular and unicarpellate with a single pendulous anatropous bitegmic ovule are preserved with Clavatipollenites-type pollen on their apices (Pedersen et al. 1991). Although Clavatipollenites is typically considered to be of chloranthaceous affinity, the anatropous ovule is inconsistent with modern Chloranthaceae and may be yet another example of mosaicism in the fossil record.

Among Hamamelididae, flowers sharing characters with those of modern Trochodendrales/Buxales have been reported from the earliest Cenomanian Mauldin Mountain locality (Spanomera: Drinnan et al. 1991; Friis et al. 1991). In one species, Spanomera mauldinensis, small pistillate and staminate flowers with undifferentiated perianths and four or five floral parts per whorl are included in diclinous inflorescences (Drinnan et al. 1991). The anthers are dorsifixed, tetrasporangiate, and borne on short filaments (Drinnan et al. 1991). Connective tissue is on the dorsal side of the anther and extends slightly (Drinnan et al. 1991). Pollen is prolate, tricolpate and semi-tectate and reticulate to striate with relatively prominent sculptural elements as in modern Tetracentron (Bogle & Philbrick 1980). Dehiscence of pollen sacs is by linear slits instead of valvate flaps (Drinnan et al. 1991).

Isolated stamens with affinities to modern Rosidae appear for the first time in the Early Cenomanian. As in most extant Rosidae (Endress & Stumpf 1991), these stamens are filamentous with tetrasporangiate anthers, little connective development, and dehiscence by longitudinal slits. There are several dispersed rosid-type stamens with typical rosid pollen and there are also small pentamerous flowers with apparent rosid affinities (Friis et al. 1991). Although this constitutes the earliest evidence of rosids based on megafossils of reproductive structures, the pollen and leaf records suggest an earlier origin of the rosid lineage (Hickey & Doyle 1977; Hickey 1978; Lidgard & Crane 1990).

There are two taxa of magnolialean fruits with associated leaves and bracts in slightly younger Cenomanian deposits in the mid-western USA (Dilcher & Crane 1984; Crane & Dilcher 1984). Stamens have not been found attached or associated with these magnolialean remains. Intact flowers of rhamnoid Rosidae have been described from the same mid-western deposits by Basinger & Dilcher (1984). These flowers have stamens with robust filaments and proportionally large anthers that dehisce by longitudinal slits (Basinger & Dilcher 1984). There is no connective extension and pollen is tricolporate with a wall structure similar to that found in some modern Rhamnales (M. Zavada, unpublished data).

Summary of the Cenomanian

The fossil record of stamens in the Cenomanian, although sparse, documents certain events and the appearance of major taxonomic groups. The earliest unequivocal evidence of Magnoliales *sensu* Cronquist suggests that specialization for beetle pollination, today involving a complex set of interactions in certain Magnoliales, may have already occurred within some taxa. Calycanthoid flowers from the mid-Albian, however, suggest that specialization for coleopteran pollination was established at an earlier time within Laurales (Crane *et al.* 1994). The Rosidae appear for the first time with well-developed petals (Basinger & Dilcher 1984). The appearance of petals is compatible with the possibility that they were taking the place of stamens and staminodes as attractants in at least the hamamelid–rosid lineage, while, if reconstructed flowers are correct, other stamens and staminodes have been modified into nectaries. Thus, three of Cronquist's subclasses of angiosperms are present in the fossil record based on flowers or dispersed floral parts by the mid-Cretaceous.

TURONIAN

A newly exploited and particularly rich set of localities from Turonian deposits on the Atlantic Coastal Plain is providing new insights into angiosperm floral (and stamen) diversity during the early part of the Late Cretaceous (e.g., Crepet *et al.* 1992; Nixon & Crepet 1993, Herendeen *et al.* 1993, 1994; Crepet & Nixon 1994). The fossils are in various stages of investigation, but may include as many as 100 taxa based on floral remains alone. In addition to the great number of newly discovered angiosperm taxa, there are taxa that have also been reported from slightly older Cenomanian deposits and those that have been reported from younger sediments in both Scania and Portugal (e.g., Friis *et al.* 1991). These Turonian fossils include all subclasses of dicotyledonous angiosperms except Asteridae (but note that recent analyses suggest that Ericales is best considered to be part of a broader asterid lineage, e.g., Anderberg 1992; Olmstead *et al.* 1993). Interestingly, there is no conclusive evidence of monocots despite a paleoclimate and local ecological setting appropriate for taxa of modern monocotyledons (Christopher 1977; G.J.

Brenner, W.L. Crepet, & K.C. Nixon, unpublished data).

In the Turonian localities of the Atlantic Coastal Plain, stamens are common both as dispersed organs or attached to flowers. As with many Cretaceous angiosperm flowers that have been described recently, preservation is by charcoalification or lignification. In charcoalified specimens, preservation is three dimensional and, depending on the degree to which the fossil was exposed to high temperature, may include detailed cellular as well as morphological preservation (see discussion in McGinnes *et al.* 1971, 1974). Lignified stamens often have well-preserved pollen while pollen is variously preserved in charcoalified flowers. This may be related to the maximum temperature reached within the anthers during the charcoalification event.

Overall, the New Jersey Turonian angiosperm flora reflects an increase in diversity in stamens representing Magnoliidae, Hamamelididae and Rosidae relative to earlier deposits on the basis of stamen, floral or vegetative remains. Perhaps the most striking taxonomic aspect of the flora is a sharp increase in diversity in groups placed by Cronquist in his subclass Rosidae and the first appearance of taxa placed in Dilleniidae. Another aspect of these fossils, consistent with most Cretaceous floral remains, is their small size relative to flowers of extant taxa to which they appear to be most closely related. Thus, virtually all the flowers we present here are less than 4 mm in either diameter or length, and most have dimensions in the range 1–2 mm.

The precise affinities of many stamen types and flowers to which they are connected are still in the process of investigation, but we have included several to provide a representative assessment of Turonian anther and stamen diversity. The affinities of several of these (as yet unpublished) fossils are referred to only in the most general terms, and without formal names, in order to avoid possible confusion in the literature. References to morphological similarity with extant families are not intended to establish taxonomic affinity, but only to provide a framework for comparsion pending more complete analyses.

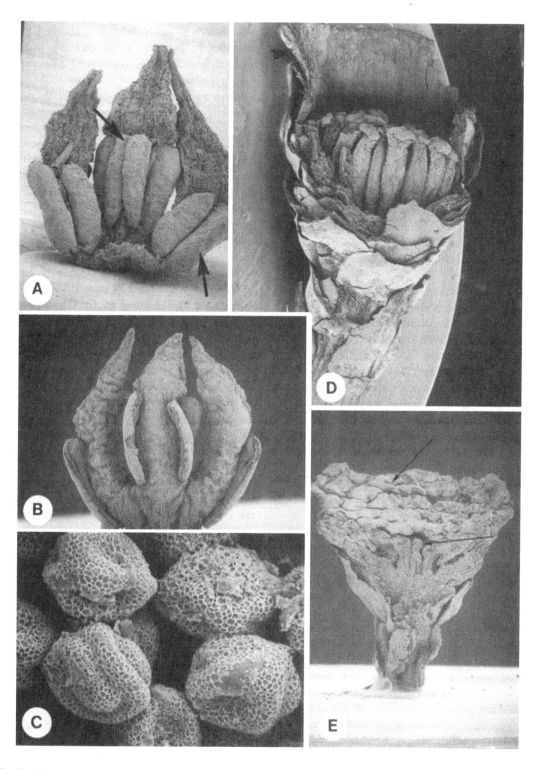

Fig. 1. A–C. *Chloranthistemon*. A. abaxial view, note pollen sacs (arrows) ×70. B. Adaxial view of A. ×70. C. Pollen. ×2000. D, E. Magnolialean taxon 1. D. Flower with abscised stamens showing carpels and cupulate receptacle (arrow). ×11.2. E. Flower at an earlier developmental stage showing spirally arranged immature carpels (arrows). ×23.

Magnoliidae

Stamens similar to those of modern genera of Chloranthaceae, e.g., *Chloranthistemon crossmanensis*, appear for the first time in these sediments (Herendeen *et al.* 1993). These stamens are dorsiventrally flattened and trilobed with two sporangia on each lateral lobe and four on the central lobe (Fig. 1A, B). They are similar to younger chloranthoid stamens from Scania described by Crane *et al.* (1989). Pollen is well preserved and is typically chloranthoid with loose reticulate ektexine. There is considerable variation in the nature of the apertures within the same pollen sacs (Fig. 1C), a phenomenon also observable in *Chloranthistemon endressii* (Crane *et al.* 1989).

Magnoliales are represented by well-preserved flowers having characters that are now found in magnoliids such as Eupomatiaceae, Himantandraceae and Calycanthaceae (Endress & Hufford 1989; Crepet & Nixon 1994). One taxon, Magnolialean taxon 1, has a cupulate floral receptacle with helically arranged carpels attached at the base (Fig. 1D, E). There are several marginally attached, distally winged seeds in each carpel of presumably mature fruits (Fig. 2A). Associated ovate stamens have four adaxial microsporangia dehiscing by valves (Fig. 2B, C). They are auriculate, and have distally incurved connective extensions with elongate hairs on their distal abaxial surfaces (Fig. 2C). Pollen is monosulcate and foveolate with occasional folds and atectate wall structure (Fig. 2D, E).

Another taxon shares eupomatioid characters. In this taxon, magnolialean taxon 2, stamens are covered by bracts that are attached at the cupule rim. Each of these bracts is narrowly attached and expands into an ovate concave structure that covers approximately one third of the flower (Fig. 2F, G). The expanded bracts overlap to close the flower much like a calyptra (in Eupomatiaceae or Himantandraceae), and they evidently abscise because specimens are preserved with exposed stamens and no covering bracts (Fig. 3A–C). The stamens themselves are linear-lanceolate with an extended connective, and four adaxial valvate sporangia (Fig. 3D). Pollen is globose with reticulate, apparently tectate columellate exine structure (Fig. 3E). Stamens are borne distally and internally on the receptacular cupule and are acutely incurved between the points of attachment and the

pollen sacs. Pollen sacs are not always evenly filled, suggesting the possibility of occasional failed pollen development. While stamens decrease in size toward the interior of the receptacular cup, all of them have pollen sacs and there are no stamens with obviously staminodal morphology (Fig. 3C, G). The receptacular cup bears no internal appendages between the stamen attachment zone and the base (Fig. 3C). At the base of the cupule are numerous carpels with marginal winged seeds and bilobed stigmatic areas (Fig. 3C, F). The carpels are surrounded by pistillodes (Fig. 3C, F). There are several different taxa represented by fossil flowers with a basic floral plan similar to that of magnolialean taxon 2. There are no morphologically definable staminodes internal to the stamens in any of these taxa.

Calycanthus-like cupulate magnoliid flowers are also present in the Turonian (Fig. 4A, B). They have urceolate cupulate receptacles with spirally arranged tomentose bracts on the outside (Fig. 4B). Near the rim of the cupule is a whorl or tight spiral of stamens (Fig. 4B). Stamens have massive, more or less deltoid connectives and pollen is reticulate and on occasion disulcate (Fig. 4D, E). Carpels are spirally arranged with elongate styles extending into the internal hairs that fill the opening of the receptacle (Fig. 4A). Carpels enclose one mature seed, and the seeds have a small marginal ridge or wing (Fig. 4C). Although the outer receptacle has numerous imbricate bracts, it is not possible to tell whether these were petaloid as in modern *Calycanthus*. Nor is there any indication of the same type of food bodies that terminate

Fig. 2. A–E. Magnolialean taxon 1. A. Carpels with winged seeds (arrow heads) in the broken carpel on the left. Note the large spherical cells (e.g. c) in the carpel epidermis. ×38.5. B. Adaxial view of stamen with adaxial sporangia. Valves (arrow heads) are recurved. ×38.5. C. Adaxial view of a stamen with an incurved connective extension (ce). Note the abaxial hairs. ×77. D. Cluster of pollen in pollen sac. ×1190. E. Two monosulcate grains. One is broken illustrating the solid wall structure (arrow). ×1533. F, G. Magnolialean taxon 2. F. Lateral view of flower showing the enclosing expanded bracts (b). ×18. G. Top view of the flower illustrated in F showing the large calyptra-like expanded bracts (b). ×25.

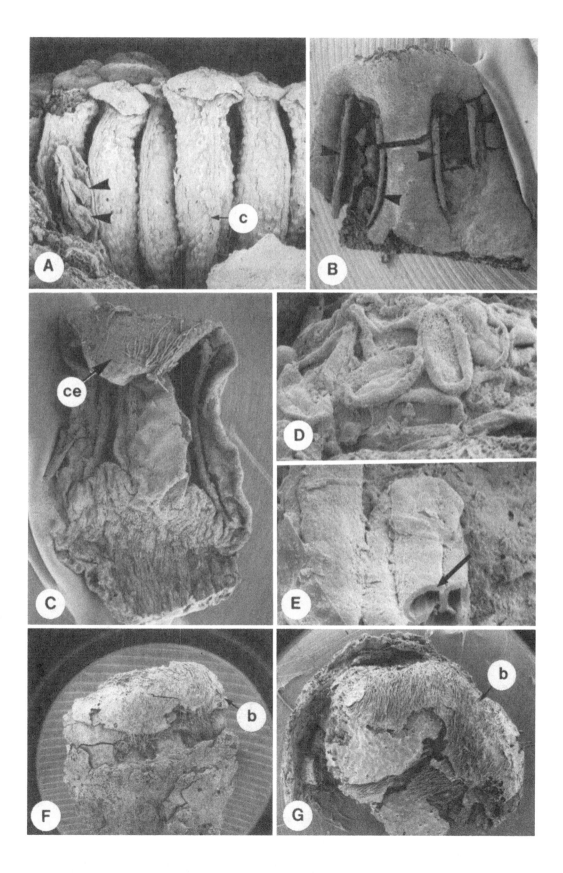

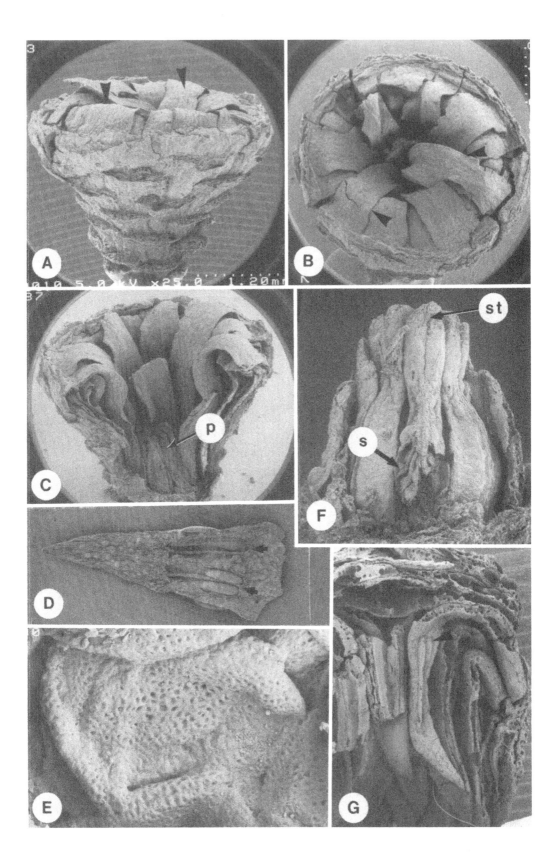

the stamen connectives in modern calycanthoids. These fossils will be treated in greater detail in a future paper.

Fossil magnoliid flowers with receptacular cups are complex and, in aggregate, have characters that, in modern taxa, suggest specialization for pollination by certain Coleoptera (possibly Nitidulidae and Curculionidae as in modern *Eupomatia* and *Calycanthus*: Grant 1950; Endress 1990; Bergstrom *et al.* 1991; and other references in Crepet & Nixon 1994b). The cupulate magnolialean fossils have large spherical cells in the carpel walls and stamen connective tissues that are suggestive of ethereal oil cells (Figs. 2A, 3D). In modern *Eupomatia*, ethereal oils play an important role in pollination (Bergstrom *et al.* 1991), suggesting that the apparent ethereal oil cells in the fossils had similar functions. Stigmatic areas in the fossils have structural features compatible with the production of stigma exudates and they might have played a role in the pollination syndromes of these fossils (see discussion by Lloyd & Wells 1992).

There are at least two taxa of Lauraceae represented by flowers in the Turonian deposits of the Atlantic Coastal Plain (Figs. 4F, 5A). One taxon has stamens dehiscing by four ventral valves in two tiers (Fig. 5A, B). Another fossil taxon, *Perseanthus crossmanensis*, has well-preserved flowers with two whorls of tepals, three cycles of stamens

and a fourth cycle of staminodes (Fig. 4F, G; Herendeen *et al.* 1994). Paired lateral substipitate appendages are borne on the filaments of the third cycle of stamens (Fig. 5C). Unfortunately, the filaments are broken and lack anthers (Fig. 4F). Pollen, however, is preserved on the filament bases (Fig. 5D). Pollen of Lauraceae is rarely preserved, presumably because there is very little sporopollenin in the exine (e.g., Hesse & Kubitzki 1983; Kubitzki 1981). In this instance, pollen preservation may have been the result of the mode of preservation. Pollen has a very thin wall (Fig. 5D) and small unornamented spines with expanded bases (Fig. 5D). Its morphology is consistent with floral morphology in suggesting affinities with the tribe Perseeae in Lauraceae (Herendeen *et al.* 1994).

A final taxon that may have either magnoliid (including ranunculiid) or even monocot affinities is represented by trimerous staminate flowers with three stamens and six more or less equal, valvate tepals (Fig. 5E, F). The anthers are essentially sessile on a flat receptacle. The dithecal anthers have well-developed connectives and abaxial pollen sacs (Fig. 5F). Dehiscence is longitudinal with the anther walls becoming strongly recurved and overlapping with anthers of adjacent stamens. There is a blunt, rounded connective extension (Fig. 5F). Pollen is trichotomosulcate and exine ornamentation is smooth with small spines (Fig. 5G). The exact affinities of this fossil taxon are still under investigation, and it would be premature to speculate further on its affinities, but the large, sessile anthers are distinctive within the known fossil assemblage. The variation in fossils of this type also suggests the possibility that more than one species of this group occurred at the locality.

Fig. 3. Magnolialean taxon 2. A. Lateral view of flower after abscission of calyptrate bracts illustrating the incurved laminar stamens (e.g. arrows heads). ×17.5. B. Top view of the specimen illustrated in A showing the laminar incurved stamens (e.g. arrow heads). ×17.5. C. Lateral view of the same specimen illustrated in A and B dissected to reveal the carpels surrounded by pistillodes (p). ×17.5. D. Adaxial view of the distal two-thirds of a stamen from the specimen illustrated in A–C. Note the four pollen sacs (each arrow points to two) and the large sperical cells in the connective tissue. ×42. E. Spherical reticulate pollen photographed within an anther. ×3150. F. A flower without the cupule or stamens illustrating carpels with bilobed stigmatic regions (st) and winged seeds (s). ×56.7. G. Specimen illustrated in Fig. 2F, G showing the stamens enclosed by expanded bracts and the pollen sacs on the adaxial surfaces of the stamens (e.g. arrow head). ×18.

Hamamelididae

One of the most interesting fossils described from this locality (Crepet *et al.* 1992) has definite affinity with modern Hamamelidaceae. Flowers are pistillate or staminate in tiny (*c.* 2 mm diameter) spherical inflorescences of about 15 flowers each (Fig. 6A). Staminate flowers have a four-lobed sepal cup and four stamens opposite the sepal cup lobes (Fig. 6B). The distal stamen connectives are incurved and meet marginally and completely seal the flowers (Fig. 6B). Alternating with the stamens are four staminodes (an interpretation based on position

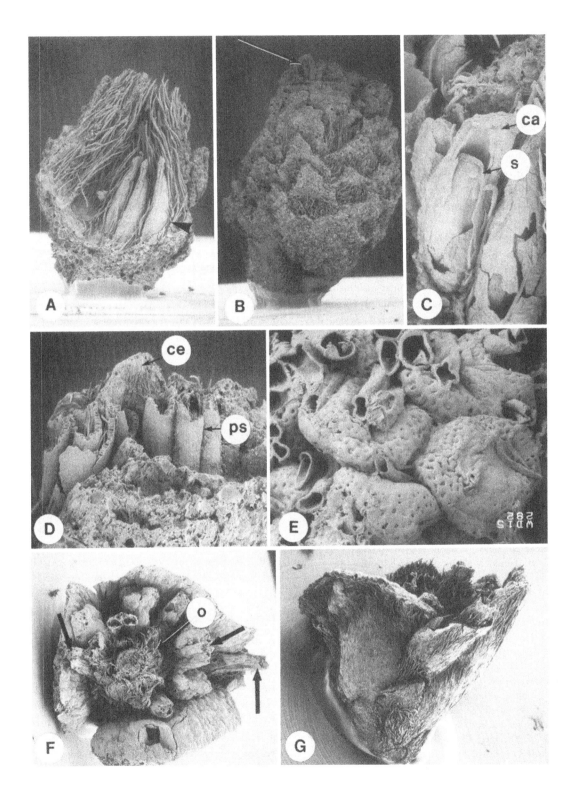

and morphology) that are also distally incurved and juxtaposed marginally (Fig. 6C). These have extremely large cells on their convex, adaxial surfaces that may be nectary cells (Fig. 6D; Crepet *et al.* 1992). Stamens have sagittate connective extensions and valvate dehiscence in four adaxial sporangia (Fig. 6E), similar to some modern hamamelidaceous stamens (Hufford & Endress 1989). Pollen is tricolpate and reticulate (Fig. 6F). Proximally, the stamens are fused with the bases of the staminodes into an hourglass-shaped staminal tube within the sepal cup (Fig. 6D). The morphology of the staminodes and their position alternate with the stamens suggests that they are homologous with petals in some modern Hamamelidaceae and by inference may represent a transitional stage between apetalous and petalous flowers. This interpretation is supported by the frequently incurved nature of petals, often with adaxial basal nectaries, in modern Hamamelidaceae (Crepet *et al.* 1992).

These fossil flowers might also be considered prehypanthial, in that they possess both a sepal cup and a short staminal tube. Adnation of the staminal tube and the sepal cup would produce a compound structure equivalent to the hypanthium of modern Hamamelidaceae and most modern Rosidae (Crepet *et al.* 1992). While this fossil has features that can be interpreted as pre-hypanthial, the broader fossil record suggests that the hamamelid/rosid hypanthium originated earlier, because rosids with petals are already known in the fossil record of the Cenomanian (Basinger & Dilcher 1984). Rosid-like leaves have been reported from yet older sediments (Doyle & Hickey 1976; Hickey & Doyle 1977; Hickey 1978).

Rosidae

In the context of older (Cenomanian) and younger (Santonian–Campanian, see below) reports of Rosidae based on floral and stamen remains, rosid taxa from the Turonian suggest that the early part of the Late Cretaceous was an interval of rapid rosid radiation. Flowers and stamens from the New Jersey deposits reveal that Rosidae reached a considerable degree of diversity and illustrate concomitant variation in stamen and staminode morphology. At least two orders of Rosidae have been tentatively identified among flower fossils from the Turonian deposits of the Atlantic Coastal Plain.

Various flowers with generalized rosid affinity are known from New Jersey. These typically have stamens with well-differentiated anthers and thin terete filaments, without a significant connective extension. The stamens are dithecal and dehisce by longitudinal slits. Two such taxa are illustrated in Fig. 7. One (Fig. 7A, B) is a generalized saxifragalean-type flower with a well-developed hypanthium, five stamens with filamentous basifixed anthers, two carpels and an apparently nectiferous disc. The other taxon has flowers with five relatively elongate sepal lobes (Fig. 7C) alternating with five narrow petals (Fig. 7F). There are two whorls of stamens with dorsifixed somewhat sagittate anthers and rugulate/perforate tricolporate pollen (Figs. 7D–H, 8A). The pistil is five-carpellate with short recurved styles and decurrent stigmas (Fig. 7D). There is a stylopodium (Fig. 7D) and there are two ovules per carpel. While the exact affinities of this latter rosid have also not yet been determined, it shares characters with various Rosales as well as with the families Araliaceae and Geraniaceae in the orders Geraniales and Apiales.

A flower of possible myrtalean affinity (Fig. 8B, C) is missing anthers but has distinctive, terminally coiled filaments (Fig. 8C, D) with massive amounts of triporate pollen on their distal surfaces (Fig. 8E). The flower has a distinctive tubular/campanulate sepal cup (Fig. 8B, C) and a bicarpellate ovary with free styles (Fig. 8C). The relative depth of the sepal cup suggests pollinators specialized for nectar-feeding through elongate mouthparts. Hymenoptera are among the most significant pollinators

Fig. 4. A–E. Calycanthoid. A. Internal view of a longitudinally broken cupulate receptacle with carpels (arrow head) and elongate hairs attached on the inside. ×37. B. External view of the specimen illustrated in A. Note the helically arranged, tomentose bracts and the stamen near the top of the specimen (arrow). ×37. C. Broken carpel (ca) illustrating a single seed (s) with a small marginal wing. ×98. D. Stamen with abaxial pollen sacs (ps) and broad, ovate connective extension (ce). ×98. E. Disulcate, reticulate pollen. ×1400. F, G. Lauraceae. F. Flower with tomentose ovary (o), six tepals, and several whorls of stamen filaments. ×35. G. Lateral view of the specimen illustrated in F showing the alteration of large and small tepals. ×35.

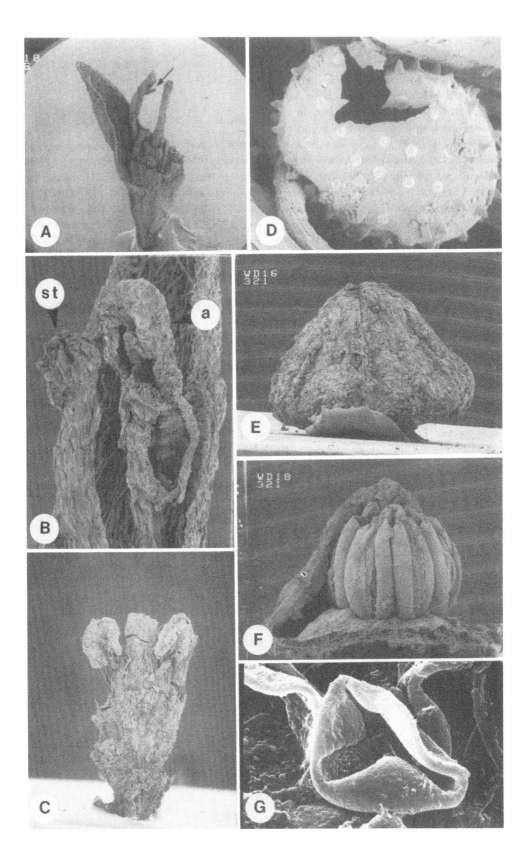

in some possibly related groups today (Faegri & van der Pijl 1971; Procter & Yeo 1972), but other nectar-feeding insects cannot be dismissed.

Dilleniidae

The Turonian deposits provide the earliest floral evidence for subclass Dilleniidae as defined by Cronquist (1981), and the diversity within this subclass is remarkable, suggesting a rapid radiation of these groups by the Turonian. Their stamens demonstrate a broad range of characters, and some are suggestive of rather specific and derived modes of insect pollination.

One series of flowers shares characters with modern Clusiaceae, placed in Theales by Cronquist (1981). Stamens are fascicled (Fig. 8F, H) and the outer fascicles, have small, short anthers that appear to be sterile (Fig. 8F, H). In some modern Clusiaceae, stamens of similar morphology produce resin as a reward for hymenopteran pollinators (Armbruster 1984). Anthers in the fossils open by slits and sometimes have strands of an amorphous substance preserved in the open slits (Fig. 8G). This substance may be preserved resin, but this is not necessarily an indication that these particular stamens produced resin exclusively. Resins in modern Clusiaceae may have a protective role even in functioning anthers (W.S. Armbruster, personal communication). Some anthers in the fossil flowers have *in situ* tricolporate pollen with perforate tectum that is similar to the pollen of some modern Clusiaceae (Fig. 9A; Erdtmann 1952).

Stigmas are expanded, distinctive (Fig. 8F, H) and also are similar to those of certain modern genera of Clusiaceae (e.g., *Clusia*, *Garcinia*), as is the general floral morphology.

Another fossil of probable dilleniid affinity *sensu* Cronquist has relatively large flowers compared with most other flowers from these localities (Fig. 9B, C). Anthers are elongate with tubular connective extensions and are borne in two whorls (Fig. 9C). Stigmas are distinctively lobed (Fig. 9C). These flowers are preserved in bud and the robust stamen filaments may reflect their immaturity. Pollen is not preserved in any of the available specimens.

Yet another probable dilleniid flower may have affinities with one or more families centered around the Capparales, but affinities are equivocal at this time. The flowers are irregular with different-sized petals and an asymmetrical calyx (Fig. 9D), and have a whorl of five stamens alternating with five distinctive setiform staminodes (Fig. 9E, F). Stamens and staminodes are fused into a staminal tube that surrounds the stipitate bicarpellate ovary (Fig. 9F). Stamens appear ditheal and have short recurved connective extensions (Fig. 9G, H). Pollen is tricolporate with rugulate-reticulate micromorphology (Fig. 9I).

Fossils with floral morphology suggestive of Ericales are diverse by the Turonian and include an array of taxa that include mosaics of modern ericalean, ebenalean and/or thealean taxa. More definite affinities of these taxa are being explored with cladistic analyses of fossil and extant taxa (Nixon & Crepet, unpublished data). The first of these has superior tricarpellate ovaries, elongate styles that are distally free, and two whorls of five stamens each inserted on the corolla tube (Fig. 10A–E). Anthers are tetrasporangiate, sagittate, and elongate, similar to those of the younger taxon *Actinocalyx* described by Friis (1985b; Fig. 10D, E). There are no signs of elongate slits that might be associated with dehiscence (Fig. 10F), but neither is there evidence of terminal pores, and flowers with mature stamens with dehisced anthers have not been discovered. Filaments are expanded near the bases (Fig. 10D) and adnate to the corolla (Fig. 10B), with distinctively sculpted epidermis distally (Fig. 10G). Sporangia are expanded near the attachment points and taper to a prolonged distal segment (Fig. 10A, E). Anther epidermis includes pyramidal processes that are most pronounced near the widest portions of the anthers (Fig.

Fig. 5. A. Another lauracean flower with one intact tepal and stamen. Note the open anther valve near the stigma (arrow). ×21. B. Higher magnification view of the flower seen in A showing the stigma (st) and anther (a). Note that three of the anther's four sporangia are visible and that the distal sporangia are smaller than the proximal ones. ×105. C. Proximal part of a stamen of Lauraceae with lateral glands/staminodes. ×35. D. Pollen of Lauraceae illustrated in Fig. 4F, G. ×1400. E–G. Trimerous staminate magnoliid flower. E. Lateral view. ×49. F. Lateral view of the flower illustrated in E, further dissected, showing the stamens with recurved overlapping valves and the rounded connectives. ×63. G. Trichotomosulcate pollen. ×4500.

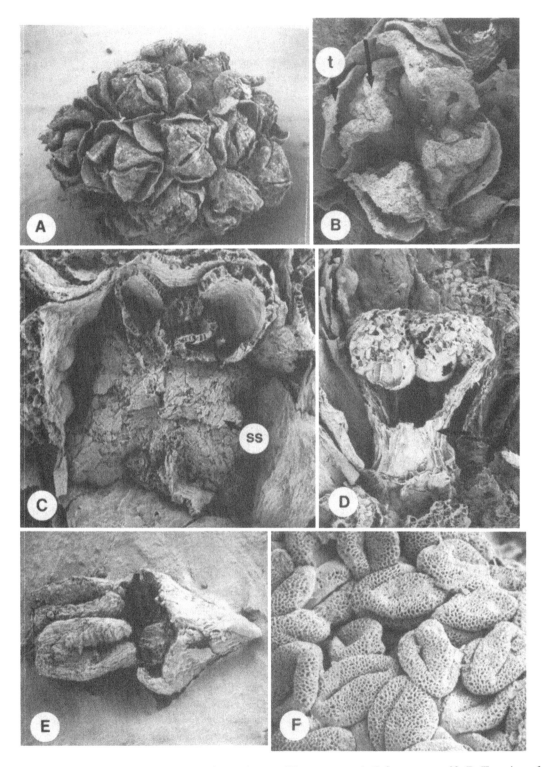

Fig. 6. Staminate inflorescence and flowers of a pre-hamamelidacean taxon. A. Inflorescence. ×32. B. Top view of a staminate flower illustrating the stamens (arrow) opposite the tepal lobes (t). ×86. C. Staminate flower in top view with most of the anther removed to reveal the sagittate incurved staminodes (ss). ×142. D. Flower in lateral view with stamens removed showing the hourglass-shaped staminal tube (arrow) and the bulbous adaxially expanded inner distal staminodes. ×140. E. A stamen with sagittate connective and pollen sacs. ×88. F. Tricolporate reticulate pollen. ×1400.

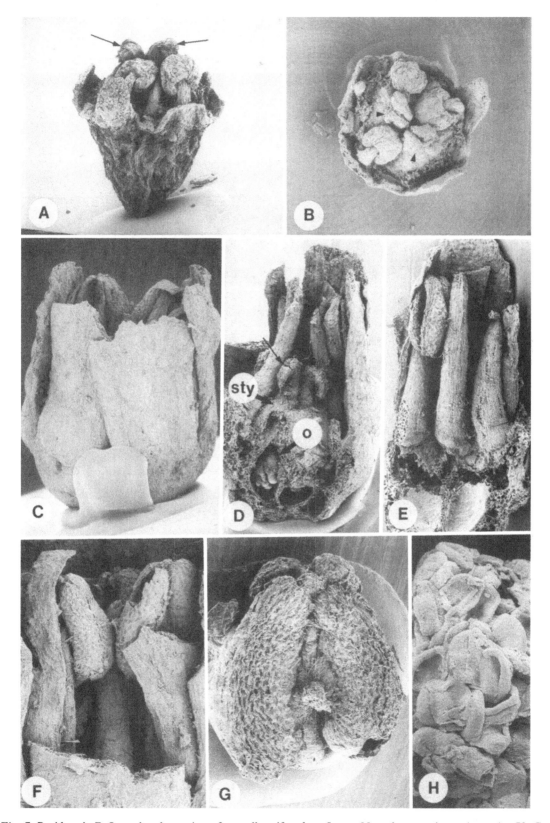

Fig. 7. Rosidae. A, B. Lateral and top view of a small saxifragalean flower. Note the two stigmas (arrows). ×70. C–H. Different views of the same rosid specimen. C. Lateral view of flower with elongate sepal lobes. ×43. D. Sepals and petals removed to expose stamens, pistil with recurved stigmas (arrow), stylopodium (sty), and two ovules/carpel (o). ×39. E. Part of the flower removed illustrating two whorls of stamens and the inner walls of two ovary locules. ×49. F. View of flower showing dorsifixed anthers and narrow petals inside sepal lobes. ×70. G. Anther with filament attachment. ×126. H. Tricolporate pollen. ×1750.

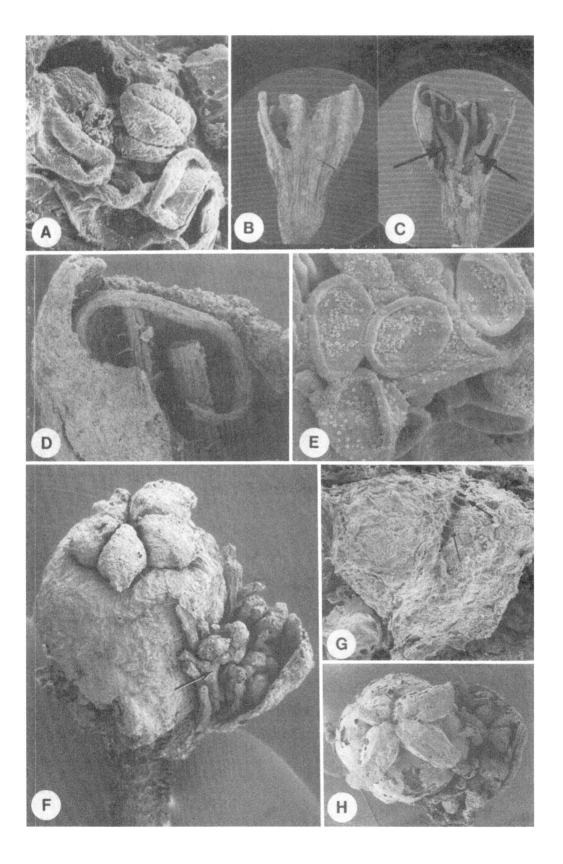

10D, E). These are epidermal cells that have endothecium-like thickenings internally (Fig. 10I). The anthers are attached to a distally flattened filament ventrally (Fig. 10E). Pollen is tricolporate with distinctive ektexine bridges in the endoaperturial regions and exine micromorphology is very finely scabrate (Fig. 10H). In view of the suite of characters, including unusual ventral versatile anther attachment, and the probable affinities of these flowers, it is possible that these anthers are inverted and that the narrow upright portions are pseudoterminal.

Other five-merous flowers of possible ericalean/ebenalean/ affinity have prominent abaxial sepal glands. There are literally hundreds of specimens of flowers representing at least five and perhaps more distinct, but similar, taxa. We consider three of these herein. In addition to the characters mentioned above, these taxa share clawed petals, more or less spherical lobed stigmas, numerous ovules per carpel, apparent staminodal nectaries alternating with the stamens, and pollen with smooth exine ornamentation and pronounced ektexine bridges in the endoaperturial regions.

One of the taxa in this complex has five upright sepals bearing two rows of abaxial glands, five clawed petals with undulating margins alternate with the sepals, and a whorl of five stamens alternating with the petals (Fig. 11A, B). Anthers have paired lateral/introrse pollen sacs and a broad connective with dorsal attachment (Fig. 11B–D). Filaments are flattened (Fig. 11B, C). Inside of and alternate with the stamen whorl is a whorl of five nectaries (Fig. 11B, E) with typical nectary cell structure

Fig. 8. A. Higher magnification view of pollen illustrated in Fig. 7H. ×3150. B. Sepal tube of myrtalean flower. ×14. C. Interior view of B illustrating the bicarpellate ovary with separate styles (arrows), and the coiled stamen filaments. ×14. D. Close-up of C illustrating a coiled stamen filament. ×70. E. Triporate pollen on stamen filament. ×7000. F. Clusioid flower with distinctive stigmas and fascicled stamens (arrow). ×75. G. Close-up of an anther showing preserved strands (arrows) stretching across the slightly open dehiscence slit. ×560. H. Top view of a different clusioid specimen, again illustrating the large stigma lobes and the more elongate stamens with larger anthers that are proximal to the ovary relative to the shorter outer stamens. × 56.

(Fig. 11F). Petal claws are adaxially grooved and juxtaposed with the opposite staminodes so tightly as to suggest that nectar may have moved into the petal grooves after secretion from the nectaries (Fig. 11E). Pollen is tricolporate and monadinous with extremely smooth exine sculpting and notable ektexine bridges in the region of the endoaperture (Fig. 11G). The pistil is composed of five fused carpels, a single style and a five-lobed stigma (Fig. 11B, C).

Another similar fossil flower type differs in several distinctive, and with respect to function, potentially significant ways. The larger stigma is more spherical and has indentations below the conjunctions of the lobes (Fig. 12A). Stamens are shorter relative to the stigmas and the anthers do not extend beyond the base of the stigma (Fig. 12A). The anthers are spurred (Fig. 12A, B) and, while the individual pollen grains are similar to those of the taxon described above, the contents of each theca are entirely interconnected by fine viscin threads (Fig. 12C, D) into polyad-like masses. The quality of preservation and ubiquity of these threads is remarkable.

A third ericalean/ebenalean taxon is distinctive in having glands widely distributed on the abaxial surfaces of the sepals (Fig. 12E) and styles covered with stellate trichomes (Fig. 12F). Anthers differ from those of the two taxa mentioned above. They do not have spurs and the distal connective is retuse (Fig. 12F, G). Fruits of this taxon retain the stamens and sepals and are consistently asymmetrical suggesting a degree of bilateral symmetry in the flowers (Fig. 12H).

A final ericalean taxon, *Paleoenkianthus*, has recently been described by Nixon & Crepet (1993; Fig. 13C). Stamens of *Paleoenkianthus* provide the first evidence of pseudoterminal dehiscence by either a pore or short aperture (occlusion of the apertures by amorphous material preserved within the anthers makes a more precise determination difficult) and also the first evidence of awns associated with the anthers (Fig. 13A, B). Pollen is monadinous and tricolporate with reticulate micromorphology but has apparent viscin threads as in some taxa of modern Ericales that have pollen in tetrads (Fig. 13D). In modern ericads, these characters have specific functions associated with pollination by bees (e.g., Knudsen & Olesen 1993).

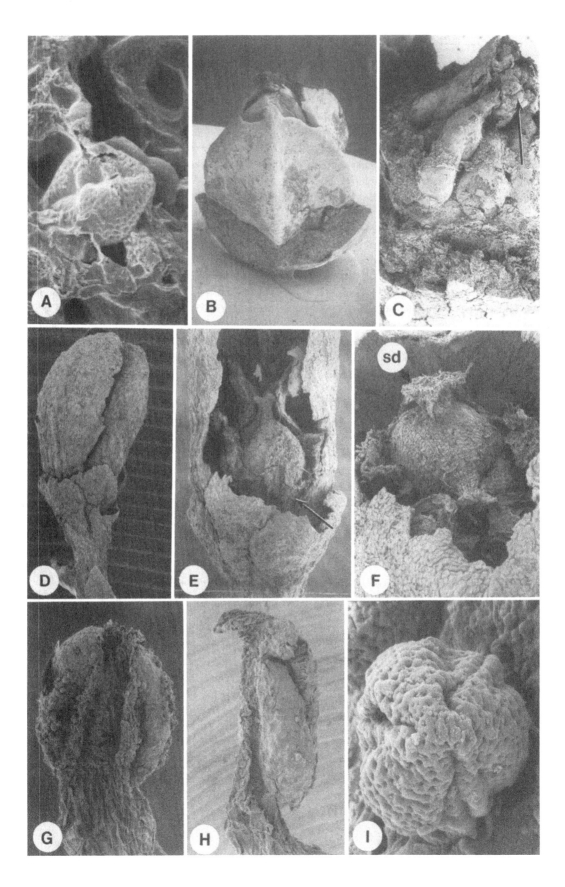

Summary/implications of the Turonian fossil record

1. The presence of magnolialean flowers (magnolialean taxa 1 and 2) with character complexes compatible with those of modern taxa specialized for beetle pollination by Cucurlionidae suggests that such specialization may have occurred in Magnoliales by the Turonian. In modern *Eupomatia*, pollen and tissues of stamens and staminodes are ingested by the weevil *Elleschodes* and ethereal oils in the tissues are an important part of the reward structure as well as serving an attractive role (Armstrong & Irvine 1990; Endress 1990; Bergstrom *et al.* 1991). In addition, in *Eupomatia* stamens serve to provide a sheltered physical space for mating and early development of these pollinators (Armstrong & Irvine 1990). Stamens therefore might be inferred to have served as protection as well as attractants and rewards in these magnolialean fossils as in modern taxa with similar stamens.

2. A rapid radiation of rosid and dilleniid taxa and the many stamen and anther types associated with them is evident by the Turonian. These sediments include the earliest fossil evidence of reproductive organs of Dilleniidae. Dilleniidae are already diverse with highly specialized taxa and stamen characters. In rosids, the frequency of what appear to be staminodal nectaries in Turonian taxa suggests that nectary rings in certain Rosidae might have been derived from reduced staminodal nectaries and underscores the importance of staminodes in floral evolution and floral reward structure within the rosid–hamamelid complex.

3. Fossil flowers similar to those of modern *Clusia* are evidence that close relatives of modern taxa now pollinated by meliponine and euglossine bees existed in the Turonian. Although the morphology of fossil stamens is suggestive of resin-bearing stamens in modern taxa, it is impossible to confirm or refute the possibility that some anthers were modified to produce resins as in modern species of *Clusia*.

4. The complex of ericalean/thealean/ebenalean flowers with a diversity of floral forms documents an early association of characters that, in related modern taxa, is indicative of rather derived modes of pollination by bees or syrphid flies. Modern analogs of these flowers include buzz-pollinated species as well as species with floral nectar rewards. One of the fossil taxa (Fig. 12A–D) has apparent nectaries, pollen in loose polyads and enlarged stigmas. These fossils represent the earliest evidence of stamens with spurs and also provide the earliest evidence of functional polyads.

The fossil record of Hymenoptera is growing rapidly and the acknowledged stem group of bees, the sphecid wasps, is known from Barremian deposits (Jarzembowski 1984). Derived bees, Meliponinae, are reported from Campanian–Maastrichtian sediments (Michener & Grimaldi 1988a, b), strongly suggesting the possibility that bees were available during the Turonian. While associations of Turonian floral characters might indicate the presence of bee pollination, it is also possible that these floral characters reflect specializations associated with other orders of anthophilous insects that have now been displaced by bees.

SANTONIAN–CAMPANIAN

Charcoalified and lignified flowers of Cretaceous age were first reported from the Santonian–Campanian deposits in Scania, Sweden (e.g., Friis & Skarby 1982;

Fig. 9. A. Pollen from the clusioid flower illustrated in Fig. 8F. ×500. B, C. Possible malvalean flower in bud. B. Flower showing keeled petal and broken fleshy sepal lobes. ×14.5. C. Bud with sepals and petals removed showing two whorls of stamens with elongate tubular anther connectives adjacent to the lobed stigma (arrow). ×19.6. D–I. Flower of dilleniid (possible capparalean) affinity. D. Unopened flower illustrating inequal petal sizes and sepals. ×42. E. The flower illustrated in D, dissected to reveal the ovary, stamens and setiform staminodes. Note the fusion of the stamen and staminode bases to form a tube (arrow) around the base of the stipitate ovary. ×77. F. Close-up of the ovary showing stipitate staminodes (sd). Stamens have been removed. ×140. G. Ventral view of dithecal anther. ×210. H. Lateral anther view. Note geniculate filament and the recurved connective extension. ×210. I. Polar view of tricolporate pollen grain. ×7700.

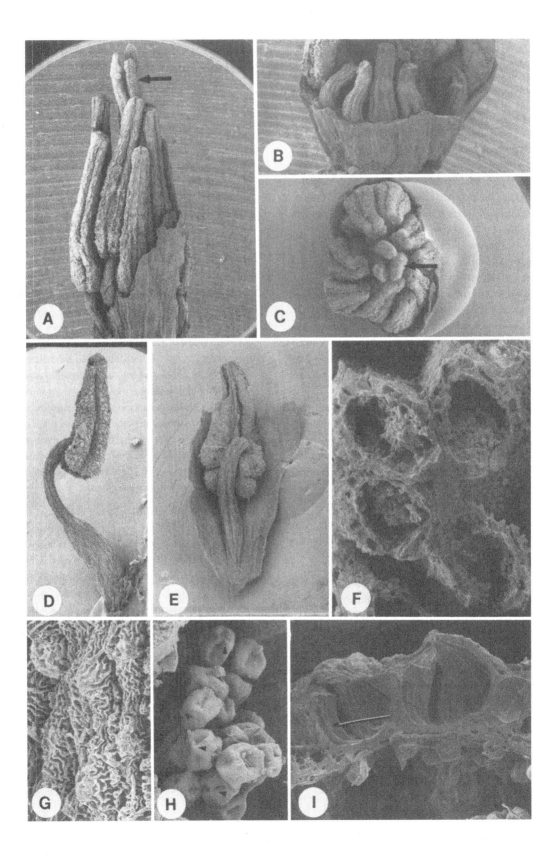

Friis 1984). Within Magnoliidae, these include stamens similar to those of modern Chloranthaceae (*Chloranthistemon*: Crane *et al.* 1989). In Hamamelididae, there are platanoid remains (*Platananthus*: Friis *et al.* 1988) and flowers of Hamamelidaceae having anthers with a single valve over each of the two thecae (*Archamamelis*: Endress & Friis 1991). These sediments also include cosexual flowers with evident rosid affinities and *Normapolles* grains preserved in their anthers (Friis 1983; Friis & Crane 1989). Other Rosidae are represented by several small, generally saxifragalean floral types with dorsifixed tetrasporangiate anthers having little connective development between the thecae and inner whorls of nectaries (e.g., Friis 1985a, 1990). Finally in Dilleniidae, possible Ericales are represented by the fossil flower *Actinocalyx* with anthers having smooth tricolporate pollen and apparent pollenkitt (Friis 1985b). Pollen in the rosid/dilleniid taxa of the Santonian–Campanian is uniformly small (10–15 µm) and smooth foveolate, tricolporate (Friis *et al.* 1991).

In general, taxa from Scania seem to represent a continuation of lineages that appeared earlier in the fossil record. A major distinguishing feature of these deposits is the association of some elements of the *Normapolles* flora with cosexual rosid flowers (Friis 1983). These fossils signal the origin of certain lineages that are now represented by wind-pollinated taxa with unisexual flowers.

CAMPANIAN–MAASTRICHTIAN

Numerous new angiosperm taxa appear suddenly in the Early Tertiary, suggesting that the uppermost Cretaceous was an important time in angiosperm radiation, but sufficient samples are not available to provide direct fossil evidence of such a radiation. The Cretaceous–Tertiary boundary is an interesting time period and a popular subject of speculation also because of the extinction of 'dinosaurs' and one or more correlated cataclysmic events that may have been causally related to more general mass extinctions in both plants and animals. It is therefore unfortunate that so few fossils of angiosperm reproductive structures are known from the time interval preceding the K–T boundary.

Although the general record is sparse, there is evidence of the appearance of important new taxa in the Campanian–Maastrichtian interval. Within Rosidae, flowers of Combretaceae are found with associated leaf fragments in the Upper Cretaceous rocks of Portugal (Friis *et al.* 1992). The oldest known monocots are Santonian palms followed by Campanian seeds (Friis 1988). By the Maastrichtian, monocots are better documented and include palm leaves and pollen (Daghlian 1981), and Maastrichtian leaves of Zingiberaceae (Hickey & Pedersen 1978). Maastrichtian pollen ascribed to grasses has been described, but the suite of characters diagnostic of grass pollen has not been fully demonstrated in these dispersed grains (see discussion in Crepet & Feldman 1991). Unfortunately, little is known about monocot stamens in this Cretaceous interval. Definitive evidence of Asteridae (discounting Ericales) is also missing from the uppermost Cretaceous.

EARLY TERTIARY

In addition to the continuing radiation of many groups that became established and diversified in the Cretaceous, a profusion of new taxa appears in the Early Tertiary. We will not attempt to summarize relevant fossil data here. However, most published accounts of stamen morphology in Tertiary fossils are descriptions of flowers/inflorescences. There are very few accounts

Fig. 10. A–H. Ericalean/Thealean flower. A. Lateral view with most of corolla missing. Note the two whorls of stamens with elongate anthers and the three styles/stigmas (arrow). ×25. B. A close-up of another specimen illustrating the corolla tube and stamen filaments. ×24.5. C. Top view of A showing stamen arrangement and stigmas (arrow). ×24.5. D. Lateral view of a stamen showing the curved filament that is proximally expanded. ×24.5. E. A stamen inserted on a corolla lobe illustrating the anther shape and ventral versatile attachment of the anther. ×28. F. A broken anther showing the four sporangia without an obvious dehiscence suture. ×126. G. Distal filament epidermis. ×1050. H. Pollen within the anther showing the finely scabrate exine micromorphology and the ektexinous bridges in the endoapertural regions (arrow heads). ×1050. I. Anther wall showing the structure of the outwardly pyramidal epidermal cells and their internal thickenings (arrow). ×560.

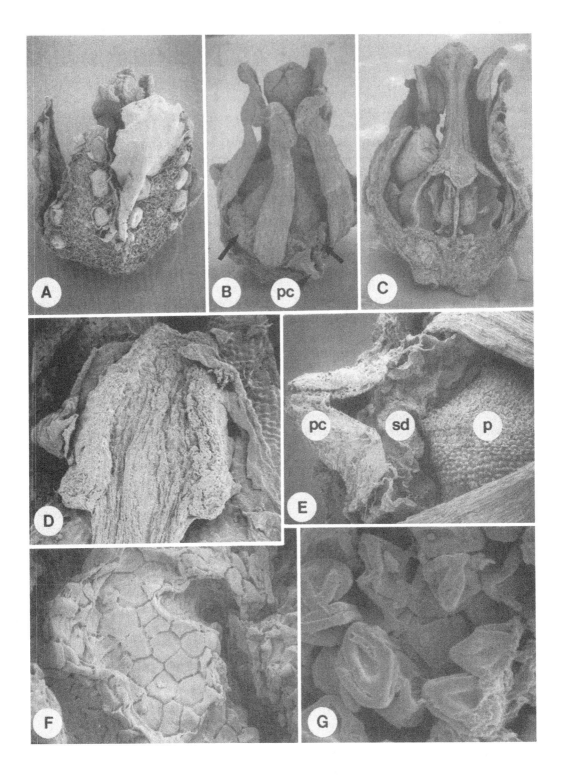

of isolated Tertiary stamens. To generalize, increasing floral diversity in the Tertiary reflects continuing specialization for various insect vectors, development of the wind pollination syndrome in 'higher Hamamelididae' (e.g., Crane 1981; Manchester & Crane 1983; Wing & Hickey 1984; Manchester 1987; Crepet & Nixon 1989) and grasses (Crepet & Feldman 1991), increasing diversity within Asteridae, and additional monocot diversity (Boyd 1992).

DISCUSSION AND FURTHER CONCLUSIONS

The fossil history of stamens and flowers in dicotyledonous angiosperms reveals the following generally observable pattern in the progression of subclasses (as outlined by Cronquist 1981):

Early Cretaceous: Magnoliidae (especially chloranthoids), 'Lower' Hamamelididae (tricolpates)

Mid/early Late Cretaceous: (tricolporates) Rosidae, including 'Higher' Hamamelididae, additional Magnoliidae, Dilleniidae (particularly Ericales/Theales)

While too sparse to provide a definitive pattern, particularly within Magnoliidae, the early fossil record of stamens reflects modifications consistent with a progression from generalized insect and wind pollination (assuming fossil platanoids were anemophilous) in the

Fig. 11. Ericalean/ebenalean flower with abaxial sepal glands. A. Flower with clawed petal, upright sepals with glands in two rows and anthers surrounding stigma. ×32.9. B. Flower with sepals and petals missing showing the flat wide filaments, dorsifixed anthers and broad abaxial anther connective, staminodes alternating with stamens (arrows), and one broken petal claw base (pc). ×34.3. C. Flower broken longitudinally. Note the cocoon on the left and the placentation. ×32.9. D. Top view of a flower with stamens, petals and distal sepals missing showing the staminodes and lobed ovary. ×133. E. Top view of a petal claw base (pc)/staminode (sd) juxtaposition. The pistil (p) is on the right and the staminode is flanked by stamen filaments. ×125. F. Close-up of presumably nectiferous cells of the staminode. ×637. G. *In situ* pollen. ×2460.

Early Cretaceous to highly specific pollination syndromes and correlated changes in stamen morphology by the mid and Late Cretaceous. The early appearance of chloranthoid stamens, anther contents and dispersed *Clavatipollenites* pollen might be interpreted as support for the primitive nature of branched (i.e., pinnate) stamens within angiosperms (see discussion in Herendeen *et al.* 1993). Concomitantly, such data might be interpreted as evidence of the antiquity of Chloranthaceae relative to other groups of angiosperms, and again this is consistent with certain hypotheses of phylogeny and with available fossil evidence (Dahlgren *et al.* 1985; Burger 1977; Nixon *et al.* 1994; see also Dilcher 1979). However, dispersed pollen with winteroid characters (Walker *et al.* 1983) and flattened latrorse stamens of the same age (Friis *et al.* 1994) are evidence that considerable diversification had already occurred by the time of the oldest fossil angiosperm flowers (and stamens). Even so, definitive magnolialean stamen fragments do not appear until the Cenomanian (Friis *et al.* 1991).

Beyond their basic function, stamens have had a critical role as floral building blocks in the evolution of floral morphology and function through time. In variously modified form, they have served both as attractants and rewards in addition to having been modified in many ways to function optimally in placing pollen on pollinators or to allow pollen dispersal by wind or other abiotic means. Through time, and within different lineages, there has been a repetition of a transformation from stamens to petals/nectaries. For example, stamens have become modified into attractants or rewards in some Magnoliales, the Hamamelididae, and Rosidae at different times in the radiation of the flowering plants. The recurrence of this transformation may be related in each instance to coevolution with a different group or groups of pollinators that, through the agencies of their particular mouthparts, behavior, vision, and nutritional requirements affected specific transformations in floral morphology. Thus, for example, evolution of staminodes to provide protection, attractants, and rewards based on the chewing and ingesting of tissues and pollen in some Magnoliales evidently took place in conjunction with Coleoptera (possibly curculionid or nititulid beetles), while the transition to petals and staminodal nectaries in the hamamelid–rosid lineage most likely involved Diptera and perhaps Hymenoptera that had mouthparts

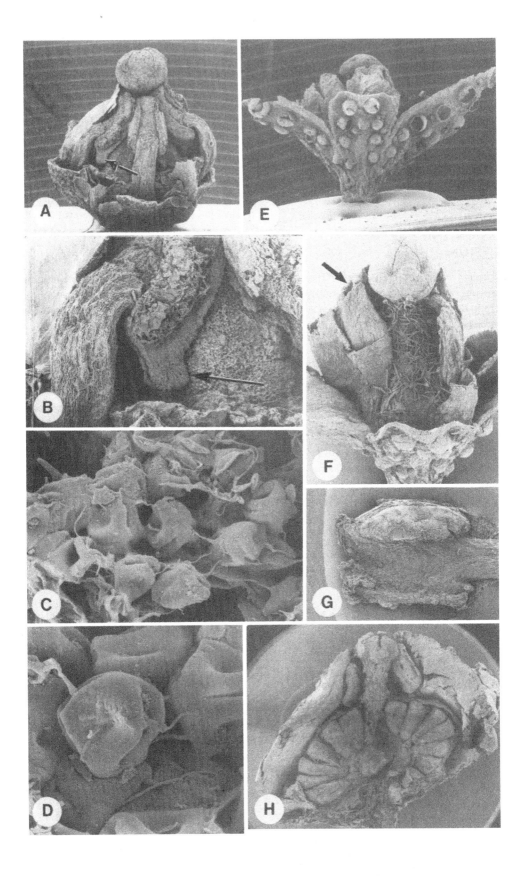

more suitable for liquid feeding. It is more difficult to evaluate the role of nectar as an attractant in the history of Magnoliidae because nectiferous surfaces on stigmas, carpels, stamen bases, or perianth are harder to identify in fossils. Lloyd & Wells (1992) have illustrated that stigmatic exudates are important attractants/rewards in fly – and beetle-pollinated *Pseudowintera* and such attractants cannot be ruled out when considering Cretaceous Magnoliales (especially in view of stigmatic epidermis characters in some taxa, e.g., magnolialean taxon 1: Crepet & Nixon 1994).

One logical prelude to transformation of stamens into either attractive or reward-providing structures involves multiplication of stamen numbers or cycles. If all stamens were converted to one of these functions, then their role in pollen production would, of course, be compromised. Thus, increases in the numbers of spirally arranged stamens in some taxa or in the number of cycles of stamens in others seem to be necessary correlates of having stamens specialized for functions in attraction and reward. Therefore, duplication of stamen cycles (or, in some Magnoliidae, simply increases in the number of spirally arranged stamens, unless the original condition included numerous stamens) must have been a critical part of floral evolution, and duplication followed by specialization of some of the new cycles of stamens in association with modes of pollination would have been a neces-

sary combination of events that might have been repeated at different times and within different monophyletic groups.

In the early chloranthoids, magnoliids and hamamelids, there was apparently no 'division of labor' in stamens. It is possible that, in these taxa, stamens that were fully functional in the production of pollen served also as attractants and rewards simultaneously (Friis *et al.* 1991). There are no showy bracts or petals in such flowers and the stamens have large amounts of sterile tissue, including extended connectives, that would have been rather obvious to potential pollinators (Friis *et al.* 1991).

In the first complete flower fossils of Magnoliales, stamens have not yet become as specialized as the stamens in modern related taxa, with some functioning in pollen production and others in the functions of attraction and reward (Crepet & Nixon 1994). Although modern platanoids have staminodes, staminodes did not exist in their early counterparts in a form suggestive of nectar secretion (Friis *et al.* 1986).

One of the most notable aspects of the fossil record of stamens in the hamamelid – rosid lineage is the profusion of taxa that have apparent staminodal nectaries in the early part of the Late Cretaceous. This phenomenon may reveal mid-Cretaceous radiation in anthophilous nectar-feeding insects and is consistent with the increasing diversity in Hymenoptera and Diptera occurring over the same interval (Michener & Grimaldi, 1988a,b; Carpenter & Rasnitsyn 1990; Crepet *et al.* 1991; Carpenter 1992).

From the perspective of the evolution of floral structure, the staminodes of pre-Hamamelidaceae from the Turonian (Crepet *et al.* 1992) are most interesting. In phylogenetic context, these suggest that Hamamelidaceae evolved from platanoid ancestors via the modification of one whorl of stamens into staminodes and the fusion of the staminal tube with a sepal cup to form a hypanthium. The character transformation from such a hypanthium-bearing early hamamelid to more typical Rosidae could then be envisioned as having included a multiplication of stamen whorls and then a 'division of labor' through specialization of some whorls as petals while others became modified as nectaries. This transition also evidently involved a shift to colporate apertures in the pollen.

The fossil record provides insights into the evolution

Fig. 12. Ericalean/ebenalean flowers. A–D. Five-merous flower with lobed stigmas and clawed petals. A. Dissected flower, with petals removed. Arrow indicates anther spur. ×32. B. Close-up of anther spur (arrow). ×105. C. *In situ* pollen with viscin threads. ×2100. D. Closer view of pollen grains showing viscin threads and smooth exine. ×4200. E–H. Another possible ericalean five-merous flower with prominent globose multicellular glands on abaxial surfaces of sepals. E. Overview of flower showing glandular sepals. ×42. F. Dissection showing stellate trichomes on ovary and style, and globose lobed stigma. Note the notched anther connectives (arrow). ×70. G. Close-up of anther, adaxial surface, showing broad connective with a distal notch. ×105. H. Longitudinal dissection showing axile placentation and numerous clavate ovules/seeds. Note the irregular shape of the flower. ×21.

Fig. 13. *Paleoenkianthus*. A, B. Stamens. Note awns in A, and inversion of pollen sacs. ×250. C. Open bud. Note the styles and anther inside. ×51.1. D. Pollen connected by fine threads. ×1540.

of character complexes related to various modes of pollination today. From the fossil record we may now identify minimal times for the origination of such character complexes, as discussed above, providing a rough pattern for the evolution of floral morphology and inferred modes of pollination through time. Early evidence of stamens and flowers suggests that the first angiosperms did not have highly attractive petals nor did they have highly modified staminal nectaries. As far as can be determined, early angiosperms appear to have had flowers with few or relatively low numbers of floral parts. Fossil evidence suggests that early modes of insect pollination were similar to insect pollination in certain extant angiosperms. For example, Winteraceae today may be pollinated by thrips, Diptera, micropterigid Lepidoptera and certain Coleoptera (Gottsberger *et al.* 1980; Thien 1980; Thien *et al.* 1985; Miller 1989). Pollination in Chloranthaceae is still incompletely understood. *Chloranthus* and *Sarcan-*

dra are considered well suited for insect pollination because of cosexual flowers and attractive stamens (Endress 1990, and references therein). *Ascarina* shows variability in floral sexuality and attractiveness of the stamens and may be pollinated by both wind and insects (Endress 1990). *Hedyosmum* flowers are well suited for wind dispersal of pollen and this genus is one of the few extant Magnoliidae to have such floral and pollen structure (Endress 1990). Early Hamamelididae may also have been insect-pollinated but illustrate no features suggestive of a high degree of specialization (e.g., Friis *et al.* 1988). The Early Cretaceous insect fauna includes flies, beetles and certain Hymenoptera as well as micropterigid moths (Jarzembowski 1984; Crepet & Friis, 1987; Crepet *et al.* 1991; Carpenter 1992); none of these can be reasonably ruled out as important pollinators in the first phases of angiosperm radiation.

Studies of the fossil record of stamens and flowers are in their infancy. The recent discoveries of Early and mid-Cretaceous fossil flowers have provided a completely revised understanding of early floral and stamen evolution. Continuing investigations in the context of improved phylogenetic analyses should make significant contributions from the fossil record to our understanding of angiosperm evolution and relationships. Increased understanding of stamen and floral structure in early angiosperms will improve this understanding and provide a basis for analysis of competing hypotheses about coevolution and adaptation of angiosperms and insect pollinators, and factors affecting early angiosperm diversification and success.

ACKNOWLEDGMENTS

We thank M.A. Gandolfo for assistance with laboratory work and literature searches; Jennifer Svitko for electron microscopy; Else-Marie Friis, Swedish Natural History Museum, Stockholm, for Fig. 5 and access to unpublished data; William G. D'Arcy, Missouri Botanical Garden, Patrick S. Herendeen, Field Museum, Chicago, and Scott L. Wing, National Museum of Natural History, Washington, DC, for their helpful comments and suggestions. Research was supported by NSF DEB92-01179 and the College of Agriculture and Life Sciences, Cornell University, Ithaca, New York.

LITERATURE CITED

Anderberg, A.A. 1992. The circumscription of the Ericales, and their cladistic relationships to other families of 'Higher' Dicotyledons. Syst. Bot. 17: 660–675.

Andrews, H.N. 1963. Studies in Paleobotany. Wiley, New York.

Archangelsky, S. & T.N. Taylor. 1993. The ultrastructure of in situ *Clavatipollenites* pollen from the Early Cretaceous of Patagonia. Amer. J. Bot. 879–885.

Armbruster, W.S. 1984. The role of resin in angiosperm pollination: ecological and chemical considerations. Amer. J. Bot. 71: 1149–1160.

Armstrong, J.E. & A.K. Irvine. 1990. Functions of staminodia in the beetle-pollinated flowers of *Eupomatia laurina*. Biotropica 22: 429–431.

Basinger, J.F. & D.L. Dilcher. 1984. Ancient bisexual flowers. Science 224: 511–513.

Bergstrom, B., I. Groth, O. Pellmyr, P.K. Endress, L.B. Thien, A. Hubener & W. Francke. 1991. Chemical basis of a highly specific mutualism: chiral esters attract pollinating beetles in Eupomatiaceae. Phytochemistry 30: 3221–3225.

Bernhardt, P. & L.B. Thien. 1987. Self-isolation and insect pollination in the primitive angiosperms: new evaluations of older hypotheses. Pl. Syst. Evol. 156: 159–176.

Bierhorst, D.W. 1971. Morphology of Vascular Plants. Macmillan, New York.

Bogle, A.L. & C.T. Philbrick. 1980. A generic atlas of hamamelidaceous pollens. Contrib. Gray Herb. Harv. Univ. 210: 29–103.

Boyd, A. 1992. *Musopsis* n. gen.: a banana-like leaf genus from the Early Tertiary of eastern North Greenland. Amer. J. Bot. 79: 1359–1367.

Brenner, G.J. 1963. The spores and pollen of the Potomac Group of Maryland. State of Maryland Dep. Geol. Mines Water Res. *Bull.* 27: 1–125.

Burger, W.C. 1977. The Piperales and the monocots. Alternate hypotheses for the origin of monocotyledonous flowers. Bot. Rev. 43: 345–393.

Carpenter, F.M. 1992. Superclass Hexapoda. Treatise on Invertebrate Paleontology (edited by R.L. Kaesler), Vols. 3, 4. Geological Society of America, University of Kansas Press, Lawrence, Kansas.

Carpenter, J.M. & A.P. Raznitsyn. 1990. Mesozoic Vespidae. *Psyche* 97: 1–20.

Chapman, J.L. 1987. Comparison of Chloranthaceae pollen with the Cretaceous 'Clavatipollenites complex': taxonomic implications for palaeopalynology. Pollen Spores 29: 249–272.

Chase, M.W., D.E. Soltis, R.G. Olmstead, D. Morgan, D.H. Les, B.D. Mishler *et al.* 1993. Phylogenetics of seed plants: an analysis of nucleotide sequences from the plastid gene *rbcL*. *Ann. Missouri Bot. Gard.* 80: 528–580.

Chlonova, A.F. & T.D. Surova. 1988. Pollen wall ultrastructure of *Clavatipollenites incisus* Chlonova and two modern species of *Ascarina* (Chloranthaceae). Pollen Spores 30: 29–44.

Christopher, R.A. 1977. Selected *Normapolles* genera and the age of the Raritan and Magothy formations (Upper Cretaceous) of Northern New Jersey. pp. 58–69 in A Field Guide to Cretaceous and Lower Tertiary Beds of the Raritan and Salisbury Embayments, New Jersey, Delaware, and Maryland (edited by J.P. Owens, N.F. Sohl & J.P. Minard). Am. Assoc. Pet./Geol. Soc. Econ. Paleontol. Minerol. Guidebook. Ann. Conv. Washington, D.C.

Couper, R.A. 1958. British Mesozoic microspores and pollen grains. Palaeontographica 103B: 75–179.

Crabtree, D.R. 1987. Angiosperms of the northern Rocky Mountains: Albian to Campanian (Cretaceous) megafossil floras. Ann. Missouri Bot. Gard. 74: 707–747.

Crane, P.R. 1981. Betulaceous leaves and fruits from the British Upper Palaeocene. Bot. J. Linn. Soc. London 83: 103–136.

Crane, P.R. 1985. Phylogenetic analysis of seed plants and the origin of angiosperms. Ann. Missouri Bot. Gard. 72: 716–793.

Crane, P.R. 1989. Paleobotanical evidence on the early radiation of non-magnoliid dicotyledons. Pl. Syst. Evol. 162: 165–191.

Crane, P.R. & D.L. Dilcher. 1984. *Lesqueria*: an early angiosperm fruiting axis from the mid-Cretaceous. Ann. Missouri Bot. Gard. 71: 384–402.

Crane, P.R., E.M. Friis & K.R. Pedersen. 1986. Angiosperm flowers from the Lower Cretaceous: fossil evidence on early radiation of dicotyledons. Science 232: 852–854.

Crane, P.R., E.M. Friis & K.R. Pedersen. 1989. Reproductive structure and function in Chloranthaceae. Pl. Syst. Evol. 165: 211–226.

Crane, P.R., E.M. Friis & K.R. Pedersen. 1994. Paleobotanical evidence on the early radiation of magnoliid angiosperms. In The early fossil record of angiosperm flowers (edited by E.M. Friis & P.K. Endress). Pl. Syst. Evol. 8 (suppl.): 51–72.

Crepet, W.L. 1974. Investigations of North American cyca-

deoids: the reproductive biology of *Cycadeoidea*. Palaeontographica 148B: 144–169.

Crepet, W.L. 1979. Insect pollination: a paleontological perspective. BioScience. 29: 102–108.

Crepet, W.L. 1983. The role of insect pollination in the evolution of the angiosperms. pp. 29–50 in Pollination Biology (edited by L. Real) Academic Press, New York.

Crepet, W.L. & G.D. Feldman. 1991. The earliest remains of grasses in the fossil record. Amer. J. Bot. 78: 1010–1014.

Crepet, W.L. & E.M. Friis. 1987. The evolution of insect pollination mechanisms in angiosperms. pp. 181–201 in (editors), The Origin of Angiosperms and their Biological Consequences (edited by E.M. Friis, W.G. Chaloner and P.R. Crane). Cambridge University Press, Cambridge.

Crepet, W.L. & K.C. Nixon. 1989. Extinct transitional Fagaceae from the Oligocene and their phylogenetic implications. Amer. J. Bot. 76: 1493–1505.

Crepet, W.L. & K.C. Nixon. 1994. Flowers of Turonian Magnoliidae and their implications. In The early fossil record of angiosperm flowers (edited by E.M. Friis & P.K. Endress). Pl. Syst. Evol. 8 (Suppl.): 73–91.

Crepet, W.L., E.M. Friis & K.C. Nixon. 1991. Fossil evidence for the evolution of biotic pollination. Phil. Trans. R. Soc. Lond. 333B: 187–195.

Crepet, W.L., K.C. Nixon, E.M. Friis & J.V. Freudenstein. 1992. Oldest fossil flowers of hamamelidaceous affinity, from the Late Cretaceous of New Jersey. Proc. Natl. Acad. Sci. USA 89: 8986–8989.

Cronquist, A. 1981. An Integrated System of Classification of Flowering Plants. Columbia University Press, New York.

Daghlian, C.P. 1981. A review of the fossil record of monocotyledons. Bot. Rev. 47: 517–555.

Dahlgren, R.M.T., H.T. Clifford & P.F. Yeo. 1985. The Families of the Monocotyledons: Structure, Evolution and Taxonomy. Springer-Verlag, New York.

Delevoryas, T. 1963. Investigations of North American cycadeoids: cones of *Cycadeoidea*. Amer. J. Bot. 50: 45–52.

Dilcher, D.L. 1974. Approaches to the identification of angiosperm leaf remains. Bot. Rev. 40: 1–157.

Dilcher, D.L. 1979. Early angiosperm reproduction: an introductory report. Rev. Palaeob. Palyn. 27: 291–328.

Dilcher, D.L. & P.R. Crane. 1984. *Archaeanthus*: an early angiosperm from the Cenomanian of the western interior of North America. Ann. Missouri Bot. Gard. 71: 351–383.

Donoghue, M.J. & J.A. Doyle. 1989. Phylogenetic studies

of seed plants and angiosperms based on morphological characters. pp. 181–193 in The Hierarchy of Life (edited by B. Fernholm, K. Bremer & H. Jörnvall). Elsevier, New York.

Doyle, J.A. 1969. Cretaceous angiosperm pollen of the Atlantic Coastal Plain and its evolutionary significance. J. Arnold Arbor. 50: 1–35.

Doyle, J.A. 1978. Origin of angiosperms. Annu. Rev. Ecol. Syst. 9: 365–392.

Doyle, J.A. & M.J. Donoghue. 1986. Seed plant phylogeny and the origin of angiosperms: an experimental cladistic approach. Bot. Rev. 52: 321–431.

Doyle, J.A. & L.J. Hickey. 1976. Pollen and leaves from the mid-Cretaceous Potomac Group and their bearing on early angiosperm evolution. pp. 139–206 in Origin and Early Evolution of Angiosperms (edited by C.B. Beek). Columbia University Press, New York.

Doyle, J.A. & C.L. Hotton. 1991. Diversification of early angiosperm pollen in a cladistic context. pp. 168–195 in Pollen and Spores: Patterns of Diversification (edited by S. Blackmore & S.H. Barnes). Clarendon Press, Oxford.

Doyle, J.A. & E.I. Robbins. 1977. Angiosperm pollen zonation of the continental Cretaceous of the Atlantic Coastal Plain and its application to deep wells in the Salisbury Embayment. Palynology 1: 43–78.

Drinnan, A.N., P.R. Crane, E.M. Friis & K.R. Pedersen. 1990. Lauraceous flowers from the Potomac Group (mid-Cretaceous) of eastern North America. Bot. Gaz. 151: 370–380.

Drinnan, A.N., P.R. Crane, E.M. Friis & K.R. Pedersen. 1991. Angiosperm flowers and tricolpate pollen of buxaceous affinity from the Potomac group (mid-Cretaceous) of eastern North America. Amer. J. Bot. 78: 153–176.

Endress, P.K. 1986. Reproductive structures and phylogenetic significance of extant primitive angiosperms. Pl. Syst. Evol. 152: 1–28.

Endress, P.K. 1987. The Chloranthaceae: reproductive structures and phylogenetic position. Bot. Jb. Syst. Pflanzengesch. Pflanzengeogr. 109: 153–226.

Endress, P.K. 1990. Evolution of reproductive structures and functions in primitive angiosperms (Magnoliidae). Mem. New York Bot. Gard. 55: 5–34.

Endress, P.K. & E.M. Friis. 1991. Archamamelis, hamamelidalen flowers from the Upper Cretaceous of Sweden. Pl. Syst. Evol. 175: 101–114.

Endress, P.K. & L.D. Hufford. 1989. The diversity of stamen structures and dehiscence patterns among Magnoliidae. J. Linn. Soc., Bot. 100: 45–85.

Endress, P.K. & S. Stumpf. 1991. The diversity of stamen structures in Lower Rosidae (Rosales, Fabales, Proteales, Sapindales). J. Linn. Soc., Bot. 107: 217–294.

Erdtmann, G., 1952: Pollen Morphology and Plant Taxonomy: Angiosperms. Almqvist & Wiksell, Stockholm.

Faegri, K. & L. van der Pijl. 1971. The Principles of Pollination Ecology, 2nd ed. Pergamon Press, Oxford.

Friis, E.M. 1983. Upper Cretaceous (Senonian) floral structures of juglandalean affinity containing Normapolles pollen. Rev. Paleobot. Palynol. 39: 161–188.

Friis, E.M. 1984. Preliminary report of Upper Cretaceous angiosperm reproductive organs from Sweden and their level of organization. Ann. Missouri Bot. Gard. 71: 403–418.

Friis, E.M. 1985a. Structure and function in Late Cretaceous angiosperm flowers. Biol. Skr. K. Danske Vidensk. Selsk. 25: 1–37.

Friis, E.M. 1985b. Actinocalyx gen. nov., sympetalous angiosperm flowers from the Upper Cretaceous of southern Sweden. Rev. Paleobot. Palynol. 45: 171–183.

Friis, E.M. 1988. Spirematospermun chandlerae sp. nov., an extinct species of Zingiberaceae from the North American Cretaceous. Tert. Res. 9: 7–12.

Friis, E.M. 1990. Silvianthemum suecicum gen. et sp. nov., a new saxifragalean flower from the Late Cretaceous of Sweden. Biol. Skr. K. Danske Vidensk. Selsk. 36: 1–35.

Friis, E.M. & P.R. Crane. 1989. Reproductive structures of Cretaceous Hamamelidae. pp. 155–174 in Evolution, Systematic and Fossil History of the Hamamelidae (edited by P.R. Crane & S. Blackmore). Oxford University Press, Oxford.

Friis, E.M. & W.L. Crepet. 1987. Time of appearance of floral features. pp. 145–179. in The Origins of Angiosperms and their Biological Consequences (edited by E.M. Friis, W.G. Chaloner & P.R. Crane). Cambridge University Press, Cambridge.

Friis, E.M. & P.K. Endress. 1990. Origin and evolution of angiosperm flowers. Adv. Bot. Research. 17:99–162.

Friis, E.M. & A. Skarby. 1982. Scandianthus gen. nov., angiosperm flowers of saxifragalean affinity from the Upper Cretaceous of southern Sweden. Ann. Bot. 50: 569–583.

Friis, E.M., P.R. Crane & K.R. Pedersen. 1986. Floral evidence for Cretaceous chloranthoid angiosperms. Nature 320: 163–164.

Friis, E.M., P.R. Crane & K.R. Pedersen. 1988. Reproductive structures of Cretaceous Platanaceae. Biol. Skr. K. Danske Vidensk Sels. 31: 1–56.

Friis, E.M., P.R. Crane & K.R. Pedersen. 1991. Stamen

diversity and *in situ* pollen in Cretaceous angiosperms. pp. 197–224 in Pollen and Spores: Patterns of Diversification (edited by S. Blackmore & S.H. Barnes). Clarendon Press, Oxford.

Friis, E.M., P.R. Crane & K.R. Pedersen. 1994. Angiosperm floral structures from the Early Cretaceous of Portugal. in, The early fossil record of angiosperm flowers (edited by E.M. Friis & P.K. Endress). Pl. Syst. Evol. 8 (suppl.): 31–49.

Friis, E.M., K.R. Pedersen & P.R. Crane. 1992. *Esgueria* gen. nov., fossil flowers with combretaceous features from the Late Cretaceous of Portugal. Biol. Skr. K. Danske Vidensk. Selsk. 41: 5–25.

Gottsberger, G. 1977. Some aspects of beetle pollination in the evolution of flowering plants. Pl. Syst. Evol. Suppl. 1: 211–226.

Gottsberger, G. 1988. The reproductive biology of primitive angiosperms. Taxon 37: 630–643.

Gottsberger, G., I. Silberbauer–Gotsberger & F. Ehrendorfer. 1980. Reproductive biology in the primitive relic angiosperm *Drimys brasiliensis* (Winteraceae). Pl. Syst. Evol. 135: 11–39.

Grant, V. 1950. The pollination of *Calycanthus occidentalis*. Amer. J. Bot. 37: 294–296.

Herendeen, P.S., W.L. Crepet & K.C. Nixon. 1993. *Chloranthus*-like stamens from the Upper Cretaceous of New Jersey. Amer. J. Bot. 80: 865–871.

Herendeen, P.S., W.L. Crepet & K.C. Nixon. 1994. Fossil flowers and pollen of Lauraceae from the Upper Cretaceous of New Jersey. Pl. Syst. Evol. 189: 29–40.

Hesse, M. & K. Kubitzki. 1983. The sporoderm ultrastructure in *Persea, Nectandra, Hernandia, Gomortega* and some other lauralean genera. Pl. Syst. Evol. 141: 299–311.

Hickey, L.J., 1978. Origin of the major features of angiospermous leaf architecture in the fossil record. Cour. Forsch. Inst. Senckenberg. 30: 27–34.

Hickey, L.J., & J.A. Doyle. 1977. Early Cretaceous fossil evidence for angiosperm evolution. Bot. Rev. 43: 3–104.

Hickey, L.J., & K.R. Pedersen. 1978. *Zingiberopsis*, a fossil genus from the ginger family from Late Cretaceous to early Eocene sediments of western interior North America. Can. J. Bot. 56: 1136–1152.

Hufford, L.D. & P.K. Endress. 1989. The diversity of stamen structures and dehiscence patterns among Hamamelididae. J. Linn. Soc., Bot. 99: 301–346.

Hughes, N.F. 1976. Paleobiology of Angiosperm Origins. Cambridge University Press, Cambridge.

Hughes, N.F., C.E. Drewry & J.F. Lang. 1979. Barremian earliest angiosperm pollen. Palaeontology 22: 513–535.

Jarzembowski, E.A. 1984. Early Cretaceous insects from Southern England. Mod. Geol. 9: 71–93.

Knudsen, J.T. & J.M. Olesen. 1993. Buzz-pollination and patterns in sexual traits in North European Pyrolaceae. Amer. J. Bot. 80: 900–913.

Kubitzki, K. 1981. The tubular exine of Lauraceae and Hernandiaceae, a novel type of exine structure in seed plants. Pl. Syst. Evol. 138: 139–146.

Lidgard, S. & P.R. Crane. 1990. Angiosperm diversification and Cretaceous floristic trends: a comparison of palynofloras and leaf macrofloras. Paleobiology 16: 77–93.

Lloyd, D.G. & M.S. Wells. 1992. Reproductive biology of a primitive angiosperm, *Pseudowintera colorata* (Winteraceae) and the evolution of pollination systems in the Anthophyta. Pl. Syst. Evol. 181: 77–95.

Loconte, H. & D.W. Stevenson. 1990. Cladistics of the Spermophyta. Brittonia 42: 197–211.

Loconte, H. & D.W. Stevenson. 1991. Cladistics of Magnoliidae. Cladistics 7: 267–296.

Manchester, S.R. 1987. The fossil history of Juglandaceae. Ann. Missouri Bot. Gard. Monogr. 21: 1–137.

Manchester, S.R. & P.R. Crane. 1983. Attached leaves, inflorescences and fruits of *Fagopsis*, an extinct genus of fagaceous affinity from the Florissant Flora of Colorado, U.S.A. Amer. J. Bot. 70: 1147–1164.

Martens, P. 1971. Les gnetophytes. Handb. Pflanzen-anat. 12: 1–295.

McGinnes, E.A., Jr., S.A. Kandeel & P.S. Szopa. 1971. Some structural changes observed in the transformation of wood into charcoal. Wood Fiber 3: 77–83.

McGinnes, E.A., P.S. Szopa, and J.E. Phelps. 1974. Use of scanning electron microscopy in studies of wood charcoal formation. Scan. Electron Microsc. 3: 469–476.

Michener, C.D. & D.A. Grimaldi. 1988a. A *Trigona* from Late Cretaceous amber at New Jersey (Hymenoptera: Apidae: Meliponinae). Amer. Mus. Novitates 2917: 1–10.

Michener, C.D. & D.A. Grimaldi. 1988b. The oldest fossil bee: apoid history, evolutionary stasis, and the antiquity of social behavior. Proc. Natl. Acad. Sci. USA 85: 6424–6426.

Miller, J.M. 1989. The archaic flowering plant family Degeneriaceae: its bearing on an old enigma. Natl. Geogr. Res. 5: 218–232.

Muller, J. 1970. Palynological evidence on early differentiation of angiosperms. Biol. Rev. 45: 417–450.

Nixon, K.C. & W.L. Crepet. 1993. Late Cretaceous fossil flowers of ericalean affinity. Amer. J. Bot. 80: 616–623.

Nixon, K.C., W.L. Crepet, D.A. Stevenson, & E.M. Friis. 1994. A reevaluation of seed plant phylogeny. Ann. Missouri Bot. Gard. 81: 484–485.

Olmstead, R.G., B. Bremer, K.M. Scott & J.D. Palmer. 1993. A parsimony analysis of the Asteridae *sensu lato* based on *rbcL* sequences. Ann. Missouri Bot. Gard. 80: 700–722.

Pedersen, K.R., P.R. Crane, A.N. Drinnan & E.M. Friis. 1991. Fruits from the mid-Cretaceous of North America with pollen grains of the Clavatipollenites type. Grana 30: 577–590.

Pellmyr, O., L.B. Thien, G. Bergstrom & I. Groth. 1990. Pollination of New Caledonian Winteraceae: opportunistic shifts or parallel radiation with their pollinators? Pl. Syst. Evol. 173: 143–157.

Procter, M & P. Yeo. 1972. The Pollination of Flowers. Collins, London.

Regal, P.J. 1977. Ecology and evolution of flowering plant dominance. Science 196: 622–629.

Taylor, T.N. 1982. The reproductive biology of early seed plants. BioScience 32: 23–28.

Taylor, T.N. & L.J. Hickey. 1990. An Aptian plant with attached leaves and flowers: implications for angiosperm origin. Science 247: 702–704.

Thien, L.B. 1980. Patterns of pollination in primitive angiosperms. Biotropica 12: 1–13.

Thien, L.B., P. Bernhardt, G.W. Gibbs, O. Pellmyr, G. Bergström, I. Groth & G. McPherson. 1985. The pollination of *Zygogynum* (Winteraceae) by a moth, *Sabatinca* (Micropterigidae): an ancient association? Science 227: 540–543.

Thomas, H.H. 1915. On *Williamsoniella*, a new type of bennettitalean flower. Phil. Trans., Ser. B 207: 113–148.

Upchurch, G.R., Jr. 1984. Cuticle evolution in Early Cretaceous angiosperms from the Potomac Group of Virginia and Maryland. Ann. Missouri Bot. Gard. 71: 522–550.

Upchurch, G.R. & D.L. Dilcher. 1990. Cenomanian angiosperm leaf megafossils, Dakota Formation, Rose Creek Locality, Jefferson County, southeastern Nebraska. U.S. Geol. Serv. Bull. 1915: 1–55.

Walker, J.W. & A.G. Walker. 1984. Ultrastructure of Lower Cretaceous angiosperm pollen and the origin and early evolution of flowering plants. Ann. Missouri Bot. Gard. 71: 464–521.

Walker, J.W., G.J. Brenner & A.G. Walker. 1983. Winteraceous pollen in the Lower Cretaceous of Israel: early evidence of a magnolialean angiosperm family. Science 220: 1273–1275.

Ward, J.V, J.A. Doyle & C.L. Hotton. 1989. Probable granular magnoliid angiosperm pollen from the Early Cretaceous. Pollen Spores. 33: 101–120.

Whalley, P. 1977. Lower Cretaceous Lepidoptera. Nature 266: 526.

Wieland, G.R. 1906. American Fossil Cycads. Carnegie Institute, Washington.

Wing, S.L. & L.J. Hickey. 1984. The *Platycarya* perplex and the evolution of Juglandaceae. Amer. J. Bot. 78: 388–411.

Wolfe J.A. & G.R. Upchurch, Jr. 1986. Vegetation, climatic and floral changes at the Cretaceous–Tertiary boundary. Nature 324: 148–152.

3 • The origin and early evolution of angiosperm stamens

LARRY HUFFORD

INTRODUCTION

Our understanding of the origin and early evolution of key angiosperm innovations, such as the carpel, bitegmic ovule, endosperm, and stamen, is quite limited. It is confounded by both the great divergence of flowering plants from other extant seed plants and by our limited knowledge of those extinct seed plants (particularly Bennettitales and Gnetales) that radiated during the Mesozoic. Recent cladistic analyses have refined our understanding of angiosperm evolution by identifying those groups of seed plants most closely related to the flowering plants (Crane 1985; Doyle & Donoghue 1986). The results of these cladistic analyses provide a powerful avenue for explorations of character state evolution (Doyle & Donoghue 1993). The origin and early evolution of angiosperm stamens are explored in this paper using the results of recent cladistic analyses. Key problems that are examined include (1) the implications of outgroup microsporophyll attributes for understanding the conditions from which stamens evolved and possible plesiomorphic states among angiosperms, (2) the implications of different rootings of angiosperm cladograms for interpreting stamen diversification, and (3) the evolution of significant aspects of stamen morphology (including laminar forms, filaments, pollen sac positions, loss of thecal construction, and bifurcated stomia that lead to valvate dehiscence). Prior to discussing these problems, I reevaluate how stamen morphology has been described in order to clarify ambiguities and identify key characters.

MATERIALS AND METHODS

Collections examined in this study are listed in Table 1. Stamens for histological and scanning electron microscopic investigation (SEM) were fixed in formalin–acetic acid–alcohol (50%). Specimens for sectioning were dehydrated in a graded tertiary butyl alcohol series, embedded in paraplast, and sectioned at 8–14 µm. Specimens for SEM were dehydrated in a graded ethanol series, critical point dried and coated with gold or palladium prior to examination at 10–20 kV.

Character state evolution was examined on the basis of cladogram topologies derived from studies of seed plants by Crane (1985, 1988) and Doyle & Donoghue (1986) and angiosperms by Donoghue & Doyle (1989), Hamby & Zimmer (1992), and Chase *et al.* (1993). Phylogenetic models of their cladogram topologies and variations of them were constructed using MacClade (Maddison & Maddison 1992). Character state distribution on the phylogenetic models was determined using the Trace Character option of MacClade (Maddison & Maddison 1992).

ASPECTS OF STAMEN MORPHOLOGY

Eames's (1961) discussion of stamen morphology is instructive. He described the typical stamen as comprising more or less distinct filament and anther regions. The anther was described as the fertile portion of the stamen which consists of microsporangia and the sterile tissue between and bearing sporangia. Eames recognized the tissue 'connecting and embedding' the microsporangia as connective. Like other basic descriptions of stamen form (e.g., Gray 1879; McLean & Ivimey-Cook 1956;

Table 1. *Collection data*

Acorus calamus L.	Walton 728, 20.7.1993
Akebia trifoliolata (Thunb.) Koidz.	Cultivated, Missouri Botanical Garden
Alisma subcordatum Raf.	Walton 738, 20.7.1993
Anemopsis californica Hook.	Hufford 555, 21.4.1992
Aristolochia elegans Mast.	Cultivated, Greenhouse, University of Minnesota–Duluth
Asarum canadense L.	Hufford 582, 8.5.1993; Hufford s.n., 7.5.1992
Brasenia schreberi Gmel.	C. Chapin s.n., 26.8.1992
Calla palustris L.	Walton 622, 20.6.1993
Caltha palustris L.	Hufford s.n., 17.5.1992
Calycanthus floridus L.	Cultivated, Greenhouse, University of Minnesota–Duluth
Dicentra cucullaria (Goldie) Walp.	Hufford s.n., 2.4.1988
Dicentra formosa (Andr.) Walp.	Hufford 427, 10.6.1990
Drimys winteri Forst.	Cultivated, University of California, Berkeley, Botanical Garden
Ephedra fragilis Fresen.	N. Friedman s.n., 13.5.1993
Eucnide aurea (Gray) Thompson et Ernst	Greenhouse, University of California, Berkeley (seed from Thompson 3584)
Fortunearia sinensis Rehder et Wilson	Cultivated, Missouri Botanical Garden
Gnetum gnemon L.	N. Friedman s.n., 13.5.1993
Hamamelis vernalis Sargent	Cultivated, Arnold Arboretum
Houttuynia cordata Thunb.	Cultivated, Greenhouse, University of Minnesota–Duluth
Lindera benzoin L.	Cultivated, Missouri Botanical Garden
Magnolia biondi Pampan	Cultivated, Arnold Arboretum
Nuphar luteum subsp. *variegatum* (Engelm. ex Clinton) Beal	P. Monson s.n., 18.6.1992
Nymphaea odorata Ait.	C. Chapin s.n., 19.7.1992
Sagittaria cuneata Sheldon	Hufford 561, 20.6.1992
Sanguinaria canadensis L.	Hufford s.n., 7.5.1992
Tacca chantrieri Andre	Cultivated, Greenhouse, University of Minnesota–Duluth
Thalictrum dioicum L.	Hufford 585, 8.5.1993
Typha latifolia L.	Hufford 560, 20.6.1992
Umbellularia californica (H. & A.) Nutt.	Cultivated, Missouri Botanical Garden
Welwitschia mirabilis Hook.	Cultivated, Greenhouse, Iowa State University

Esau 1977; Gifford & Foster 1989; Weberling 1989), his does not attempt to define terms such as filament and anther using morphological criteria (such as position, development, or unique structural attributes).

The conventional uses of terms such as 'anther,' 'connective,' and 'filament' are ambiguous and, generally, provide little insight into stamen morphology. For example, 'anther' is usually defined in terms of function as the stamen region where pollen forms and in terms of form as not being filament (e.g., Gray 1879; McLean & Ivimey-Cook 1956; Eames 1961; Esau 1977; Guedes 1979; Gifford & Foster 1989; Weberling 1989). Some workers (e.g., Takhtajan, 1991) use the term anther in a fashion interchangeable with thecae. Coulter &

Chamberlain (1912) described the term 'anther' as 'one of convenience, but represents a morphological complex made up of sporangia and more or less sporophyll tissue.' Terms of convenience and definitions based on function or the absence of attributes are adequate for neither homologizing character states nor hypothesizing synapomorphy. In order to discuss the evolution of form, definitions must be grounded in morphological attributes.

I propose that we use 'anther' to refer to the level or region of the microsporophyll in which microsporangia are positioned (Fig. 1). This definition significantly focuses on the basic structural attribute inherent to most discussions of anthers: the presence of microsporangia in which pollen develops. All fertile stamens, thus, have an anther. This observation, based on a morphological definition of 'anther,' contrasts with descriptions of some angiosperms with laminar stamens as lacking both filament and anther (e.g., Bailey & Smith 1942; Canright 1952; Gifford & Foster 1989). It is consistent with Takhtajan's (1991) discussion of *Degeneria* as bearing anthers despite the absence of a distinct filament. The proposed definition of 'anther' is important because it facilitates homologizing regions among diverse stamens, including laminar with nonlaminar forms, those with and without sterile, distal protrusions, and those in which the conservative symmetrical positioning of paired pollen sacs is lost.

Most stamens bear four microsporangia which are positioned in two pairs. If the microsporangia of each pair are adjacent for most of their length and share a common region of dehiscence (a stomium), then they form a unit called a theca. Most stamens have two thecae. Thecal 'construction' may be lost if microsporangia are not positioned directly adjacent to one another for all or nearly all of their length and each microsporangium has its own separate stomium. Lauraceae, for example, are characterized by the loss of thecae. Those Lauraceae with tetrasporangiate stamens have pollen sacs that are not adjacent and dehisce independently (Fig. 43). Loss of thecal construction may also occur with evolutionary increase or decrease of microsporangial number per stamen. For example, if a stamen has only one or two unpaired microsporangia (i.e., not adjacent for most of their length and not sharing a common stomium such that each dehisces independently), then it lacks thecae. Lauraceae, such as *Lindera benzoin* (Figs. 45, 46), demonstrate this change which is also found in various other angiosperm families (Hufford & Endress 1989; Endress & Stumpf 1990). Similarly, evolutionary elaboration of microsporangia, resulting in numerous sporangia, each dehiscing independently (e.g., *Polyporandra* and *Viscum*, Endress & Stumpf 1990) is associated with loss of thecal construction. Myristicaceae and Canellaceae demonstrate the loss of thecal construction associated with extensive synorganization of the androecium (Wilson 1966; Wilson & Maculans 1967; Armstrong & Wilson 1978). Deviation from the formation of four microsporangia is not always associated with loss of thecal

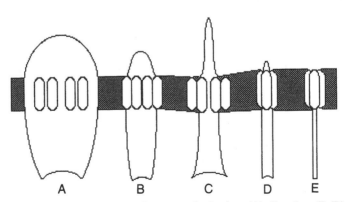

Fig. 1. Stamen form and anther position. Stamen forms may be laminar (A), lingulate (B–D), or filamentous (E) (see text for distinguishing attributes). The anther, defined by the position of the microsporangia, is shown by background shading on illustrations of stamens with different forms.

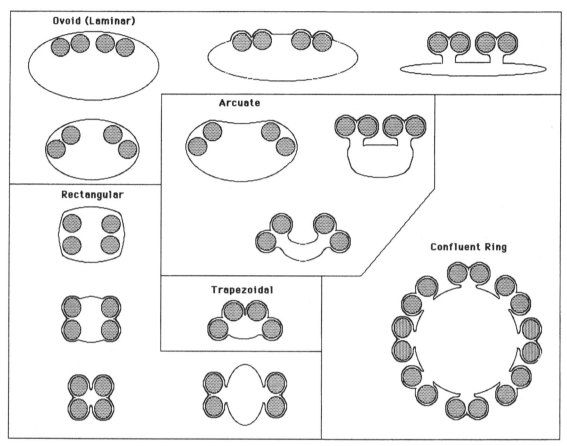

Fig. 2. The diversity of anther transectional shapes. Ovoid (laminar) anthers have facially positioned microsporangia (or microsporangia positioned at the corners of either the abaxial or adaxial surfaces and the lateral sides of the stamen) and a width that exceeds breadth. Arcuate anthers have laterally positioned microsporangia and either a concave or convex shape relative to the floral axis. Trapezoidal anthers are similar to arcuate but have a highly restricted connective (such that the curvature of the connective characteristic of an arcuate anther may not be visible) and both the microsporangia on either abaxial or adaxial sides of the stamen will be directly adjacent to each other. Rectangular anthers have laterally positioned microsporangia. A line drawn through the centers of both microsporangia of one theca will approximately parallel a similar line drawn through the other theca.

construction. For example, monothecate stamens have only two microsporangia, but these are paired such that they are adjacent for most of their length and share a common stomium (e.g., *Eucnide aurea*, Figs. 3, 4; Hufford 1988).

Stamens vary in the longitudinal and latitudinal position of thecae/microsporangia. The longitudinal position of thecae/microsporangia can be characterized simply as abaxial, adaxial, or lateral (marginal). Dorsiv-

entrally flattened stamens that have either abaxial or adaxial thecae/microsporangia might be said to exhibit facial positioning. We can characterize latitudinal position as either terminal (Fig. 1E; *Sanguinaria*, Fig. 52; *Sagittaria*, Fig. 63) or subterminal (Fig. 1A–D; *Houttuynia*, Fig. 26; *Nuphar*, Fig. 37). Stamens with subterminal anthers vary extensively in the extent and form of the sterile distal tissue. Most stamens with subterminal anthers have only a slight distal protrusion (Figs. 26, 54);

however, it is much more extensive in a some taxa. Parkin (1951) has argued effectively that the hyper-elongation of sterile distal tissue of stamens has evolved independently a number of times. In some Magnoliaceae, sterile distal tissue may serve as a hook that anchors abscised stamens in the floral vicinity and, perhaps, lengthens the time of pollen presentation (Howard 1948; Canright 1952; Endress 1994. Stamens with long, pointed distal tips (e.g., as in some Aristolochiaceae, Fig. 28; Dioscoreaceae; and Stemonaceae) have also been implicated in trap-flower pollination syndromes.

Anthers may have various transectional shapes. For this discussion of stamen evolution, I formulated an artificial classification (Fig. 2) of transectional shapes that uses position of microsporangia in conjunction with shapes of the stamen surfaces to identify categories. Ovoid (laminar) anthers have facially positioned microsporangia (or microsporangia at the corners of either the abaxial or adaxial surfaces and the lateral sides of the stamen). Ovoid anthers may have variously shaped (e.g., convex, concave, or flat) adaxial and abaxial surfaces, but characteristically have a width that exceeds their breadth. Arcuate anthers have laterally positioned microsporangia and either a convex or concave shape relative to the floral axis. Trapezoidal anthers are similar to arcuate, but the pollen sacs on one face of the stamen are much closer than on the other. Consequently, one face of trapezoidal anthers is much narrower than the other. Rectangular anthers have laterally positioned pollen sacs and two planes of symmetry. A line drawn through the centers of both microsporangia of one theca will approximately parallel a similar line drawn through those of the other theca.

Introrse, latrorse, and extrorse are terms to describe directions of pollen release from dehiscing thecae, which is important in pollination biology. The terms are commonly used to convey the positions of pollen sacs, but they are inappropriate for this purpose. For example, arcuate anthers with laterally positioned pollen sacs may have either introrse or extrorse dehiscence, depending upon the direction the stamen is curved. Functional changes from introrse to extrorse dehiscence may be more easily mediated than changes from a developmentally lateral position. Coding a labile functional orientation rather than morphological position in the course of a phylogenetic analysis could produce erroneous results.

I restrict usage of the term 'connective' to refer only to tissue of the anther region located between thecae of stamens with laterally positioned microsporangia. Although the tissue of transectionally ovoid anthers may not differ in histology from the connective of those that are arcuate, rectangular or trapezoidal, it is generally far more extensive and not limited simply to 'connecting' the opposing thecae.

Stamens differ in the degree to which their microsporangia protrude as major surface forms. The microsporangia of *Degeneria* and *Galbulimima* are embedded in the tissues of the abaxial side of the stamen and are not protrusive (Endress & Hufford 1989); whereas, those of the laminar *Nuphar* are protrusive (Fig. 37). In contrast, some stamens have microsporangia that appear to protrude along the length of the thecal face. This often results from, or is accentuated by, invagination of the stomium in the area between the two pollen sacs of the theca (Fig. 27). Generally, microsporangial protrusion is associated with limited sterile tissue in the areas directly surrounding the pollen sacs; whereas, greater amounts of sterile tissue around individual microsporangia tend to mask their form, and they often appear less protrusive (Fig. 7). Microsporangia may also be highly protrusive when offset from the axis of the stamen. For example, the ovoid (laminar) anther of *Austrobaileya* has thecae offset on the adaxial face of the stamen (Endress 1980b; Endress & Hufford 1989). The arcuate anthers of *Asarum* (Figs. 28, 29) and *Akebia* (Figs. 47, 48) have laterally positioned thecae that are offset toward the abaxial side of the stamen.

I follow the treatment of Hufford & Endress (1989) in which stomium refers to a dehiscence region. Prior to dehiscence, a stomium has a variety of structural attributes, including most notably a furrow on the surface of the thecae or pollen sac (Hufford & Endress, 1989). The stomium may have various shapes or locations relative to pollen sacs and thecae (Endress & Hufford 1989; Hufford & Endress 1989; Endress & Stumpf 1990, 1991). The tetrasporangiate stamens of most flowering plants have a single linear, unbifurcated stomium positioned longitudinally between the pair of pollen sacs of a theca (Fig. 27). Linear stomia with distal and/or proximal bifurcations are common in some groups, especially Magnoliales (Endress & Hufford 1989) and 'lower' Hamamelidae (Hufford & Endress 1989). For example *For-*

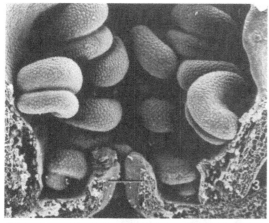

Figs. 3, 4. Stamens of *Eucnide aurea*. Fig. 3. Monothecal anthers, oriented transversely, are inserted on the corolla tube. The lower microsporangium of each theca is somewhat narrower than the upper. Fig. 4. Flower section showing transverse sections of two anthers. The anther on the left is monothecal, and its two pollen sacs and filament are visible. The sectioned teratological anther on the right (T) demonstrates a reversion to the tetrasporangiate form present in most *Eucnide*. Scale bars represent 100 μm.

tunearia (Hamamelidaceae, Fig. 7) has both proximal and distal stomial bifurcations, resulting in a valvate form of dehiscence. A stomium may also extend around the perimeter of a theca (e.g., *Noahdendron* of Hamamelidaceae; Hufford & Endress, 1989) or around individual pollen sacs. The latter is found in those taxa which diverge from

tetrasporangiate anthers (e.g., *Hamamelis* of Hamamelidaceae; Figs. 5, 6) or have vertically offset pollen sacs (e.g., Lauraceae; Figs. 43, 44). These curvilinear stomia that extend around the perimeter of thecae or pollen sacs have also been associated with valvate dehiscence (i.e., the outward, flap-like, opening of part of the pollen sac wall at dehiscence). Hence, it is important to recognize that valvate dehiscence is a functional attribute that has evolved from different, nonhomologous stomial patterns among the flowering plants. So-called porate dehiscence has also evolved independently in various angiosperm groups. Porate dehiscence may be associated with either highly restricted linear stomia (often positioned at the thecal apex) or ring-like stomia.

'Filament,' like 'anther,' is one of the most ambiguous floral terms. Esau (1977), for example, considered the filament to be a single-veined stalk. Weberling (1989) described the filament as a filiform stalk. Broad, bifacial filaments that are not, however, very filiform are characteristic of some Loasaceae, such as *Cevallia* and *Fuertesia* (L. Hufford, unpublished data). Similarly, the bulbously inflated filaments of Hamamelidaceae, such as *Fothergilla* (Weaver 1969), diverge from the filiform. In this chapter, stamens are regarded as having a filament if the region subjacent to the anther is morphologically distinct from that of the anther region, such that one would not regard the shape of the basal zone as continuous with that of the anther. Saururaceae (Figs. 26, 27) and *Sagittaria* (Fig. 63) might be considered to have stamens that demonstrate well-differentiated filaments. *Asarum* (Fig. 28) and *Akebia* (Fig. 47) are good examples of genera with stamens that demonstrate a continuity of the basal region of the stamen with the axis of the anther that would lead me to describe each as lacking a filament. It may be difficult to determine whether some stamens have a filament because the basal region may have little distinction from the anther or because it is unclear whether it is continuous with the anther. Both Lauraceae and Nymphaeales, in which a wide range of stamen forms may be found, include taxa in which the identification of the filament may be equivocal. Filaments may have various forms, but generally serve as a pedestal for the anther. For example, *Tacca chantrieri* (Fig. 8) has a broad, V-shaped pedestal that is quite different from the conventional interpretation of a filament but should be characterized as such.

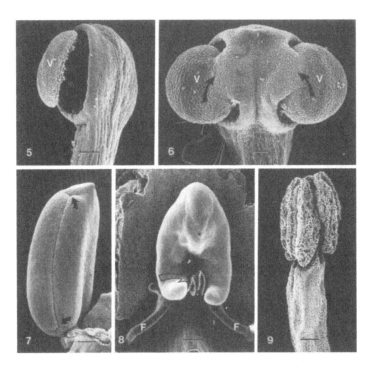

Figs. 5–9. Stamens of *Hamamelis*, *Fortunearia*, *Tacca*, and *Drimys*. Figs. 5, 6. *Hamamelis vernalis*. Fig. 5. Lateral view of anther showing a single valve (V) that opened via a curved stomium. Fig. 6. Ventral view of stamen showing both valves (V) of the dehisced anther. Each valve forms from the outer wall of the single microsporangium on each side of the anther. A curvilinear stomium circumscribes all but the ventral (hinge at arrow) side of the microsporangium. Fig. 7. Lateral view of *Fortunearia sinensis*, showing stomial bifurcations (arrow-heads) at the proximal and distal ends of the theca. Fig. 8. Ventral view of *Tacca chantrieri*, showing the broad V-shaped filament (F), highly protrusive, offset thecae (arrow) and the massive upper portion of the stamen axis (asterisk). Fig. 9. *Drimys winteri* near dehiscence. Scale bars represent 100 μm in Figs. 4, 5, 8; 300 μm in Fig. 6; and 500 μm in Fig. 7.

Three terms are used in this chapter to describe overall form of stamens: laminar, lingulate, and filamentous. Laminar stamens are broad and typically dorsiventrally flattened (Fig. 1A). The pollen sacs of laminar stamens have facial positions that are distinctly submarginal and subterminal. Lingulate stamens are narrower than laminar stamens (Fig. 1B–D). Pollen sacs of lingulate stamens may be facially or laterally positioned. If their pollen sacs are facially positioned, then they will be more or less at the stamen margin. The axis of a lingulate stamen usually extends distal to the anther. Lingulate stamens may range from dorsiventrally flattened to having a radial thickness that is as great as or greater than their tangential width. Filamentous stamens (Fig. 1E) have a more or less filiform axis with a small diameter and little, if any, extension of the axis beyond the anther (i.e., the anther is terminal or nearly so).

PRIOR HYPOTHESES ON PLESIOMORPHIC ATTRIBUTES

Although various hypotheses on the form of plesiomorphic stamens have been proposed, most may be grouped in two general categories: (1) stamens were originally branched or pinnatifid with terminal sporangia, or

(2) stamens were originally laminar with subterminal microsporangia embedded superficially on one face.

Wilson (1937) advocated the idea that stamens are axial, not foliar, structures and represent modifications of sporangiate telomes. He envisioned the two thecae of a typical stamen as arising via the phylogenetic fusion of two pairs of telomes, each of which bore a single terminal sporangium in an earlier fern-like ancestor. Wilson hypothesized that fasciculate stamens, such as those of Dilleniaceae, Hypericaceae, and Malvaceae, represent a single axis with terminal branches, each of which bears sporangia at its tip. Wilson (1937) emphasized the vascular system in flowers of these families to hypothesize that clusters of stamens were single branches. In some members of these families, clustered stamens are vascularized by bundles that diverge from a common 'trunk bundle' in the receptacle. Thus, despite the numerous stamens in each flower, they have only a few trunk bundles. Among the extant angiosperms, flowers of Engler's Parietales and Malvales, that included the families identified above, were considered by Wilson (1937) to represent plesiomorphic androecial attributes.

Stebbins (1974) also advocated the idea that clusters of stamens are homologous to a single microsporangiate organ, but unlike Wilson (1937) he did not attempt to derive the original stamen from telomes. Stebbins (1974) suggested that the original angiosperm androecium may have been composed of a small number of branches (axes) that had terminal sporangiate ramifications or ramified sporophylls (i.e., foliar not axial structures). Like Wilson (1937), Stebbins cited the trunk bundles that vascularize clusters of stamens in flowers of some angiosperms as evidence that they are parts of compound organs (of either axial or foliar homology). Stebbins (1974) emphasized Dilleniaceae as exemplars of plesiomorphic androecia. He also observed that trunk bundles vascularize androecia of various Magnoliidae, including groups such as Degeneriaceae, Annonaceae, and Magnoliaceae that are often cited as exhibiting plesiomorphic attributes.

Most contemporary assessments of stamens follow the interpretation of Arber & Parkin (1907) that they are simple sporophylls with two synangia. Arber & Parkin's investigation of seed plant strobili, especially those of Bennettitales, led them to deduce that the plesiomorphic attributes of angiosperm stamens included a short filament and subterminal synangia with a prominent extension of sterile sporophyll tissue (a 'connective protrusion') beyond the synangia, such as exemplified by many Magnoliaceae (e.g., Fig. 31). They hypothesized that stamens with more pronounced sporangiophores (filaments) and synangia with simpler connectives were derived among angiosperms.

Recent workers (especially Canright 1952; Takhtajan 1969, 1991; Cronquist 1988) focused on *Degeneria* as exemplary of plesiomorphic stamen attributes that include: a broad (largely laminar) form, subterminal sporangia positioned in two parallel, bisporangiate synangia (thecae), synangia that are embedded in the superficial tissue of the abaxial side of the stamen, and vascularization by three vascular bundles that enter the stamen from the receptacle. From this perspective, the stamens of most angiosperms share derived attributes that include a thin, basal filament, more-or-less terminally positioned microsporangia, lateral thecae, and thecal separation by a narrow connective. In contrast, Gottsberger (1974, 1977, 1988) has argued that unifacial stamens with terminally positioned microsporangia, such as those of *Drimys* (Fig. 9), best exemplify the plesiomorphic attributes of angiosperms.

These alternative hypotheses on plesiomorphic attributes of androecia and stamen form may be best evaluated in the context of the explicit phylogenetic hypotheses for flowering plants (and the more inclusive seed plant group) and outgroup analysis that follow.

OUTGROUP ANALYSIS

Recent phylogenetic analyses of seed plants have differed little in their placement of angiosperms in contrast to the range of earlier, noncladistic hypotheses (Crane 1988). Crane's (1985) analysis placed the angiosperms as the sister taxon of the Gnetales in a clade that in turn formed the sister group of Bennettitales and *Pentoxylon*. The subsequent analysis of Doyle & Donoghue (1986) placed the angiosperms as the sister group of a clade consisting of *Pentoxylon*, Bennettitales, and Gnetales. Doyle & Donoghue (1986) called attention to this clade that consisted of angiosperms, Bennettitales, Gnetales, and *Pentoxylon* by designating it the anthophyte clade. The closest relatives of the anthophytes are unclear (Crane 1988), and this may limit attempts to reconstruct the deeper

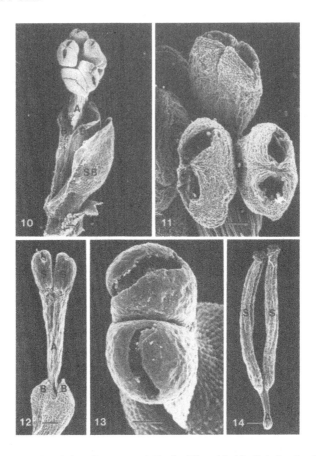

Figs. 10–14. Microsporophylls of *Ephedra, Gnetum,* and *Typha*. Figs. 10, 11. *Ephedra fragilis*. Fig. 10. Male strobilus with subtending bract (SB). The strobilus consists of an androphore (A) surrounded by two concrescent bracts (B). Numerous synangia are positioned near the terminus of the androphore. Fig. 11. Closer view of the bisporangiate synangia shown in Fig. 10. Figs. 12, 13. *Gnetum gnemon*. Fig. 12. Male strobilus showing an androphore (A) surrounded by two concrescent bracts (B). The androphore bears two terminal sporangia (S). Fig. 13. Distal view of two, dehisced sporangia. Fig. 14. Androecium from *Typha latifolia*. The androecium of *Typha* parallels that found in *Ephedra* and *Gnetum* in forming an androphore (A) terminated by synangia (S). Scale bars represent 1000 μm in Fig. 10; 100 μm in Figs. 11, 13; and 200 μm in Figs. 12, 14.

history of microsporophyll diversification. The recognition of the monophyletic anthophyte group, however, focuses attention on *Pentoxylon*, Bennettitales, and Gnetales as the best outgroups to examine in order to understand the androecial and staminal character states important in the early evolution of angiosperms.

Pentoxylon remains a poorly known fossil taxon from the Jurassic and Cretaceous. Its male strobilus (*Sahnia nipaniensis*) had a broad, conical receptacle that bore a series of sterile bracts subjacent to the fertile region.

Each male strobilus had numerous filamentous, pinnatifid microsporophylls. Unilocular microsporangia were borne at the tips of stalks (pinnae). Vishnu–Mittre (1952) described the microsporophylls as basally connate, forming a shallow cup-like region around the strobilus apex, but Rao (1981) suggested that more recent studies showed that they were free and helically arranged.

Bennettitales are first known from the Upper Triassic (Crane 1988), although Middle Jurassic and Cretaceous forms are the best reconstructed. The bisexual strobili of *Williamsoniella* (Middle Jurassic) and *Cycadeoidea* (Upper Jurassic–Lower Cretaceous) have received extensive attention in discussions of angiosperm origins. Both had strobili that were flower-like in having a whorl of bracts subtending the microsporophylls and, distally, the receptacle bore ovules and interseminal scales (Crane 1985). The microsporophylls of both *Williamsoniella* and *Cycadeoidea* were lobed (pinnatifid) and bore synangia along the lobes (Crepet 1974; Crane 1985). The position of the synangia on the lobes has not been addressed, but reconstructions show that they may be positioned on margins rather than on abaxial or adaxial faces. Synangia of *Cycadeoidea* had eight to 20 tubular microsporangia (Crepet 1974).

Weltrichia, the microsporangiate strobili associated with *Williamsonia* of the Middle Jurassic, consisted of up to 30 microsporophylls that were distally free but proximally united to form a shallow cup-like region (Crane 1985). The free distal part of each microsporophyll of *Weltrichia sol* was adaxially 'branched,' and those pinnae bore two-valved synangia that included numerous microsporangia. *Weltrichia pecten* and *W. whitbiensis* were similar to *W. sol* but did not have 'branched' microsporophylls; instead, they had sessile synangia positioned adaxially in two rows (Crane 1985).

The microsporophylls of Upper Triassic Bennettitales differ from those of the later taxa described above. *Bennettistemon amblum* and *B. ovatum* are laminar microsporophylls that bear separate adaxial sporangia, not synangia (Crane 1986). Microsporophylls of *B. amblum* have sporangia localized on costae that diverge from the midrib. Sporangia are borne over the entire adaxial surface of *B. ovatum* microsporophylls. *Haitingeria krasseri* from the Upper Triassic of Austria is also laminar but is pinnatifid. Unlike the pinnae of the later *Weltrichia*, *Williamsonia*, and *Cycadeoidea*, those of *H. krasseri* are not oriented toward the floral axis (Crane 1986). Each pinna of the *H. krasseri* microsporophyll bears up to six pairs of microsporangia, but it is not clear whether they are synangial. *Leguminanthus*, a microsporophyll also from the Upper Triassic of Austria, has entire margins, but it is adaxially conduplicate and bears numerous rows of microsporangia on its adaxial surface (Crane 1986). Some pollen sacs of *Leguminanthus* appear clustered, but there is no evidence that they were united into synangia (Crane 1986). Crane (1986) hypothesized that *Leg-*

uminanthus, *Haitingeria*, and *Bennettistemon* may have been borne in helices along strobilus axes, rather than in whorls like the microsporophylls of Jurassic and Cretaceous Bennettitales.

Interpretations of Gnetalean strobili have long been controversial. Their construction, however, is based on an opposite and decussate phyllotaxy. The microstrobilus of most *Ephedra* (Fig. 10) species consists of a bilobed 'perianth' that is subjacent to and alternating with an 'androecium' of two microsporophylls (Eames 1952; Martens 1971). These microsporophylls are generally connate, such that the 'androecium' consists of an androphore terminated by a cluster of sporangia (Figs. 10, 11; the sporangia may be more or less stalked in some species: Eames 1952; Martens 1971). Among extant *Ephedra* species, however, various degrees of connation may be found between the two microsporophylls, including those that are completely separate and have a brief receptacle projecting upward between them (Eames 1952). In those species with separate microsporophylls, each bears a cluster of terminal synangia (Eames 1952 – although he interprets the 'synangia' as simple sporangia with one or more longitudinal septa). Synangia in *Ephedra* are usually bisporangiate (Figs. 10, 11), but may include up to four sporangia.

The microstrobilus of *Gnetum* is similar to that of *Ephedra* in consisting of a bilobed 'perianth' that is subjacent to the 'androecium' (Fig. 12). The 'androecium' consists of an androphore that terminates in two separate sporangia (Figs. 12, 13; Vasil 1959).

Androphore evolution, such as found in *Ephedra* and *Gnetum*, may be facilitated by evolutionary shifts to unisexual flowers. For example, we find the parallel evolution of similar androphores among angiosperms with unisexual flowers in Myristicaceae (Wilson & Maculans 1967; Armstrong & Wilson 1978), *Ruscus* (Dahlgren *et al.* 1985), and *Typha* (Fig. 14; Dahlgren *et al.* 1985).

Welwitschia 'microstrobili' have a distally positioned ovule that is putatively nonfunctional; ovules located in the compound strobilus or 'cone' of *Gnetum* have been long reported to be sterile, but some have been shown recently to function in reproduction (W.E. Friedman, personal communication.) Unlike the microstrobili of *Ephedra* and *Gnetum*, that of *Welwitschia* has two pairs of bracts below the microsporophylls (Fig. 15; Martens 1971; Crane 1985). The 'androecium' consists of a

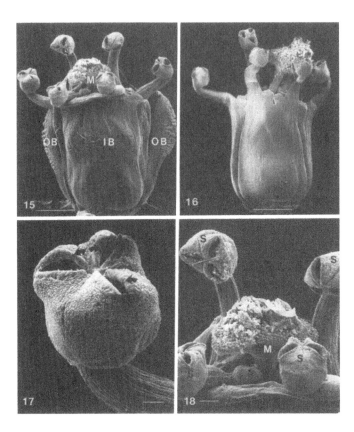

Figs. 15–18. *Welwitschia mirabilis.* Fig. 15. Male strobilus showing six synangia projecting above the outer (OB) and inner (IB) bracts. The discoid micropyle (M) of the 'nonfunctional' ovule is positioned between the synangia. The discoid micropyle is covered with pollen. Fig. 16. 'Androecium' of *Welwitschia* showing concrescent basal region with distal filamentous projections that bear the synangia. All floral bracts have been removed. Fig. 17. Synangium showing the regions (arrows) of longitudinal dehiscence for the three sporangia. Fig. 18. Higher magnification of the micropyle shown in Fig. 15. Pollen from the synangia (S) is deposited on the concave surface of the discoid micropyle (M) where it mixes with ovular secretions (the hardened white flakes). The micropyle with the pollen load may function in secondary pollen presentation. Scale bars represent 1000 μm in Figs. 15, 16; 100 μm in Fig. 17; and 200 μm in Fig. 18.

basally tubular (or cup-like) region with six free lobes that bear terminal synangia (Fig. 16). The number and position of the microsporophylls in this 'androecium' have been richly debated. Most workers interpret the 'androecium' to consist of either (1) two microsporophylls, each with three lobes, or (2) six microsporophylls (with each synangium-bearing lobe corresponding to an individual sporophyll). Two observations may support the interpretation that the 'androecium' consists of two lobed microsporophylls: (1) an opposite and decussate phyllotaxy is pervasive among all other Gnetalean strobili, and two opposite microsporophylls that are decussate with the pair of bracts just below them is consistent with this phyllotactic pattern; and (2) developmental study has indicated that 'androecial' inception begins with the initiation of two opposite primordia, positioned decussate to the subjacent bract pair (Martens 1971). The developmental study of the *Welwitschia* microstrobilus should be repeated using scanning electron microscopy and histological preparations to determine the extent and position of these initial 'androecial' primordia and whether they are common primordia from which all synangium-bearing lobes develop or if some of the lobes develop from primordia that are initiated on the receptacle separate from the first 'androecial' primordia. The synangia of *Welwitschia* contain three microsporangia, each of which dehisces independently via a short apical slit (Fig. 17; Church 1914).

Specimens of *Welwitschia* examined for this study have shown that the abaxial and especially adaxial pair of microsporophylls tend to release pollen on the discoid micropyle (Figs. 15, 16, 18). Pearson (1929) previously noted that pollen tended to lie in a mass at the summit of the male strobilus. The direct deposition of pollen on the discoid stigma suggests that it has a role in secondary pollen presentation. Church (1914) noted that a sticky substance is exuded through the discoid micropyle of male strobili. A hardened substance, presumably the remnant of this sticky secretion, was often found with the pollen on the micropyle in scanning electron microscope preparations (Fig. 18). The sticky micropylar exudate has been suggested to serve as an attractant for animal pollinators, and the presentation of the pollen on the discoid micropyle is also expected to be associated with biotic pollination. Pollination by hemipterans has been suggested (Pearson 1929), although Bornman (1972) has discounted these ideas with the proposal of wind pollination. No thorough investigation of reproductive biology of *Welwitschia* has been conducted.

Gnetalean megafossils are relatively scarce (Crane & Upchurch 1987). The few fossil male strobili and microsporophylls attributed to Gnetales usually bear *Ephedripites* pollen which has a polyplicate exine similar to that of extant *Ephedra* and *Welwitschia*, but unlike that of other seed plants. *Maculostrobus clathratus* from the Upper Triassic of Arizona is a strobilus with helically arranged microsporophylls (Ash 1972). Each microsporophyll has a slender stalk and a broad head with an acuminate apex. A few pendant, elliptical microsporangia, from which *Ephedripites* pollen have been isolated, are attached to the adaxial surface and lower edge of the sporophyll head. *Maculostrobus* is distinctive, if Gnetalean, because the 'inflorescence' is simple rather than compound and its sporophylls are neither tightly nested nor associated with bracts. *Piroconites kuespertii* from the early Jurassic of Germany is a laminar microsporophyll

with the adaxial surface covered by synangia (van Konijnenberg-van Cittert 1992). Each synangium consists of three microsporangia that bear *Ephedripites* pollen. *Piroconites* are dispersed sporophylls, although associated with a subtending bract, and their arrangement in a strobilus is not known.

The nonanangiospermous anthophytes display a fairly high level of microsporophyll diversity that may reflect differences in pollen presentation and pollination processes. The diversity complicates reconstruction of the character suite from which angiosperm stamens evolved. Recent phylogenetic studies have provided support for the monophyly of Bennettitales and Gnetales (Crane 1985, 1986, 1988; Doyle & Donoghue 1986). If true, then only the plesiomorphic attributes of these groups need to be used in reconstructing representative taxa for an examination of microsporophyll character evolution among anthophytes. Attributes of Triassic Bennettitales (*Bennettistemon*, *Haitingeria*, and *Leguminanthus*) and Early Gnetales (*Maculostrobus* and *Piroconites*) have been selected to represent their respective monophyletic groups (Table 2). *Caytonia* and *Glossopteris* have been selected as outgroups based on results from Doyle & Donoghue (1986; the sister taxon of the anthophytes shown in Crane's 1985 investigation is a large, highly diverse group of extinct and extant seed plants). The characters examined and their states are provided in Table 2.

Mapping microsporophyll attributes on the Doyle & Donoghue (1986) topology for anthophytes showed that they share derived laminar microsporophylls that are entire, and that this state is retained by angiosperms (Fig. 19). The Crane (1985) topology for anthophyte relationships shows that the basal attribute for anthophytes is equivocal because of variability within the group but is consistent with the idea that laminar stamens are a plesiomorphic attribute retained by angiosperms (Fig. 20). The diversity within each of the anthophyte groups implies that microsporophyll form has been modified along various avenues. If we add derived taxa for Bennettitales and Gnetales to the anthophyte cladograms, increasing the variability, the ancestral form for angiosperm microsporophylls becomes equivocal using the Doyle & Donoghue (1986) topology (Fig. 21). As noted above, however, if Bennettitales, Gnetales, and *Pentoxylon* are monophyletic groups, then the addition of

Table 2. *Character states used to evaluate the evolution of microsporophylls among anthophytes*

Taxon	Form	Longitudinal position of sporangia	Synangia
Caytonia	Filamentous pinnate	Scattered	Present
Glossopteris	Filamentous pinnate	?	Absent
Pentoxylon	Filamentous pinnate	Scattered	Absent
Triassic Bennettitales	Laminar entire or pinnate	Scattered	Absent
Late Bennettitales	Laminar pinnate	Scattered	Present
Early Gnetales	Laminar entire or peltate	Scattered or subterminal	Present or absent
Extant Gnetales	Synorganized androphore pinnate or entire[a]	Terminal	Present or absent
Angiosperms	Laminar entire or filamentous unbranched	Terminal or localized subterminal	Present

[a] Extant Gnetales are difficult to interpret. For example, the androphore of *Ephedra* may be homologous with pinnate microsporophylls since the synangia are often stalked. Some *Ephedra* and *Gnetum* in which the sporangia are not stalked may be homologous to synorganized laminar sporophylls that are entire. The different interpretations offered for *Welwitschia* (see text) also add to the difficulties in coding extant Gnetales.

the later taxa is superfluous and may provide misleading equivocation rather than informing us of appropriate limitations.

The anthophyte cladograms show that microsporangia may have been scattered largely over the length of the microsporophyll in the angiosperm ancestor rather than localized in either a terminal or subterminal position as is characteristic of most extant angiosperms (Fig. 22). Deducing whether sporangia were in facial or lateral positions in the angiosperm ancestor is complicated by the limited data that may be derived from fossils. Early Bennettitales and Gnetales (although *Maculostrobus* is difficult to interpret) appear to have had a facial position of microsporangia that might be homologous with the pollen sac placement of some Magnoliales that have ovoid anther transections. It is equivocal whether the angiosperm stamen was derived from a microsporophyll that had synangia or discrete sporangia (Fig. 23). Both are present among early Gnetales, but *Pentoxylon* and the early Bennettitales had discrete sporangia. Later Bennettitales were synangial. The outgroup *Caytonia* was synangial, but *Glossopteris* was not. Among the early anthophytes, homoplastic gains of synangia may be more common than changes from synangia to discrete microsporangia, although the latter occurs in some angiosperms (e.g., Lauraceae and Hamamelidaceae). The repeated gains among the anthophytes may be associated with the importance of synangia in effective pollen release and display in groups with insect pollination. None of the other anthophyte groups display the great conservatism of the angiosperm pairing of pollen sacs and pairing of thecae. A basic analysis such as this demonstrates, however, that greater knowledge of extinct anthophytes, especially those of the angiosperm stem lineage (Doyle & Donoghue 1993), and their relationships will be required to reconstruct effectively the microsporophyll attributes that preceded the flowering plant stamen.

This outgroup analysis provides insights for deciding among the prior hypotheses elaborated above. The results of this analysis are more consistent with hypotheses that the first angiosperms had laminar stamens rather than the highly branched forms advocated by Wilson (1937) and Stebbins (1974). The stamen trunk bundles of Dilleniaceae, Hypericaceae, and Malvaceae emphasized as a primitive attribute by Wilson (1937) and Stebbins (1974) commonly occur among polystaminate members of the former Dilleniidae (*sensu* Takhtajan 1980; Cronquist 1981). Ronse Decraene & Smets (1992b)

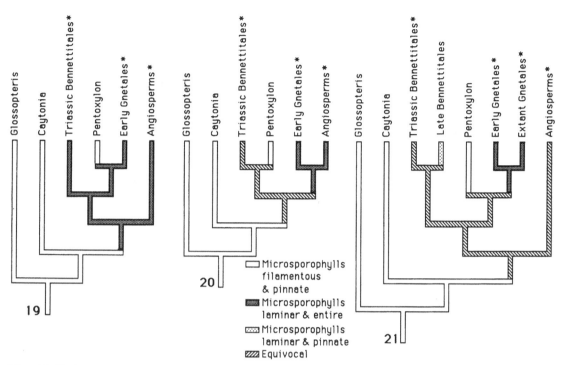

Figs. 19–21. Alternative cladograms for microsporophyll evolution among anthophytes. Character states for anthophytes and the two outgroups, *Caytonia* and *Glossopteris*, are provided in Table 1. Asterisks indicate polymorphism within taxa (see Table 1). Fig. 19. Topology based on results from Doyle & Donoghue (1986). Taxon exemplars represent only the earliest attributes for each anthophyte group. Fig. 20. Topology based on results from Crane (1985). Taxon exemplars represent only the earliest attributes for each anthophyte group. Fig. 21. Same topology as Fig. 19 but both early and late exemplars are included for Bennettitales and Gnetales.

have shown that stamen trunk bundles are associated with androecia that form a few common primordia (from which numerous individual stamens subsequently arise) and flowers that lack hypanthia. These attributes are typical of the groups Wilson (1937) and Stebbins (1974) considered to be primitive angiosperms. Earlier investigators had also hypothesized that stamen trunk bundles were derived among different groups of angiosperms (Corner 1946; Sporne 1958; Tucker 1972). Results from recent phylogenetic analyses (Crane 1985; Doyle & Donoghue 1986) do not support Wilson's contention that angiosperm stamens are derived from telomes. These phylogenetic analyses have placed angiosperms among highly derived seed plants far removed from telomic levels of construction (see also Baum 1950, 1952; Canright 1952). The angiosperm groups, such as Dillenia-

ceae, Hypericaceae, and Malvales, that Wilson (1937) and Stebbins (1974) emphasized as demonstrating plesiomorphic androecial attributes have been shown to be derived among the higher dicots, rather than placed among basal angiosperms, in recent phylogenetic analyses (Hufford 1992; Chase *et al.* 1993). Hence, it is unlikely that their attributes will bear directly on our understanding of plesiomorphic angiosperm attributes.

IMPLICATIONS OF ALTERNATIVE ANGIOSPERM ROOTINGS

The phylogenetic analysis of major clades of angiosperms using cladistic methods is in its infancy. Current results are weakened by as yet unaddressed issues about taxon sampling and methodological problems inherent to ana-

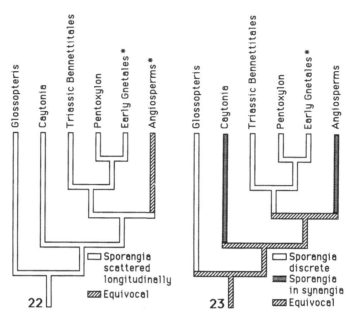

Figs. 22, 23. Microsporophyll attributes of anthophytes mapped on a topology based on Doyle & Donoghue (1986). Character states for anthophytes and the two outgroups, *Caytonia* and *Glossopteris*, are provided in Table 1. Asterisks indicate polymorphism within taxa (see Table 1). Fig. 22. Position of sporangia. Fig. 23. Sporangia discrete or synangial.

lyzing large data sets. Hence, recent results must be used cautiously but are the best devices available for exploring character evolution. The results of recent phylogenetic analysis of basal angiosperms place very different groups among the lowermost branches of cladograms. This is a key concern when we consider patterns of character evolution among basal angiosperms.

The analysis of Donoghue & Doyle (1989) using structural data placed Magnolialean groups at the base of the most parsimonious cladogram. They emphasized, however, that some alternative rootings were almost as parsimonious. For example, placing Nymphaeales as the sister group to the rest of the angiosperms, which results in the shift of Magnoliales from the base of the tree, is only one step less parsimonious than the shortest trees. The Nymphaeales in the most parsimonious cladograms found by Donoghue and Doyle (1989) are part of a group they called the paleoherbs. In their most parsimonious cladograms, this monophyletic paleoherb group consisted of the monocots, their nymphaealean sister group, Piperaceae, Saururaceae, Lactoridaceae, and Aristolochi-

aceae. Rerooting the Donoghue and Doyle (1989) topology near Nymphaeales, among the so-called paleoherbs, foreshadows the results obtained by Hamby & Zimmer (1992) using rRNA data. Hamby and Zimmer's (1992) most parsimonious trees placed Nymphaeales as the sister group of other angiosperms and other paleoherbs, such as Piperales and monocots, among the lowermost branches. The *rbc*L trees of Chase *et al.* (1993) also favored the early branching of paleoherbs; however, unlike the results of Hamby & Zimmer (1992), their trees placed the unusual aquatic plant *Ceratophyllum* as the sister group of the rest of the angiosperms, and at the next node the eudicots (nonmagnoliid dicots) diverged from a clade that consisted of other paleoherbs, Magnoliales, and Laurales. These scenarios imply very different patterns of stamen evolution during early angiosperm diversification. The taxa sampled and pattern of ingroup relationships found by Donoghue & Doyle (1989) provide a useful tool for examining implications of alternative phylogenetic scenarios.

If Magnoliales are the sister group to the rest of the

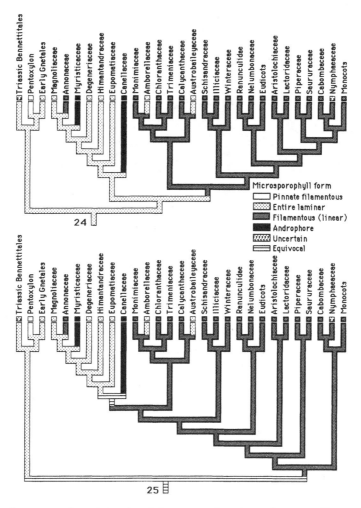

Figs. 24, 25. Stamen form mapped on alternative cladogram rootings for angiosperms. Taxa and topologies are based on results from Donoghue & Doyle (1989). The character state 'filamentous' denotes stamens that are neither lingulate nor laminar, but have a prominent linear axis; it does not imply that stamens have a differentiated filament. Fig. 24. Topology rooted with Magnoliales as sister group of the rest of the angiosperms. Fig. 25. Topology rooted among paleoherbs with Nymphaeales and monocots forming the sister group of the rest of the angiosperms.

angiosperms, then flowering plants may be interpreted to have retained stamens with a laminar or lingulate form from their anthophyte ancestors (Fig. 24). Even this rooting, however, indicates that more-or-less filamentous (denoting stamens that are not lingulate or laminar, but those with a prominent linear axis; filamentous in this context does not connote a stamen with a well-differentiated filament) stamens evolved early in magnoliid diversification. That is, we find shifts from laminar/lingulate forms via diminished tissue bulk outside of the anther region and the distinctive positioning of this region near the stamen terminus, although a short distal protrusion (restricted in width and breadth to the margins of the anther region) is generally present. For

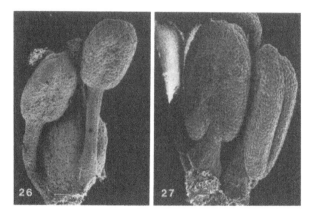

Figs. 26, 27. Stamens of Saururaceae. Fig. 26. *Houttuynia cordata.* Fig. 27. *Anemopsis californica.* Scale bars represent 100 μm.

example, within Magnoliaceae some stamens have terminal anther regions, slightly protrusive thecae, and a cylindrical basal filament (Canright 1952). Similarly in Laurales, Trimeniaceae have stamens with cylindrical bases and anther regions with prominent thecal protrusion (Endress & Sampson 1983; Endress & Hufford 1989). The placement of Magnoliales as the sister group of the rest of the angiosperms indicates that reversals to laminar forms would have occurred in at least three clades, including Amborellaceae, Austrobaileyaceae, and Nymphaeaceae.

The placement of *Ceratophyllum* as the sister group of angiosperms in cladograms of Chase *et al.* (1993), like the similar placement of Magnoliales in Donoghue & Doyle (1989), also implies that the earliest angiosperms may have retained or departed little from the laminar stamens of anthophyte ancestors. The lingulate form, ovoid transectional shape, and abaxial–lateral position of embedded microsporangia that are characteristic of *Ceratophyllum* (Endress & Hufford 1989) are generally similar to the basic attributes of laminar stamens of Magnoliales. The derivation of such a stamen form or its transformation into those found among other angiosperms would require no changes beyond those that would be envisioned for stamens of Magnoliales. The absence of a distinct stomium and endothecium (Shamrov 1983; Endress & Hufford 1989) in *Ceratophyllum* may be specializations for an aquatic habitat.

If angiosperms are rooted among certain paleoherbs

(Fig. 25), then stamen evolution may have proceeded quite differently from the interpretation with a Magnoliales- or *Ceratophyllum*-based phylogenetic hypothesis. Paleoherbs such as Aristolochiaceae (Figs. 28–30; Leins & Erbar 1985; Sugawara 1987; Leins *et al.* 1988) and Lactoridaceae (Carlquist 1964, Lammers *et al.* 1986) have thick, fleshy stamen axes, distinctly subterminal anthers with arcuate transections and offset, highly protrusive thecae. *Aristolochia* is somewhat different from other Aristolochiaceae because of the coalescence of the androecium and gynoecium (Fig. 30; Johri & Bhatnagar 1955; Nair 1962). Among piperalean paleoherbs, Saururaceae (Figs. 26, 27) have a filamentous rather than lingulate-laminar stamen form with a well-differentiated filament, slightly subterminal anther with rectangular transection, and moderately protrusive thecae. Piperaceae display little filament differentiation, but sometimes have distinctively terminal anthers (Bornstein 1991; Endress, 1994). Nymphaeales, another paleoherb group among Magnoliidae (Donoghue & Doyle 1989), have very diverse stamens. This diversity includes laminar forms in *Nuphar* (Figs. 37–39) and *Victoria* (Endress & Hufford 1989) and laminar to lingulate forms in *Nymphaea* (Fig. 36), but less lingulate to filamentous forms in *Brasenia* (Figs. 34, 35) and *Cabomba* (figured in Moseley 1958). Gottsberger (1988) argued that laminar stamens were derived within Nymphaeaceae in association with dynastid scarab beetle pollination. Dynastid scarab beetles have a fossil record extending back only to the

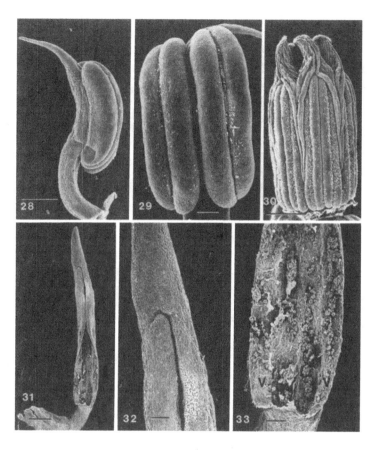

Figs. 28–33. Stamens of *Asarum*, *Aristolochia*, and *Magnolia*. Figs. 28, 29. *Asarum canadense*. Fig. 28. Side view of stamen, showing the highly protrusive thecae that are offset on the abaxial side. Fig. 29. Facial view of thecae. Fig. 30. Flower of *Aristolochia elegans* with perianth removed, showing the cowl-like gynoecium surrounding the synorganized androecium. Figs. 31–33. *Magnolia biondi*. Fig. 31. Lateral view of stamen. Fig. 32. Distal portion of theca, showing curved end of stomium. Fig. 33. Proximal portion of theca, showing two valves (V) that result from the stomial bifurcation at this end of the theca. Scale bars represent 500 μm in Figs. 28, 31; 200 μm in Fig. 29; 1000 μm in Fig. 30; and 100 μm in Figs. 32, 33.

Tertiary, well after the early diversification of angiosperms. If paleoherbs such as Piperales or the Nymphaealean *Brasenia* (Figs. 34, 35) or *Cabomba* are representative of the earliest angiosperms, then the plesiomorphic stamen form for flowering plants may be a more-or-less filamentous form with a well-differentiated filament. This scenario implies that all laminar stamens would have been derived within flowering plants.

Multiple origins of laminar stamens are indicated by all of the recent phylogenetic analyses of basal angio-

sperms. A paleoherb rooting (outside of Nymphaeaceae with laminar stamens) requires only one more instance of laminar stamen evolution than does the Magnolialean rooting demonstrated by Donoghue & Doyle (1989). The paleoherb rooting, however, helps to emphasize the rather facile shifts in angiosperm stamen evolution that parallel those in microsporophylls of other anthophytes. For example, outside of the angiosperms, Gnetales may demonstrate microsporophyll evolution from laminar forms (e.g., *Piroconites*) to syncrescent androphores

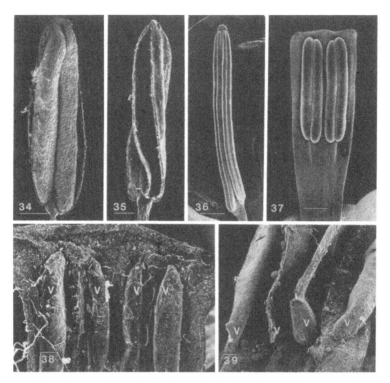

Figs. 34–39. Stamens of Nymphaeales. Figs. 34, 35. *Brasenia schreberi.* Fig. 34. Face view of dehisced stamen. Fig. 35. Lateral view of dehisced stamen, showing slight twist. Figs. 36. *Nymphaea odorata.* Figs. 37–39. *Nuphar luteum* subsp. *variegatum.* Fig. 37. Abaxial view of stamen, showing laminar form with protrusive pollen sacs. Fig. 38. Distal portion of dehisced theca, showing valves (V) resulting at dehiscence. Fig. 39. Proximal portion of dehisced theca, showing valves (V) resulting at dehiscence. Scale bars represent 300 μm in Figs. 34, 35; 1000 μm in Fig. 36; 500 μm in Fig. 37; and 100 μm in Figs. 38, 39.

(*Ephedra* and *Gnetum*) and back to a basally unified laminar form in *Welwitschia.* Among Bennettitales, laminar stamens have shifted from entire to pinnatifid (Crane 1986) and from separate pinnae to concrescence of basal pinnae with the distal (Crepet 1974).

The paleoherb rooting also calls attention to specializations of Magnolialean laminar stamens. For example, the embedded pollen sacs of Magnoliales with laminar stamens have been identified often as a plesiomorphic attribute (Canright 1952; Takhtajan 1969, 1991; Cronquist 1988). The outgroup analysis demonstrated that embedded pollen sacs are not present among nonangiospermous anthophytes or most other seed plants; instead, they are derived in the Magnoliales (Fig. 40 shows this feature mapped on the magnolialean rooted cladogram of

Donoghue & Doyle 1989). Similarly, even when Magnoliales are placed as the sister group of the rest of the angiosperms, the abaxial position of microsporangia that is common in this clade is shown to be a derived attribute (Fig. 41).

Previous authors have proposed that laminar stamens may be part of a specialized syndrome for cantharophily (Carlquist 1969; Gottsberger 1974, 1977, 1988). Gottsberger (1977, 1988) has observed that flowers of Magnoliales with laminar stamens often have fruity or aminoid scents that may serve as a deceit to attract beetles with fruit associations. These tend to be among the more derived beetles of Curculionidae and Scarabinaeidae. As flower visitors, they forage on the floral appendages as well as pollen, and this may have served as a selective

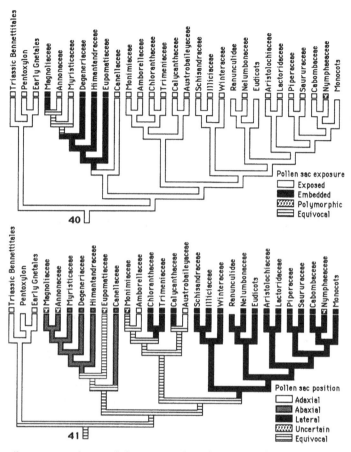

Figs. 40, 41. Stamen attributes mapped on a cladogram topology from Donoghue & Doyle (1989). Fig. 40. Pollen sac exposure. Fig. 41. Pollen sac position.

force for larger appendage size (with excess sterile tissue surrounding pollen sacs of stamens) and dimorphism in the androecium (staminodes in addition to fertile stamens) (Gottsberger 1977, 1988). The large stamens and staminodes may serve not only as food rewards to these pollinators, but may also provide protection for the beetles from predators and daily climatic changes (Gottsberger 1977, 1988). Staminodes are specialized in *Eupomatia* and *Galbulimima* to provide a sticky exudate for pollinators (Endress 1977, 1984a, b; Armstrong & Irvine 1990). Armstrong & Irvine (1990) have observed that beetles forage primarily on the apices of the staminodes in *Eupomatia*. Gottsberger (1974, 1977, 1988) has argued that a less specialized cantharophily than that

present among Magnoliales may have characterized the earliest angiosperms. He envisioned the earliest angiosperms as lacking the fruity scents of Magnoliales and having smaller flowers with exposed stamens and carpels (not hidden by large staminodes). These earliest angiosperms, Gottsberger (1977, 1988, Gottsberger *et al.* 1980) has proposed, were visited by primitive beetles that foraged only on pollen and not by the later, more destructive curculionids and scarabaeids. *Degeneria*, which has often been proposed as displaying plesiomorphic attributes for angiosperms, differs from the primitive form envisioned by Gottsberger (1974, 1977, 1988) in having large, laminar stamens and fragrant staminodia that present a sticky exudate, but also appears to

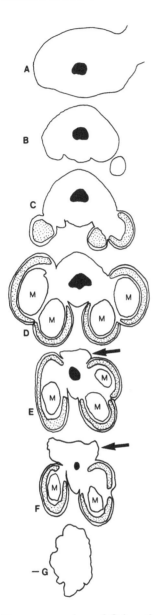

Fig. 42. Transverse sections of *Calycanthus floridus* stamen. Arrows indicate brief laminar expression of stamen axis. A, B. Stamen base. C–F. Anther region. G. Distal tip of stamen. Vascular tissue is represented by solid black, endothecium is stippled. M, microsporangia. Scale bar represents 40 μm.

have a range of 'licking' floral visitors (including various lepidopterans) as well as nitidulid beetles that feed only on pollen (and not floral appendages) (Bailey & Smith 1942; Thien 1980; Miller 1989).

DIVERSIFICATION AMONG MAGNOLIIDAE

Among the Magnoliidae, Laurales may have the most diverse stamens. It is difficult to assess the character state transformations that have been important in this diversification because we have no reasonable or comprehensive phylogenetic hypotheses for Laurales. Core Laurales (equivalent to Monimiaceae in Donoghue & Doyle 1989) might be considered to include Lauraceae, Gomortegaceae, Hernandiaceae (including *Gyrocarpus*), and Monimiaceae (*sensu* Philipson 1987). Various other groups have sometimes been allied with these core Laurales in the order. The phylogenetic analysis of Donoghue & Doyle (1989) placed Amborellaceae, Trimeniaceae, Chloranthaceae, Austrobaileyiaceae, and Calycanthaceae as part of a monophyletic group that included core Laurales. This contrasts with the recent *rbc*L-based phylogeny of Magnoliidae by Qiu *et al.* (1993) in which core Laurales (Gomortegaceae was not sampled) formed a monophyletic group in which Calycanthaceae was nested. The Qiu *et al.* results placed *Amborella*, *Austrobaileya*, and Chloranthaceae in a clade with Nymphaeales and Illiciales, a group positioned closer to Magnoliales than to core Laurales.

Diversity in the broad Laurales of Donoghue & Doyle (1989) can be categorized in three groups: (1) laminar stamens (including *Austrobaileya*, *Amborella*, and Calycanthaceae; although members of the latter family are not broadly laminar, they demonstrate similarities with *Amborella*), (2) nonlaminar lingulate stamens (Chloranthaceae and Trimeniaceae), and (3) a core Laurales association in which stamens have a fairly well-differentiated filament and anthers in which thecal form and stomium position are highly labile.

Mature stamens of *Austrobaileya* and *Amborella* are clearly laminar (i.e., have an ovoid anther transverse section, Fig. 2; sections illustrated in Endress & Hufford, 1989). Endress (1983) has shown that the sterile marginal areas of the anther region form after thecal definition

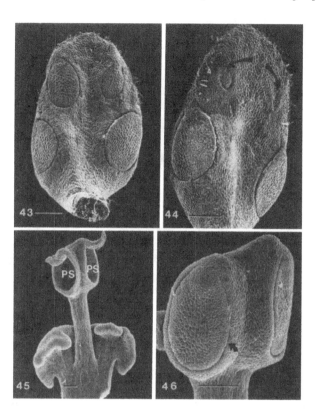

Figs. 43–46. Stamens of Lauraceae. Figs. 43, 44. *Umbellularia californica* stamens from the same plant. Fig. 43. Stamen with four pollen sacs. Fig. 44. Stamen with two functional pollen sacs and two rudimentary pollen sacs (arrows). Figs. 45, 46. *Lindera benzoin.* Fig. 45. Dehisced stamen with glandular appendages. *L. benzoin* is bisporangiate. Each pollen sac (PS) opens via a valve that lifts upward. Fig. 46. Predehiscence stamen, showing the curvilinear stomium (arrow) of one pollen sac. Scale bars represent 200 μm.

during stamen development of *Austrobaileya.* Unfortunately, developmental data are not available for *Amborella* or most Magnoliales with laminar stamens. Lateral expansion of the anther margin of laminar stamens of *Nuphar* appears contemporaneous with thecal definition (Moseley 1971) rather than following as in *Austrobaileya.* Similarities between stamens of *Calycanthus* and *Amborella* have not been previously noted. Over most of its length, the anther of *Calycanthus* is arcuate in transection (Fig. 42); however, the distal end of the anther is slightly ovoid (laminar) because of a limited marginal expression on the adaxial side of the thecae. A single cross-section of the anther of *Calycanthus* during early stamen development shows a unifacial primordium, similar to that in

higher dicots, at the time of thecal formation (Dengler 1972); however, this datum has little application without information on shapes over the length of the anther and on the shape transformations during floral ontogeny. *Amborella* is also similar to *Calycanthus* in the restriction of the anther terminus to a small knob (figs. 54, 77, 79 in Endress & Hufford 1989). The knobby apex of the *Calycanthus* stamen has been shown to have protein-rich superficial cells (Rickson 1979) that are eaten by beetles during pollination (Grant 1950), although this is not characteristic of *Chimonanthes* (Nicely 1965). The structure of the apical knob and its potential role in pollination of *Amborella* have not been investigated. Bailey & Swamy (1948) reported that the apical knobs of outer

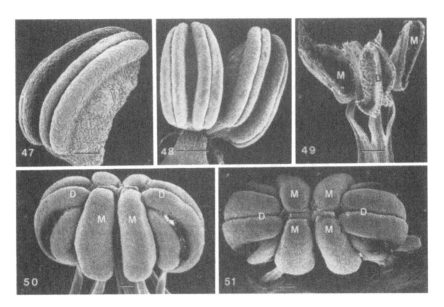

Figs. 47–51. Stamens of Lardizabalaceae and Papaveraceae. Figs. 47, 48. *Akebia trifoliolata* (Lardizabalaceae). Fig. 47. Lateral view of stamen, showing highly protrusive, curved thecae. Fig. 48. Male flower, showing abaxial view of surface of thecae. Figs. 49–51. *Dicentra* (Papaveraceae). Fig. 49. One androecial unit of *D. cucullaria*. The androecial unit consists of one dithecal anther (D) and two monothecal anthers (M). All thecae dehisced on this androecial unit. Figs. 50, 51. *D. formosa* showing paired androecial units tightly appressed to stigma as found in prepollination flower. M, monothecal anthers; D, dithecal anthers. Fig. 50. Lateral view of anthers of two androecial units. Fig. 51. Top view of anthers of two androecial units. Scale bar represent 300 μm in Figs. 47, 48; and 200 μm in Figs. 49–51.

stamens of *Amborella* form distinctive papillae similar to those of *Calycanthus*.

Phylogenetic relationships among core Laurales are poorly understood. Recognition of the two large families, Lauraceae and Monimiaceae, continues to be supported (Philipson 1987), but questions about monophyly and the derived character states that may support particular groups have not been addressed. Most authors have accepted the placement of *Atherosperma*, *Laureliopsis*, and *Siparuna* in Monimiaceae, but the placement of the divergent Hernandiaceae and Gomortegaceae in relationship to the two larger families remains a problem. Although we can outline basic aspects of stamen diversity, exploring important character state transformations and homoplastic evolution must be deferred until reasonable phylogenetic hypotheses are formulated.

The most interesting variation in stamens of core Laurales involves compromised thecal construction. As noted earlier, most angiosperm stamens are tetrasporangiate with pollen sacs paired on each side of the anther. If the pollen sacs of each pair are positioned adjacently and share a common stomium, then they may be considered to form a theca. Thecal construction is lost among core Laurales because of (1) vertical displacement and independent dehiscence of microsporangia and (2) loss of one microsporangium from each pair.

The tight developmental specification for the adjacent positioning of pollen sac pairs on each side of the stamen to form a theca is highly conserved among angiosperm clades but has been lost in the Lauraceae. Lauraceae that possess tetrasporangiate stamens have vertically displaced pollen sacs, each of which dehisces independently via a curved stomium (Fig. 43). Associated with this change is a heightened lability in the relative sizes of pollen sacs within anthers (sizes are generally similar within stamens of most angiosperms). For example, size

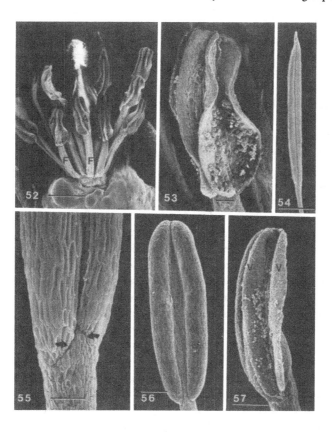

Figs. 52–57. Stamens of Papaveraceae and Ranunculaceae. Figs. 52, 53. *Sanguinaria canadensis*. Fig. 52. Flower with perianth removed to show androecium. Stamens have well-differentiated filaments (F) and terminal anthers (A). Fig. 53. Dehisced anther. Figs. 54, 55. *Thalictrum dioicum*. Fig. 54. Lateral view of stamen showing subterminal anther region and stomium that is unbifurcated distally but bifurcated proximally. Fig. 55. Higher magnification of stomium base from Fig. 54, showing bifurcation of stomium (arrows). Figs. 56, 57. *Caltha palustris*. Fig. 56. Facial view of stamen, showing well-differentiated filament and terminal anther. Fig. 57. Dehisced anther, showing valves (V). Scale bars represent 1000 μm in Figs. 52, 54; 100 μm in Figs. 53, 55; and 300 μm in Figs. 56, 57.

differences among the pollen sacs within stamens and among stamens of individual plants of *Umbellularia* are evident (Figs. 43, 44). Size differences among pollen sacs are one of the reasons why anthers of some Lauraceae are longitudinally asymmetrical (e.g., figures in Endress & Hufford 1989). Compromised thecal construction, resulting in greater functional independence of pollen sacs, may also have relaxed selection on the number of pollen sacs. Many Lauraceae have only two pollen sacs per anther (Figs. 45, 46; Endress & Hufford 1989;

Endress & Stumpf 1990). Gomortegaceae, Hernandiaceae and many Monimiaceae (including *Atherosperma* and *Monimia*) are similar to those Lauraceae that possess stamens with only two pollen sacs (one on each side of the anther). Indeed, among core Laurales, only Monimiaceae (including *Hortonia*, *Peumus*, and *Xymalos*) have stamens in which pollen sacs are paired in adjacent positions and share a common stomium (Endress & Hufford 1989). Monimiaceae, such as *Siparuna* and some *Tambourissa*, demonstrate additional modification in the adjacent positioning of pollen sacs from opposite sides of the anther (thecal coalescence) and their dehiscence via a single stomium (Endress & Lorence 1983; Endress & Hufford 1989). Endress (1980a) noted that Monimiaceae with a single stomium for pollen sacs on opposing sides of the anther tend to be those with globular flowers.

In addition to Monimiaceae among Magnoliidae, *Zygogynum* and *Bubbia* (Winteraceae) also sometimes form a single stomium that is continuous across both thecae. *Bubbia* and *Zygogynum* are similar to those *Tambourissa* that demonstrate this attribute in having microsporangia that meet at the stamen apex or are confluent (Endress & Hufford 1989). Indeed, the pollen sacs of these Winteraceae have a largely apical position. The apical to near apical position of the pollen sacs in these two genera and *Tasmannia*, *Exospermum*, and some *Drimys* results in the convergence of the thecae (i.e., the thecal apices are closer together than are their bases). This is unusual among the Magnoliidae, which generally have parallel thecae. The brachiate stamens of *Kadsura* (Schisandraceae; figured in Endress & Hufford 1989), which have pollen sacs laterally offset on arm-like outgrowths, have slightly divergent thecae with the apices farther apart than are the bases. Endress & Stumpf (1991) have shown that most lower rosid groups have anthers with convergent thecae (sagittate in their terminology). Convergent thecae are probably common among the higher dicots and associated with the shift from basifixed to predominately dorsifixed anthers and the restriction of connective tissue along the length of the thecae.

Stamen forms among Ranunculales (*sensu* Cronquist 1981) have not been broadly surveyed, except among Berberidaceae (Endress & Hufford 1989). Berberidaceae have diverse dehiscence patterns which may be associated with size differences between dorsal and ventral pollen sacs in each theca. In most examined members of the family (e.g., *Berberis*, *Caulophyllum*, *Epimedium*, *Jeffersonia*, *Mahonia*, and *Vancouveria*; Endress & Hufford 1989) which have pollen sacs of different sizes the stomium extends longitudinally between the two pollen sacs of each theca, around the base of the dorsal pollen sac and upward along the dorsal side of the theca. This permits the wall of the dorsal pollen sac to curve upward at dehiscence. The ventral pollen sac wall is not circumscribed by stomium and does not open outward. *Podo-*

phyllum has a stomium that extends along the proximal, dorsal, and distal margins of each theca and hinges outward at dehiscence along the ventral thecal margin. Other Berberidaceae have only a conventional longitudinal stomium positioned between the pollen sacs of each theca (e.g., *Nandina*). The diversity of dehiscence patterns in Berberidaceae is similar to that of Hamamelidaceae among the eudicots but is greater than that in most other families of Magnoliidae.

Androecial evolution within Fumarioideae (Papaveraceae) has been richly debated (summarized in Ronse Decraene & Smets 1992a). Flowers of most Fumarioideae have two androecial units. Each unit consists of a centrally positioned dithecal anther and two laterally positioned monothecal anthers (Figs. 49–51). These androecial units are particularly interesting because each has a form similar to the 'stamen' of *Chloranthus* (Chloranthaceae). In both *Chloranthus* and Fumarioideae, the origin of the androecial units bearing one dithecal anther and two lateral monothecal anthers is unclear. Endress (1987) hypothesized that the *Chloranthus* androecial unit evolved by fractionation of the thecae of a single stamen and the separation of the lower thecal fractions on separate lateral lobes. Payer (1857), Eichler (1865), and Buchenau (1866) have advocated a similar explanation for the origin of the androecial units of Fumarioideae. Comparative developmental studies indicate, however, that the monothecal anthers of the androecial units of Fumarioideae were coopted from separate stamens in a different androecial whorl (Ronse Decraene & Smets 1992a). Significant progress could probably be made toward resolving this dilemma by mapping the data on androecial development on a robust phylogeny for Fumarioideae and their relatives. In *Dicentra* (Figs. 50, 51), the anthers of the androecial units tightly enclose the lobed stigma. Macior (1970) observed that self-pollen is regularly deposited on the stigma, but that it seldom results in fertile seeds. The tight positioning of the anthers around the stigma is associated with pollination by *Bombus* species that must pry apart the tightly appressed petal/anther complex to reach nectar in spurs at the base of the flower (Macior 1970, 1978).

Most Ranunculales have stamens with well-differentiated filaments and terminal to subterminal anthers (e.g., Figs. 54, 56). Stamens of some Lardizabalaceae (Figs. 47, 48; particularly those of *Sargentodoxa*,

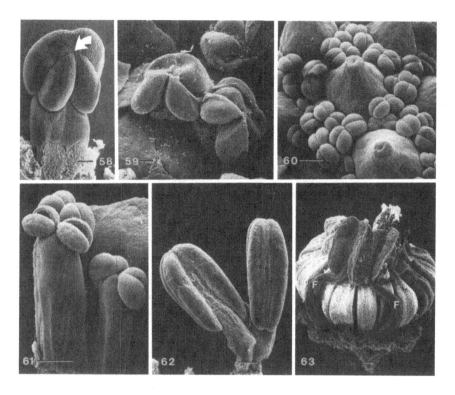

Figs. 58–63. Stamens of monocotyledons. Figs. 58, 59. *Acorus calamus* (Araceae). Fig. 58. Ventral view of stamen showing ventral pollen sacs smaller than dorsal and continuity of stomium across thecae (arrow). Fig. 59. Portion of flower, showing two dehisced stamens. Figs. 60, 61. *Calla palustris.* Fig. 60. Flowers with undehisced stamens protruding just above the surface of the spadix. Fig. 61. Stamens showing thick, linear axis and distally positioned thecae. Fig. 62. Facial and lateral views of two stamens of *Alisma subcordatum.* Fig. 63. Flower of *Sagittaria cuneata* with perianth and some stamens removed. Dehisced stamens show well-differentiated filament (F) and longitudinally restricted connective that results in basally protrusive thecae. Scale bars represent 100 μm in Figs. 58, 59, 63; 500 μm in Figs. 60, 61; and 200 μm in Fig. 62.

Stapf 1926) among Ranunuculales are most similar to those of paleoherbs and Illiciales in lacking filament differentiation (they have a thick, fleshy axis that is continuous along the stamen length) and possessing a prominent distal protrusion. Most Ranunculales with subterminal anthers generally have only a slight apical protrusion with a highly restricted width and breadth relative to the anther. Anthers may be straight or curved, but are predominantly basifixed. Most Ranunculales have linear, unbifurcated stomia. Ranunculaceae are interesting because they appear to be one of the groups in which bifurcated stomia and valvate dehiscence have evolved (Table 3; Figs. 55, 57). Although the family has not been

sampled extensively, the few examined taxa typically have proximal (but not distal) stomial bifurcations that facilitate valvate dehiscence.

Endress and coworkers (Endress & Hufford 1989; Hufford & Endress 1989; Endress & Stumpf 1990, 1991) have elaborated the details of stamen dehiscence patterns among Magnoliidae and basal groups of eudicots. One result of these surveys has been the description of branched stomia that are associated with valvate dehiscence. Valvate dehiscence results when a portion of the pollen sac wall opens outward, flap-like, to present pollen (which often adheres to the surface of the opened wall; Fig. 33) at the time of pollination. Valvate

Table 3. *Families and genera known to possess stomial bifurcations and valvate dehiscence. Sampling in the large families has been limited*

Family	Genus	Bifurcation (proximal, distal or both)
Degeneriaceae	*Degeneria*	Both
Himantandraceae	*Galbulimima*	Both
Eupomatiaceae	*Eupomatia*	Both
Magnoliaceae	*Magnolia*	Proximal, no consistent distal bifurcation
Annonaceae	*Asimina*	Both
	Cananga	Distal
	Monodora	Distal
	Polyalthia	Both
Monimiacae	*Peumus*	Both
Chloranthaceae	*Sarcandra*	Both
	Choranthus	Both
Nymphaeaceae	*Nuphar*	Both
Piperaceae	*Piper*	Distal
Ranunculaceae	*Caltha*	Proximal
	Thalictrum	Proximal
	Trauvetteria	Proximal
Trochodendraceae	*Trochodendron*	Both
Tetracentraceae	*Tetracentron*	Both
Cercidiphyllaceae	*Cercidiphyllum*	Proximal
Eupteleaceae	*Euptelea*	Proximal
Myrothamnaceae	*Myrothamnus*	Proximal
Hamamelidaceae	*Altingia*	Both
	Corylopsis	Both
	Distylium	Proximal
	Eustigma	Both
	Fortunearia	Both
	Fothergilla	Both
	Loropetalum	Both
	Maingaya	Both
	Matudaea	Both
	Molinadendron	Both
	Parrotiopsis	Both
	Rhodoleia	Both
	Sinowilsonia	Both
Platanaceae	*Platanus*	Distal
Nothofagaceae	*Nothofagus*	Distal

Based on data derived from Endress (1987, 1994), Endress & Hufford (1989), Hufford & Endress (1989), and this chapter.

dehiscence has been well known in core Laurales (Fig. 45, 46) and *Hamamelis* (Hamamelidaceae; Figs. 5, 6) where it is associated with a curvilinear stomium that extends around the margin of a pollen sac. In contrast, a stomium that extends longitudinally between the adjacent pollen sacs of a theca may be bifurcated at proximal and distal ends, permitting two valves to form. Hufford & Endress (1989) suggested that bifurcated stomia are generally associated with nonprotrusive pollen sacs that are embedded in sterile tissue, particularly around their proximal and distal shoulders. Bifurcated stomia are common among Magnoliidae (Table 3). Among the extant groups of basal angiosperms, bifurcated stomia are most common among Magnoliales in which stamens have embedded or only slightly protrusive pollen sacs. Bifurcated stomia appear to have evolved separately in the Laurales (*Peumus* of Monimiaceae: Endress & Hufford 1989; and *Chloranthus* and *Sarcandra* of Chloranthaceae: Endress 1987), Piperales (*Piper angustum* – distal bifurcation only; Endress, 1994), Nymphaeales (*Nuphar*, Figs. 38, 39) and Ranunculales (*Thalictrum*, Fig. 55, *Trauvetteria*, and perhaps *Caltha*, Fig. 57, of Ranunculaceae; Endress & Hufford, 1989). The proximal and distal bifurcations of stomia clearly evolve separately (or may be lost separately; Hufford & Endress, 1989). Magnoliaceae (Endress & Hufford 1989; Figs. 31–33) and Ranunculaceae (Endress & Hufford 1989, Figs. 54, 55) include taxa with proximal bifurcations but not distal.

Bifurcated stomia may be more common among Magnoliidae than higher dicots and monocots. No monocots have been reported to have bifurcated stomia, but their independent evolution has been documented among eudicots (Table 3). The higher dicots with stomial bifurcations are all basal members of the Hamamelidae, except *Xanthoceras* (Sapindaceae) which Endress & Stumpf (1991) suggested has only a basal bifurcation, but it does not open to create valvate dehiscence. Many Hamamelidaceae have bifurcated stomia (Hufford & Endress 1989). This may reflect the plesiomorphy of bifurcated stomia in the family and retention of the attribute through the evolutionary radiation of the entomophilous taxa.

EUDICOTS

The phylogenetic placement of eudicots (nonmagnoliid dicots) among Magnoliidae is unresolved. The results of Donoghue & Doyle's (1989) phylogenetic analysis placed eudicots as the sister group of Ranunculales (although neither group was well sampled in their study). The eudicot–Ranunculales clade was placed as the sister group of paleoherbs. Studies of reproductive structures (e.g., Williams *et al.* 1993) have emphasized a placement of eudicots near Illiciales. The results of Chase *et al.* (1993) were distinctive in showing a very deep origin of eudicots among Magnoliidae. Their results showed the eudicots and Ranunculales diverged from the rest of the flowering plants at the second node from the base of the angiosperms. The results of Chase *et al.* (1993) imply that putatively derived attributes of eudicot flowers would have originated much earlier than expected. Similarly, derived stamen attributes must be envisioned as evolving much earlier and more rapidly in the Chase *et al.* scenario than has been conventionally accepted.

Stamens of most eudicots have well-differentiated filaments and terminal to subterminal anthers. Those with subterminal anthers tend to have only a very brief sterile, distal projection. The width and breadth of the sterile distal tip is generally highly restricted relative to the distal end of the anther. The sterile distal tip is sometimes displaced from a central position toward either the abaxial or adaxial side of the anther (e.g., *Loropetalum* and *Maingaya* of Hamamelidaceae; Hufford & Endress 1989). Hence, among the eudicots the sterile distal tip may have derived novel positions. Basal members of the eudicot group have basifixed anthers, a connective that extends the full length of the thecae, and nonprotrusive microsporangia. In general form, stamens of the basal eudicot groups, such as *Trochodendron*, *Tetracentron*, *Cercidiphyllum*, and other 'lower' hamamelids, are quite similar to those found among Ranunculales, especially Ranunculaceae, and Illiciales.

Relatively early in eudicot evolution we find a more-or-less lateral fixation of the filament evolving repeatedly. These so-called versatile anthers probably facilitate pollination by the larger, more specialized pollinators (such as bees) that rake stamens along their torsos when probing flowers. Along with the shift in filament fixation, we find longitudinally restricted connectives, resulting in sagittate or X-forming anthers, and more protrusive microsporangia. These derived attributes tend to characterize the rosid grade (Hufford 1992) which evolved early in the diversification of eudicots. Although the stamens of many eudicots, particularly those of 'lower' hamamelids, are similar to forms found among Ranunculales and paleoherbs, they tend to differ markedly from those of other Magnoliidae.

Dehiscence patterns tend to be less variable among eudicots than Magnoliidae. As discussed above, valvate dehiscence via bifurcated stomia is common among 'lower' Hamamelidae that appear to be among the earliest groups of eudicots (Hufford 1992; Chase *et al.* 1993) but is absent in the diversification of eudicots. Eudicots display porate forms of dehiscence that are not present among Magnoliidae. Among the eudicots, various means of porate dehiscence appear to have evolved (Endress & Stumpf 1990, 1991). Simple linear stomia positioned between the microsporangia of each theca are by far the most common means of dehiscence among eudicots. Variations in pollen presentation among eudicots are more commonly achieved via adnation of stamens to the corolla and by great developmental control of filament growth and orientation in contrast to patterns among Magnoliidae.

MONOCOTYLEDONS

Recent phylogenetic analyses have placed the monocots among magnoliid paleoherbs (as the sister group of Nymphaeales: Donoghue & Doyle 1989; or the sister group of Aristolochiaceae: Bharathan & Zimmer 1993) or as their sister group (Chase *et al.* 1993). Duvall *et al.* (1993) identified *Acorus* as the sister group of the rest of the monocots in which Alismatales, Arales, and Dioscoreales form basal clades (which is consistent with the results of Bharathan & Zimmer 1993). *Acorus* has a terminal anther and broad, bifacial filament (Fig. 58) Its ventral pollen sacs are smaller than the dorsal, and the stomium is confluent across the two thecae (Figs. 58, 59). The basal groups of monocots all have stamens with filaments and anthers with at least slightly protrusive pollen sacs that are similar to those common among paleoherbs. The diversity of stamens among the monocots parallels that of eudicots. For example, various monocots show trends toward filament connation, longitudinal restriction of the connective that results in basally and distally protrusive thecae (sagittate and X-forming anthers in the terminology of Schaeppi 1939), dorsifixed

and ventrifixed anthers, reductions in pollen sac number, and restriction of the stomium that results in porate dehiscence. Few specific stamen forms present among Magnoliidae are replicated among monocots, although examples of two presumed parallelisms are discussed below.

Many Araceae (e.g., *Calla*, Fig. 61) have stamens that resemble those of Winteraceae. Members of both groups have a thick stamen axis and apical or nearly apical thecae that often have distally positioned stomia (e.g., *Bubbia* of Winteraceae; Bailey & Nast 1943; Vink 1983, 1985). The similar 'packing' and orientation of the stamens in both families may influence the thecal positions and direction of dehiscence. *Bubbia*, for example, has numerous tightly packed stamens on the floral axis. Its stamens separate little at anthesis; hence, the distal positions of the thecae with their 'apical' stomia may have been advantageous for adequate pollen presentation. The dense packing of flowers on inflorescences of some Araceae (e.g., *Calla*, Fig. 60) presents a similar problem. Among these Araceae, the stamen axes elongate very little and the anthers with their convergent, distally positioned thecae protrude just slightly above the surface of the spadix.

Members of Dioscoreales, especially *Stemona*, *Stenomeris*, *Tacca*, and *Trichopus*, have stamens with a form resembling that of *Asarum* (Aristolochiaceae). These taxa have stamens with a thick axis that tapers distally well beyond the anther. The anthers of these taxa share a longitudinal curvature, arcuate transection, and thecae distinctly offset from the axis. This stamen form has been implicated as part of a trap-flower pollination syndrome; however, few of the taxa have been investigated in detail, and *Asarum* (Kugler 1934; Werth 1951; Lu 1982; Mesler & Lu 1993) has been shown to be largely self-pollinating (although other Aristolochiaceae do appear to have a trap-flower pollination syndrome; Brantjes 1980). The stamens of *Stenomeris* (Dioscoreaceae) are pendant and their elongate apices extend to rest on the stigma (Burkill 1960; Dahlgren *et al.* 1985; Heel 1992). Burkill (1960) suggested that these distal prolongations may serve as bridges for pollinators to access the stigma. *Afrothismia* (Burmanniaceae) has similar pendant stamens with recurved tips forming a cone over the stigma (Dahlgren *et al.* 1985). Stamens of *Tacca* also contact the stigma; however, this is through elaborations of the anther axis rather than through a distal prolongation (Fig. 8; Heel 1992).

The very slender, long anthers that are common among lilioid monocots are far less prevalent among eudicots and uncommon among Magnoliidae. The thecae of these anthers have linear stomia. At dehiscence the entire anther twists, resulting in a wide opening of each theca. Such twisting is permitted by the narrow connective and small diameter of the filament relative to the anther. Most Magnoliidae have a stamen axis that is far too massive to permit such twisting without more extensive formation of mechanical tissue than is typically present (e.g., see data in Endress & Hufford 1989). Only those relatively derived Magnoliidae (e.g., *Brasenia*, Fig. 35; and some Papaverales and Ranunculales) with well-differentiated filaments that have a restricted diameter relative to the anther demonstrate similar twisting at dehiscence.

CONCLUSIONS

Anthophytes have diverse microsporophylls. The fossil record currently shows that the earliest Bennettitales and Gnetales (except *Maculostrobus*) tend to have broad, laminar microsporophylls with sporangia or synangia on their surface. Current data support the proposal that laminar or lingulate stamen forms were retained by primitive angiosperms from their anthophyte ancestors; however, this hypothesis is only modestly robust. Better supported hypotheses of character evolution are needed but require greater knowledge about relationships among early angiosperms as well as about nonangiospermous anthophyte diversification. Additional information about the evolution and diversity of the angiosperm stem lineage (Doyle & Donoghue 1993) is particularly important for resolving many questions about early flowering plants, including stamen diversification. Although diversification in the different anthophyte groups has proceeded along distinct avenues, simpler linear to filamentous forms and even highly pinnatifid forms have commonly evolved from the early thick, laminar microsporophylls. The reason for the parallelism in each of the groups is unclear. Highly pinnatifid forms may have been one way of increasing number of sporangia, and hence pollen, with minimal resource investment in microsporophylls. Simpler filamentous to linear forms may

have evolved in association with changes in strobilus architecture. For example, they may be associated with shifts from helical to whorled arrangements of appendages and the tight arrangement of these parts inside of a protective perianth. Changes to simpler stamens may have also been mediated by novel pollination strategies, especially shifts away from coleopterans that indiscriminately forage on floral parts as well as pollen.

The results of recent phylogenetic analyses of angiosperms indicate that basal clades do not consist entirely of taxa with laminar or lingulate stamens. Evidence is mounting that Magnoliales may not be the basal group or sister group of the rest of the angiosperms. If recent phylogenetic hypotheses that show these modified placements for Magnoliales, and particularly those that root angiosperms among paleoherbs, are correct, then stamen evolution may need to be viewed as more labile or changing more rapidly during the early diversification of angiosperms than has been previously appreciated. If stamen form did diversify earlier and more rapidly than conventionally accepted, then this may mean that concepts of floral evolution in general will need to be modified. Investigations of pollination biology (Thien 1980; Thien *et al.* 1985) provide some basis for suggesting that pollinator diversity was greater among early angiosperms than was formerly thought, and this may be an indication that morphological evolution was also more diverse.

The large flowers of Magnoliales with heteromorphic androecia that include both fleshy staminodes and fertile stamens appear specialized for pollination by some coleopteran groups. The movements of the stamens and staminodes and the rewards they offer pollinators are part of their syndrome of specialization for beetle pollination. It is unclear, however, whether these specializations arose at the base of the angiosperms or somewhat later. Most of the peculiar attributes of Magnolialean androecia are not found among eudicots or monocots, groups in which floral architecture and reproductive biology are sufficiently different from those of Magnoliales that they have diversified within a different range of constraints.

ACKNOWLEDGMENT

This research was funded in part by the Olga Lakela Herbarium, University of Minnesota-Duluth. I thank G.

Walton, P. Monson, W. Friedman, and C. Chapin for providing specimens and A. Winbauer for technical assistance. I thank W. D'Arcy, W. Friedman, R. Keating, and S. Manchester for helpful reviews of the manuscript.

LITERATURE CITED

Arber, E.A.N. & J. Parkin. 1907. On the origin of angiosperms. J. Linn. Soc. Bot. 22: 29–80.

Armstrong, J.E. & A.K. Irvine. 1990. Functions of staminodia in the beetle-pollinated flowers of *Eupomatia laurina*. Biotropica 22: 429–431.

Armstrong, J.E. & T.K. Wilson. 1978. Floral morphology of *Horsfieldia* (Myristicaceae). Amer. J. Bot. 65: 441–449.

Ash, S.R. 1972. Late Triassic plants from the Chinle Formation in northeastern Arizona. Palaeontology 15: 598–618.

Bailey, I.W. & C.G. Nast. 1943. The comparative morphology of the Winteraceae. I. Pollen and stamens. J. Arnold Arbor. 24: 340–346 + 3 pls.

Bailey, I.W. & A.C. Smith. 1942. Degeneriaceae, a new family of flowering plants from Fiji. J. Arnold Arbor. 23: 356–365 + 5 pls.

Bailey, I.W. & B.G. Swamy. 1948. *Amborella trichopoda* Baill., a new morphological type of vesselless dicotyledon. J. Arnold Arbor. 29: 245–254 + 5 pls.

Baum, H. 1950. Unifaziale und subunifaziale Strukturen im Bereich der Blütenhülle und ihre Verwendbarkeit für die Homologisierung der Kelch- und Kronblätter. Österr. Bot. Z. 97: 1–43.

Baum, H. 1952. Die Bedeutung der diplophyllen Übergangsblätter für den Bau der Staubblätter. Österr. Bot. Z. 99: 228–243.

Bharathan, G. and E. Zimmer. 1993. Phylogenetic relationships of basal monocots – rDNA sequence data. Amer. J. Bot. 80 (Suppl.): 132.

Bornman, G.H. 1972. *Welwitschia mirabilis*: paradox of the Namib Desert. Endeavour 31: 95–99.

Bornstein, A.J. 1991. The Piperaceae in the Southeastern United States. J. Arnold Arbor., Suppl. Ser. 1: 349–366.

Brantjes, N.B.M. 1980. Flower morphology of *Aristolochia* species and the consequences for pollination. Acta Bot. Neerl. 35: 521–528.

Buchenau, F. 1866. Bemerkungun über den Blüthenbau der Fumariaceen und Cruciferen. Flora 49: 39–46.

Burkill, I.H. 1960. The organography and the evolution of Dioscoreaceae, the family of the yams. J. Linn. Soc. (Bot.) 56: 319–412.

Canright, J.E. 1952. The comparative morphology and relationships of the Magnoliaceae. I. Trends of specialization in the stamens. Amer. J. Bot. 39: 484–496.

Carlquist, S. 1964. Morphology and relationships of Lactoridaceae. Aliso 5: 421–435.

Carlquist, S. 1969. Toward acceptable evolutionary interpretations of floral anatomy. Phytomorphology 19: 332–363.

Chase, M.W., D.E. Soltis, R.G. Olmstead, D. Morgan, D.H. Les, B.D. Mishler, M. R. Duvall, R.A. Price, H.G. Hills, Y.-L. Qiu, K.A. Kron, J.H. Rettig, E. Conti, J.D. Palmer, J.R. Manhart, K.J. Systsma, H.J. Michaels, W.J. Kress, K.G. Karl, W.D. Clark, M. Hedren, B.S. Gaut, R.K. Jansen, K.-J. Kim, C.F. Wimpee, J.F. Smith, G.R. Furnier, S.H. Strauss, Q.-Y. Xiang, G.H. Plunkett, P.S. Soltis, S.M. Swensen, S.E. Williams, P.A. Gadek, C.J. Quinn, L.E. Eguiarte, E. Golenberg, G.H. Learn, Jr., S.W. Graham, S.C. H. Barrett, S. Dayanandan, & V.A. Albert. 1993. Phylogenetics of seed plants: an analysis of nucleotide sequences from the plastid gene *rbc*L. Ann. Missouri Bot. Gard. 80: 528–580.

Church, A.H. 1914. On the floral mechanism of *Welwitschia mirabilis*. Phil. Trans. R. Soc. Lond., B 205: 115–151.

Corner, E.J.H. 1946. Centrifugal stamens. J. Arnold Arbor. 27: 423–437.

Coulter, J.M. & C.J. Chamberlain. 1912. Morphology of Angiosperms. Appleton and Co., New York.

Crane, P.R. 1985. Phylogenetic analysis of seed plants and the origin of angiosperms. Ann. Missouri Bot. Gard. 72: 716–793.

Crane, P.R. 1986. The morphology and relationships of Bennettitales. pp. 163–175 in Systematic and Taxonomic Approaches in Paleobotany (edited by R.A. Spicer & B.A. Thomas). Clarendon Press, Oxford.

Crane, P.R. 1988. Major clades and relationships in the 'higher' gymnosperms. pp. 218–272 in Origin and Evolution of Gymnosperms (edited by C.B. Beck). Columbia University Press, New York.

Crane, P.R. & G.R. Upchurch. 1987. *Drewria potomacensis* gen. et sp. nov., an early Cretaceous member of Gnetales from the Potomac Group of Virginia. Amer. J. Bot. 74: 1722–1736.

Crepet, W.L. 1974. Investigations of North American Cycadeoids: the reproductive biology of *Cycadeoidea*. Palaeontographica, B 148: 144–169.

Cronquist, A. 1981. An Integrated System of Classification

of Flowering Plants. Columbia University Press, New York.

Cronquist, A. 1988. The Evolution and Classification of Flowering Plants, 2nd ed. New York Botanical Garden, New York.

Dahlgren, R.M.T., H.T. Clifford, & P.F. Yeo. 1985. The Families of the Monocotyledons. Springer-Verlag, Berlin.

Dengler, N.G. 1972. Ontogeny of the vegetative and floral apex of *Calycanthus occidentalis*. Can. J. Bot. 50: 1349–1356.

Donoghue, M.J. & J.A. Doyle. 1989. Phylogenetic analysis of angiosperms and the relationships of Hamamelidae. pp. 17–45 in Evolution, Systematics, and Fossil History of the Hamamelidae, Vol. 1. Introduction and 'Lower' Hamamelidae (edited by P.R. Crane & S. Blackmore). Clarendon Press, Oxford.

Doyle, J.A. & M.J. Donoghue. 1986. Seed plant phylogeny and the origin of angiosperms: an experimental cladistic approach. Bot. Rev. 52: 321–431.

Doyle, J.A. & M.J. Donoghue. 1993. Phylogenies and angiosperm diversification. Paleobiology 19: 141–167.

Duvall, M.R., M.T. Clegg, M.W. Chase, W.D. Clark, W.J. Kress, H.G. Hills, L.E. Equiarte, J.F. Smith, B.S. Gaut, E.A. Zimmer, & G.H. Learn, J.R. 1993. Phylogenetic hypotheses for the monocotyledons constructed from *rbc*L sequence data. Ann. Missouri Bot. Gard. 80: 607–619.

Eames, A.J. 1952. Relationships of the Ephedrales. Phytomorphology 2: 79–100.

Eames, A.J. 1961. Morphology of the Angiosperms. McGraw-Hill, New York.

Eichler, A.W. 1865. Über den Blüthenbau der Fumariaceen, Cruciferen und einiger Capparideen. Flora 48: 433–444.

Endress, P.K. 1977. Über Blütenbau und Verwandtschaft der Eupomatiaceae und Himantandraceae (Magnoliales). Ber. Dtsch. Bot. Ges. 90: 83–103.

Endress, P.K. 1980a. Ontogeny, function and evolution of extreme floral construction in Monimiaceae. Pl. Syst. Evol. 134: 79–120.

Endress, P.K. 1980b. The reproductive structures and systematic position of the Austrobaileyaceae. Bot. Jahrb. Syst. 101: 393–433.

Endress, P.K. 1983. The early floral development of *Austrobaileya*. Bot. Jahrb. Syst. 103: 481–497.

Endress, P.K. 1984a. The flowering process in the Eupomatiaceae (Magnoliales). Bot. Jahrb. Syst. 104: 297–319.

Endress, P.K. 1984b. The role of inner staminodes in the floral display of some relic Magnoliales. Pl. Syst. Evol. 146: 269–282.

Endress, P.K. 1987. The Chloranthaceae: reproductive structures and phylogenetic position. Bot. Jahrb. Syst. 109: 153–226.

Endress, P.K. 1994. Shapes, sizes and evolutionary trends in stamens of Magnoliidae. Bot. Jahrb. Syst. 115: 429–460.

Endress, P.K. & L. Hufford. 1989. The diversity of stamen structures and dehiscence patterns among Magnoliidae. Bot. J. Linn. Soc. 100: 45–85.

Endress, P.K. & D.H. Lorence. 1983. Diversity and evolutionary trends in the floral structure of *Tambourissa* (Monimiaceae). Pl. Syst. Evol. 143: 53–81.

Endress, P.K. & F.B. Sampson. 1983. Floral structure and relationships of the Trimeniaceae (Laurales). J. Arnold Arbor. 64: 447–473.

Endress, P.K. & S. Stumpf. 1990. Non-tetrasporangiate stamens in the angiosperms: structure, systematic distribution and evolutionary aspects. Bot. Jahrb. Syst. 112: 193–240.

Endress, P.K. & S. Stumpf. 1991. The diversity of stamen structures in 'lower' Rosidae (Rosales, Fabales, Proteales, Sapindales). Bot. J. Linn. Soc. 107: 217–293.

Esau, K. 1977. Anatomy of Seed Plants, 2nd ed. Wiley, New York.

Gifford, E.M. & A.S. Foster. 1989. Morphology and Evolution of Vascular Plants. W.H. Freeman, New York.

Gottsberger, G. 1974. The structure and function of the primitive angiosperm flower – a discussion. Acta Bot. Neerl. 23: 461–471.

Gottsberger, G. 1977. Some aspects of beetle pollination in the evolution of flowering plants. Plant Syst. Evol., Suppl. 1: 211–226.

Gottsberger, G. 1988. The reproductive biology of primitive angiosperms. Taxon 37: 630–643.

Gottsberger, G., I. Silberbauer–Gottsberger, & F. Ehrendorfer. 1980. Reproductive biology in the primitive relic angiosperm *Drimys brasiliensis* (Winteraceae). Pl. Syst. Evol. 135: 11–39.

Grant, V. 1950. The pollination of *Calycanthus occidentalis*. Amer. J. Bot. 37: 294–297.

Gray, A. 1879. Botanical Textbook, Vol. 1. Structural Botany. American Book Co., New York.

Guédès, M. 1979. Morphology of Seed Plants. Cramer, Vaduz.

Hamby, R.K. & E.A. Zimmer. 1992. Ribosomal RNA as a phylogenetic tool in plant systematics. pp. 50–91 in Molecular Systematics of Plants (edited by P.S. Soltis,

D.E. Soltis, & J.J. Doyle). Chapman and Hall, New York.

Heel, W.A. Van. 1992. Floral morphology of Stemonaceae and Pentastemonaceae. Blumea 36: 481–499.

Howard, R.A. 1948. The morphology and systematics of the West Indian Magnoliaceae. Bull. Torrey Bot. Club 75: 335–357.

Hufford, L. 1988. Roles of early ontogenetic modifications in the evolution of floral forms of *Eucnide* (Loasaceae). Bot. Jahrb. Syst. 109: 289–333.

Hufford, L. 1992. Rosidae and their relationships to other nonmagnoliid dicotyledons: A phylogenetic analysis using morphological and chemical data. Ann. Missouri Bot. Gard. 79: 218–248.

Hufford, L. & P.K. Endress. 1989. The diversity of anther structures and dehiscence patterns among Hamamelididae. Bot. J. Linn. Soc. 99: 301–346.

Johri, B.M. & S.P. Bhatnagar. 1955. A contribution to the morphology and life history of *Aristolochia*. Phytomorphology 5: 123–137.

Konijnenburg–Van Cittert, J.H.A. Van. 1992. An enigmatic Liassic microsporophyll, yielding *Ephedripites* pollen. Rev. Palaeobot. Palynol. 71: 239–254.

Kugler, H. 1934. Zur Blütenökologie von *Asarum europaeum*. Ber. Dtsch. Bot. Ges. 52: 348–354.

Lammers, T.G., T.F. Stuessy & M. Silva O. 1986. Systematic relationships of the Lactoridaceae, an endemic family of the Juan Fernandez Islands, Chile. Pl. Syst. Evol. 152: 243–266.

Leins, P. & C. Erbar. 1985. Ein Beitrag zur Blütenentwicklung der Aristolochiaceen, einer Vermittlergruppe zu den Monokotylen. Bot. Jahrb. Syst. 107: 343–368.

Leins, P., C. Erbar & W.A. Van Heel. 1988. Floral development of *Thottea* (Aristolochiaceae). Blumea 33: 357–370.

Lu, K.L. 1982. Pollination biology of *Asarum caudatum* (Aristolochiaceae) in Northern California. Syst. Bot. 7: 150–157.

Macior, L.W. 1970. The pollination ecology of *Dicentra cucullaria*. Amer. J. Bot. 57: 6–11.

Macior, L.W. 1978. Pollination interactions in sympatric *Dicentra* species. Amer. J. Bot. 65: 57–62.

Maddison, W.P. & D.R. Maddison. 1992. MacClade, Version 3. Sinauer Associates, Sunderland.

Martens, P. 1971. Les Gnetophytes. Borntraeger, Berlin.

McLean, R.C. & W.R. Ivimey–Cook. 1956. The Textbook of Theoretical Botany, Vol. 2. Longman Green, London.

Mesler, M.R. & K.L. Lu. 1993. Pollination biology of *Asarum hartwegii* (Aristolochiaceae): an evaluation of Vogel's mushroom fly hypothesis. Madroño 40: 117–125.

Miller, J.M. 1989. The archaic flowering plant family Degeneriaceae: its bearing on an old enigma. Nat. Geog. Res. 5: 218–231.

Moseley, M.F. 1958. Morphological studies of the Nymphaeaceae. I. The nature of the stamens. Phytomorphology 8: 1–29.

Moseley, M.F. 1971. Morphological studies of Nymphaeaceae. VI. Development of flower of *Nuphar*. Phytomorphology 21: 253–283.

Nair, N.C. 1962. Studies on the Aristolochiaceae. I. Nodal and floral anatomy. Proc. Natl. Inst. Sci. India 28: 211–227.

Nicely, K.A. 1965. A monographic study of the Calycanthaceae. Castanea 30: 38–81.

Parkin, J. 1951. The protrusion of the connective beyond the anther and its bearing on the evolution of the stamen. Phytomorphology 1: 1–8.

Payer, J.B. 1857. Traité d'Organogénie Comparée de la Fleur. Masson, Paris.

Pearson, H.H.W. 1929. Gnetales. Cambridge University Press, Cambridge.

Philipson, W.R. 1987. A classification of the Monimiaceae. Nord. J. Bot. 7: 25–29.

Qiu, Y.-L., M.W. Chase, D.H. Les & C.R. Parks. 1993. Molecular phylogenetics of the Magnoliidae: cladistic analysis of nucleotide sequences of the plastid gene *rbc*L. Ann. Missouri Bot. Gard. 80: 587–606.

Rao, A.R. 1981. The affinities of the Pentoxyleae. Palaeobotanist 28–29: 207–209.

Rickson, F.R. 1979. Ultrastructural development of the beetle food tissue of *Calycanthus* flowers. Amer. J. Bot. 66: 80–86.

Ronse Decraene, L.P. & E.F. Smets. 1992a. An updated interpretation of the androecium of the Fumariaceae. Can. J. Bot. 70: 1765–1776.

Ronse Decraene, L.P. & E.F. Smets 1992b. Complex polyandry in the Magnoliatae: definition, distribution and systematic value. Nord. J. Bot. 12: 621–649.

Shamrov, I.I. 1983. Anthecological investigations of three species of the genus *Ceratophyllum* (Certophyllaceae). Bot. Zhurn. 68: 1357–1365.

Schaeppi, H. 1939. Vergleichend-morphologische Untersuchungen an den Staubblättern der Monocotyledonen. Nova Acta Leopold. n. F. 6: 389–447.

Sporne, K.R. 1958. Some aspects of floral vascular systems. Proc. Linn. Soc. London 169: 75–84.

Stapf, O. 1926. *Sargentodoxa cuneata*. Curtis's Bot. Mag. 151, pls. 9111–9112.

Stebbins, G.L. 1974. Flowering Plants, Evolution Above the Species Level. Belknap, Cambridge.

Sugawara, T. 1987. Taxonomic studies of *Asarum* sensu

latB. III. Comparative floral anatomy. Bot. Mag. Tokyo 100: 335–348.

Takhtajan, A. 1969. Flowering Plants: Origin and Dispersal. Oliver and Boyd, Edinburgh.

Takhtajan, A. 1980. Outline of the classification of flowering plants (Magnoliophyta). Bot. Rev. 46: 225–359.

Takhtajan, A. 1991. Evolutionary Trends in Flowering Plants. Columbia University Press, New York.

Thien, L.B. 1980. Patterns of pollination in the primitive angiosperms. Biotropica 12: 1–13.

Thien, L.B., P. Bernhardt, G.W. Gibbs, O. Pellmyr, G. Bergstrom, I. Groth & G. McPherson. 1985. The pollination of Zygogynum (Winteraceae) by the moth, Sabatinca (Micropterigidae): an ancient association? Science 227: 540–543.

Tucker, S.C. 1972. The role of ontogenetic evidence in floral morphology. pp. 359–369 in Advances in Plant Morphology (edited by Y.S. Murty, B.M. Johri, H.Y. Mohan Ram & T.M. Varghese) Sarita Prakashan, Nauchandi.

Vasil, V. 1959. Morphology and embryology of Gnetum ula Brongn. Phytomorphology 9: 167–215.

Vink, W. 1983. The Winteraceae of the Old World. IV. The Australian species of Bubbia. Blumea 28: 311–328.

Vink, W. 1985. The Winteraceae of the Old World. V. Exospermum links Bubbia to Zygogynum. Blumea 31: 39–55.

Vishnu–Mittre. 1952. A male flower of the Pentoxyleae with remarks on the structure of the female cones of the group. Palaeobotanist 2: 75–84 + 5 pls.

Weaver, R.E. 1969. Studies in the North American genus Fothergilla (Hamamelidaceae). J. Arnold Arbor. 50: 599–619.

Weberling, F. 1989. Morphology of Flowers and Inflorescences. Cambridge University Press, New York.

Werth, E. 1951. Asarum europaeum, ein permanenter Selbstbefruchter. Ber. Dtsch. Bot. Ges. 64: 287–294.

Williams, E.G., T.L. Sage & L.B. Thien. 1993. Functional syncarpy by intercarpellary growth of pollen tubes in a primitive apocarpous angiosperm, Illicium floridanum (Illiciaceae). Amer. J. Bot. 80: 137–142.

Wilson, C.L. 1937. The phylogeny of the stamen. Amer. J. Bot. 24: 686–699.

Wilson, T.K. 1966. The comparative morphology of the Canellaceae. IV. Floral morphology and conclusions. Amer. J. Bot. 53: 336–343.

Wilson T.K. & L.M. Maculans. 1967. The morphology of the Myristicaceae. I. Flowers of Myristica fragrans and M. malabarica. Amer. J. Bot. 54: 214–220.

4 · Diversity and evolutionary trends in angiosperm anthers

PETER K. ENDRESS

INTRODUCTION

As in other plant organs, the specific structure of anthers is dependent upon the phylogenetic history (bauplan), on architectural factors and on ecological factors at both the morphological and histological level. Although it would be important to know more about them for evolutionary reconstructions, the relationships of these aspects have been poorly studied. The aim here is to show examples of such interdependencies.

BASAL ANTHER ORGANIZATION AND DEVIATIONS THEREFROM

Angiosperm anthers have a remarkably uniform organization. They have four sporangia that are arranged pairwise in two lateral thecae. Each theca opens by a longitudinal slit (stomium) between the two pollen sacs and by partial obliteration of the septum between the two pollen sacs (Fig. 1.1). This situation probably occurs in more than 90% of the angiosperm species (Endress & Stumpf 1990). An exact figure is difficult to assess at present, since a large number of orchids probably have more than four sporangia; however, detailed comparative studies are wanting for that family.

In a few groups each theca opens with two instead of one longitudinal slit. This is the case in some Rutaceae and Meliaceae among dicots (Endress & Stumpf 1991), in *Strelitzia* (Endress 1994a), *Freycinetia javanica* (Pandanaceae) (Huynh 1992); *Xerophyta minima*, *X. plicata* (but not in other species of *Xerophyta*) (Velloziaceae) (de Menezes 1988), and in Brazilian *Vellozia* species (de Menezes 1980). Rarely, the septum is reduced (*Hydrangea*) (Endress & Stumpf 1991). In many

magnoliids and lower hamamelids each stomium bifurcates at the lower and upper end. As a result, instead of a simple longitudinal slit, two valves are formed as opening devices by a bifurcation of the stomium at both ends (Endress & Hufford 1989; Hufford & Endress 1989) (Fig. 9.12–13). In some Lauraceae the two pollen sacs of a theca open with two separate valves (Fig. 9.17–18). Valvate dehiscence is associated with the presence of a thick connective but not always present in such anthers. It is most probably derived from simple longitudinal dehiscence (see Endress & Hufford 1989; Hufford & Endress 1989). Poricidal anthers usually open with two apical pores. In most cases the pores are a result of restriction of the longitudinal slits to the apical part of the thecae. Rarely the pores may be formed by circular abscission of a small part of the anther wall (some Ericaceae) (Addicott 1982).

A more profound deviation is increase or decrease in number of sporangia. An increase happens by longitudinal and lateral subdivision of the original four sporangia, whereby the opening mechanism with longitudinal slits is retained (Fig. 1.2–3). This is known from about 30 families, mainly dicots (see Endress & Stumpf 1990). A decrease from four to two sporangia is known from representatives of at least 60 families.

In addition to an increase in number of sporangia, thecal organization (i.e., the presence of two pollen sacs on each side of the anther with a dehiscence region between them) may be lost, and each of the numerous microsporangia may open by a pore of its own (Fig. 1.5). This is known only from Icacinaceae and Viscaceae. Another athecal anther form occurs in *Rafflesia* with numerous microsporangia converging toward the anther apex, where pollen of all sporangia is released by a single

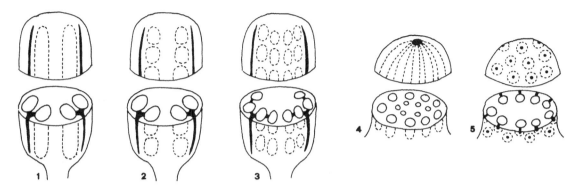

Fig. 1. Organization of anthers. Surface view and transverse section. Dehiscence region indicated by thick black lines; sporangia hatched in surface view, encircled in transverse section. 1–3. Anthers with thecal organization. 4, 5. Anthers with athecal organization. 1. Anther with basal organization. 2. Anther with transverse subdivision of pollen sacs. 3. Anther with transverse and longitudinal subdivision of pollen sacs. 4. Anther with numerous convergent sporangia dehiscing by a single apical pore. 5. Anther with numerous sporangia, each dehiscing by a separate pore (after Endress & Stumpf (1990)).

pore (Fig. 1.4). In all these athecal stamens a filament is absent. The occurrence of these different anther forms has been reviewed by Endress & Stumpf (1990).

ANTHER DIVERSITY

In spite of the comparative stability of the basic organiz- ation of angiosperm anthers there is diversity in many other respects. It is not intended here to give an exhaus- tive review on anther diversity but to show some salient features and problems.

Anther size

Anther size varies considerably. In *Strelitzia* anthers are 5 cm long; in *Lepianthes* (Piperaceae) they are less than 0.2 mm, thus more than 250 times shorter (Endress 1994b) (Fig. 2.1–2). Nevertheless, both have the same organization. In some cleistogamous flowers, reduced but functional anthers may be even smaller, and the number of pollen grains produced may be very low. Upper sizes are probably limited by functional constraints. They may be physiological: in anther development there are crucial phases with synchrony in the sporogenous tissues (e.g., Heslop–Harrison 1972). Synchrony might not be poss- ible if the sporogenous tissue were too massive. It is of interest in this respect that increase in anther size is

mainly expressed as an increase in length and less so as an increase in thickness. In this way the ratio of spor- ogenous tissue to tapetum does not decrease. There may also be biophysical constraints on large sizes: in *Strelit- zia*, with exceedingly long anthers, the anthers are sup- ported by two petals in which they are enclosed (Endress 1994a).

Among dicots, large anthers are known from *Rothman- nia* (Rubiaceae), where they attain a length of 4 cm (Robbrecht & Puff 1986). In *Demosthenesia* (Ericaceae) they may be more than 4 cm long, but part of their length consists of sterile areas (Luteyn 1978).

Different behavior of dehisced anthers

After dehiscence, angiosperm thecae generally gape widely, and they remain in this condition through anthesis by a desiccatory process (e.g., Schmid 1976; Keijzer 1987). Many anthers may more or less close again when wet. However, there are not many studies on this phenomenon. It has been mentioned for the anthers of Lauraceae that open by valves (Kerner 1905) and it also occurs in the valvate anthers of Hamamelida- ceae (personal observation). Reversible anther opening has been studied for the longitudinally dehiscent anthers of *Lilium philadelphicum* (Edwards & Jordan 1992). The anthers were 90% closed within 15–20 minutes of being

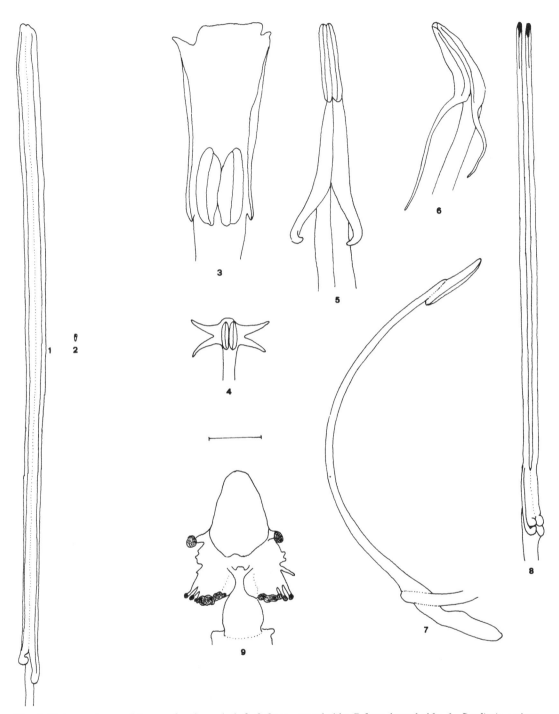

Fig. 2. Extreme sizes and forms of anthers. 1–6, 8, 9 from ventral side; 7 from lateral side. 1. *Strelitzia reginae* (Strelitziaceae), E 9571.* 2. *Lepianthes peltata* (Piperaceae), E s.n., 1991. 3. *Costus igneus* (Costaceae), E 9668. 4. *Globba winitii* (Zingiberaceae), E 9771. 5. *Roscoea purpurea* (Zingiberaceae), E 7522. 6. *Thunbergia mysorensis* (Acanthaceae), E 9665. 7. *Salvia patens* (Lamiaceae), E 8001. 8. *Demosthenesia cordifolia* (Ericaceae) (after Luteyn 1978). 9. *Psychopsis papilio* (Orchidaceae), E 9742, gynostemium (stamen fused with gynoecium). Scale bar represents 4 mm.

* Collection numbers by the author are abbreviated to E throughout the legends in this chapter.

sprinkled with water. In the reclosed anthers the epidermal and endothecial cells are fully expanded, seemingly rehydrated. Reversible opening may provide protection for pollen in rainy periods, as is the case for pollen in poricidal anthers (see next section).

Apart from the examples shown in the next section, correlations between anther opening mode and particular endothecium structure have scarcely been worked out, although the diversity and potential macrosystematic significance of endothecium differentiation have been discussed in general (Noel 1983) and for several families or orders, such as Asteraceae (Dormer 1962; Thiele 1988), Araceae (French 1985a, b, 1986), Iridaceae (Manning & Goldblatt 1990; Manning, this volume), Orchidaceae (Freudenstein 1991), and Poales/Restionales (Manning & Linder 1990).

Examples of functional aspects of special endothecium differentiation are some poricidally dehiscent Araceae (e.g. *Spathicarpa*, *Xanthosoma*). They have slimy pollen that is extruded through the pore by the contraction of the dehisced anther by an inversely oriented endothecium; the flowers are pollinated by flies (Troll 1928). The combination of poricidal anthers, slimy pollen and fly-pollination is also present in Rafflesiaceae, Hydnoraceae, and Viscaceae (Endress & Stumpf 1990). Poricidal anther opening combined with dry pollen, buzz-pollination and special endothecium differentiation is discussed in the next section. In addition, it has long been known that submersed flowers, such as *Ceratophyllum*, do not produce an endothecium. In *Callitriche* emersed flowers have an endothecium, whereas submersed ones do not (Strasburger 1902).

POLLINATION BIOLOGY AND ANTHER SHAPE

Particular modes of pollination may considerably influence the shape of anthers. Here, two examples are shown.

Nectar-flowers that are pollinated by large animals, seesaw mechanism, narrow opening of thecae, viscin threads

Owing to imprecise pollination, nectar-flowers that are pollinated by large animals (such as butterflies, hawk-moths, birds, bats) tend to have large, relatively broad anthers with the stomia facing the same direction, extremely introrse (in monosymmetric flowers), slit-like and not widely opening (Fig. 3). The narrow opening may arise because, after dehiscence, the two parts of the thecal wall curve inwards rather than outwards (Fig. 4). In combination with mobility of a seesaw mechanism such anthers work like a rubber stamp to transfer pollen onto bird feathers or butterfly wings. The anther openings are passively directed towards the pollinator's body at the slightest touch, whereby pollen is transferred onto it, e.g. in *Delonix*, *Amherstia* (Caesalpiniaceae), *Erythrina* (Fabaceae), *Melianthus* (Melianthaceae), *Cleome* (Capparaceae), *Aeschynanthus* (Gesneriaceae), *Pancratium* and *Crinum* (Amaryllidaceae). In the more polysymmetric flowers of *Gloriosa* (Colchicaceae) the narrow opening slits of the anthers are latrorse. The anatomical differentiations accompanying the narrow opening slits have apparently not been comparatively studied.

The presence of viscin threads is concentrated in flowers that are pollinated by large animals. With the help of the viscin threads pollen grains hang together in loose aggregates, which enhances the chances of pollination by the enlarged area of the pollen mass. Viscin threads occur in some Annonaceae (*Porcelia*), Caesalpiniaceae (*Delonix*, *Bauhinia*), Passifloraceae (*Tetrastylis*), Onagraceae, Ericaceae (*Rhododendron*), Balsaminaceae, Strelitziaceae (*Strelitzia*) (Cruden & Jensen 1979; Graham *et al.* 1980; Kronestedt & Bystedt 1981; Hesse 1981, 1984, 1986; Waha 1984; Vogel & Cocucci 1988; Morawetz & Waha 1991; Buzato & Franco 1992). Viscin threads are differentiations of the pollen sacs that arise in various ways. In most groups they originate inside the pollen sacs and contain sporopollenin (Fig. 5.1–2); in orchids they are lipidic (Schill & Wolter 1986). In contrast, in *Impatiens* (Vogel & Cocucci 1988) and *Strelitzia* (Kronestedt & Bystedt 1981) they are formed by epidermal cells of the thecae and are cellulosic (Fig. 5.3–6). In plants with viscin threads the pollen:ovule ratio is comparatively low (Cruden & Jensen 1979). The clumps of pollen, loosely held together by viscin threads, may be compared to some extent with the 'search vehicles' of water-pollinated plants (Cox & Knox 1988).

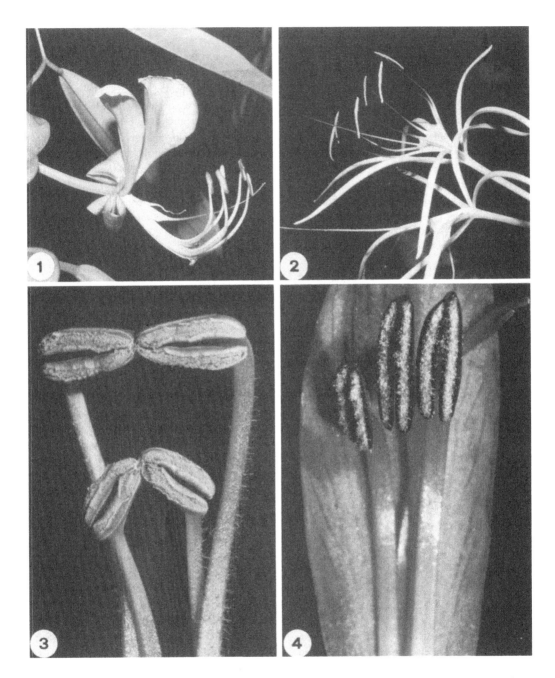

Fig. 3. Anthers of nectar-flowers that are pollinated by large animals, in the living state. 1. *Amherstia nobilis* (Caesalpiniaceae), E 9357 (×0.7). 2. *Hymenocallis expansa* (Amaryllidaceae), E s.n. (×0.7). 3. *Aeschynanthus speciosus* (Gesneriaceae), E s.n. (×20). 4. *Antholyza aethiopica* (Iridaceae), E 7672 (×20).

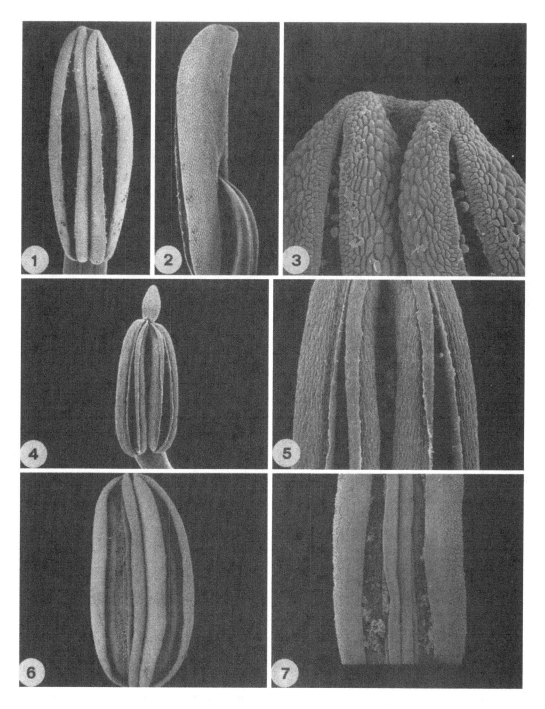

Fig. 4. Anthers of nectar-flowers that are pollinated by large animals, fixed material. 1–3. *Antholyza aethiopica* (Iridaceae), E 7672. 1. From ventral side (×13). 2. From lateral side (×13). 3. Apex from ventral side (×70). 4, 5. *Anigozanthos flavidus* (Haemodoraceae), E 6655, from ventral side. 4. Anther (×15). 5. Middle part of anther (×45). 6. *Delonix regia* (Caesalpiniaceae), E 6412, from ventral side (×13). 7. *Amherstia nobilis* (Caesalpiniaceae), E 9357, middle part of anther from ventral side (×13).

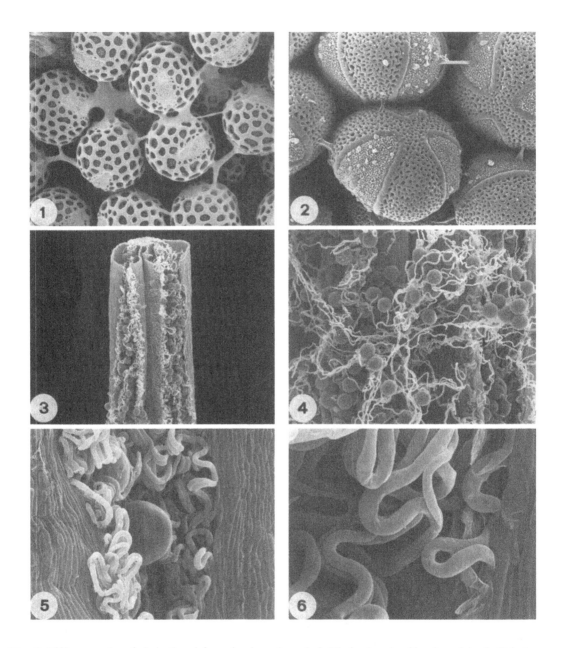

Fig. 5. Different modes of viscin thread formation in anthers. 1, 2. Viscin threads of locular origin. 1. *Delonix regia* (Caesalpiniaceae), E 9279 (×500). 2. *Caesalpinia sepiaria* (Caesalpiniaceae), E 5313 (×1000). 3–6. Viscin threads of epidermal origin in *Strelitzia reginae* (Strelitziaceae), E 9571. 3. Anther with upper part cut off, from ventral side (×18). 4. Pollen grains with viscin threads (×40). 5. Opening stomium with a pollen grain and viscin threads originating from epidermal cells (×150). 6. Viscin threads originating by isolation and elongation of epidermal cells (×400).

'Solanum-type' flowers, buzz-pollination, and poricidal anthers, heteranthery and synanthery

One of the most obvious relationships between anther structure and pollination biology is buzz-pollination and 'Solanum-type' flowers with dry pollen and poricidal

anthers e.g., Vogel 1978; Buchmann 1983) (Fig. 6). Another tendency of taxa with this syndrome is to exhibit larger anthers and shorter filaments than closely related groups without the syndrome (Halsted 1890). 'Solanum-type' flowers are widespread among dicots and monocots.

Among monocots they are represented in many liliids

Fig. 6. 'Solanum-type' flowers in the living state. 1. *Lycianthes rantonnei* (Solanaceae), E. s.n. (×1). 2. *Dichorisandra speciosa* (Commelinaceae), E. s.n. (×3). 3, 4. Flowers of two Loasaceae in the same magnification: 'Solanum-type' (4), as contrasted with 'non-Solanum-type' (3). 3. *Mentzelia lindleyi*, E 7691 (×20). 4. *Schismocarpus pachypus*, E 5363 (×20).

and commelinids (Fig. 7). Many of these taxa have poricidal anthers, e.g. *Cyanella* (Tecophilaeaceae), *Cyanastrum* (Cyanastraceae), *Dichorisandra* (Commelinaceae),
Mayaca (Mayacaceae), and Rapateaceae (Tiemann 1985).
This is, however, not always the case. A number
of taxa mentioned by Dahlgren *et al.* (1985) and Gerenday & French (1988) for poricidal anthers have normal
longitudinal dehiscence (Fig. 8). The following 'Solanum-type' flowers and related forms have longitudinal,
and not poricidal, anther dehiscence: *Dianella*
(Phormiaceae), *Heteranthera* (Pontederiaceae), *Xiphidium*
(Haemodoraceae), *Xerophyta elegans* (Velloziaceae), *Eustrephus* and *Geitonoplesium* (Luzuriagaceae). The distinction between poricidal and longitudinally dehiscent
anthers is not always clear, because in some cases
dehiscence begins apically with a pore and later
extends along the entire length of the anther, e.g.
Monochoria (Pontederiaceae) (Cook 1989), *Galanthus*
(Amaryllidaceae). Among dicots there are many families
containing some representatives with 'Solanum-type'
flowers and poricidal anthers.

In poricidal anthers of buzz-pollinated flowers the
endothecium is often lacking or is modified in some way.
For monocots, Gerenday & French (1988) found only a
weak association between poricidal dehiscence and lack
of an endothecium. However, some of the taxa mentioned by them as having a normal endothecium are not
poricidal (see previous paragraph). In fact, the association is closer. This is also the case in dicots. Examples
are Dilleniaceae (Rao 1961: wall thickenings only in the
epidermis), Ericaceae, Melastomataceae (Matthews &
Knox 1926; Matthews & Maclachlan 1929), *Tetratheca*
(Tremandraceae) (personal observation), and Gesneriaceae (Huber 1953). Conversely, it is noteworthy that a
normal endothecium is also absent in certain taxa that
are non-porate but closely related to porate groups
(Xyridaceae: Gerenday & French 1988; Epacridaceae:
Endress & Stumpf 1990). Are these evolutionary reversals or is it not poricidal dehiscence but buzz-pollination
that influences the histological change? The entire biological situation makes it probable that the use of
vibrations for pollen collecting came first, since it is
widely used in flowers with dry pollen (Buchmann 1985;
Proença 1992), not only in those with tubular anthers
where it was first described. The differentiation of long,

tubular, poricidal anthers was a response to this method
of pollen-collecting. Better protection of the pollen and
portionwise, parsimonious release from the anthers must
have been an advantage for the plants. Renner (1989a)
found a particularly high proportion of buzz-pollinated
plants on the high Tepuis of Venezuela (cited in Endress
1994a).

As a specialization in buzz-pollinated flowers there
may be heteranthery, and there may be functional differentiation into feeding and pollinating stamens (e.g.,
Vogel 1978). In many groups pollen of both feeding and
pollinating stamens is fertile (Renner 1989b). Heteranthery is known from many dicots and monocots. Among
monocots, it has evolved in parallel in a number of liliids
and commelinids. Among liliids it occurs in *Sternbergia*
(Amaryllidaceae) (Dafni & Werker 1982), *Cyanella*
(Tecophilaeaceae) (Dulberger & Ornduff 1980), *Monochoria* (Pontederiaceae) (Cook 1989, 1990), and *Dilatris*,
Haemodorum, *Schiekia*, and *Xiphidium* (Haemodoraceae)
(Simpson 1990). It is common in many Commelinaceae,
with extreme forms in *Tripogandra* and *Cochliostema*.
According to the position of the different stamen morphs
in flowers of Commelinaceae two patterns may be distinguished, one in monosymmetric flowers, one in polysymmetric flowers (Rohweder 1956). In extreme cases
only a single larger stamen is present, while the other
stamens are absent or present as mere rudiments (e.g.,
Solanum sect. *Androceras*: Whalen 1978; *Rhynchanthera*
species, Melastomataceae: Renner 1990; *Eurystemon*,
Heteranthera, *Monochoria*, Pontederiaceae: Cook 1989,
1990; *Cyanella*, Tecophilaeaceae: Dulberger & Ornduff
1980; *Pyrrorhiza*, Haemodoraceae: Simpson 1990).

Another specialization of buzz-pollinated flowers
results in more or less (postgenitally) connate anthers
that form a unified apparatus, such as in *Walleria*
(Tecophilaeaceae) (Dahlgren & van Wyk 1988), *Conanthera* (Tecophilaeaceae) (Dahlgren *et al.* 1985), *Echeandia*
(Anthericaceae) (Bernhardt & Montalvo 1977), *Eustrephus* (Luzuriagaceae) (personal observation), *Sollya*
(Pittosporaceae) (Endress & Stumpf 1991; personal
observation), *Lycopersicon* (Solanaceae) (Bonner &
Dickinson 1989; Endress 1994a).

In these postgenitally connate anthers, the poricidal
dehiscence tends to be secondarily lost, and they open
by longitudinal slits. A functional aspect of this corre-

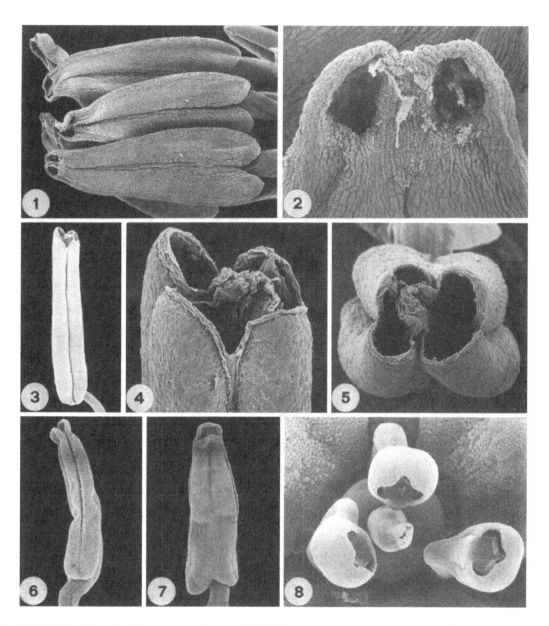

Fig. 7. Poricidal anthers of 'Solanum-type' flowers. 1, 2. *Dichorisandra speciosa* (Commelinaceae), E 6571. 1. Androecium from abaxial side, all anthers directed to the same side by torsion of filaments of the stamens of the abaxial half of the flower (×13). 2. Anther apex, from ventral side (×90). 3–5. *Cyanastrum bussei* (Cyanastraceae), E 9395. 1. Anther from the side (×13). 4. Anther apex from the side (×80). 5. Anther apex from above (×60). 6, 7. *Cyanella orchidiformis* (Tecophilaeaceae), E 9776. 8. *Mayaca* sp. (Mayacaceae) E 9504, floral center with three anthers from above (× 40).

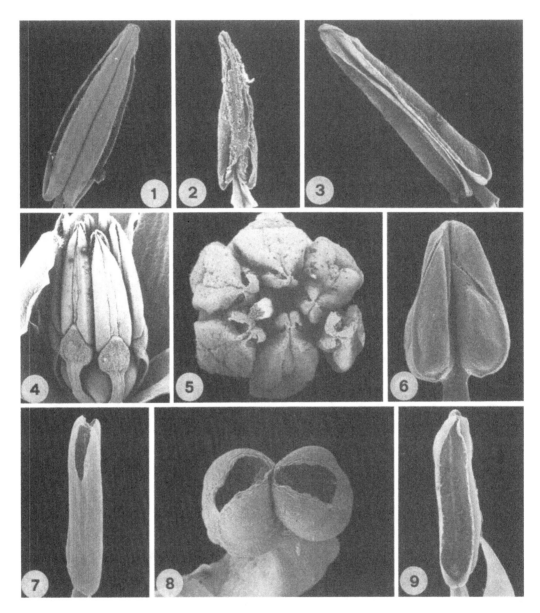

Fig. 8. Anthers with longitudinal dehiscence of 'Solanum-type' flowers. 1. *Geitonoplesium cymosum* (Luzuriagaceae), E 6154, from ventral side (×13). 2. *Eustrephus latifolius* (Luzuriagaceae), E 6638, from ventral side (×15). 3. *Xerophyta elegans* (Velloziaceae), E 7241, from ventral side (×13). 4, 5. *Dianella coerulea* (Phormiaceae), E 7364, androecium. 4. From the side (×13). 5. From above (×25). 6. *Helmholtzia acorifolia* (Philydraceae), E 4217, from ventral side (×18). 7–9. *Monochoria vaginalis* (Pontederiaceae), C.D.K. Cook & R. Frey 4735. 7. Small anther, from the side (×20). 8. Small anther, from above (×50). 9. Large anther, from the side (×13).

lation is that pollen is well protected by the anther fusion, because the opening slits are on the inner side of the tube resulting from anther fusion.

ANTHER SHAPE AND FLORAL CONSTRUCTION

In flowers with an elaborate construction the anthers show especially well their potential for evolutionary plasticity (Fig. 2.3–9). Anther shape may then be understood only in the framework of the entire flower structure. Examples are anthers with a lever mechanism that functions only in the intact flower. In monosymmetric flowers with stamen pairs the anthers may be highly asymmetric. In bee-pollinated *Salvia* species (Lamiaceae) with two lateral stamens, the connective is extremely broad. One of the thecae of both stamens is sterile but forms a frontally expanded plate that acts as a lever to expose the fertile theca (e.g., Werth 1956; Fig. 2.7). In *Roscoea* (Zingiberaceae), which has a single (symmetrical) median stamen, two basal extensions, one from each theca, form a functionally analogous lever (Troll 1951; Fig. 2.5). In *Thunbergia* (Acanthaceae), which has four (asymmetric) lateral stamens, similar basal extensions serve as levers to open the otherwise closed thecae (Endress 1994a; Fig. 2.6).

ANTHER SHAPE AND ANTHER HISTOLOGY (INDEPENDENT OF POLLINATION BIOLOGY)

Anthers with a basal pit (centrifixed anthers) (see Endress & Stumpf 1991) occur in a number of rosids, liliids and alismatids. The pit sometimes has a somewhat dorsal position. Schaeppi (1939) mentions it only for some Liliaceae (*Tulipa*, *Gagea*, *Erythronium*) and Butomaceae (*Butomus*), but the feature seems to be widespread in Liliales (Huber 1969; see also Krause 1930). It also occurs in some Asparagales, e.g., *Doryanthes* (Doryanthaceae) (Patil & Pai 1981), Strumariinae (Amaryllidaceae) (D. & U. Müller–Doblies 1985), and in some Melanthiales, such as *Veratrum* (Melanthiaceae) (Shamrov 1990). It has even been mentioned for bamboos, *Anomochloa*, (Arber 1934). In *Doryanthes* (personal observation) and *Veratrum* (Shamrov 1990) the endothecium-like tissue is extraordinarily extensive and is present in almost the entire connective. This is also the case in similar anthers among dicots, where they occur more rarely (Crassulaceae, Saxifragaceae) (Endress & Stumpf 1991). This parallel makes it probable that the association of a basal pit and an extensive endothecium-like tissue is an architectural feature and does not represent two independent features that would point to a close relationship among those liliids where they occur.

ANTHER SHAPE AND SYSTEMATICS

In view of the great plasticity of anther shapes related to pollination biology, are there traits that characterize large taxa at all?

This is, of course, the case in groups with anthers that are functionally associated with each other and/or with the gynoecium and, therefore, with architectural (functional) constraints on anthers. For example, in Asteraceae the anthers are relatively uniform (Bremer 1987; Thiele 1988).

Anthers with valvate dehiscence are known only in magnoliids and hamamelidids (Endress & Hufford 1989; Hufford & Endress 1989). A special type with a single valve but two pollen sacs per theca occurs exclusively in Dicoryphinae of Hamamelidaceae (Endress 1989).

Extrorse anthers are somewhat concentrated in primitive dicots and primitive monocots and are rare in more advanced groups (Figs. 9, 10). *Ceratophyllum*, perhaps the most basal clade of the angiosperms, has extrorse anthers, and so do about half of the magnoliid families (Endress 1994b). Among the basal monocots, extrorse anthers are concentrated in the alismatids and Araceae, although *Acorus*, which is claimed to be the most basal monocot by some authors (e.g., Duvall *et al.* 1993; S. Mayo, personal communication, J.C. French, personal communication), has introrse anthers (Endress 1993).

At the middle evolutionary level of the dicots, anther features are not highly distinctive at the macroevolutionary level, although certain characteristics may be concentrated at the family level or lower, as shown for the lower rosids (Endress & Stumpf 1991). As an example, among the rosids, anthers with basal pits occur in Crassulaceae and Saxifragaceae; however, they are more common in monocots (many liliids) (see above).

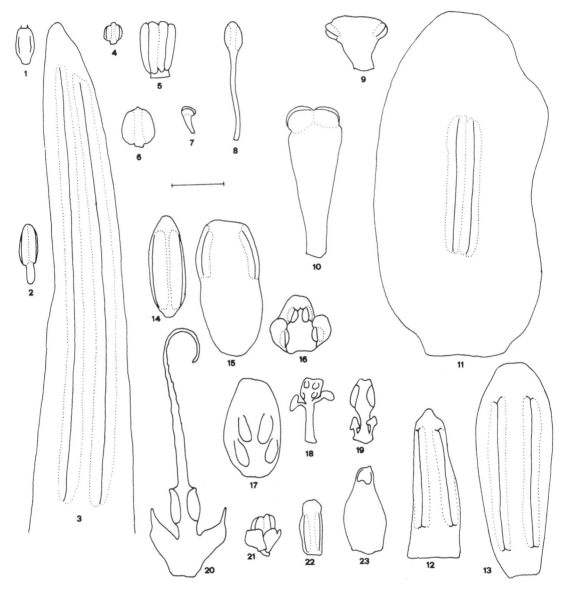

Fig. 9. Anthers (stamens) of basal angiosperms (dicotyledons). 1–2, 4, 10, 13, 21, 22 from dorsal side; 3, 5–9, 11, 12, 14–20, 23 from ventral side. 1. *Ceratophyllum demersum* (Ceratophyllaceae), E 9755. 2. *Cabomba aquatica* (Cabombaceae), E 9407. 3. *Nymphaea nouchalii* (Nymphaeaceae), E 9718. 4. *Lactoris fernandeziana* (Lactoridaceae) (after Carlquist 1964). 5. *Saruma henryi* (Aristolochiaceae), H.-N. Qin s.n., 1992. 6. *Macropiper excelsum* (Piperaceae), E 6356a. 7. *Peperomia caperata* (Piperaceae), E 7238. 8. *Saururus cernuus* (Saururaceae), E 7267. 9. *Kadsura japonica* (Schisandraceae), E 4530. 10. *Exospermum stipitatum* (Winteraceae), A. Juncosa s.n., 1981. 11. *Austrobaileya scandens* (Austrobaileyaceae), E 4265. 12. *Eupomatia bennettii* (Eupomatiaceae), E 5197. 13. *Degeneria vitiensis* (Degeneriaceae), E. Zogg s.n., 1985. 14. *Ascarina rubricaulis* (Chloranthaceae), E 6012. 15. *Sarcandra chloranthoides* (Chloranthaceae), K.U. Kramer 6582. 16. *Chloranthus spicatus* (Chloranthaceae), E 6740. 17. *Ocotea botrantha* (Lauraceae), E 1118. 18. *Litsea* sp. (Lauraceae), H.U. Stauffer 5821a. 19. *Gomortega keule* (Gomortegaceae), M.F. Doyle III-6-86/1. 20. *Doryphora sassafras* (Monimiaceae), E 2256. 21. *Hortonia angustifolia* (Monimiaceae), E. Frey & S.T. Keppetipola s.n. 1978. 22. *Levieria acuminata* (Monimiaceae), B.P.M. Hyland 10165. 23. *Siparuna andina* (Monimiaceae), E 1096. Scale bar represents 2 mm. After Endress (1994b).

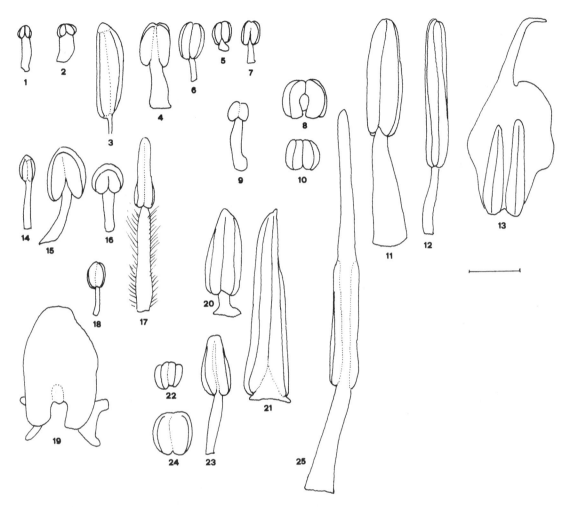

Fig. 10. Anthers (stamens) of basal monocotyledons. 1, 3, 14, 17–25 from ventral side; 2, 4–13, 15, 16 from dorsal side. 1. *Acorus calamus* (Acoraceae), E 5180. 2. *Anthurium scandens* (Araceae), E 7825. 3. *Scheuchzeria palustris* (Scheuchzeriaceae), E 6701. 4. *Sagittaria sagittifolia* (Alismataceae), E 9715. 5. *Sagittaria subulata* (Alismataceae), E 9667. 6. *Caldesia parnassifolia* (Alismataceae), E 6562. 7. *Alisma lanceolata* (Alismataceae), E 6552. 8. *Potamogeton natans* (Potamogetonaceae), E 5179. 9. *Aponogeton distachyus* (Aponogetonaceae), E 7650. 10. *Triglochin maritimum* (Juncaginaceae), E 5184. 11. *Butomus umbellatus* (Butomaceae), E 7244. 12. *Hydrocleys nymphoides* (Limnocharitaceae), E 9719. 13. *Posidonia oceanica* (Posidoniaceae), F. Markgraf s.n., 9.11.75. 14. *Tofieldia calyculata* (Melanthiaceae), E 6708. 15. *Zigadenus glaberrimus* (Melanthiaceae), E 9727. 16. *Veratrum album* (Melanthiaceae), E 9850. 17. *Narthecium ossifragum* (Melanthiaceae), E 9690. 18. *Dioscorea hemicrypta* (Dioscoreaceae), E 9542. 19. *Tacca aspera* (Taccaceae), E 9530. 20. *Convallaria majalis* (Convallariaceae), E 7647. 21. *Ophiopogon jaburan* (Convallariaceae), E 9704. 22. *Aspidistra stipulacea* (Convallariaceae), E 9547. 23. *Reineckea carnea* (Convallariaceae), E 9751. 24. *Rohdea japonica* (Convallariaceae), E 9836. 25. *Paris quadrifolia* (Trilliaceae), E 7329. Scale bar represents 2 mm.

MAJOR EVOLUTIONARY TRENDS?

There are few distinctive macroevolutionary progressions in anther shape. In the monocots and at the middle evolutionary level of the dicots, anther structure became more economical by a major trend from basifixed forms with massive connective to (mostly) dorsifixed forms with thin, narrow and short connective and bulging pollen sacs or thecae of a more parsimonious architecture. The ratio of polliniferous tissue to nonpolliniferous tissue in an anther has increased. This seems at least partly a consequence of the fact that by the evolutionary advent of an elaborate perianth, the stamens were freed from protective and attractive functions, which they still exert in many extant magnoliids that lack

an elaborate perianth (Endress 1994a). On a smaller scale, however, attractivity of anthers for pollinators has evolved secondarily in various groups of higher dicots and monocots. Examples are 'Solanum-type' flowers (see above), where parts of the anthers have become sterile and optically attractive (e.g., *Keraudrenia*, Sterculiaceae, or *Arthropodium*, Anthericaceae) and where anthers produce scent or even perfume, as in some Solanaceae (Sazima & Vogel 1989; D'Arcy *et al.* 1990).

The anthers of basal angiosperms (magnoliids, hamamelids) often have sterile apices (connective protrusions) (Parkin 1951; Endress & Hufford 1989; Hufford & Endress 1989; Endress 1994b) (Fig. 9.11–13). However, conspicuous sterile apices may also occur in the most specialized and advanced groups, such as Mim-

Fig. 11. Elaborate anthers with sterile apical extension (marked with asterisks) of advanced dicotyledons and monocotyledons. 1. *Prosopis pubescens* (Mimosaceae), E 9640, from dorsal side (×70). 2. *Centaurea crocodylium* (Asteraceae), E 9847, apical part of three anthers of an androecium, from dorsal side (×35). 3. *Asclepias curassavica*, (Asclepiadaceae), E 7368, two anthers of an androecium, from dorsal side, before anthesis (×60). 4. *Oncidium ornithorrhynchum* (Orchidaceae), E 9515, gynostemium from the side (×13).

osaceae, Asclepiadaceae, Asteraceae, Orchidaceae (Fig. 11), where they have particular functional roles. This shows that such structures may easily reappear again if needed. In some Mimosaceae the sterile anther apices are secretory or nutritive and may provide food for pollinating beetles (P. Bernhardt, personal communication). In Asclepiadaceae they are curved over the top of the style head; they may reinforce stability of the guide rails formed by the flanks of the anthers (e.g., Endress 1994a). In some Campanulaceae and in Asteraceae they may be involved in pollen portioning (Brantjes 1983; Thiele 1988). In some Orchidaceae the sterile apex covers and protects the stalk of the pollinarium (e.g. Endress 1994a). Other examples of connective protrusion in advanced groups with highly specialized flowers include *Anigozanthus* (Haemodoraceae) (Fig. 4.4), and a number of Rubiaceae (e.g., *Bertiera*, *Retiniphyllum*, *Tricalysia*) (Robbrecht 1988). Even in the magnoliids excessive elongation of the connective protrusion may be an extreme specialization, such as reverse anther exhibition after stamen abscission in some *Magnolia* species (Howard 1948).

In conclusion, although we know a great deal about the anthers of a few model organisms (such as, e.g., *Nicotiana* or *Gasteria*; Koltunow *et al.* 1990; Keijzer 1987), the diversity of anthers is much underexplored. The relationships between structure and function and, consequently, between structure and systematics should be studied much more closely.

ACKNOWLEDGMENT

For valuable material I am grateful to L. Bütschi, C.D.K. Cook, M.F. Doyle, U. Eggli, E. Frey, B.P.M. Hyland, A.M. Juncosa, K.U. Kramer, H.-N. Qin, D. Supthut, and E. Zogg. For assistance with microtome sections I thank R. Siegrist, for scanning electron micrographs I thank U. Jauch, and for photographic work I thank A. Zuppiger. I am grateful to W.G. D'Arcy, D.L. Dilcher and D.W. Stevenson for valuable comments on the manuscript.

LITERATURE CITED

Addicott, F.T. 1982. Abscission. University of California Press, Berkeley.

Arber, A. 1934. The Gramineae: a Study of Cereal, Bamboo and Grass. Cambridge University Press, Cambridge.

Bernhardt, P. & E.A. Montalvo. 1977. The reproductive phenology of *Echeandia macrocarpa* Greenm. (Liliaceae) with a reexamination of the floral morphology. Bull. Torrey Bot. Club 104: 320–323.

Bonner, L.J. & H.G. Dickinson. 1989. Anther dehiscence in *Lycopersicon esculentum* Mill. I. Structural aspects. New Phytol. 113: 97–115.

Brantjes, N.B.M. 1983. Regulated pollen issue in *Isotoma*, Campanulaceae, and evolution of secondary pollen presentation. Acta Bot. Neerl. 32: 213–222.

Bremer, K. 1987. Tribal interrelationships of the Asteraceae. Cladistics 3: 210–253.

Buchmann, S.L. 1983. Buzz pollination in angiosperms. pp. 73–113 in Handbook of Experimental Pollination Biology (edited by C.E. Jones & R.J. Little). Scientific and Academic Editions, New York.

Buchmann, S.L. 1985. Bees use vibration to aid pollen collection from non-poricidal anthers. J. Kansas Entomol. Soc. 58: 517–525.

Buzato, S. & A.L.M. Franco. 1992. *Tetrastylis ovalis*: a second case of bat-pollinated passionflower (Passifloraceae). Pl. Syst. Evol. 181: 261–267.

Cook, C.D.K. 1989. Taxonomic revision of *Monochoria* (Pontederiaceae). pp. 149–184 in The Davis & Hedge Festschrift (edited by K. Tan). Edinburgh University Press, Edinburgh.

Cook, C.D.K. 1990. Aquatic Plant Book. SPB Academic Publishing, The Hague.

Cox, P.A. & R.B. Knox. 1988. Pollination postulates and two-dimensional pollination in hydrophilous monocotyledons. Ann. Missouri Bot. Gard. 75: 811–818.

Cruden, R.W. & K.G. Jensen. 1979. Viscin threads, pollination efficiency and low pollen-ovule ratios. Am. J. Bot. 66: 875–879.

D'Arcy, W.G., N.S. D'Arcy & R.C. Keating. 1990. Scented anthers in the Solanaceae. Rhodora 92: 50–53.

Dafni, A. & E. Werker. 1982. Pollination ecology of *Sternbergia clusiana* (Ker–Gawler) Spreng. (Amaryllidaceae). New Phytol. 91: 571–577.

Dahlgren, R. & E. van Wyk. 1988. Structures and relationships of families endemic to or centered in Southern Africa. Monogr. Syst. Bot. Missouri Bot. Gard. 25: 1–94.

Dahlgren, R., H.T. Clifford & P.F. Yeo. 1985. The Families of the Monocotyledons. Springer–Verlag, Berlin.

Dormer, K.J. 1962. The fibrous layer in the anthers of Compositae. New Phytol. 61: 150–153.

Dulberger, R. & R. Ornduff. 1980. Floral morphology and

reproductive biology of four species of *Cyanella* (Tecophilaeaceae). New Phytol. 86: 45–56.

Duvall, M.R., M.T. Clegg, M.W. Chase, W.D. Clark, W.J. Kress, H.G. Hills, L.E. Eguiarte, J.F. Smith, B.S. Gaut, E.A. Zimmer & G.H. Learn, Jr. 1993. Phylogenetic hypotheses for the monocotyledons constructed from *rbc*L sequence data. Ann. Missouri Bot. Gard. 80: 607–619.

Edwards, J. & J.R. Jordan. 1992. Reversible anther opening in *Lilium philadelphicum* (Liliaceae): a possible means of enhancing male fitness. Am. J. Bot. 79: 144–148.

Endress, P.K. 1989. Aspects of evolutionary differentiation of the Hamamelidaceae and the Lower Hamamelididae. Pl. Syst. Evol. 162: 193–211.

Endress, P.K. 1993. Major evolutionary traits of monocot flowers. p. 5 in Monocotyledons (edited by Wilkin, P., S. Mayo & P. Rudall). Royal Botanic Gardens, Kew.

Endress, P.K. 1994a. Diversity and Evolutionary Biology of Tropical Flowers. Cambridge University Press, Cambridge.

Endress, P.K. 1994b. Shapes, sizes and evolutionary trends in stamens of Magnoliidae. Bot. Jahrb. Syst. 115: 429–460.

Endress, P.K. & L.D. Hufford. 1989. The diversity of stamen structures and dehiscence patterns among Magnoliidae. Bot. J. Linn. Soc. 100: 45–85.

Endress, P.K. & S. Stumpf. 1990. Non-tetrasporangiate stamens in the angiosperms: structure, systematic distribution and evolutionary aspects. Bot. Jahrb. Syst. 112: 193–240.

Endress, P.K. & S. Stumpf. 1991. The diversity of stamen structures in 'lower' Rosidae (Rosales, Fabales, Proteales, Sapindales). Bot. J. Linn. Soc. 107: 217–293.

French, J.C. 1985a. Patterns of endothecial wall thickenings in Araceae: subfamilies Pothoideae and Monsteroideae. Am. J. Bot. 72 472–486.

French, J.C. 1985b. Patterns of endothecial wall thickenings in Araceae: subfamilies Calloideae, Lasioideae, and Philodendroideae. Bot. Gaz. 146: 521–533.

French, J.C. 1986. Patterns of endothecial wall thickenings in Araceae: subfamilies Colocasioideae, Aroideae and Pistioideae. Bot. Gaz. 147: 166–179.

Freudenstein, J.V. 1991. A systematic study of endothecial thickenings in the Orchidaceae. Am. J. Bot. 78: 766–781.

Gerenday, A. & J.C. French. 1988. Endothecial thickenings in anthers of porate monocotyledons. Am. J. Bot. 75: 22–25.

Graham, A., G. Barker & M. Freitas da Silva. 1980. Unique pollen types in the Caesalpinioideae (Leguminosae). Grana 19: 79–84.

Halsted, B.D. 1890. Notes upon stamens of Solanaceae. Bot. Gaz. 15: 103–106.

Heslop-Harrison, J. 1972. Sexuality of angiosperms. pp. 133–289 in Plant Physiology: A Treatise 6 C: (edited by F.C. Steward). Academic Press, New York.

Hesse, M. 1981. Pollenkitt and viscin threads: their role in cementing pollen grains. Grana 20: 145–152.

Hesse, M. 1984. An exine architecture model for viscin threads. Grana 23: 69–75.

Hesse, M. 1986. Nature, form and function of pollen-connecting threads in angiosperms. pp. 109–118 in Pollen and Spores: Form and Function, (edited by S. Blackmore & I.K. Ferguson). Academic Press, London.

Howard, R.A. 1948. The morphology and systematics of the West Indian Magnoliaceae. Bull. Torrey Bot. Club 75: 335–357.

Huber, E. 1953. Beitrag zur anatomischen Untersuchung der Antheren von *Saintpaulia*. Sber. Akad. Wiss. Wien, Math.–Nat. Kl. Abt. I, 162: 227–234.

Huber, H. 1969. Die Samenmerkmale und Verwandtschaftsverhältnisse der Liliiflorae. Mitt. Bot. Staatssamml. München 8: 219–538.

Hufford, L.D. & P.K. Endress. 1989. The diversity of anther structures and dehiscence patterns among Hamamelididae. Bot. J. Linn. Soc. 99: 301–346.

Huynh, K.-L. 1992. The flower structure in the genus *Freycinetia*, Pandanaceae (part 2) – early differentiation of the sex organs, especially of the staminodes, and further notes on the anthers. Bot. Jahrb. Syst. 114: 417–441.

Keijzer, C.J. 1987. The process of anther dehiscence and pollen dispersal. I. The opening mechanism of longitudinally dehiscing anthers. New Phytol. 105: 487–498.

Kerner, A. 1905. Pflanzenleben 2, 2nd ed. Bibliographisches Institut, Leipzig.

Koltunow, A.M., J. Truettner, K.H. Cox, M. Wallroth & R.B. Goldberg 1990. Different temporal and spatial gene expression patterns occur during anther development. Pl. Cell 2: 1201–1224.

Krause, K. 1930. Liliaceae. pp. 227–386 in Die natürlichen Pflanzenfamilien 15a, 2nd ed. (edited by A. Engler & K. Prantl). Engelmann, Leipzig.

Kronestedt, E. & P.-A. Bystedt. 1981. Thread-like formations in the anthers of *Strelitzia reginae*. Nord. J. Bot. 1: 523–529.

Luteyn, J.L. 1978. Notes on neotropical Vaccinieae (Ericaceae). VI. New species from the Cordillera Vilcabamba and adjacent Eastern Peru. Brittonia 30: 426–439.

Manning, J.C. & P. Goldblatt. 1990. Endothecium in Iridaceae and its systematic implications. Am. J. Bot. 77: 527–532.

Manning, J.C. & H.P. Linder. 1990. Cladistic analysis of patterns of endothecial thickenings in the Poales/Restionales. Am. J. Bot. 77: 196–210.

Matthews, J.R. & E.M. Knox, 1926. The comparative morphology of the stamen in the Ericaceae. Trans. Bot. Soc. Edinburgh 29: 243–281.

Matthews, J.R. & C.M. Maclachlan. 1929. The structure of certain poricidal anthers. Trans. Proc. Bot. Soc. Edinburgh 30: 104–122.

de Menezes, N.L. 1980. Evolution in Velloziaceae, with special reference to androecial characters. pp. 117–138 in Petaloid Monocotyledons. Horticultural and Botanical Research (edited by C.D. Brickell, D.F. Cutler & M. Gregory) Academic Press, London.

de Menezes, N.L. 1988. Evolution of the anther in the family Velloziaceae. Bol. Bot. Univ. Sao Paulo 10: 33–41.

Morawetz, W. & M. Waha 1991. Zur Entstehung und Funktion pollenverbindender Fäden bei *Porcelia* (Annonaceae). Beitr. Biol. Pfl. 66: 145–154.

Müller-Doblies, D. & U. Müller-Doblies. 1985. De Liliifloris notulae. 2. De taxonomia subtribus Strumariinae (Amaryllidaceae). Bot. Jahrb. Syst. 107: 17–47.

Noel, A.R.A. 1983. The endothecium – a neglected criterion in taxonomy and phylogeny? Bothalia 14: 833–838.

Parkin, J. 1951. The protrusion of the connective beyond the anther and its bearing on the evolution of the stamen. Phytomorphology 1: 1–8.

Patil, D.A. & R.M. Pai. 1981. The floral anatomy of *Doryanthes excelsa* Corr. (Agavaceae). Ind. J. Bot. 4: 5–9.

Proença, C.E.B. 1992. Buzz pollination – older and more widespread than we think? J. Trop. Ecol. 8: 115–120.

Rao, A.N. 1961. Fibrous thickenings in the anther epidermis of *Wormia burbidgei* Hook. Curr. Sci. 31: 426.

Renner, S.S. 1989a. Floral biological observations on *Heliamphora tatei* (Sarraceniaceae) and other plants from Cerro de la Neblina in Venezuela. Pl. Syst. Evol. 163: 21–29.

Renner, S.S. 1989b. A survey of reproductive biology in neotropical Melastomataceae and Memecylaceae. Ann. Missouri Bot. Gard. 76: 496–518.

Renner, S.S. 1990. Reproduction and evolution in some genera of neotropical Melastomataceae. Mem. New York Bot. Gard. 55: 143–152.

Robbrecht, E. 1988. Tropical Woody Rubiaceae. Opera Bot. Belg. 1: 1–271.

Robbrecht, E. & C. Puff. 1986. A survey of the Gardenieae and related tribes (Rubiaceae). Bot. Jahrb. Syst. 108: 63–137.

Rohweder, O. 1956. Die Farinosae in der Vegetation von El Salvador. Abh. Geb. Auslandkunde 61: 1–197.

Sazima, M. & Vogel, S. 1989. *Cyphomandra* species (Solanac.) visited by euglossine bees: floral fragrances as a reward for pollen dispersal. 9 Symposium Morphologie, Anatomie und Systematik, Vienna, Abstr. 52.

Schaeppi, H. 1939. Vergleichend-morphologische Untersuchungen an den Staubblättern der Monocotyledonen. Nova Acta Leop., n.F., 6: 390–447.

Schill, R. & Wolter, M. 1986. Ontogeny of elastoviscin in the Orchidaceae. Nord. J. Bot. 5: 575–580.

Schmid, R. 1976. Filament histology and anther dehiscence. Bot. J. Linn. Soc. 73: 303–315.

Shamrov, I.I. 1990. Melanthiaceae. pp. 59–63 in Comparative Embryology of Flowering Plants. Monocotyledones (edited by T.B. Batygina & M.S. Yakovlev). Nauka, Leningrad.

Simpson, M. 1990. Phylogeny and classification of the Haemodoraceae. Ann. Missouri Bot. Gard. 77: 722–784.

Strasburger, E. 1902. Ein Beitrag zur Kenntniss von *Ceratophyllum submersum* und phylogenetische Erörterungen. Jahrb. Wiss. Bot. 37: 477–526.

Thiele, E.-M. 1988. Bau und Funktion des Antheren-Griffel-Komplexes der Compositen. Diss. Bot. 117. Cramer, Berlin.

Tiemann, A. 1985. Untersuchungen zur Embryologie, Blütenmorphologie und Systematik der Rapateaceen und der Xyridaceen-Gattung *Abolboda* (Monocotyledoneae). Diss. Bot. 82. Cramer, Vaduz.

Troll, W. 1928. Ueber *Spathicarpa sagittifolia* Schott. Flora 123: 286–316.

Troll, W. 1951. Botanische Notizen II. Abh. Math.-Nat. Kl. Akad. Wiss. Lit., Mainz 1951: 25–80.

Vogel, S. 1978. Evolutionary shifts from reward to deception in pollen flowers. pp. 89–96 in The Pollination of Flowers by Insects (edited by A.J. Richards). Academic Press, London.

Vogel, S. & A. Cocucci. 1988. Pollen threads in *Impatiens*: their nature and function. Beitr. Biol. Pfl. 63: 271–287.

Waha, M. 1984. Zur Ultrastruktur und Funktion pollenverbindender Fäden bei Ericaceae und anderen Angiospermenfamilien. Pl. Syst. Evol. 147: 189–203.

Werth, E. 1956. Zur Kenntnis des Androecums der Gattung *Salvia* und seiner stammesgeschichtlichen Wandlung. Ber. Dtsch. Bot. Ges. 69: 381–386.

Whalen, M.D. 1978. Reproductive character displacement and floral diversity in *Solanum* section *Androceras*. Syst. Bot. 3: 77–86.

5 · Are stamens and carpels homologous?

WILLIAM BURGER

INTRODUCTION

A central paradigm in angiosperm morphology has been the assumption that the fertile parts of flowers are leaf-like structures. These ideas received support from Goethe's analysis of the developmental changes that occur as a simple herb grows to maturity, producing leaves in early stages and floral parts in later stages (Goethe 1790; Arber 1946). Goethe's concepts of developmental metamorphoses have been superseded by views in which the postulated homologies are those of evolutionary history (Arber & Parkin 1907; Celakovsky 1901; Eyde 1975). The leaf-like nature of stamens and carpels remains a fundamental tenet in the writing of many contemporary authors (Guedes 1966; Fahn 1982; Gifford & Foster 1989; Takhtajan 1991). The underlying assumption in these writings is that stamens and carpels are homologous in the sense that both were derived directly from leaf-like organs bearing sporangia or cupules. However, angiosperm stamens are not very leaf-like, and other origins for the stamen have been proposed (Wilson 1937; Zimmermann 1957). Why has the notion of a leaf-like stamen been so widely accepted among modern authors? The answer, I believe, had much to do with a focus on the primitive nature of the Magnoliales and the discovery of a relictual genus, *Degeneria*, on a remote island (Bailey & Smith 1942; Miller 1989).

The broad laminar stamens of *Degeneria*, *Himantandra*, *Austrobaileya* and Nymphaeaceae have been interpreted as primitive and illustrated in many texts (Canright 1952; Eames 1961; Raven *et al.* 1981; Fahn 1982; Gifford & Foster 1989; Takhtajan 1991). These laminar stamens have convinced many botanists that stamens, the least leaf-like element of the flower, are also leaf homologs. However, there are problems with such an interpretation. Takhtajan (1991, p. 84) finds it 'morphologically puzzling' that among the so-called primitive stamens of Magnoliales there are some genera with adaxial anthers and other genera with abaxial anthers. Others have used this anomaly to question the interpretation of these laminar stamens as primitive.

In addition, laminar stamens are found only in flowers that are relatively large with thick damage-resistant parts, which has led to the interpretation that leaf-like stamens may be specialized adaptations for beetle pollination (Carlquist 1969; Sauer & Ehrendorfer 1970; Gottsberger 1974, 1988; Stebbins 1974: 226). These authors argue that the laminar stamens of Magnoliales, Laurales and Nymphaeales represent special adaptations, and should not be interpreted as primitive leaf-like structures (see also Loconte & Stevenson 1991: 291). However, these counter-arguments have not replaced the general consensus that the laminar stamens found in some genera of Magnoliales are strong evidence for leaf-like stamens in the origin of angiosperms. I believe this problem deserves further discussion and that a short review of the question of stamen-carpel homology may help clarify the issue.

The question of stamen and carpel homology is more general than the question of whether two similar character states in related lineages represent a monophyletic synapomorphy. Stamens and carpels often occur in the same flower; they are important elements in a large suite of characteristics that define angiosperms. Rather than a question of monophyly, this is a question of similar antecedent structure. Nevertheless, the criteria used to evaluate homology in phylogenetic studies should be useful in addressing the question of carpel and stamen homology. The following discussion considers this question from the perspective of five general criteria of homology: general similarity, teratology, fossil data, sister-group similarities, and ontogeny. More precise criteria for strict morphological homology as have been summarized by Kaplan (1977, 1984) are, I believe, not appropriate to the more general evolutionary question posed here.

GENERAL SIMILARITY

Stamens are usually easily recognized, the great majority have a slender stalk supporting anthers that are rounded in cross-section (cf. Endress & Hufford 1989; Hufford & Endress 1989; Endress & Stumpf 1991). The carpel has been interpreted as a 'phylogenetically folded megasporophyll that has enclosed one or more ovules' (Bold 1973: 577). It was defined earlier as a 'simple pistil, or element of a compound pistil, answering to a single leaf' (Jackson 1950). The interpretation of carpel elements in compound pistils often makes the carpel a more abstract concept; for the purposes of this discussion I shall limit the comparisons to free independent stamens and simple monocarpellate pistils. The simple pistil (with single locule, style and stigma) or carpel is generally interpreted as a single leaf-like structure. This contrasts with more complex gynoecia that are interpreted as the product of the union of two or more carpels. Despite the fact that stamens can also become fused into multistaminate androecia, there is little similarity between stamens and carpels. Stamens are almost always stalked structures bearing saccate anthers. Even when fused into tubes or fascicles, stamens usually retain their characteristic stalked anthers. The exceptions, such as the stamens of orchids and sunflowers, are clearly derived. Gynoecia, with their ovuliferous locules, bear little resemblance to stamens, despite their great diversity. The stamens of *Degeneria* or *Chloranthus*, considered primitive and leaf-like by many authors, do not resemble carpels. Stamens and carpels fail the simplest criterion of homology: they do not share a general similarity.

TERATOLOGY

Abnormal structures, and especially teratological transformation series, have often been used as an argument for stamen–carpel homology (Heslop–Harrison 1952;

Wagner 1968). These arguments fail to consider the developmental potential of plant cells, both in the meristems and in more differentiated tissue. Even cells taken from the cortex of a carrot have been able to recreate a new plantlet. This potential makes the interpretation of abnormal structures as atavistic highly speculative. The fact that stamens are borne where carpels should have formed (or vice versa) may imply nothing more than a failure of developmental controls. Likewise, the occurrence of stamen–carpel chimeras may be the consequence of developmental anomalies, having nothing to do with evolutionary history.

Abnormal development can also be used to argue against stamen–carpel homology. For example, the replacement of anthers for ovules within the ovary of onion flowers (Gohil & Kaul 1981) or within the ovary of tobacco flowers (Malmberg & McIndoo 1983) could be used as evidence for stamen–ovule homology. Unlike higher animals, where cells lose their developmental plasticity early in differentiation, arguments for homology from plant teratology should be given little credence.

FOSSIL EVIDENCE

A diversity of flowers have now been recovered from the Cretaceous (Friis & Endress 1990; Friis *et al.* 1991). *Archaeanthus*, of the middle Cretaceous (Dilcher & Crane 1984), is very similar in general form to modern flowers of the Magnoliaceae, with stamen-scars that have a narrow base, as do modern representatives of this family. *Chloranthus*-like stamens from the upper Cretaceous are narrower and less leaf-like than those of contemporary species (Friis & Endress 1990, fig. 10). Virtually all the stamens ascribed to angiosperm flowers from the early Cretaceous have narrow basal filaments and elongate anthers (Friis *et al.* 1991). None appear to be as leaf-like as *Degeneria* or *Himantandra* (Friis & Endress 1990, p. 130), and they weaken the argument that these contemporary genera retain an ancient character state.

Cornet (1986, 1989) has made a strong case for the angiosperm-like nature of the Triassic *Sanmiguelia*. Unfortunately, the reproductive material of *Sanmiguelia* is very poorly preserved, and Cornet's interpretation of stamen and carpel similarity should be confirmed with better material. As in the case of abnormal structures, I believe the fossil record does not provide us with substantial evidence regarding stamen and carpel homology.

SISTER-GROUP SIMILARITIES

A consensus has developed in recent years that, among living plants, the Gnetales represent the sister group to angiosperms. These views are based on patterns of chromosomal differentiation (Ehrendorfer 1976), cladistic analyses of morphology (Crane 1985; Doyle & Donoghue 1986, 1993), similarities in fertilization (Friedman 1992), and phylogenies based on DNA sequences (Chase *et al.* 1993; Hamby & Zimmer 1992).

Immediately apparent in any angiosperm–Gnetales comparison is the fact that the pollen-producing organs are saccate structures borne on narrow stalks. Even in *Welwitschia*, where the microsporangiophores are united to form a broad tube, the pollen sacs are borne on short stalks. The usually bisporangiate pollen sacs of *Ephedra* and *Gnetum* are superficially similar to those of angiosperms. These sister-group similarities argue against the contention that flattened leaf-like anthers are primitive in angiosperms.

Carpels have no clear counterpart in the Gnetales. While the ovules of Gnetales may be enclosed within a tubular outer tissue, these tissues do not form a closed locule and are not involved in fertilization. These outer coverings of the ovule are probably the product of fusion of decussate bracts (Crane 1985, fig. 3); they are not equivalent to the angiosperm carpel.

While comparison of sister-groups is a strong criterion for homology (Watrous & Wheeler 1981), it is compromised by our ignorance of the extinct intermediate lineages which may have had an important role in early diversification. Nevertheless, the general similarities of pollen-bearing organs in Gnetales and angiosperms make a strong case for a primitive angiosperm stamen with rounded anthers and slender stalk. The lack of a carpel-like gynoecium in Gnetales supports the contention that the carpel is a defining autapomorphy for angiosperms (Crane 1985; Friis & Endress 1990). It is also worth noting that no one has argued for any general homology between the male and female elements in the 'flowers' of Gnetales. This, also, weakens the argument for such a homology among angiosperms.

ONTOGENY

Ontogeny has played an important role in the development of evolutionary hypotheses, especially in zoology (some recent discussions are: Kraus 1988; Rieppel 1989; Wheeler 1990). I believe ontogeny provides the most convincing data we possess as regards the question of stamen–carpel homology. Much more attention has been given the ontogeny of carpels than has been afforded stamens, but this may be due to the greater diversity of gynoecia and placentation-types.

Simple (monocarpellate) pistils usually begin from primordia that quickly develop lateral flanks (Eames 1961, p. 227). This is identical to the primordia of most angiosperm leaves (Hagemann 1970). A comparison of scanning electron micrographs of the early *Magnolia grandifolia* leaf primordium (Postek & Tucker 1982) and the early development of simple pistils (Heel 1981) shows a striking similarity. Likewise, the near-identity of ontogeny of the tubular leaf-traps of *Sarracenia* and some simple pistils (Franck 1975) gives further evidence of carpel–leaf homology. Such studies have covered many taxa and provide compelling evidence for the claim that the carpel is leaf-like in origin. A similar convergent ontogeny can be seen in the early development of some aroid spadices and their subtending spathe (Uhlarz 1983) where, again, a single subtending leaf has become an enclosing protective structure.

The early ontogeny of stamens begins with a small rounded elevation (Foster & Gifford 1959, fig 18–16). This proceeds to elongate in the development of independent stamens. I have seen no evidence that the flanking base, characteristic of the early ontogeny of leaves and carpels, has ever been seen in the early development of independent stamens. Flanking bases and lateral fusion do occur, but this is associated with the development of staminal tubes in lineages with such androecia (Sattler 1973: 181). Takhatajan (1991: 91) cites the strong ontogenetic evidence for the leaf-like nature of carpels, but he fails to discuss its lack in stamen development. Instead, he focuses on the laminar stamens in a few genera of Magnoliales as evidence of ancestral leaf-like form. However, a study of *Austrobaileya* by Endress (1983) found that the laminar margins of the stamens develop late in ontogeny, and the author concluded they did not represent a primitive condition.

Despite these differences in early ontogeny, imaginative botanists have developed arguments for homology in the later ontogeny of stamens and carpels. Troll (references in Just 1939) and his followers have elaborated a theory that sees strong similarities between stalked (peltate/ascidiate) carpels and the development of certain stamens (Baum & Leinfellner 1953; Leinfellner 1956; Heel 1983). While these authors present much evidence for leaf-like carpel development, their claim for leaf-like stamens appears to be based on derived or teratological structures. The carpel–stamen similarities they illustrate often develop later in ontogeny, after the base has narrowed into a stalk-like portion. The generally later development of the narrowed base in ascidiate carpels suggests that this is a derived condition.

While early histogenesis may be quite variable, there are reports that stamens originate in deeper layers of the meristem than do sepals, petals and carpels (Barnard 1958; Richardson 1969) or in a stem-like fashion (Satina & Blakeslee 1941). Such reports are consistent with the discovery of a gene in *Antirrhinum* affecting early development of bracts, sepals, petals and carpels but not stamens (Coen & Meyerowitz 1991). Taken together, the evidence of early ontogeny and histogenesis appears to be consistent. There is strong evidence for a leaf-like origin of the carpel, but none for a similar origin of the stamen.

DISCUSSION

Viewed from the perspective of these five general criteria we should conclude that the stamen and carpel are not homologous. The ontogenetic evidence for the leaf-like nature of the carpel is especially strong. This conclusion implies that the carpel has incorporated a leaf-like element into its structure; a configuration not found in the Gnetales. Evidence from the sister group also suggests that neither stamens nor carpels were derived directly from fertile (sporangia- or cupule-bearing) leaves. Such a conclusion implies that the leaf incorporated into the carpel must have been a subtending bract or sterile leaf. This subtending leaf (or leaves) came to enclose the ovules or ovules (cf. Crane 1985, fig. 21; Stewart 1983, fig. 27.21; Stevenson 1993) and, subsequently, the ovules have become borne on the leaf-like portion of the gynoecium in many lineages.

Interpreting the carpel as a compound structure (ovule or ovule-cluster with enclosing sterile leaf) clarifies many problems. It explains why some angiosperm gynoecia have been called acarpellate, based on a definition of the carpel as an ovule-bearing leaf. In such gynoecia, the ovules develop directly from the shoot-apex and are not 'leaf-borne' (Macdonald 1974; Sattler 1974). A good example is seen in *Illicium floridanum* Ellis where the ovule originates directly from the floral axis and is axillary to the carpel (Robertson & Tucker 1979). In addition, sterile subtending bracts (as origin of the leaf-like constituents of modern angiosperm gynoecia) make it easier to understand the fusion of lateral margins in multicarpellate gynoecia; something that is more difficult to envision if the primitive carpel were ascidiform or cupulate *and* ovule-bearing. Viewing the carpel as a compound structure also helps explain the diversity of angiosperm gynoecia and why their interpretation has been the subject of so much acrimonious debate (cf. Cresens & Smets 1989).

A rejection of stamen–carpel homology also leads to the conclusion that the angiosperm stamen is a simpler structure than is the carpel. Microsporangia borne on leaf-like structures may have existed in the ancestry of angiosperms, but the sister group comparison indicates that the common ancestor of Gnetales and angiosperms had lost this feature. While the question of the origin of the basic dithecal-tetrasporangiate angiosperm anther cannot be answered at present (Endress & Stumpf 1990, p. 232), eliminating a leaf-like element from its recent history will facilitate comparisons with the pollen-producing organs of gymnosperms (cf. Wilson 1937, Zimmerman 1956). These conclusions also make clear that the stamen is an older structure than the carpel, sharing fundamental similarities with the pollen-producing organs of Gnetales.

LITERATURE CITED:

Arber, A. 1946. Goethe's botany. Chronica Bot. 10: 63–124.
Arber, E.A. & J. Parkin. 1907. On the origin of Angiosperms. J. Linn. Soc. Bot. 38: 29–80.
Bailey, I.W. & A.C. Smith. 1942. Degeneriaceae, a new family of flowering plants from Fiji. J. Arnold Arbor. 23: 355–365.

Barnard, C. 1958. Floral histogenesis in the monocotyledons. III. The Juncaceae. Aust. J. Bot. 6: 285–298.

Baum, H. & W. Leinfellner. 1953. Die ontogenetischen Abänderungen des diplophyllen Grundebaues der Staubblätter. Öst. Bot. Z. 100: 593–600.

Bold, H.C. 1973. Morphology of Plants, 3rd ed. Harper & Row, New York.

Canright, J.E. 1952. The comparative morphology and relationships of the Magnoliaceae. 1. Trends of specialization in the stamens. Amer. J. Bot. 39: 487–497.

Carlquist, S. 1969. Toward acceptable evolutionary interpretations of floral anatomy. Phytomorphology 19: 332–362.

Celakovsky, L.J. 1901. Ueber den phylogenetischen Entwicklungsgang der Blüthe und über den Ursprung der Blumenkrone. II. Sitzungsber. Königl. Böhm. Ges. Wiss. Prag, Math. -Naturwiss. Cl. 1900, 3: 1–221.

Chase, M.W., D.E. Soltis, R.G. Olmstead, D. Morgan, D.H. Les, B.D. Mishler, M.R. Duvall, R.A. Price, H.G. Hills, Y.-L. Qui, K.A. Kron, J.H. Rettig, E. Conti, J.D. Palmer, J.R. Manhart, K.J. Sytsma, H.J. Michaels, W.J. Kress, K.G. Karol, W.D. Clark, M. Hedrén, B.S. Gaut, R.K. Jansen, H.-J. Kim, C.F. Wimpee, J.F. Smith, G.R. Furnier, S.H. Strauss, Q.-Y. Xiang, G.M. Plunkett, P.S. Soltis, S. Swensen, S.E. Williams, P.A. Gadek, C.J. Quinn, L.E. Eguiarte, E. Golenberg, G.H. Learn, Jr., S.W. Graham, S.C.H. Barrett, S. Dayanandan, & V. A. Albert. 1993. Phylogenies of seed plants: an analysis of nucleotide sequences from the plastid gene rbcL. Ann. Missouri Bot. Gard. 89: 528–580.

Coen, E.S. & E.M. Meyerowitz. 1991. The war of the whorls: genetic interactions controlling flower development. Nature 353: 31–37.

Cornet, B. 1986. The leaf venation and reproductive structures of a late Triassic angiosperm, Sanmiguelia lewisii, Evol. Theory 7: 231–309.

Cornet, B. 1989. The reproductive morphology and biology of Sanmiguelia lewisii, and its bearing on angiosperm evolution in the Late Triassic. Evol. Trends Pl. 3: 25–51.

Crane, P. 1985. Phylogenetic analysis of seed plants and the origin of Angiosperms. Ann. Missouri Bot. Gard. 72: 716–793.

Cresens, E.M. & E.F. Smets. 1989. The carpel – a problem child of floral morphology and evolution. Bull. Jard. Bot. Etat. 59: 377–409.

Dilcher, D.L. & P.R. Crane, 1984. Archaeanthus: an early angiosperm from the Cenomanian of the western interior of North America. Ann. Missouri Bot. Gard. 71: 351–383.

Doyle, J.A. & M.J. Donoghue 1986. Relationships of angiosperms and Gnetales: a numerical cladistic analysis. pp. 177–198 in Systematics and Taxonomic Approaches in Palaeobotany (edited by R.A. Spicer & B.A. Thomas). Syst. Assoc. Spec. Vol. 31. Clarendon Press, Oxford.

Doyle, J.A. & M.J. Donoghue. 1993. Phylogenies and angiosperm diversification. Paleobiology 19: 141–167.

Eames, A.J. 1961. Morphology of Angiosperms. McGraw-Hill, New York.

Ehrendorfer, F. 1976, Evolutionary significance of chromosomal differentiation patterns in gymnosperms and primitive angiosperms. pp. 220–240 in Origin and Early Evolution of Angiosperms (edited by C.B. Beck). Columbia University Press, New York.

Endress, P.K. 1983. The early floral development of Austrobaileya. Bot. Jahrb. Syst. 103: 481–497.

Endress, P.K. & L.D. Hufford. 1989. The diversity of stamen structures and dehiscence patterns among Magnoliidae. Bot. J. Linn. Soc. 100: 45–85.

Endress, P.K. & S. Stumpf. 1990. Non-tetrasporangiate anthers in angiosperms. Bot. Jahrb. Syst. 112: 193–240.

Endress, P.K. & S. Stumpf. 1991. The diversity of stamen structures in 'lower' Rosidae (Rosales, Fabales, Proteales, Sapindales). Bot. J. Linn. Soc. 107: 217–293.

Eyde, R.H. 1975. The foliar theory of the flower. Amer. Sci. 63: 430–437.

Fahn, A. 1982. Plant Anatomy, 3rd ed. Pergamon Press. Oxford.

Foster, A.S. & E.M. Gifford, Jr. 1959. Comparative Morphology of Vascular Plants, W.H. Freeman & Co., San Francisco.

Franck, D.H. 1975. Early histogenesis of the adult leaves of Darlingtonia californica (Sarraceniaceae) and its bearing on the nature of epiascidiate foliar appendages. Amer. J. Bot. 62: 116–132.

Friedman, W.E. 1992. Evidence of a pre-angiosperm origin of endosperm: implications for the evolution of flowering plants. Science 255: 336–339.

Friis, E.M. & P. Endress. 1990. Origin and evolution of angiosperm flowers. Adv. Bot. Res. 17: 99–162.

Friis, E.M., P.R. Crane & K.R. Pedersen. 1991. Stamen diversity and in situ pollen of Cretaceous angiosperms. pp. 197–224 in Pollen and Spores (edited by S. Blackmore & S.H. Barnes). Syst. Assoc. Spec. Vol. 44. Clarendon Press, Oxford.

Gifford, E.M. & A.S. Foster. 1989. Morphology and Evolution of Vascular Plants, 3rd ed. W.H. Freeman, New York.

Goethe, J.W. von, 1790. Versuch die Metamorphose der Pflanzen zu erklären. Gotha.

Gohil, R.N. & R. Kaul, 1981. On the occurrence of ovular anthers in *Allium cepa*. Naturwiss. 68: 331.

Gottsberger, G. 1974. The structure and function of the primitive angiosperm flower – a discussion. Acta Bot. Neerl. 23: 461–471.

Gottsberger, G. 1988. The reproductive biology of primitive angiosperms. Taxon 37: 630–643.

Guedes, M. 1966. Homologie de carpelle et de l'étamine chez *Tulipa gesneriana*. Öst. Bot. Z. 113: 47–83.

Hagemann, W. 1970. Studien zur Entwicklungsgeschichte der Angiospermenblätter. Bot. Jahrb. Syst. 90: 297–413.

Hamby, R.K. & E.A. Zimmer. 1992. Ribosomal RNA as a phylogenetic tool in plant systematics. pp. 50–91 in Molecular Systematics of Plants (edited by P.S. Soltis, D.E. Soltis & J.J. Doyle). Chapman & Hall, New York.

Heel, W.A. van. 1981. A SEM investigation on the development of free carpels. Blumea 27: 499–522.

Heel, W.A. van. 1983. The ascidiform early development of free carpels, a SEM investigation. Blumea 28: 231–270.

Heslop-Harrison, J. 1952. A reconsideration of plant teratology. Phyton 4: 19–34.

Hufford, L.D. & P.K. Endress. 1989. The diversity of anther structure and dehiscence patterns among Hamamelididae. Bot. J. Linn. Soc. 99: 301–346.

Jackson, B.D. 1950. A Glossary of Botanic Terms, 4th ed. Hafner, New York.

Just, T. 1939. The morphology of the flower: the typological approach to the nature of the flower. Bot. Rev. 5: 115–131.

Kaplan, D.R. 1977. Morphological status of the shoot systems of Psilotaceae. Brittonia 29: 30–53.

Kaplan, D.R. 1984. The concept of homology and its central role in the elucidation of plant systematic relationships. pp. 51–70 in Cladistics: Perspective on the Reconstruction of Evolutionary History (edited by T. Duncan & T.F. Stuessy). Columbia University Press, New York.

Kraus, F. 1988. An empirical evaluation of the use of the ontogeny polarization criterion in phylogenetic inference. Syst. Zool. 37: 106–141.

Leinfellner, W. 1956. Die blattartig flachen Staubblätter und ihre gestaltlichen Beziehungen zum Bautypus der Angiospermen-Staubblattes. Öst Bot. Z. 103: 247–290.

Loconte, H. & D.W. Stevenson. 1991. Cladistics of the Magnoliidae. Cladistics 7: 267–296.

Macdonald, A.D. 1974. Theoretical problems of interpreting floral organogenesis of *Laportea canadensis*. Can. J. Bot. 52: 639–644.

Malmberg, R.L. & J. McIndoo. 1983. Abnormal floral development of a tobacco mutant with elevated polyamine levels. Nature 305: 623–625.

Miller, J.M. 1989. The archaic flowering plant family Degeneriaceae: its bearing on an old enigma. Natl. Geogr. Res. 5: 218–231.

Postek, M.T. & S.C. Tucker, 1982. Foliar ontogeny and histogenesis in *Magnolia grandiflora* L.I. Apical organization and early development. Amer. J. Bot. 69: 556–569.

Raven, P.H., R.F. Evert & H. Curtis. 1981. Biology of Plants, 3rd ed. Worth, New York.

Richardson, F.C. 1969. Morphological studies of the Nymphaeaceae. IV. Structure and development of the flower of *Brasenia schreberi* Gmel. Univ. Calif. Publ. Bot. 47: 1–101.

Rieppel, O. 1989. Ontogeny, phylogeny, and classification. Abh. Verh. Nat. Ver. Hamburg 28: 63–82.

Robertson, R.E. & S.C. Tucker, 1979. Floral ontogeny of *Illicium floridanum*, with emphasis on stamen and carpel development. Amer. J. Bot. 66: 605–617.

Satina, S. & A.F. Blakeslee, 1941. Periclinal chimeras in *Datura stramonium* in relation to development of leaves and flower. Amer. J. Bot. 28: 862–871.

Sattler, R. 1973. Organogenesis of Flowers: A Photographic Text-atlas. University of Toronto Press, Toronto.

Sattler, R. 1974. A new approach to gynoecial morphology. Phytomorphology 24: 22–34.

Stebbins, G.L. 1974. Flowering Plants. Evolution above the Species Level. Harvard University Press, Cambridge, Mass.

Stevenson, D. 1993. Homology of the seed and associated structures in spermatophytes (Abstract). Amer. J. Bot. 80 (6, suppl.): 125.

Sauer, W. & F. Ehrendorfer. 1970. Chromosomen, Verwandschaft und Evolution tropisher Holzpflanzen. II. Himantandraceae. Öst. Bot. Z. 118: 38–54.

Stewart, W.N. 1983. Paleobotany and the Evolution of Plants. Cambridge University Press, Cambridge.

Takhtajan, A. 1991. Evolutionary Trends in Flowering Plants. Columbia University Press, New York.

Uhlarz, H. 1983. Typologische und ontogenetische Untersuchungen an *Spathicarpa sagittifolia* Schott (Araceae): Wuchsform und Infloreszenz. Beitr. Biol. Pfl. 57: 389–429.

Wagner, W.H. 1968. Teratological stamens and carpels of a willow from northern Michigan. Michigan Bot. 7: 113–120.

Watrous, L.E. & Q.D. Wheeler, 1981. The outgroup comparison method of character analysis. Syst. Zool. 30: 1–11.

Wheeler, Q.D. 1990. Ontogeny and character phylogeny. Cladistics 6: 225–268.

Wilson, C.L. 1937. The phylogeny of the stamen. Amer. J. Bot. 24: 686–699.

Zimmermann, W. 1957. Phylogenie der Blüte. Phyton 7: 162–182.

6 · Heterochrony in the anther

JEFFREY P. HILL

INTRODUCTION

The morphological and cellular bases of development of the angiosperm anther have been the subject of study for a diverse range of research interests. Morphological and histological aspects of stamen initiation and subsequent anther differentiation have formed the basis of broad discussions concerning the homology of vegetative and floral organs (Arber 1937; Wilson 1937; Satina & Blakeslee 1941; Boke 1949; Eames 1961; Kaplan 1968). Systematic classifications of microsporangial development in the anther have been compiled (Davis 1966), providing character states for taxonomic use. The anther continues to be recognized as a model organ for addressing basic and applied questions concerning plant cell biology, morphogenesis, breeding, and evolution (Erickson 1948; Cheng *et al.* 1979; McHughen 1980; Gould & Lord 1988; Kaul 1988; Lord *et al.* 1989; Dawe & Freeling 1992; Mascarenhas 1992; Sawhney 1992).

Not surprisingly, general aspects of anther differentiation have recently come under intensive study utilizing the methods of plant molecular biology (Schwartz–Sommer *et al.* 1990; Coen 1991; Coen & Meyerowitz 1991; Scott *et al.* 1991; Goldberg *et al.* 1993). These recent advances have created new opportunities for assessing old questions about the evolution of anthers as well as other plant organs. For example, manipulation of the floral homeotic (organ identity) mutants to show that leaves are the developmental 'ground state' of floral organs clearly has implications for the study of morphological evolution, as does the suggestion that these floral organ identity genes might also be expressed in the organs of non-flowering vascular plants (Doyle 1993). Notwithstanding these approaches, comparative studies

of cellular and morphological development continue to play an important dual role in establishing basic details of how plant organs develop, and how ontogenetic patterns at these levels of organization have been modified phylogenetically (Tucker 1992).

The desire to explain patterns of morphological evolution in mechanistic terms is leading to an increasing integration of organismal, molecular, and evolutionary biology that was considered to be lacking not too long ago (Gould 1977; Bonner 1982; Diggle 1992). In this regard, heterochrony, or changes in ontogenetic rates or timing during evolution, has emerged as a paradigm for many phylogenetically oriented studies of morphology and development (Raff & Kaufman 1983; Lord & Hill 1987; Guerrant 1988; Raff 1990; Dengler 1992). For the developmental biologist, efforts to elucidate heterochrony embody a desire to link heritable variation in the rates or timing of developmental events to changes in morphology.

The formalism proposed to describe different styles of morphological heterochrony (Alberch *et al.* 1979) is based on size and shape; the element of developmental timing or rates is incorporated into these models by including a measure of ontogenetic time. Although the need to explicitly incorporate chronological time in studies of heterochrony has been debated (Blackstone 1987; Strauss 1987), comparative documentation of the absolute timing of developmental events in the organisms under study is clearly a requirement if mechanistic explanations of heterochrony are sought (Raff & Wray 1989). In the absence of this restricted use, heterochronic models serve mainly as descriptive explanations of patterns of morphological change (Raff & Wray 1989; Hall 1990).

Studies of morphological heterochrony that incorporate time can still pose problems of interpretation when cellular and molecular levels of control related to the heterochrony are also characterized. Some changes that represent an alteration in timing at the morphological level will arise incidentally from non-heterochronic changes at other levels (Raff & Wray 1989). The importance of this fact does not lie in its use as a criterion for disqualifying certain examples of variation in morphology as 'real' cases of heterochrony. Instead, it underscores the need for greater scrutiny of comparative development in specific cases so that a realistic picture of how morphology changes phylogenetically can emerge. Furthermore, we can expect to find that some cases of heterochrony observed at the organismal level are actually due to changes in timing at cellular and subcellular levels of organization.

The purpose of this chapter is to examine critically whether heterochrony, defined in the restricted sense as an evolutionary change in absolute timing of development, provides a satisfactory explanation for differences observed in anther ontogeny at the morphological level. In so doing, I will first review what is known about the cellular and temporal patterns of early anther differentiation in general. This will be followed by a discussion of the salient results of previous studies that suggest a heterochronic origin for reduced anther forms that have evolved in association with the evolution of self-pollination. In light of the observed developmental changes, the role that controls over cell cycle time and cell size at division might play in heterochrony of the anther will be considered. The emphasis throughout the discussion is on the early phase of stamen and anther development prior to the initiation of microsporocyte meiosis.

CELLULAR PATTERNS OF STAMEN INITIATION

The details of cell division patterns during stamen initiation from the floral apex in several dicotyledons indicate that organ initiation is characterized by some combination of: (1) anticlinal cell divisions in the outer tunica (L_1) layer; (2) mainly periclinal divisions in the corpus (L_3); and (3) variable (periclinal or anticlinal) divisions in the second tunica layer (L_2), depending on the species (Fig. 1A; Satina & Blakeslee 1941; Boke 1949;

Cheung & Sattler 1967; Kaplan 1968; Dengler 1972; Hicks 1973; Hill & Lord 1989). During the ensuing growth of the stamen primordium, cells of the protoderm continue to divide anticlinally. In *Datura*, anticlinal divisions also predominate in the subepidermal layer, so that neither L_1 nor L_2 cells invade the more internal tissues of the developing organ during this early phase of growth (Satina & Blakeslee 1941). In species where periclinal divisions in L_2 have been observed, the exact timing of these divisions relative to the timing of primordial emergence is variable. If periclinal divisions in L_2 are not immediately evident at the time of stamen initiation, the likelihood of such divisions in L_2 at the tip of a primordium appears to increase as the primordium enlarges in size; based on histological analyses, this period is described as a phase of apical growth of the stamen.

Aside from their possible role as subapical initials, it is not clear to what extent cells in L_2 generally participate in the enlargement of the stamen primordium prior to the differentiation of the microsporangia (Fig. 1B–D; Boke 1949; Kaplan 1968). Earlier research conducted under the premise that stamens were leaf homologues advanced evidence to support the notion that cells in L_2 regularly contribute to more internal layers from hypodermal marginal meristems (Kaussmann 1941; cited in Eames 1961). In other cases, it appears that cell division activity in L_2 during this period of primordial development remains anticlinal (Satina & Blakeslee 1941). Some of this variation may be attributed to differences among species in the extent of marginal expansion of the stamen primordium relative to the timing of microsporangial initiation (Dengler 1972).

DEVELOPMENT OF THE MICROSPORANGIUM

The initiation of microsporangia in the anther is characterized anatomically by a round of periclinal divisions in archesporial initials that differentiate in the L_2 layer of the stamen primordium. The inner derivative of this periclinal division is the primary sporogenous cell, and the outer one is the primary parietal cell (Fig. 2A; Maheshwari 1950; Davis 1966; Gifford & Foster 1989). The origin of microsporangia in the L_2 layer of the primordium has been corroborated by clonal analysis (Satina & Blakeslee 1941; Dawe & Freeling 1990), and

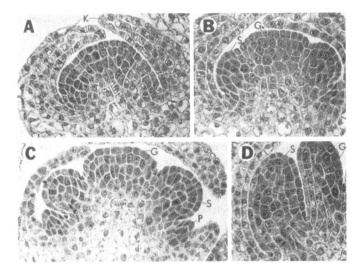

Fig. 1. Early stamen development in *Arabidopsis thaliana*. All sections are longitudinal. A. Stamen initiation from the floral apex. The primordium beneath the left sepal (K) shows cell walls resulting from periclinal divisions in L_2 and L_3. B. Early emergence of stamen primordia (S). Cell walls from recent periclinal divisions in L_2 are evident near the tip of each stamen. The gynoecium (G) has also been initiated. C. Stamen primordial enlargement during the proliferative meristematic phase of growth. Variable lengths of anticlinal walls in L_2 cells suggest periclinal divisions have recently occurred near the tip of the primordium. P, petal primordium. D. Stamen primordium near the end of the proliferative growth phase. Periclinal divisions in L_2 are evident apically and in an archesporial initial about half way down the adaxial face of the stamen. A and C are reprinted from Hill & Lord (1989), with permission.

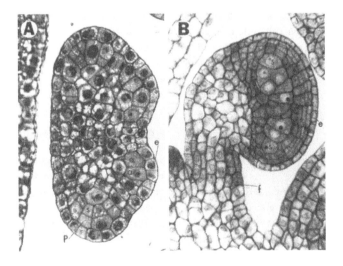

Fig. 2. Microsporangium development in the anther. A. Transverse section through an anther of *Collomia*. In the front locules, archesporial initials (a) expanded in the radial direction are evident beneath the epidermal layer (e). p, primary parietal cell; s, primary sporogenous cell. B. Longitudinal section through a stamen of *Arenaria*. Most of the cellular development of the microsporangium is complete. Files of parietal cells beneath the epidermis have arisen by periclinal divisions of the primary parietal cells. s, sporogenous tissue; f, stamen filament. A is reprinted from Lord *et al.* (1989), with permission.

appears to be a general feature of angiosperms. Analyses of periclinal chimeras demonstrate that L_1 cells occasionally displace L_2 cells after a periclinal division occurs in L_1. The inner daughter cell of the periclinal division assumes the L_2 position and subsequently contributes to normal microsporangium and pollen formation (Stewart & Dermen 1979). Cells in L_2 of the developing stamen function based on their hypodermal position at the time microsporangium development begins. How cells come to occupy the L_2 position based on previous cell division patterns is not relevant.

Within this general framework, the cytological details that culminate with division of archesporial initials in the hypodermal layer are variable. In many cases, the differentiation of archesporial initials prior to their division makes these cells cytologically distinct from surrounding tissues. Initials can be densely cytoplasmic with large nuclei, and/or the cells are larger than neighboring cells due to expansion of their anticlinal cell walls prior to periclinal division (Fig. 2A; Boke 1949; Kaplan 1968; Raghavan 1988; Lord et al. 1989). In other cases, initials are not very different in appearance compared with neighboring cells. Archesporial initials can be difficult to identify in Lilium (Gould & Lord 1988).

Morphologically, the differentiation and division of archesporial initials occurs close to the time that the adaxial face of the anther begins to expand radially in two zones that run along most of its length. These zones later develop into the front locules of the organ. In the relatively large stamen primordia of lily (Gould & Lord 1988), however, changes in anther shape are less evident at these stages of archesporial cell development.

Variation in the cellular patterns of anther wall formation form the basis of Davis's (1966) systematic classification of different wall types. This scheme is based on the observed cell division activity in the secondary parietal cell layers formed by periclinal divisions of the primary parietal cells. After two secondary parietal cell layers form, periclinal divisions may occur in both layers, the cells of the inner layer only, the cells of the outer layer only, or be completely suppressed in both layers. The utility of this classification scheme has been questioned because the first cell division of the primary parietal cell may not be periclinal (Gifford & Foster 1989). It is noteworthy, however, that examples can be found where nearly all cell divisions of the secondary parietal cells are periclinal (Fig. 2B). Although cell division patterns may not strictly follow Davis's (1966) scheme, there are clearly some systematic differences among angiosperms with respect to the cell partitioning patterns in the anther wall layers.

Generally, the innermost parietal cell layer adjacent to the developing sporogenous tissue will differentiate into tapetum and the outermost layer will form the endothecium. The origin of tapetal cells from sporogenous tissue has also been reported (Eames 1961; Vasil 1967; Dengler 1972).

The pollen mother cells or microsporocytes develop from the primary sporogenous cells that originate via archesporial cell division. Primary sporogenous cells may function directly as microsporocytes, or there may be a period of mitotic proliferation prior to microsporocyte differentiation (Erickson 1948; Eames 1961; Vasil 1967; Bennett et al. 1973; Raghavan 1988). In cases where mitotic activity occurs, the number of pre-meiotic divisions in the sporogenous tissue may be relatively constant within taxa, although details about this aspect of anther development have received less formal attention than the development of the wall layers (Davis 1966; and see below). During pre-meiotic interphase, microsporocytes become very nearly synchronized in their cell cycles (Erickson 1964; Bennett et al. 1973; Yoshioka et al. 1981). With the exception of pollen development, most of the cells of the mature anther are present by the time microsporocytes enter meiosis.

TEMPORAL PATTERNS OF ANTHER DIFFERENTIATION

The description of cellular patterns of microsporangial development in pre-meiotic anthers is usually based on anatomical observations, documenting first the planes of cell division, followed by the patterns of cell differentiation. These reports are often made without any reference to time. Certainly, the time course of events in the anther may be superfluous to particular studies. Additionally, the morphological circumstances in which anthers of particular species develop may preclude any effective means of measuring time (Raghavan 1988). These early phases of anther development commonly occur in meristematic primordia that cannot be observed without disrupting either the growth of the plant in general, or the growth of floral organs in particular. It is worth examining a few representative studies where satisfactory methods have been devised to allow the timing of stamen and anther development to be inferred.

The basic timing of stamen development can be approximated using a conventional method of plant growth analysis – harvesting plants or flowers from a population at known intervals and recording their relative developmental progress (Hunt 1982). Such an approach may be particularly useful in cases where the stages of pre-meiotic anther development last several months or more (Steeves et al. 1991). This approach may overestimate the duration of various phases of development, because slow-growing and fast-growing individuals are not distinguished; a phase of ontogeny will be interpreted to begin when the fastest individuals reach that stage, and will end when the slowest individuals finish it.

Erickson (1948) showed that nondestructive growth measurements of floral buds can be used with a high degree of accuracy to infer the growth of developing anthers inside the bud. In *Lilium* anthers, organ growth was biphasic, consisting of an early period of rapid expansion followed by a period of slower growth. The decline in growth rate occurred about midway between the end of microsporocyte meiosis and the start of microspore mitosis.

Combining data on anther growth rate with observations on cellular development inside the organ led to the conclusion that the sporogenous tissues in lily anthers are characterized by a series of up to four nearly equally spaced rounds of nuclear division (i.e., two pre-meiotic mitoses, meiosis, and microspore mitosis) (Erickson 1948). The reasonably high degree of cell division synchrony in meiosis and microspore mitosis seen in lily anthers is a feature of anther development that can be generalized to other angiosperms (Erickson 1964; but see Burns 1972; Gould & Lord 1988; Schrauwen et al. 1988). However, the suggestion that mitotic divisions in sporogenous tissues preceding meiosis may also be partially synchronized has not been very thoroughly investigated. The results from lily suggest that archesporial cell division at the beginning of anther differentiation may be when some degree of synchrony is first established. These early phases of anther development were not included in Erickson's (1948) work, since floral

buds containing anthers at these stages of development were inaccessible for growth studies.

Correlations between the development of neighboring flowers within a spike have been used to determine the timing of reproductive events in *Triticum* (Bennett et al. 1973). Adjacent spikelets within a developing spike were harvested up to 5 days apart and the developmental changes in anthers from the first to the second sampling times were analyzed. Early development of the sporogenous tissue in each anther lobe produces a single file of cells about 10 cells in length. An analysis of sporogenous cell division activity leading to the formation of approximately 100 microsporocytes showed (1) that approximately three successive rounds of cell division occur prior to meiosis; (2) that the cell cycle time in premeiotic sporogenous cells increases with each successive division; (3) that the volume of individual sporogenous cells increases throughout the period of sporogenous cell development; and (4) that pre-meiotic sporogenous tissue exhibits low-amplitude mitotic peaks similar to those observed in lily.

The method of analysis for the study of *Triticum* sporogenous cell development was based on direct observations of these cells after columns of cells were extruded from ruptured anthers (Bennett et al. 1973). The timing of earlier phases of anther development, including archesporial cell division and early sporogenous cell division, was not quantified.

The temporal patterns of anther development can also be assessed with the plastochron index (Erickson & Michelini 1957). A plastochron is the amount of time between the initiation of successive organs during a shoot's growth. The plastochron index method combines nondestructive observations of shoot and/or inflorescence growth on one set of plants while another set of plants growing under the same conditions are sacrificed for microscopic study. This allows the age of successively smaller buds in an inflorescence to be inferred (Erickson 1976; Hill & Lord 1990a; Lord et al. 1994).

In tobacco, the flower plastochron can be estimated from observations on the timing of corolla expansion in successive flowers in a cyme (Table 1; Hill & Malmberg 1991). Previous analysis indicated that variation in the plastochron along the cyme was due primarily to variation in the rate of flower initiation, not the rate of flower growth. The tendency for the plastochron to increase in

Table 1. *The average plastochron in two cymes of* Nicotiana tabacum

Flower position	Plastochron (days)
2	1.82±0.23
3	1.31±0.35
4	1.51±0.11
5	1.83±0.01
6	1.56±0.34
7	1.53±0.22
8	1.84±0.59
9	2.48±0.15

Growth measurements used for calculations were from cymes on different plants. Flowers are numbered sequentially in the order of their appearance in each cyme. Plants were grown at 23 °C under continuous illumination. Plastochron values are in days±S.D.

later flower positions is probably related to seed set in earlier flowers.

Although knowledge of the age difference between successive flowers in the tobacco cyme allows the rate of early stamen and anther development to be inferred, providing useful estimates of developmental timing in primordia becomes more complicated if accurate correlations between morphological and anatomical rates of development are sought. Samples that must be sacrificed for morphological study using the scanning electron microscope (SEM) may not be at exactly the same developmental stage as those sacrificed for histological study. This difficulty can be solved by combining both types of observations on the same tissue sample. The mold cast procedure (Green and Linstead 1990) can be utilized to prepare epoxy samples for SEM (Fig. 3), and the same tissue samples can be subsequently processed for conventional histology (Fig 4A–C).

A representative example shows that when the ninth flower of a cyme has reached the point of stamen initiation (Figs. 3A, 4A, Table 2), the eighth flower which is 2.5 days older (Fig. 3B) has five stamen primordia approximately 110 μm in length. Differentiation of the microsporangia is not yet evident (Fig. 4B, Table 2). Stamens from the seventh flower are 1.8 days older and have reached approximately 360 μm in length. The

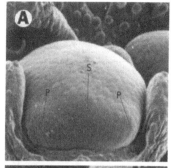

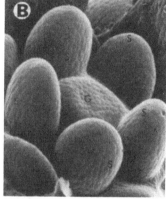

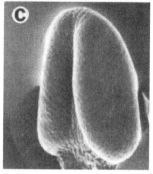

Fig. 3. Morphological development of stamen primordia in *Nicotiana tabacum*. All scanning electron micrographs are of epoxy casts made from flower primordia or individual stamens from consecutive flowers within a single cyme. A. Ninth flower produced in the cyme at the time of stamen primordial emergence (S) from the floral meristem, alternate to the petal lobes (P). B. Eighth flower produced in the cyme. Five stamen primordia are present and the gynoecium (G) has begun to develop. C. A single anther from the seventh flower produced in the cyme. The front locules are clearly differentiated.

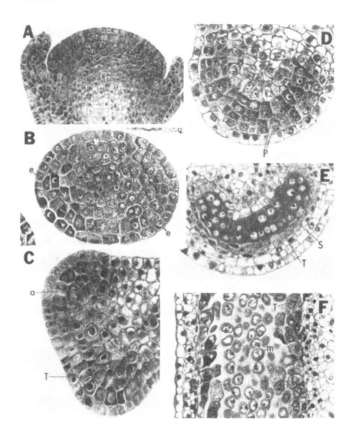

Fig. 4. Histological development of six consecutive flower/stamen primordia within a single tobacco cyme. All sections stained with toluidine blue. B–E are transverse views and the front (adaxial) side of the stamen is towards the bottom; A and F are longitudinal. A. Section through the floral apex at the stamen initiation phase. The same flower shown in Fig. 3A. B. Section through a stamen primordium. L_2 cells beneath the labelled epidermal cells (e) have divided periclinally. The flower with this stamen is shown in Fig. 3B. C. Tobacco anther with archesporial initials (a) present. Adjacent cells in L_3 will develop into the inner tapetum (T). The same anther is shown in Fig. 3C. D. Sixth flower produced in the cyme. The anther wall, excluding the epidermis, is two cell layers deep (P), as is the sporogenous tissue (s). The inner tapetum (T) has become evident. E. The fifth flower produced in the cyme. Sporogenous tissue (S) is near the end of mitotic proliferation and the outer tapetum (T) is now evident. The anther wall is now generally four or five layers thick. F. The fourth flower produced in the cyme. Microsporocytes (m) have entered the first meiotic division, and tapetal cells (T) are binucleate.

anthers of this seventh flower primordium have begun to develop the characteristic shape of the tetrasporangiate organ (Fig. 3C). Within these organs, archesporial initials are evident in the L_2 layer (Fig. 4C, Table 2). The data from this cyme indicate the total interval between stamen initiation and the start of meiosis in the anther is about 7.4 days (Table 2). Sampling additional cymes with flower ages that differ from those described here will provide a more continuous picture of the temporal pattern of anther development; this work is in progress.

Table 2. *Stages of premeiotic stamen and anther development at successive flowers in a cyme in* Nicotiana tabacum

Flower position	Stamen age (days)	Length (μm)	Developmental stage
9	0	—	Petal primordia initiated; stamen initiation just beginning
8	2.48	111	Stamen primordial development in progress; radial expansion of adaxial (front) anther locules not yet evident morphologically
7	4.32	360	Archesporial initials evident in L_2
6	5.85	672	Anther wall (excluding epidermis) two cell layers wide; sporogenous layer two cell layers wide; inner tapetum differentiating
5	7.41	1221	Anther wall four cell layers wide; sporogenous tissue two or three cell layers wide, and leptonema of first meiotic prophase in progress; inner and outer tapetal layers present
4	9.24	2300	First meiotic division in microsporocytes in progress; tapetum cells binucleate

Flower ages were determined with the plastochron index.

DIRECT OBSERVATION OF ANTHER GROWTH

The approaches described thus far for evaluating the timing of anther development rely mainly on correlated patterns of plant growth. Stamen growth is never actually observed. In this context, anatomical details describing planes of cell division have been given a disproportionate role in identifying regions of organ growth (Poethig 1984). The aforementioned phase of apical growth, corresponding to the appearance of periclinal divisions in L_2, represents an example of inferring growth patterns from anatomical patterns. Correspondence between diagnostic anatomy and zones of growth are possible when organ growth is indeterminate and steady, such as in roots (Silk *et al.* 1989). The complex morphology of anthers, along with their determinate growth pattern, suggests that the relationship between organ expansion and cellular development will change as the organ matures.

The problem of obtaining direct observations on anther growth has been approached in lily, which has extremely large stamens (Gould & Lord 1988). By removing part of the perianth and applying marks to the surface of developing anthers, the growth of individual organs was viewed for as many as four consecutive days. The results indicated that the anther is characterized by at least one, and sometimes more than one, peak in growth rate along its length; in older anthers, waves of growth were observed to propagate down the organ. In pre-meiotic anthers, troughs in growth may correspond to peaks in mitotic activity, while peaks in growth are due to expansion of interphase cells (Gould & Lord 1988). A distinction should be made between the spatial propagation of mitotic peaks discussed here and the low amplitude mitotic peaks mentioned in earlier studies (Erickson 1948; Bennett *et al.* 1973). Those studies refer to the average amount of mitotic activity throughout the sporogenous tissue over time, without any reference to spatial variation along the length of an anther.

The smallest anther size class analyzed was from 1.1 mm to 2.9 mm in length (Gould & Lord 1988). Secondary parietal and sporogenous tissues are already present at the low end of this size range. Although the peak in relative growth was most frequently located near the tip of these anthers, local maxima in mitotic divisions occurred at different locations along sectioned organs. This indicates that the propagation of cell division/

expansion waves down the organ may be important at the earliest stages of development examined, which includes the final phases of mitotic activity in the parietal and sporogenous cell lineages. It is not clear to what extent these patterns observed in lily will generalize to other species where stamen primordia are substantially smaller in size.

As in Erickson's (1948) previous study of anther development in lily, earlier phases were inaccessible for growth analysis. The mold-cast method, which allows growing meristematic tissues to be viewed sequentially (Williams & Green 1988; Hernandez *et al.* 1991), has potential for assessing stamen and anther growth during these early phases by means of direct observation. The disposition of tobacco stamen primordia on the floral meristem (Fig. 3B; Rosenberg & Bonnett 1983) relative to the onset of anther histogenesis (Fig. 4, Table 2) suggests that surface growth data can be collected from these organs while archesporial cell divisions are occurring. Preliminary results show that stamen development can be tracked for several days after the organs initiate from the floral meristem (Fig. 5A, B). Inspection of epidermal cell patches on the stamen surface indicate substantial spatial differences exist in cell division patterns (Fig. 5C, D) and surface expansion patterns. Further work to characterize these phases of development is in progress. The results will allow rates of early stamen development to be quantified by direct observation instead of by inference using the plastochron index. More accurate estimates of developmental rates will help to indicate conclusively whether mitotic divisions during early anther development become synchronized prior to meiosis, as suggested for lily. This technique will also show whether waves of cell division and expansion, similar to those seen in later stages of lily anther development, characterize the early phases of stamen ontogeny.

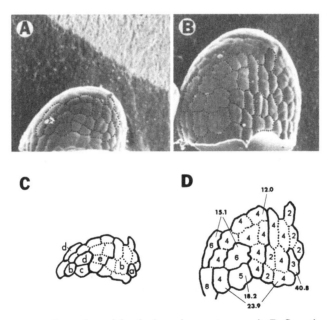

Fig. 5. Cell division patterns on the surface of developing tobacco stamens. A, B. Scanning electron micrographs of epoxy casts of the same stamen made one day apart. Identical cell patches are outlined in each figure; in A, patches initially contain one or two cells. C. Same cell patches shown in A. Heavy lines enclose regions, identified by lower case letters, which subsequently showed similar cell cycling rates. D. Same cell patches shown in B, and the same heavy lines depicted in C. Numbers inside each cell patch indicate the final cell number when the second cast was made. Numbers outside the map indicate the average cell cycle time in hours within regions a through e. The most rapid cell cycling occurred in regions d (15.1 hours) and e (12.0 hours).

ANTHER EVOLUTION: A TEST FOR HETEROCHRONY

The extended, often characteristic patterns of cell division leading to the formation of microsporangia in the anther provide an excellent set of natural developmental markers during anther ontogeny in angiosperms. The conservative nature of these events, reflecting their fundamental importance in reproductive biology, make them well suited to unambiguous phylogenetic comparisons. In spite of their conservative nature, it is also quite clear that anther histogenesis is subject to evolutionary change. Research on plant mating systems and pollination biology has played an important part in identifying significant patterns of variation in anther form within and among individuals of natural populations (Ornduff 1969; Lloyd & Webb 1986; Webb & Lloyd 1986; Barrett 1988; Wyatt 1988; Young & Stanton 1990; Svensson 1992). Variation in pollen output in the context of breeding system evolution is expected on theoretical grounds (Charlesworth & Charlesworth 1981), and observed empirically across a wide range of angiosperm taxa. Furthermore, variation in pollen output is often modulated by heritable changes in pollen production per anther, indicating that aspects of pre-meiotic anther differentiation have been modified (Richards & Barrett 1984, 1992).

Detailed comparative studies of anther ontogeny have been conducted on several cleistogamous species (reviewed in Lord & Hill 1987). In all cases, the anther in self-pollinating cleistogamous flowers is considered to have evolved from a larger organ found in chasmogamous flowers. The point of developmental divergence between stamens in cleistogamous and chasmogamous flowers differs in three species that have been analyzed. In *Viola odorata*, which has the most divergent floral dimorphism, the differences in development are evident before stamens are initiated (Mayers & Lord 1983a). A growth analysis suggests that all phases of development in the cleistogamous flower of *Viola* are speeded up compared with the chasmogamous floral forms (Mayers & Lord 1983b). *Collomia grandiflora* represents an intermediate case. The cleistogamous anther differs in shape from its chasmogamous counterpart, and does not initiate the two front (adaxial) locules (Lord *et al.* 1989). Stamen primordial development is indistinguishable in

the two flower types until a precocious onset of archesporial division occurs in the closed flower form. Acceleration of the onset of this differentiation event was estimated to be about 2 days (Minter & Lord 1983; Lord & Hill 1987). In *Lamium amplexicaule*, the developmental differences between anther forms also arise during premeiotic anther development; after archesporial cell division, the cleistogamous anther produces fewer pollen mother cells and about one fourth as much pollen as chasmogamous anthers (Lord 1979).

Although the two anther forms in *Lamium* have diverged by the start of pollen mother cell meiosis, no differences in timing leading up to this divergence could be detected. In *Collomia*, there is a larger (10-fold) difference in pollen output between the two anther forms than is seen in *Lamium*. The relatively modest decline in pollen number in the cleistogamous *Lamium* anther may be driven by equally modest changes in developmental timing that could not be detected.

Populations of *Arenaria uniflora* exhibit changes in floral form in association with the evolution of self-pollination that are not unlike the morphological distinctions between cleistogamous and chasmogamous flowers found on a single plant (Wyatt 1984; Hill & Lord 1990b). Given the existing evidence to suggest a heterochronic origin of cleistogamous floral forms from chasmogamous ones, a comparative study of flower development was conducted to assess whether the morphological divergence among these populations could also be attributed to heterochrony (Hill 1989). With respect to the cleistogamous species, phenotypic modifications of anthers in *Arenaria* most closely resemble those in *Lamium*: the organs from self-pollinating flowers have a slight (2-to 3-fold) reduction in pollen output and morphologically resemble dwarfs of the anthers of outcrossing flowers.

A comparison of flower development in representative self-pollinating and outcrossing populations showed that the early development of floral meristems was the same in terms of the size and shape (Hill 1989). In particular, no early differences in stamen primordial size or shape were detected (Hill & Lord 1990b). As noted above, stamen initiation in general is followed by an early phase of cellular development characterized by proliferative mitotic activity that enlarges the primordium above the floral apex. Histologically, the end of this phase is marked by the differentiation and subsequent division of archesporial initials in L_2. The point of ontogenetic divergence between anthers from self-pollinating and outcrossing anthers of *Arenaria* was first evident at the end of this proliferative growth stage. Anthers from self-pollinating flowers reached the archesporial cell division stage when stamen primordia were smaller than comparable primordial stages in outcrossing floral morphs.

The onset of archesporial cell division in smaller stamen primordia in the self-pollinating flowers suggested that a key modification responsible for the evolution of the small anthers from large ones was a shift in the time when microsporangial differentiation begins in relation to the overall pattern of primordial growth. However, these observations alone cannot resolve whether the derived ontogeny of the small anthers represents (1) a temporal acceleration of archesporial cell differentiation to an earlier time (i.e., progenesis), or (2) a reduced rate of stamen primordial growth such that stamen primordia in self-pollinating flowers are smaller when the phylogenetically unaltered time for archesporial cell division arrives (i.e., neoteny). Nor can it be assumed that these two features are mutually exclusive events. Such distinctions, which would help to clarify the possible mechanistic cause(s) in ontogeny responsible for the derived phenotype, require knowledge of the relationship of anther development to time in the two flower forms.

Growth analysis of floral buds from outcrossing and self-pollinating populations of *Arenaria* using the plastochron index indicated that the self-pollinating forms grew more slowly than the outcrossing forms throughout their development (Hill *et al.* 1992). However, the use of the plastochron index to estimate developmental times of meristematic growth in *Arenaria* could not provide conclusive evidence of the exact nature of timing differences for two principle reasons. First, the high variance of the plastochron in the inflorescences self-pollinating populations meant that the estimation of rates of flower development in microscopic primordia was prone to substantial errors. Second, if a shift in the timing of archesporial division was responsible for the evolution of the small anther, the absolute difference in timing that now exists is small, and therefore difficult to detect. In this respect, *Arenaria* resembles *Lamium*, where temporal differences in development could not be detected until after morphological differences were visible.

Current problems in anther development

Important gaps remain in our understanding of anthers concerning first, the features of pre-meiotic development in general terms, and secondly, the evidence for heterochrony. For example, molecular and cellular characterizations of stamen- and anther-specific gene expression have provided a wealth of information primarily on late stages of organ development, usually beginning with pollen mother cells present (Koltunow *et al.* 1990; Mascarenhas 1992; Scott *et al.* 1991; Goldberg *et al.* 1993). An equivalent characterization of molecular aspects of development between the time of stamen initiation and pollen mother cell differentiation has not been forthcoming. This probably reflects some combination of (1) the practical limitations of acquiring sufficient volumes of extremely small tissue samples for analysis of these early stages; (2) an explicit research interest in later phases of microsporogenesis; and (3) the tendency for the greatest degree of organ specific gene expression to occur at later phases of development when a variety of anther cells undergo terminal differentiation. Nevertheless, at least a small number of unique polypeptides appear in association with the initiation and early development of stamens (Rembur *et al.* 1992; Wang *et al.* 1992), and modified expression of plasma membrane arabinogalactan proteins are known that anticipate the differentiation of germ line cells in pea (Pennell & Roberts 1990). Stamen-specific mitochondrial gene functions have also been implicated in pre-meiotic anther development (Hicks *et al.* 1977; Rosenberg & Bonnett 1983; Bonnett *et al.* 1991).

Adequate characterizations of temporal patterns of early stamen and anther development have also been difficult to obtain, both in the context of basic studies of organ development and in the context of comparative studies to document heterochrony. Again, perhaps due mainly to practical limitations, characterizations of real-time development of the earliest phases of growth have not been readily available. Attempts to utilize correlations between floral bud growth and anther development have fallen short, because buds cannot be observed during the early microscopic phases. In the examples of heterochrony cited above, these are precisely the stages that may be subject to temporal change. Thus far, the plastochron index has not provided the precision necessary to evaluate the nature of these small changes in timing during anther evolution.

It is also noteworthy that the histological details of early anther differentiation show significant variation between species, and some ambiguities still exist as to exactly how the microsporangium develops in certain cases (Kaussmann 1941; Brunkener 1975; Bhandari & Khosla 1982; Johri 1984). Differences at these early stages of differentiation may affect how anther ontogeny changes phylogenetically in different species.

POSSIBLE MECHANISMS FOR HETEROCHRONY IN THE ANTHER

From both developmental and evolutionary perspectives, a key component of the modification of anthers in association with the evolution of self-pollination is a reduction in cell number (e.g., pollen grains) in the organ. Irrespective of the absolute changes in developmental timing that may be involved, the details of comparative studies of anther ontogeny already show that at the cell and tissue level, either the rate of cell division, or the rate of organ expansion, or the timing of differentiation relative to cell division and expansion, has changed. Post-meiotic abortion of gametophytic cells is not a mechanism for reducing pollen output in the cases studied thus far. These facts suggest that unambiguous conclusions about the mechanisms leading to heterochrony in anthers must be based on more accurate information about rates of cell partitioning and organ expansion during pre-meiotic phases of development. The most likely means to obtain these data will utilize a species like tobacco that allows accurate comparisons of organ growth over time (Fig. 5) to be made on stamen primordia that have evolved divergent ontogenies in association with breeding system variation (Breese, 1959).

To show the relationship between different levels of control over pollen production (P) in a flower, Richards & Barrett (1992) have developed the equation

$$P = (ALS)\,(2^n)\,(4),$$

where A is the number of anthers per flower, L is the number of pollen locules per anther, S is the number of primary sporogenous cells per locule, and n is the number of times each sporogenous cell divides mitotically preceding meiosis. The product of these vari-

ables is increased by a factor of four because each microspore mother cell produces four pollen grains.

In many species, $(4AL)$ is essentially constant, so most of the variation in pollen production arises from changes in the number of archesporial initials and/or the number of pre-meiotic mitoses of sporogenous cells. Thus, the number of microsporocytes (M) produced per anther locule is simply

$$M = 2^n S.$$

If the duration of sporogenous cell divisions is known, the element of time can be incorporated to estimate developmental rates. Using functional notation, the mean absolute rate ($\overline{G}$) at which sporogenous cell number increases within a single locule is

$$\overline{G} = \frac{(2^{n_{t_f}})\,(S_0) - (S_0)}{(t_f - t_0)},$$

where t_0 is the time archesporial cell division ends, t_f is the time sporogenous cell mitosis ends, S_0 is the number of primary sporogenous cells at t_0, and n_{t_f} is the number of times each sporogenous cell divided mitotically during the interval $t_f - t_0$. This absolute growth rate can be converted to a relative growth rate by taking logarithms (Hunt, 1982). The mean relative rate of sporogenous cell increase in a single locule ($\overline{R}$) is

$$\overline{R} = \frac{(n_{t_f})\,\ln(2)}{(t_f - t_0)}.$$

This equation shows that the relative rate of sporogenous cell increase is independent of S_0, and depends instead on sporogenous cell cycle time and the total duration of sporogenous cell division. More appropriate rate functions could be developed to reflect periodicity in sporogenous cell divisions and differences in duration of successive division cycles of sporogenous cells.

These equations point out that changes in the number of primary sporogenous cells can alter the number of microsporocytes, but alterations in mitotic patterns or the duration of cell division during sporogenous tissue development will have a greater impact on pollen production (Richards & Barrett 1992). To model sporogenous tissue development more realistically, it may be useful to quantify S and n separately in different regions within a single locule (e.g., the anther tip versus the base). Also, non-integer values of n can represent mitotic activity that is not evenly distributed among developing sporogenous cells. For example, an n value of 1.585 implies each primary sporogenous cell produces three microsporocytes. This could occur if division of each primary sporogenous cell produced one secondary sporogenous cell that divided again and another one that did not. Quantitative studies defining these parameters of early anther development will clarify which of these aspects of cellular development are most likely to change phylogenetically.

Cell cycle mutations: a source of heterochrony at the cellular level

The cell cycle mutations of fission yeast (Lee & Nurse 1988) have been proposed as a clear case where heterochrony observed at the phenotypic level is due directly to a genetic mechanism involving timing controls during development (Raff & Wray 1989; Lord et al. 1994). Single gene mutations cause yeast cells to divide at smaller or larger than normal cell sizes. Since mitosis essentially constitutes reproductive maturity in these single-celled organisms, alterations in the size at which yeast cells undergo division have been interpreted as progenesis (division at an earlier time than normal) and hypermorphosis (division at a later time than normal).

The significance of these heterochronic mutations in yeast relates to the demonstration of a heritable dissociation between cell growth and cell division. The cell division cycle and cell growth cycle are normally coupled during development (Barlow 1969; Mitchison 1971; Baserga 1984; Cuadrado et al. 1989; Baroni et al. 1992). For example, nutrient-shift experiments show that fast-growing yeast cells are relatively large in size at division when nutrient availability is high. When nutrients are depleted, cells slow their growth and also begin dividing at a new, smaller cell size (Fantes & Nurse 1977). A dissociation between growth and division is seen in the wee 1–50 temperature sensitive mutant, where cells divide precociously at a smaller size but grow at the same rate as wild type strains (Nurse 1975).

In the wee 1–50 mutant, the duration of the cell cycle that is in progress at the time the cells are temperature-shifted is shortened. Interestingly, this is the only cell cycle that is shortened; subsequent cell cycle times of small mutant cells maintained at the shifted temperature are the same duration as wild type cells. The shortened cell cycle time is a transitory effect because the mutation alters cell size at division, and not cell cycle time or the rate of cell growth, per se. The features of this mutant nicely illustrate how one small change in timing can contribute to a new, stable phenotype.

In plant tissues, cell division and growth can be associated in a variety of ways, thereby affecting the size of cells observed both within meristematic and nonmeristematic tissues (Green 1976; Kaplan & Hagemann 1991). In higher plants, cell size has been observed to be correlated with the probability that meristematic cells will initiate differentiation (Cuadrado et al. 1987). Studies with root meristems indicate that cell division, expansion, and differentiation can be coordinated to produce a determinate organ as a normal course of development (Gunning et al. 1978). In indeterminate roots, the normal balance between growth and the timing of cell differentiation can be altered by mutation (Harte & Maek 1976), sometimes leading to a new determinate growth pattern (Benfey et al. 1993).

Comparative morphological studies of heteroblastic variation in homologous organs produced sequentially along a shoot have repeatedly shown that differences in final organ form often can be attributed to changes in timing of the onset of differentiation events relative to patterns of cell division and organ expansion (Foster 1935; Kaplan 1973; Richards 1983; Dengler 1992). Heteroblastic differences can be attributed in some part to the developmental age of an individual shoot (Allsopp 1967). However, interpretations of heteroblastic differences due to differences in shoot age are often complicated by significant changes in apical size that accompany changes in organ form (Richards 1983). The changing relationship of cell division, expansion, and differentiation patterns behind the meristem of *Azolla* roots (Gunning et al. 1978) is significant in this respect, since the root tip does not change in size with increasing age. Thus, it is reasonable to expect that mutations that alter the relationships between cellular development and differentiation could produce changes in ontogeny between individuals (e.g., the evolution of the anther in self-pollinating flowers) that are similar to the kinds of developmental variation seen ontogenetically within heteroblastic shoots (e.g., the evolution of the anther in cleistogamous flowers).

The importance of cell cycle controls in plant development continues to be emphasized in the context of increasing knowledge about cell cycle genes (reviewed in Jacobs 1992; Francis 1992). The most extensive studies of cell cycle changes in relation to morphogenesis are those on the floral transition (Bernier 1988). Although a partial synchronization of cells in the shoot apex is frequently observed, flowering can occur in the absence of such changes. A possible explanation is that the synchronization normally produces a population of cells competent to respond to the floral stimulus at the apex, but compensatory changes can occur that still allow flowering to proceed if cell division synchrony is prevented (Bernier 1988).

Following the floral transition, the stimulation of cell division just prior to floral morphogenesis may reflect a general activation of cells that subsequently become involved in the formation of new organs. Alternatively, these cell divisions may be a more specific means of delineating groups of cells as subpopulations that will function as organ initials (Francis 1992). For example, Francis & Herbert (1993) have recently suggested an interaction between floral homeotic genes and cell cycle genes that could produce the temporal sequence of primordium initiation at the flower apex. Cell cycling times and coordinated changes in cell size at division may show oscillations in successive organ whorls of the flower primordium. Seeking evidence for this hypothesis will require attention to the aforementioned relationships between cell division, cell growth, and time. As indicated by yeast, small cells need not have a persistently shorter cell cycle time compared with large cells (Nurse 1975). Similarly, oscillations in cell size (growth without division; Green 1976) need not imply cell cycle time periodicity as well. Cell cohorts could therefore arise in a number of ways, and may be an important feature of many aspects of development in both plants and animals.

CONCLUSIONS

As investigations into specific examples of morphological heterochrony seek mechanistic explanations at cellular and molecular levels, many examples will not be attributable to explicitly temporal controls over development (Raff & Wray 1989; Dengler 1992). For example, absolute changes to cell cycle time may indirectly alter the relative expression patterns of genes based on their size, if large genes generally require a longer interval prior to mitosis to be completely transcribed (Shermoen & O'Farrell 1991). There are inherent difficulties associated with applying the concept of heterochrony across different levels of developmental organization because many modifications to development may lead incidentally to changes in ontogenetic timing at the organismal level. Nevertheless, efforts to analyze the mechanistic causes of morphological heterochronies have already served to better integrate results derived from molecular, developmental, and evolutionary biology, and to re-focus attention on the importance of time in ontogeny.

Research on the plant cell cycle is likely to play an important part in the search for underlying causes for the heterochronic shifts in anther ontogenies seen at the morphological level. Direct changes to the cell cycle are expected either to produce correlated changes in the growth and differentiation of meristematic tissue, or to create dissociations between these different aspects of development. The evolution of microsporangial differentiation in the anther may be ideal for characterizing the changing relationship between cell division, expansion and differentiation. In many ways, the development of the microsporangium resembles morphogenetic events seen in organogenesis at the shoot apex (Lyndon 1983), including an increased rate of cell division and an apparent change in the polarity of growth. Change in cell size, which may reflect one means of isolating a specific cohort of dividing cells for subsequent development, is commonly observed in archesporial initials. The difficulties of characterizing the cell cycle behavior of a large population of cells in the shoot apex during the floral transition are mitigated since the sporangial cell population is known to originate generally in L_2 of the stamen primordium. As more information about the genetic control of the plant cell cycle becomes available, it should be possible to manipulate cell division experimentally during critical phases of stamen morphogenesis to test hypotheses about growth, the timing of differentiation, and heterochrony in the angiosperm anther.

ACKNOWLEDGEMENTS

I thank E. Lord for helpful discussions and V. Raghavan for comments. I especially thank J. Richards for her thoughtful input, P. Green for valuable advice regarding the mold-cast method, and V. Winston for exhuming the SEM. This research was partially supported by Idaho State Board of Education Grant S 94–059 and a grant from the University Research Committee, Idaho State University, Pocatello, Idaho.

LITERATURE CITED

Allsopp, A. 1967. Heteroblastic development in vascular plants. Adv. Morphogenesis 6: 127–171.

Alberch, P., S.J. Gould, G.F. Oster & D.B. Wake. 1979. Size and shape in ontogeny and phylogeny. Paleobiology 5: 296–317.

Arber, A. 1937. The interpretation of the flower: A study of some aspects of morphological thought. Biol. Rev. 12: 157–184.

Barlow, P.W. 1969. Cell growth in the absence of division in a root meristem. Planta 88: 215–223.

Baroni, M.D., P. Monti, G. Marconi & L. Alberghina. 1992. cAMP-mediated increase in the critical cell size required for the G1 to S transition in *Saccharomyces cerevisiae*. Exp. Cell. Res. 201: 299–306.

Barrett, S.C.H. 1988. Evolution of breeding systems in *Eichhornia* (Pontederiaceae): a review. Ann. Missouri Bot. Gard. 75: 741–760.

Baserga, R. 1984. Growth in size and cell DNA replication. Exp. Cell. Res. 151: 1–5.

Benfey, P.N., P.J. Linstead, K. Roberts, J.W. Schiefelbein, M-T. Hauser & R.A. Aeschbacher. 1993. Root development in *Arabidopsis*: four mutants with dramatically altered root morphogenesis. Development 119: 57–70.

Bennett, M.D., M.K. Rao, J.B. Smith & M.W. Bayliss. 1973. Cell development in the anther, the ovule, and the young seed of *Triticum aestivum* L. var. *Chinese Spring*. Phil. Trans. R. Soc. Lond. B 266: 39–81.

Bernier, G. 1988. The control of floral evocation and morphogenesis. Annu. Rev. Plant Physiol. Plant Molec. Biol. 39: 175–219.

Bhandari, N.H. & R. Khosla. 1982. Development and histo-chemistry of anther in *Triticale* cv Tri-1. I. Some new aspects in early ontogeny. Phytomorphology 32: 18–27.

Blackstone, N.W. 1987. Size and time. Syst. Zool. 36: 211–215.

Boke, N.H. 1949. Development of the stamens and carpels in *Vinca rosea* L. Amer J. Bot. 36: 535–547.

Bonner, J.T. 1982. Evolution and Development. Dahlem Workshop on Evolution & Development Springer-Verlag, New York.

Bonnett, H.T., W. Kofer, G. Hakansson & K. Glimelius. 1991. Mitochondrial involvement in petal and stamen development studied by sexual and somatic hybridization of *Nicotiana* species. Plant Sci. 80: 119–130.

Breese, E.L. 1959. Selection of differing degrees of out-breeding in *Nicotiana rustica*. Ann. Bot. 23: 331–344.

Brunkener, L. 1975. Beitrage zur Kenntnis der fruhen Mikrosporangienentwicklung der Angiospermen. Svensk Bot. Tidskr. 69: 1–27.

Burns, J.A. 1972. Preleptotene chromosome contraction in *Nicotiana* species. J. Heredity 63: 175–178.

Charlesworth, D. & B. Charlesworth. 1981. Allocation of resources to male and female functions in hermaphrodites. Biol. J. Linn. Soc. 15: 57–74.

Cheng, P.C., R.I. Greyson & D.B. Walden. 1979. Comparison of anther development in genic male-sterile (ms10) and in male-fertile corn (*Zea mays*) from light microscopy and scanning electron microscopy. Can J. Bot. 57: 578–596.

Cheung, M. & R. Sattler. 1967. Early floral development of *Lythrum salicaria*. Can. J. Bot. 45: 1609–1618.

Coen, E.S. 1991. The role of homeotic genes in flower development and evolution. Annu. Rev. Plant Physiol. Plant Molec. Biol. 42: 241–279.

Coen, E.S. 1991. The role of homeotic genes in flower development and evolution. Annu. Rev. Plant Physiol. Plant Molec. Biol. 42: 241–279.

Cuadrado, A., J.L. Canovas & M.H. Navarrete. 1987. Influence of cell size on differentiation of root meristem cells. Envir. Exp. Bot. 27: 273–277.

Cuadrado A., M.H. Navarrete & J.L. Canovas. 1989. Cell size of proliferating plant cells increases with temperature: implications in the control of cell division. Exp. Cell. Res. 185: 277–282.

Davis, G.L. 1966. Systematic Embryology of the Angiosperms. Wiley, New York.

Dawe, R.K. & M. Freeling. 1990. Clonal analysis of the cell lineages in the male flower of maize. Dev. Biol. 142: 233–245.

Dawe, R.K. & M. Freeling. 1992. The role of initial cells in maize anther morphogenesis. Development 116: 1077–1085.

Dengler, N.G. 1972. Ontogeny of the vegetative and floral apex of *Calycanthus occidentalis*. Can. J. Bot. 50: 1349–1356.

Dengler, N.G. 1992. Patterns of leaf development in aniso-phyllous shoots. Can. J. Bot. 70: 676–691.

Diggle, P. 1992. Development and the evolution of plant reproductive characters. In Ecology and Evolution of Plant Reproduction: New Approaches (edited by R. Wyatt). Chapman and Hall, New York.

Doyle, J. 1993. Cladistic and paleobotanical perspectives on the origin of angiosperm organs. J. Cell. Biochem. Suppl. 17B: 8.

Eames, A.J. 1961. Morphology of the Angiosperms. McGraw-Hill, New York.

Erickson, R.O. 1948. Cytological and growth correlations in the flower bud and anther of *Lilium longiflorum*. Amer. J. Bot. 35: 729–739.

Erickson, R.O. 1964. Synchronous cell and nuclear division in tissues of the higher plants. pp. 11–37 in Synchrony in Cell Division and Growth (edited by E. Zeuthen). Wiley Interscience, New York.

Erickson, R.O. 1976. Modeling of plant growth. Annu. Rev. Plant Physiol. 27: 407–434.

Erickson, R.O. & F.J. Michelini. 1957. The plastochron index. Amer. J. Bot. 44: 297–305.

Fantes, P. & P. Nurse. 1977. Control of cell size at division in fission yeast by a growth-modulated size control over nuclear division. Exp. Cell Res. 107: 377–386.

Foster, A.S. 1935. A histogenetic study of foliar determination in *Carya buckleyi* var. *arkansana*. Amer. J. Bot. 22: 88–147.

Francis, D. 1992. The cell cycle in plant development. New Phytol. 122: 1–20.

Francis, D. & R.J. Herbert. 1993. Regulation of cell division in the shoot apex. pp. 201–210 in Molecular and Cell Biology of the Plant Cell Cycle (edited by J.C. Ormrod & D. Francis) Plant Growth Regulation, Supplement. Kluwer, Dordrecht.

Gifford, E.M. & A. Foster. 1989. Morphology and Evolution of Vascular Plants, 3rd ed. W.H. Freeman, New York.

Goldberg, R.B., T.P. Beals & P.M. Sanders. 1993. Anther development: basic principles and practical applications. Plant Cell 5: 1217–1229.

Gould, K.S. & E.M. Lord. 1988. Growth of anthers in *Lilium longiflorum*. Planta 173: 161–171.

Gould, S.J. 1977. Ontogeny and Phylogeny. Harvard University Press, Cambridge, Mass.

Green, P.B. 1976. Growth and cell pattern formation on an axis: critique of concepts, terminology, and mode of study. Bot. Gaz. 137: 187–202.

Green, P.B. & P. Linstead. 1990. A procedure for SEM of complex shoot structure applied to the inflorescence of snapdragon (*Antirrhinum*). Protoplasma 158: 33–38.

Guerrant, E.O. 1988. Heterochrony in plants: the intersection of evolution, ecology, and ontogeny. pp. 111–133 in Heterochrony in Evolution (edited by M.L. McKinney) Plenum Press, New York.

Gunning, B.E.S., J.E. Hughes & A.R. Hardham. 1978. Formative and proliferative cell divisions, cell differentiation, and developmental changes in the meristem of *Azolla* roots. Planta 143: 121–144.

Hall, B.K. 1990. Heterochrony in vertebrate development. Semin. Dev. Biol. 1: 237–244.

Harte, V.C. & A. Maek. 1976. Genabhangigkeit des Wachstums pflanzlicher Meristeme untersucht an der Entwicklung isolierter Wurzeln verschiedener Genotypen von *Antirrhinum majus* L. Biol. Zbl. 95: 267–299.

Hernandez, L.F., A. Havelange, G. Bernier & P.B. Green. 1991. Growth behavior of single epidermal cells during flower formation: sequential scanning electron micrographs provide kinematic patterns for *Anagallis*. Planta 185: 139–147.

Hicks, G.S. 1973. Initiation of floral organs in *Nicotiana tabacum*. Can. J. Bot. 51: 1611–1617.

Hicks, G.S., J. Bell & S.A. Sand. 1977. A developmental study of the stamens in a male-sterile tobacco hybrid. Can. J. Bot. 55: 2234–2244.

Hill, J.P. 1989. Homeosis, heterochrony, and the evolution of floral form. Ph.D. Dissertation, University of California, Riverside.

Hill, J.P. & E.M. Lord. 1989. Floral development in *Arabidopsis thaliana*: a comparison of the wild type and the homeotic *pistillata* mutant. Can. J. Bot. 67: 2922–2935.

Hill, J.P. & E.M. Lord. 1990a. A method for determining plastochron indices during heteroblastic shoot growth. Amer. J. Bot. 77: 1491–1497.

Hill, J.P. & E.M. Lord 1990b. The role of developmental timing in the evolution of floral form. Semin. Dev. Biol. 1: 281–287.

Hill, J.P. & R.L. Malmberg. 1991. Rates of corolla growth in tobacco determined with the plastochron index. Planta 185: 472–478.

Hill, J.P., E.M. Lord & R.G. Shaw. 1992. Morphological and growth rate differences among outcrossing and self-pollinating races of *Arenaria uniflora* (Caryophyllaceae). J. Evol. Biol. 5: 559–573.

Hunt, R. 1982. Plant Growth Curves. Edward Arnold, London.

Jacobs, T. 1992. Control of the cell cycle. Dev. Biol. 153: 1–15.

Johri, B.M. 1984. Embryology of the Angiosperms. Springer-Verlag, Berlin.

Kaplan, D.R. 1968. Histogenesis of the androecium and gynoecium in *Downingia bacigalupii*. Amer. J. Bot. 55: 933–950.

Kaplan, D.R. 1973. Comparative developmental analysis of the heteroblastic leaf series of axillary shoots of *Acorus calamus* L. (Araceae). Cellule 69: 253–290.

Kaplan, D.R. & W. Hagemann. 1991. The relationship of cell and organism in vascular plants. BioScience 41: 693–703.

Kaul, M.L.H. 1988. Male Sterility in Higher Plants. Monogr. Theor. Appl. Genet. 10, Springer-Verlag, New York.

Kaussmann, V.B. 1941. Vergleichende Untersuchungen uber die Blattnatur der Kelch-, Blumen- und Staubblätter. Bot. Archiv. 42: 503–572.

Koltunow, A.M., J. Truettner, K.H. Cox, M. Wallroth, & R.B. Goldberg. 1990. Different temporal and spatial gene expression patterns occur during anther development. Plant Cell 2: 1201–1224.

Lee, M.P. Nurse. 1988. Cell cycle control genes in fission yeast and mammalian cells. Trends Genet. 4: 287–290.

Lloyd, D.G. & C.J. Webb. 1986. The avoidance of interference between the presentation of pollen and stigmas in angiosperms I. Dichogamy. N.Z. J. Bot. 224: 135–162.

Lord, E.M. 1979. The development of cleistogamous and chasmogamous flowers in *Lamium amplexicaule* (Labiatae): An example of heteroblastic inflorescence development. Bot. Gaz. 140: 39–50

Lord, E.M. & J.P. Hill. 1987. Evidence for heterochrony in the evolution of plant form. pp. 47–70 in Development as an Evolutionary Process (edited by R.A. Raff & E. Raff). MBL Lectures in Biology Series. Alan R. Liss, New York.

Lord, E.M., K.J. Eckard & W. Crone. 1989. Development of the dimorphic anthers in *Collomia grandiflora*; evidence for heterochrony in the evolution of the cleistogamous anther. J. Evol. Biol. 2: 81–93.

Lord, E.M., W. Crone & J.P. Hill. 1994. Timing of events during flower organogenesis: *Arabidopsis* as a model system. Curr. Topics Dev. Biol. 29: 325–356.

Lyndon, R.F. 1983. The mechanism of leaf initiation. pp. 3–24 in The Growth and Functioning of Leaves (edited

by J.E. Dale & F.L. Milthorpe). Cambridge University Press, Cambridge.

Maheshwari, P. 1950. An Introduction to the Embryology of Angiosperms. McGraw-Hill, New York.

Mascarenhas, J.P. 1992. Pollen gene expression: Molecular evidence. Int. Rev. Cytol. 140: 3–18.

Mayers, A.M. & E.M. Lord. 1983a. Comparative flower development in the cleistogamous species *Viola odorata* II. An organographic study. Amer. J. Bot. 70: 1556–1563.

Mayers, A.M. & E.M. Lord. 1983b. Comparative flower development in the cleistogamous species *Viola odorata*. I. A growth rate study. Amer. J. Bot. 70: 1548–1555.

McHughen, A. 1980. The regulation of tobacco floral organ initiation. Bot. Gaz. 141: 389–395.

Minter, T.C. & E.M. Lord. 1983. A comparison of cleistogamous and chasmogamous floral development in *Collomia grandiflora* Dougl. ex Lindl. (Polemoniaceae). Amer. J. Bot. 70: 1499–1508.

Mitchison, J.M. 1971. The Biology of the Cell Cycle. Cambridge University Press, London.

Nurse, P. 1975. Genetic control of cell size at cell division in yeast. Nature 256: 547–551.

Or(n)duff, R. 1969. Reproductive biology in relation to systematics. Taxon 18: 121–133.

Pennell, R.I. & K. Roberts. 1990. Sexual development in the pea is presaged by altered expression of arabinogalactan protein. Nature 344: 547–549.

Poethig, S. 1984. Cellular parameters of leaf morphogenesis. pp. 235–259 in Contemporary Problems in Plant Anatomy (edited by R.A. White and W.C. Dickison). Academic Press, New York.

Raff, R.A. (editor). 1990. Heterochronic Changes in Development. Semin. Dev. Biol. 1. Saunders, Philadelphia.

Raff, R.A. & T.C. Kaufman. 1983. Embryos, Genes, and Evolution. Macmillan, New York.

Raff, R.A. & G.A. Wray. 1989. Heterochrony: developmental mechanisms and evolutionary results. J. Evol. Biol. 2: 409–434.

Raghavan, V. 1988. Anther and pollen development in rice (*Oryza sativa*). Amer. J. Bot. 75: 183–196.

Rembur, J., A. Nougarede, P. Rondet & D. Francis. 1992. Floral-specific polypeptides in *Silene coeli-rosa*. Can. J. Bot. 70: 2326–2333.

Richards, J.H. 1983. Heteroblastic development in the water hyacinth *Eichhornia crassipes* Solms. Bot. Gaz. 144: 247–259.

Richards, J.H. & S.C.H. Barrett. 1984. The developmental

basis of tristyly in *Eichhornia paniculata* (Pontederiaceae). Amer. J. Bot. 71: 1347–1363.

Richards, J.H. & S.C.H. Barrett. 1992. The development of heterostyly. pp. 85–127 in Evolution and Function of Heterostyly (edited by S.C.H. Barrett). Monogr. Theor. Appl. Genet. 15, Springer-Verlag, Berlin.

Rosenberg, S.M. & H.T. Bonnett. 1983. Floral organogenesis in *Nicotiana tabacum*: a comparison of two cytoplasmic male-sterile cultivars with a male-fertile cultivar. Amer. J. Bot. 70: 266–275.

Satina, S. & A.F. Blakeslee. 1941. Periclinal chimeras in *Datura stramonium* in relation to development of leaf and flower. Amer. J. Bot. 28: 862–871.

Sawhney, V.K. 1992. Floral mutants in tomato: development, physiology, and evolutionary implications. Can. J. Bot. 70: 701–707.

Schrauwen, J.A.M., M.W.M. Derks, P.F.M. de Groot, W.H. Reijnen, M.M.A. van Herpen & G.J. Wullems. 1988. Differential gene-expression during microsporogenesis with *Nicotiana tabacum*. pp. 3–7 in Sexual Reproduction in Higher Plants (edited by M. Cresti, P. Gori & E. Pacini). Springer-Verlag, Berlin.

Schwartz-Sommer, Z.P., W. Huijser, H. Nacken, H. Saedler & H. Sommer. 1990. Genetic control of flower development by homeotic genes in *Antirrhinum majus*. Science 250: 831–836.

Scott, R., R. Hodge, W. Paul & J. Draper. 1991. The molecular biology of anther differentiation. Plant Sci. 80: 167–191.

Shermoen, A.W. & P.H. O'Farrell. 1991. Progression of the cell cycle through mitosis leads to abortion of nascent transcripts. Cell 67: 303–310.

Silk, W.K., E.M. Lord & K.J. Eckard. 1989. Growth patterns inferred from anatomical records. Plant Physiol. 90: 708–713.

Steeves, T.A., M.W. Steeves & A.R. Olson. 1991. Flower development in *Amelanchier alnifolia* (Maloideae). Can. J. Bot. 69: 844–857.

Stewart, R.N. & H. Dermen. 1979. Ontogeny of monocotyledons as revealed by studies of the developmental anatomy of periclinal chimeras. Amer. J. Bot. 66: 47–58.

Strauss, R.E. 1987. On allometry and relative growth in evolutionary studies. Syst. Zool. 36: 72–75.

Svensson. L. 1992. Estimates of hierarchical variation in flower morphology in natural populations of *Scleranthus annus* (Caryophyllaceae), an inbreeding annual. Pl. Syst. Evol. 180: 157–180.

Tucker, S.C. 1992. The role of floral development in studies of legume evolution. Can. J. Bot. 70: 692–700.

Vasil, I.K. 1967. Physiology and cytology of anther development. Biol. Rev. 42: 327–373.

Wang, C., L.L. Walling, K.J. Eckard & E.M. Lord. 1992. Patterns of protein accumulation in developing anthers of *Lilium longiflorum* correlate with histological events. Amer. J. Bot. 79: 118–127.

Webb, C.J. & D.G. Lloyd. 1986. The avoidance of interference between the presentation of pollen and stigmas in angiosperms. II. Herkogamy. N. Z. J. Bot. 24: 163–178.

Williams, M.H. & P.B. Green. 1988. Sequential scanning electron microscopy of a growing plant meristem. Protoplasma 147: 77–79.

Wilson, C.L. 1937. The phylogeny of the stamen. Amer. J. Bot. 24: 686–699.

Wyatt, R. 1984. The evolution of self-pollination in granite outcrop species of *Arenaria* (Caryophyllaceae). I. Morphological correlates. Evolution 38: 804–816.

Wyatt, R. 1988. Phylogenetic aspects of the evolution of self-pollination. pp. 109–131 in Plant Evolutionary Biology (edited by L.D. Gottlieb & S.K. Jain) Chapman & Hall, New York.

Yoshioka, M., M. Maeda & M. Ito. 1981. The time and duration of the premeiotic interphase in microsporocytes of *Trillium kamtschaticum*. Bot. Mag. Tokyo 94: 371–378.

Young, H.J. & M.L. Stanton. 1990. Temporal patterns of gamete production within individuals of *Raphanus sativus* (Brassicaceae). Can. J. Bot. 68: 480–486.

7 • Diversity of endothecial patterns in the angiosperms

JOHN C. MANNING

INTRODUCTION

The first record of wall thickenings in anther cells appears to be that of Meyen (1828), who reported their discovery in a few genera of dicotyledons and monocotyledons and correctly supposed that they were generally distributed in the flowering plants. He observed that the anther epidermal cells lacked such thickenings, but considered that all other anther cells had them. He was content to 'quietly pass over the function of these fibres'. Shortly after this Purkinje (1830) published the results of a much more extensive investigation into the occurrence of the anther thickenings. He is responsible also for the term 'endothecium', coined to describe the inner layer of the mature anther with wall thickenings (as opposed to the exothecium which described the outer, unthickened layer which is the epidermis). From this work it emerged that endothecial thickenings were indeed very generally distributed among angiosperms (they are secondarily absent in very few instances), and more importantly, that the particular form of the endothecial thickenings was highly variable between taxa. Some discussion on the functional and taxonomic significance of the endothecial patterns was included. Mohl (1830) responded immediately with a critique, indicating that certain observations which he had made on endothecial patterns differed from those made by Purkinje, and that it was sometimes difficult to describe the patterns as simply as Purkinje suggested.

Purkinje (1830) looked at representatives from some 75 families (including some 15 families of monocotyledons), providing drawings of his observations. He did not define any thickening types, and although some of the thickening patterns are clearly described and illustrated, some are inadequately interpreted, apparently because adequate microscopy techniques were not then in existence. Other researchers were stimulated into furthering the investigation into the diversity of endothecial wall thickenings, but they were few (Chatin 1870; Le Clerc du Sablon 1885; Kuhn 1908). The work of Chatin (1870) is very unreliable, possibly due to the use of sectioned material, often of immature anthers. Le Clerc du Sablon (1885) described and illustrated thickening patterns for 20 angiosperm species, including just four monocotyledon species from two families. The first attempt to investigate thickening patterns and types systematically is that of Kuhn (1908), who examined representatives from some 183 angiosperm families, including about 25 monocotyledon families. The work of Kuhn (1908) is particularly important, not only in the extent of his survey, but also both in his appreciation of the need to consider the three-dimensional nature of the thickenings, and in his rationalization of the diversity into fewer categories. Surveys previous to his were often content to describe the appearance of the thickened endothecium, usually in the plane of the inner periclinal wall, without considering the three-dimensional shape of the thickenings.

A preliminary investigation of endothecial thickenings in the monocotyledons was carried out by Untawale & Bhasin (1973), who looked at representatives from 16 families and established two categories to accommodate the observed variation. Subsequent studies were preliminary and less ambitious. These include studies in the Asteraceae (Dormer 1962; Nordenstam 1978; Wetter 1983) and Apiaceae (Arora & Tiagi 1977).

All these surveys were carried out before the refinement of modern cladistic thought. Consequently the intellectual philosophy which guided the sampling was not directed by any phylogenetic bias and coverage of taxa was scattered. Thorough studies directed at assessing the systematic and phylogenetic significance of endothecial diversity in selected groups have only recently been initiated, mostly at the family level (Dormer 1962; French 1985a, b, 1986; Vincent & Getliffe 1988; Manning & Goldblatt 1990; Manning & Linder 1990; Freudenstein 1991; Manning & Stirton 1994).

One of the reasons for the tardy exploration of endothecial patterns as a source of phylogenetically informative data must be a lack of appreciation of the diversity in their pattern, despite the publication of the surveys listed above. One cause of this is that the endothecium is usually mentioned as an adjunct to studies on gametogenesis and anther wall development which, being based largely on sectioned material, fails to alert the worker to the true nature of the thickenings. Furthermore there may be some variation in thickening pattern within any anther making precise characterization difficult. Finally, endothecial patterns are highly prone to convergence, and until the taxon concerned is thoroughly investigated, information is seldom significant in postulating a phylogeny.

Despite this, the synthesis by Dahlgren & Clifford (1982) of available, and very inadequate, data on endothecial thickening patterns in the monocotyledons did reveal that the distribution of types was more regular than expected. This observation may have prompted the subsequent detailed studies in various families. The variation in pattern of the endothecial thickenings has clearly been overlooked until recently as a source of phylogenetically useful data, and consequently the potential of this character to provide such data has not been fully investigated.

In this review I am concerned only with the precise form of the wall thickenings in the endothecial cells, and not with the distribution of thickened cells, endothecial or otherwise, within the anther. By assessing current knowledge on structural variation in the endothecial thickenings, and analysing the type of information which is emerging, I hope this review will provide a conceptual framework for future investigations.

MATERIALS AND METHODS

The data used are drawn both from published accounts and from present investigation. Dehisced anthers from herbarium specimens are maintained in 60% aqueous lactic acid at 90°C until soft, usually some 2–4 hours. The anthers are then mounted in glycerine jelly, to which has been added a little basic fuchsin, and viewed using differential interference contrast optics. Normal optics are inadequate. The locules of at least one anther should be carefully separated and flattened intact in order to view the arrangement of endothecial cells, while the remaining anthers should be finely macerated with needles to separate the cells.

New data for the monocotyledons are presented in Appendix 2, together with available published records of adequate detail, drawn from the following accounts: Araceae (French 1985a, b, 1986); Iridaceae (Manning & Goldblatt 1990); Cyperaceae, Restionaceae, Ecdeiocoleaceae, Anarthriaceae, Centrolepidaceae, Joinvilleaceae, Flagellariaceae, Poaceae, Xyridaceae and Eriocaulaceae (Manning & Linder 1990); Orchidaceae (Freudenstein 1991); Rapateaceae (Venturelli & Bouman 1988); and some scattered observations (Untawale & Bhasin 1973). The novel observations published in Dahlgren & Clifford (1982) were not used because these authors did not distinguish between the various non-baseplate types of thickening, and I could not corroborate some of the reports of baseplates.

For the dicotyledons most of the data have been drawn from unpublished theses by Page (1980) and Queckenberg (1988), with various adjustments and omissions. Observations published in the older literature (Purkinje 1803; Chatin 1870; Le Clerc du Sablon 1885; Kuhn 1908) were not used because the reliability of the observations is uncertain in some cases. Modern published accounts used for the dicotyledons are: Mimosaceae, Caesalpiniaceae, Fabaceae, Rosaceae, Chrysobalanaceae, Connaraceae, Sapindaceae, Krameriaceae and Polygalaceae (Manning & Stirton 1994); Asteraceae (Dormer 1962; Nordenstam 1978; Wetter 1983; Vincent & Getliffe 1988) and Apiaceae (Arora & Tiagi 1977).

With such a limited data set, the conclusions which follow may be modified when more extensive information is available, but it is unlikely that major adjustments will be necessary since the available data are from taxonomically diverse groups. The classification of Cronquist (1981) is followed for the dicotyledons (class Magnoliopsida) and that of Dahlgren *et al.* (1985) for the monocotyledons (class Liliopsida), with the following exceptions: the Geosiridaceae is included in the Iridaceae following Goldblatt (1990), and the orchids are considered to comprise a single family following Freudenstein (1991). Phylogenetic diagrams used are those published by Dahlgren (1980) for the dicotyledons and Dahlgren & Clifford (1982) for the monocotyledons. In Figs. 1–31 the illustrations, from left to right, represent respectively views of thickenings in the inner periclinal, outer periclinal and anticlinal walls; the cells are oriented as if normal to the long axis of the anther.

TERMINOLOGY

The cells of the endothecium are basically cuboidal in shape, and the walls usually develop discrete thickenings of characteristic form. Individual cells may be fusiform, rectangular, isodiametric or radially elongate. Orientation is typically with the longer periclinal axis of the cell normal to the anther axis (Figs. 32, 33). The Asteraceae are a noteworthy exception to this and have endothecial cells elongated parallel to the anther axis. In cells with more or less discrete bars of thickening (U-shaped or helical) the bars almost invariably run normal to the long axis of the endothecial cell and not the anther (Fig. 35), and exceptions are very rare (some *Krameria* spp.).

It appears that the endothecial thickenings are usually lignified (Whatley 1982; Freudenstein 1991) although some are reported to be cellulosic and unlignified (de Fossard 1969) or even to contain callose (Kenrick & Knox 1979). The fine structure and development of the endothecium in *Phaseolus* has been investigated by Whatley (1982).

The wall thickenings are described in isolation from the unthickened portions of the cells walls. A cell is considered to possess a baseplate (Figs. 11–23) if the inner periclinal wall bears a more or less plate-like patch of thickening. The baseplate is considered perforate (Figs. 10B, 11, 14, 47, 48) if it contains small unthickened areas. If the plate is mostly continuous with the anticlinal walls (Figs. 11–14, 27) it is termed entire. If the plate is restricted to the central region of the inner periclinal wall and communicates with the anticlinal walls by finger-like

extensions of thickening, it is termed tympanate (Figs. 15, 16, 21, 22, 45, 46); with further reduction, palmate (Figs. 17–20, 25, 42, 44) in isodiametrical cells or rachial (Figs. 28, 43) in elongated cells. If the thickening is restricted to the margin of the inner periclinal wall, the baseplate is peripheral (Figs. 29, 49). Thickening of the anticlinal walls usually takes the form of rib-like extensions from the baseplate which almost invariably extend shortly onto the periphery of the outer periclinal wall (Figs. 21, 44) but may be continuous across it with a bar of thickening from an opposite anticlinal wall (Figs. 22, 46).

In cells in which a baseplate is absent, the inner periclinal wall usually has discrete bars of thickening extending across it (usually normal to the longer axis). These may extend across the anticlinal walls, terminating on the periphery of the outer periclinal wall thus forming a number of discrete U-shapes (Figs. 1–3, 32, 33, 35–37) which may be simple or, less commonly, branched (Fig. 38), in which case the outer periclinal wall bears more rib endings than the number of basal bars on the inner wall. If the rib endings are continuous across the outer periclinal wall, the resulting thickenings are annular (Figs. 5, 6, 55–57) in shape. Absence of bars on both periclinal faces results in columns (Figs. 26, 40) of thickening on the anticlinal walls. Discrete U-shapes characteristically occur in rectangular and fusiform cells (Figs. 32, 33). With shortening of the longer axis the number of U-shapes decreases and basal anastomosis between the bars traversing the inner periclinal wall occurs, resulting in a more or less complete palmate baseplate (Figs. 7–9, 41, 42). In another major variant, the thickenings are in the form of a helix (or very rarely a double helix) (Figs. 30, 53, 54). Shortening of the long axis in helically thickened cells favours disruption of the helix into individual gyres, or pseudannuli (Figs. 31, 56, 57), and in extreme cases the cells may bear a combination of short helices, pseudannuli, annuli and U-shapes. Extensive branching and anastomosis of individual bars of thickening results in a reticulate (Figs. 24, 50) thickening. In rare instances thickenings are actually absent.

In practice there may be some variation in the thickening type between cells in an individual anther, particularly if there is variation in the cell shape. This is usually in species with more or less isodiametric cells. For instance, intermediates between palmate baseplates and

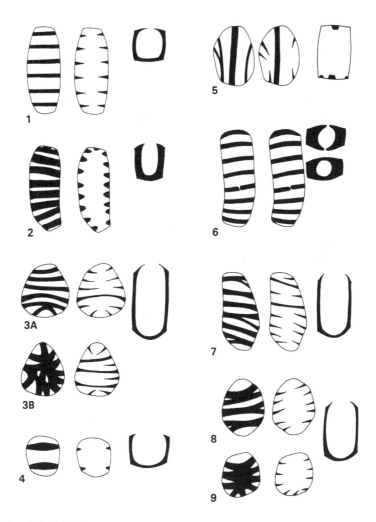

Figs. 1–6. Patterns of endothecial thickening. 1. *Aristea spiralis* (U-shapes); 2. *Dracaena hookeri* (massive U-shapes); 3. *Parinari cratellifolium*: A, typical cell (U-shapes with some basal anastomosis); B, Cell near connective (perforate, tympanate baseplate). 4. *Huttonaea pulchra* (U-shapes). 5. *Krameria spartioides* (U-shapes and annulus). 6. *Leersia hexandra* (annuli). **Figs. 7–9.** Endothecial thickenings in Sapindaceae illustrating phyletic shortening of the cell, accompanied by the formation of a baseplate from discrete U-shapes. 7. *Deinbollia oblongifolia* (U-shapes with some basal anastomosis). 8. *Atalaya alata* (U-shapes with basal anastomosis). 9. *Blighia unguiculata* (tympanate baseplate).

discrete U-shapes can occur, or helixes can be variously disrupted and incomplete. For descriptive purposes a species may thus have more than one thickening pattern, but for phylogenetic purposes the most typical or prevalent pattern alone should be used to characterize it. In particular the cells near the connective, at the anther base

or apex, and along the stomium, should not be considered as they are often anomalously thickened, or the thickenings may not be strictly endothecial in origin and thus not homologous with the others (Figs. 50–52).

At the species level very fine characterizations of the thickening type may be necessary, but for higher level

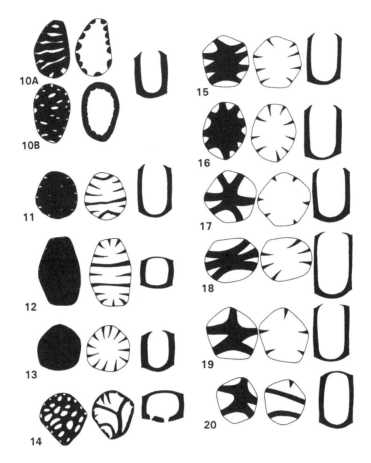

Figs. 10–20. Patterns of endothecial thickening. 10. *Cassia fistula*: A, cell near anther apex (massive U-shapes with some basal anastomosis); B, cell near porose anther base (perforate baseplate). 11. *Pentaclethra macrophylla* (perforate baseplate). 12. *Dichrostachys cinerea* (baseplate). 13. *Acacia baileyana* (baseplate). 14. *Walleria mackenzii* (perforate baseplate, bars branched and continuous across outer periclinal wall). 15. *Lonchocarpus capassa* (tympanate baseplate). 16. *Dumasia villosa* (tympanate baseplate). 17. *Canavalia rosea* (palmate baseplate). 18. *Erythrina latissima* (palmate baseplate). 19. *Galactia tenuiflora* (palmate baseplate). 20. *Pseudarthria hookeri* (palmate baseplate).

phylogenetic studies some of these distinctions are confusing or misleading. From a phylogenetic view, for example, it is pertinent to recognize that a particular palmate baseplate may be directly derived from a U-shaped type through shortening of the long axis of the cell and basal anastomosis between the bars. It is therefore allied to U-shaped thickenings, rather than related to other baseplates of potentially independent origin.

TYPES OF THICKENING

Once the thickening pattern present in any species has been described, further use of the information for phylogenetic analysis requires comparison with the thickening patterns in other species. Because of the variety of patterns that do occur such comparison is facilitated if the particular patterns can be grouped into fewer types. The

Figs. 21–31. Patterns of endothecial thickening. 21. *Dichidanthera penduliflora* (tympanate baseplate). 22. *Limnophyton obtusifolium* (tympanate baseplate). 23. *Lilium candidum* (basal anastomosis). 24. *Leucomphalos capparidoides* (reticulum). 25. *Barnhartia floribunda* (basal anastomosis). 26. *Stromanthe coccinea* (columnar). 27. *Senecio* sp. (unequally tympanate baseplate). 28. *Patersonia umbrosa* (rachial baseplate). 29. *Ecdeiocolea monostachya* (peripheral baseplate). 30. *Moraea angusta* (helix). 31. *Vellozia retinervis* (annuli and pseudannulus).

first attempt at grouping the observed variation in pattern into more inclusive categories was made by Kuhn (1908). He recognized six arbitrary categories designed merely to accommodate the observed variation. These were Bankzellen (baseplate present), Griffzellen (baseplate absent but thickening bars on inner periclinal wall branching/anastomosing), Netzfasern (thickenings reticulate), Ringfasern (thickenings annular), Spiralfasern (thickenings helical) and U-Klammern (thickenings U-shaped). These categories adequately cover most of the structural variation in the endothecium, but there are some rather unusual patterns which fall outside them, including columnar and peripheral baseplate types. Untawale & Bhasin (1973) in their very limited survey of monocotyledons distinguished only two, much more inclusive thickening types. The spiral type was described

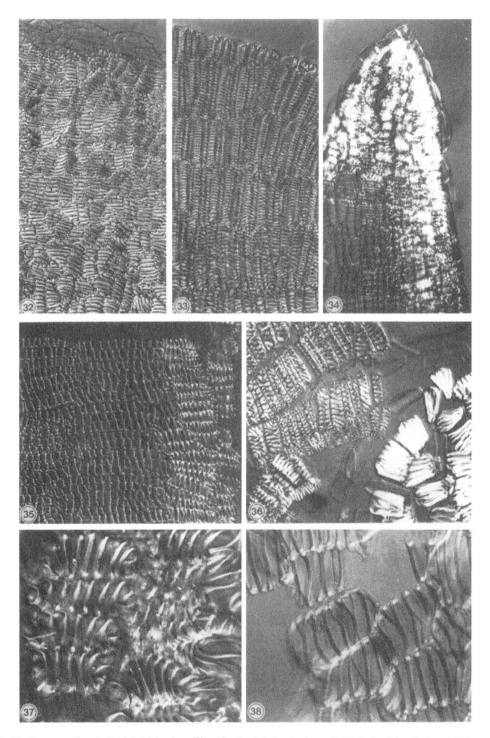

Figs. 32–38. Patterns of endothecial thickening. Fig. 32. *Apodolirion buchananii*. Endothecial cells irregularly arranged. Fig. 33. *Commelina africana*. Endothecial cells regularly arranged. Fig. 34. *Zea mays*. Endothecial cells heavily thickened at anther pore. Fig. 35. *Cyanotis speciosa*. Thickenings U-shaped. Fig. 36. *Zea mays*. Thickenings massive, U-shaped. Fig. 37. *Tulbaghia ludwigiana*. Thickenings U-shaped with some anastomosis at the cell apices. Fig. 38. *Ottelia exserta*. Thickenings U-shaped with branching/anastomosis.

as including helical, pseudannular and annular patterns and the girdle type accommodated baseplate patterns. However, genera which indisputably have U-shaped thickenings were included in the spiral category, and it is clear that these authors either failed to recognize the occurrence of U-shaped thickenings or to distinguish between them and the helical pattern. The most recent survey of patterns present in the dicotyledons is that of Page (1980), in which she distinguished seven categories. These were Thunbergioid (thickenings lacking), Physaloid (helical), Spathodoid (annular and pseudannular), Dianthoid (U-shaped and annuli in one cell), Kalancooid (U-shaped), Hylacoid (central baseplate and including basal anastomosis) and Callitricoid (peripheral baseplate). This system differs from that of Kuhn in the relative distinction that is awarded the various forms of U-shaped thickenings and also includes various degrees of baseplate development within a single category.

The more refined the distinctions between categories, the more accurately can the structural organization be characterized. This is desirable from a descriptive standpoint, but not always from a phylogenetic one, and there may be significant incongruence between the two. In an attempt to provide a system for describing endothecial patterns adequately, Noel (1983) developed a comprehensive modular system of description and nomenclature for the patterns of endothecial thickening, stressing that this was to facilitate the study of the structural variation observed, and that the taxonomic or phylogenetic significance of the thickening patterns was not at issue at that stage. This system seems satisfactory for this purpose.

Following more recent and comprehensive studies on circumscribed groups (usually at the family level), it appears that most of the phylogenetically significant variation can be captured by assigning the patterns to one of only three types: baseplate, U-shaped and helical. In this connection deviations from the typical variation found in each of these types are also informative. This broad distinction seems to reflect most of the phylogenetically significant variation although it does not describe adequately the structural variation at species level.

RELATIONSHIPS BETWEEN TYPES OF THICKENING

In the light of more intensive investigations (Manning & Linder 1990; Manning & Goldblatt 1990; Freudenstein 1991; Manning & Stirton 1994) it is emerging that not all of the categories or types of thickening that have been recognized in the past are of equal significance in a phylogenetic context. Some patterns are readily and repeatedly derived from others, and are phylogenetically insignificant above the species level, whereas other differences in endothecial patterns indicate real phylogenetic discontinuities.

The distinction between U-shapes and basal anastomosis does not seem to be highly significant in a broader context. The latter can be achieved through shortening of the long axis of the cell and develops on numerous occasions within any group in which this difference in cell shape occurs (Manning & Stirton 1994). Similarly, disruption of the helix accompanies shortening of the cell in species with helical thickenings, resulting in a variety of fragmented helices, pseudannuli and annuli. Columnar thickenings easily develop from U-shapes, again usually in isodiametric cells. Other uncommon thickening types, including reticulate and some annuli, are clearly derived but their origin can only be inferred once the family concerned is thoroughly investigated and the primitive state determined.

The only significant inclusive categories which can be distinguished are baseplate, U-shaped (includes most basal anastomosis) and helical. The most important distinction is that between U-shaped and helical thickenings. Results from the Fabales (Manning & Stirton 1994) suggest that both complete and palmate baseplates, and basal anastomosis are derived from U-shaped thickenings in the order. In the Orchidaceae (Freudenstein 1991) basal anastomosis is derived from discrete U-shapes. In Iridaceae (Manning & Goldblatt 1990) basal anastomosis, tympanate baseplates and helices are also all apparently derived from U-shaped thickenings.

The development of both helical thickenings and complete baseplates appears to be a transformation which functions at various hierarchical levels. In Iridaceae helical thickenings are characteristic of subfamily Ixioideae and of part of Iridoideae recognized as monophyletic on other grounds. In the Fabales a complete baseplate seems

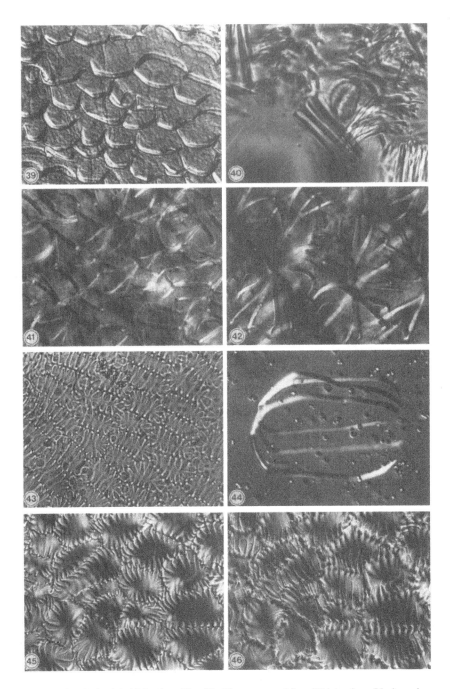

Figs. 39–46. Patterns of endothecial thickening. Fig. 39. *Huttonaea pulchra*. Thickenings U-shaped, one or two per cell. Fig. 40. *Sauromatum guttatum*. Thickenings columnar. Fig. 41. *Platylepis glandulosa*. Thickenings with irregular basal anastomosis. Fig. 42. *Lilium candidum*. Thickenings with regular basal anastomosis forming palmate baseplate. Fig. 43. *Patersonia umbrosa*. Thickenings rachial baseplate. Fig. 44. *Lilium candidum*. Anticlinal walls showing bars of thickening issuing from palmate baseplate on inner periclinal wall. Fig. 45. *Limnophyton obtusifolium*. Thickenings tympanate baseplate. Plane of focus on inner periclinal wall. Fig. 46. *Limnophyton obtusifolium*. Plane of focus on outer periclinal wall showing some bars of thickening continuous across the wall.

to be an autapomorphy for the Mimosaceae, whereas a palmate baseplate is characteristic of Fabaceae–Mellittieae and all tribes considered to fall in that clade.

THICKENING PATTERN AND FUNCTIONAL CONSTRAINTS

The endothecium is traditionally considered to promote the rupture of the stomium by differential shrinkage when the anther dries (Esau 1977). The adaptive significance, if any, of the various patterns of endothecial thickening is unknown.

There is no direct correlation between dehiscence type and thickening pattern. U-shaped thickenings occur in species with introrse, extrorse, longitudinal, transverse and poricidal dehiscence, in anthers which are versatile or basifixed. Gerenday & French (1988) found endothecial thickenings to be present in nine of 11 families of poricidal monocotyledons examined. Clearly, taxa with porose dehiscence do usually have the endothecium thickened, if only partially. In Araceae the anther is thickened throughout. In *Cassia* (Caesalpiniaceae) the endothecial thickenings are free from one another at the apical end of the anther but undergo increasing fusion basipetally, culminating in a complete, perforate baseplate at the pore.

In *Bauhinia* (Caesalpiniaceae) and *Coix* and *Zea* (Poaceae) endothecial thickenings are restricted to the pore region. In fact it seems that the thickenings in porose anthers, where present, are unusually well developed, but no particular thickening pattern is correlated with this dehiscence type. Page (1980) attempted to correlate thickening pattern with anther size. With the scant data at her disposal, the baseplate type of thickening was present in just one of 11 species of diverse taxonomic affinity which had large anthers (6 mm long or more), but was more frequent (six out of 15) in taxa with very small anthers (0.5 mm long or less).

In Orchidaceae, Freudenstein (1991) commented in ecological terms on the apparent correlation between degree of endothecial thickening (and therefore to some extent pattern) and plant habit. Epiphytic taxa were characterized by relatively reduced thickenings. This correlation appears spurious. All the epiphytic examples which he studied are placed in the Epidendroideae, for which subfamily reduced thickenings are apparently a

synapomorphy (Freudenstein 1991) and are possesed also by various terrestrial members of the subfamily. The correlation is phylogenetic and not ecological.

THICKENING PATTERN AND PHYLOGENETIC CONSTRAINTS

Thickening type appears mainly to reflect phylogenetic constraints. Within any one dehiscence type it seems possible to ignore ecological or functional constraints. Because endothecial patterns are apparently insensitive to ecological influences, they may provide data which are largely unaffected by many of the shifts in adaptive modes which influence other aspects of floral morphology. They may be highly informative in groups where either extreme floral divergence or extreme floral reduction has occurred.

The class and order

Available data for the dicotyledons are inadequate for analysis in depth, and only some groups of taxa in the monocotyledons have been surveyed with any degree of thoroughness. In fact most families are poorly known and less than a third have been sampled at all. Nevertheless, from the available data (Appendix 1) it is clear that a particular thickening type or pattern is seldom if ever characteristic of taxonomic categories at a level higher than that of the family. All orders with more than one family from which at least two species have been examined, display more than one thickening type (Figs. 58, 59). In the dicotyledons the majority of species examined have U-shaped and baseplate types (Table 1). These are scattered throughout the group. Helices are uncommon in the dicotyledons, but may be more common in the Caryophyllidae and Solanales than elsewhere. In the monocotyledons, however, helices and U-shapes are the most common types, although baseplates occur in a number of orders, particularly in the Alismatales and Poaceae (Table 1). There is thus clearly a difference in the proportional representation of the baseplate and helical thickening types in the two classes of angiosperms, while U-shaped thickenings occur in about half of the genera in both.

From the few thorough studies available, it is clear that the baseplate type is derived from the basal anasto-

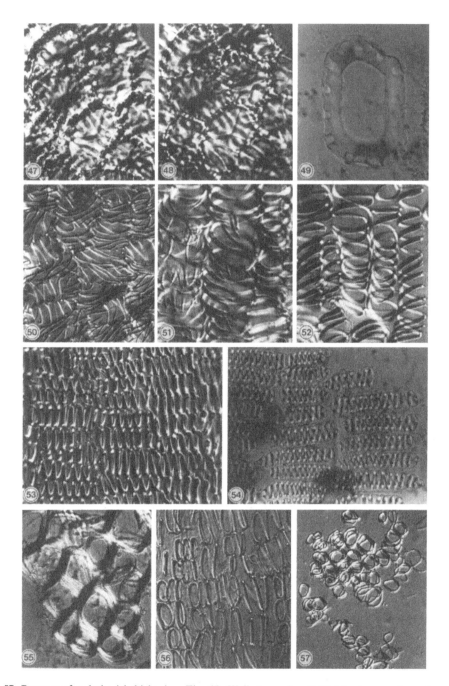

Figs. 47–57. Patterns of endothecial thickening. Fig. 47. *Walleria mackenzii*. Thickenings perforate baseplate. Plane of focus on inner periclinal wall. Fig. 48. *Walleria mackenzii*. Plane of focus on outer periclinal wall showing some branched bars of thickening continuous across the wall. Fig. 49. *Ecdeiocolea monostachya*. Thickenings peripheral baseplate. Fig. 50. *Sandersonia aurantiaca*. Thickenings reticulate, cells from outer half of anther. Fig. 51. *Sandersonia aurantiaca*. Transition between outer and connective cells. Fig. 52. *Sandersonia aurantiaca*. Thickenings helical and doubly helical, cells near connective. Fig. 53. *Empodium monophyllum*. Thickenings helical. Fig. 54. *Cyperus esculentus*. Thickenings helical. Fig. 55. *Anigozanthus viridis*. Thickenings annular. Fig. 56. *Vellozia retinervis*. Thickenings annular and pseudannular. Fig. 57. *Corymborkis corymbosa*. Thickenings annular and pseudannular.

Table 1. *Percentage representation of endothecial thickening types in angiosperm families*

	U-shaped	Helical	Baseplate	Other
Monocotyledons	46	29	14	11 (54%)
Dicotyledons	50	11	48	1 (28%)
All items	48	12	42	3 (28%)

Numbers in parentheses indicate percentage of families sampled.

mosis of U-shaped thickenings in at least some families (Iridaceae, Fabales, and probably Poaceae and Sapindaceae). It is probably derived in most families in which it is present but may be basic in some instances. A similar argument holds for the helical type. It is apparently the derived condition in the Iridaceae, for example, but may be basic for the Cyperaceae and Restionaceae. Present data are inadequate for further speculation. The validity of such claims as those of Noel (1983) that heavily thickened endothecial cells (baseplate and lateral thickening) may characterize the Magnoliidae and Ranunculidae must await extensive sampling and a better appreciation of the basic condition in each family.

Endothecial pattern is very prone to convergence and is inadequate alone as a character for phyletic analysis. The basic condition for each family, or other clade, must be determined, and this seems to be possible only by inference from existing theories of relationships within the group. Endothecial data are thus largely corroborative.

The family

Some families are conservative in the range of thickening patterns present while others contain a number of types. The Cyperaceae are remarkably uniform in thickening pattern, and a single type also characterizes the Restionaceae and Araceae among the larger families. Conversely, the much smaller families Haemodoraceae and Tecophilaeaceae contain a remarkable diversity of thickening types. In both the monocotyledons and dicotyledons available evidence indicates that in all larger families in which baseplates occur at least some species have U-shaped patterns.

Thickening patterns are not usually unique and recur in groups of diverse taxonomic affinity, but there are a few that are highly distinctive and may be diagnostic. These are mostly forms of baseplate. The peripheral baseplate is known only from Ecdeiocoleaceae (Fig. 49) and Callitrichaceae and each is different. A tympanate baseplate with bars continuous across the outer periclinal wall is known in the monocotyledons only in Alismataceae (Figs. 45, 46, 43) (*Burnatia*, *Limnophyton* and *Saggitaria*), Pontederiaceae (*Eichhornia*) and Iridaceae (*Patersonia*). There are two other types of baseplate with restricted occurrence in the monocotyledons. *Wachendorfia* has a palmate baseplate with few anticlinal bars and *Walleria* an entire, perforate baseplate with thickened anticlinal walls and some bars branching and continuous across the outer periclinal wall (Figs. 47, 48). The palmate baseplate is common also in the Fabaceae. The perforate baseplate appears to be much less common, but is also known in *Quisqualis*. Somewhat similar perforate baseplates, mostly without outer periclinal bars, are known in some other dicotyledons (*Cnestis*, *Ochna*, *Bougardia* and *Clematis* and some Mimosaceae). This recurrence of a very unusual thickening type in genera of such disparate phylogenetic affinity illustrates the extreme convergence shown by endothecial thickenings.

Non-baseplate thickenings seem to be less distinctive although double helical (*Triglochin*, *Hopkinsia* and some Iridaceae and Solanaceae) and reticulate (Fig. 50) (*Sandersonia* and *Chlorophytum*) patterns are relatively uncommon. In the U-shaped type there are a number of variants which can be distinguished and certain taxa may characteristically have one of these. Such variants usually occur in more or less isodiametric cells and are difficult to characterize because of variation within each. The few

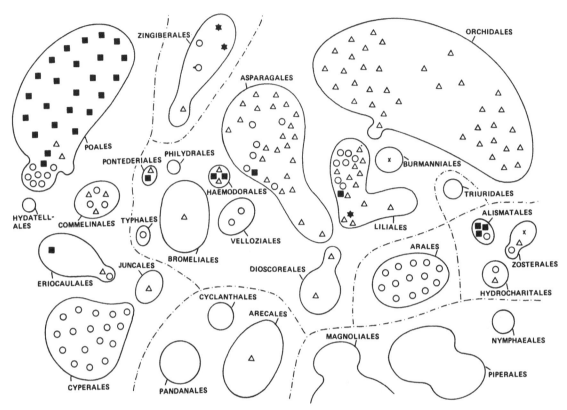

Figs. 58. Distribution of thickening types within the orders of the monocotyledons (density of symbols approximately reflects density of sampling). Square, baseplate; triangle, U-shape; circle, helix; cross, absent.

annular or U-shaped bars characteristic of the Orchidaceae – Orchidoideae are such an example (Fig. 39).

In general, therefore, variation in endothecial patterns is most usefully employed for phylogenetic studies within the family, and to date this is the only level at which it has contributed significantly. The term 'significantly' in this connection is invariably used to denote a perceived congruence between the distribution of endothecial thickening patterns and an existing phylogenetic hypothesis, and such conditions are at present mostly encountered at the family level.

The genus and species

Precise characterization of the pattern present in a species is hindered by variation in the size and shape of individual endothecial cells within an anther and variation between cells in the thickenings themselves (number of bars, degree of fusion, etc.). Endothecial patterns are unlikely to prove useful at the species level. Arora & Tiagi (1977) claim that the size and proportion of the palmate baseplate pattern in the Apiaceae may prove diagnostic for at least the genera, but no indication is given of the sample size and hence variability of the thickening patterns. The fact that *Angelica archangelica* L. is placed in two different categories of pattern suggests at least some substantial variation. Similarly, although Wetter (1983) found considerable intraspecific variation in the extent and placement of ribs in the endothecial cells of some cacaloid and senecioid genera of Asteraceae, Vincent & Getliffe (1988) were unable to distinguish any significant intraspecific variation in *Senecio*.

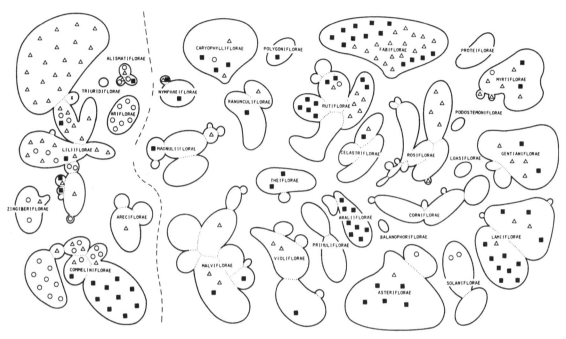

Fig. 59. Distribution of thickening types in angiosperm families. Square, baseplate; triangle, U-shape; circle, helix; cross, absent.

Accordingly, unless there are statistically adequate sampling and statistically significant differences, claims that endothecial patterns may be interspecifically diagnostic should be distrusted. The situation in *Krameria* where there are significant differences in endothecial patterns between species (Manning & Stirton 1994) is exceptional. Diagnostic patterns are more common at the generic level, particularly in very distinct genera with highly unusual thickening patterns (e.g., Callitrichaceae: *Callitriche*, Ecdeiocoleaceae: *Ecdeiocolea*).

Case studies

This section covers those families in which variation in endothecial patterns seems to reflect aspects of their phylogeny, providing an indication of the degree to which endothecial characters vary within families, and in what direction.

Araceae

Only one type of thickening occurs in the Araceae. Although considerable structural diversity is apparent, these forms are variations on a basically annular-helical pattern. The only exception is *Acorus*, which has bars of thickening radiating from the centre of the inner periclinal wall (basal anastomosis or palmate baseplate). This is clearly a derived condition. Within the family some thickening patterns are characteristic of genera, and the absence of endothecial thickenings may be a derived state linking some of the genera. The variation in endothecial thickenings can only be analyzed adequately in the light of existing ideas on relationships within the family (French 1985a, b, 1986).

Poaceae

Thickenings in the Poaceae are either baseplate (sometimes disrupted into U-shapes) or annular. Comparing the distribution of these two thickening types with existing phylogenetic hypotheses suggests that the baseplate is the basic condition for the family. The lack of baseplates in the majority of Bambusoideae is therefore apparently secondary (Manning & Linder 1990).

Iridaceae

Three types of endothecial thickenings are present in the Iridaceae. The baseplate type is clearly derived within the family. By comparing the distribution of endothecial thickenings with an existing phylogeny for the family, it appears that the U-shaped type is basic for the Iridaceae. Basal anastomosis and baseplates unite the genera with secondary thickening of the stem or rhizome (*Klattia*, *Nivenia*, *Witsenia* and *Patersonia*), while helices are derived in both the Ixioideae and the putatively monophyletic clade in the Iridoideae containing the tribes Irideae, Mariceae and Tigrideae. Some thickening types are characteristic of genera (Manning & Goldblatt 1990).

Orchidaceae

Thickening patterns in this family are variations on an annular-helical pattern, with basal anastomosis in some. Four types can be distinguished. Type 1 has numerous discrete U-shapes, annuli or helical segments; type 2 has few, usually one or two, widely spaced U-shapes or annuli; type 3 has some basal anastomosis of U-shapes or annuli; and type 4 has a dense mass of columns or bars, apparently without any higher order of organization. Intermediates between types 1 and 2 occur, but it appears that type 1 is basic for the family. Type 2 is distinctive of the Orchidoideae. Type 3 is scattered throughout the family. In other families this thickening type is derived from the pattern classed here as type 1, and the scattered distribution in the Orchidaceae is probably the result of similar parallelism. Type 4 may be distinctive for the Epidendroideae, although it is also present in some of the Spiranthoideae. Since the type is rather reduced and indistinct and difficult to homologize, its occurrence in these two subfamilies may be convergent. Within types 3 and 4 a number of variations can be distinguished, some of which correspond to tribal or subtribal associations. The Oncidiinae may be characterized by an apparent absence, or at most very weak development, of thickenings (Freudenstein 1991).

Fabales

Thickening patterns in the Fabales are mostly based on some degree of basal anastomosis of U-shaped thickenings, leading to the formation of baseplates in some instances. The Mimosaceae are largely characterized by a complete baseplate, often with bars of thickening continuous across the outer periclinal wall, but a partially disrupted baseplate occurs in the Parkieae. The Caesalpiniaceae lack a baseplate (apart from specialized exceptions in genera with porose dehiscence) and the thickenings are more or less free U-shapes, usually with some basal anastomosis. In the Fabaceae U-shapes occur in a number of tribes, but palmate baseplates are present in the remaining tribes. Significantly these are all thought to comprise a monophyletic assemblage. Comparing endothecial characters with existing ideas on phylogenies within the family suggests that U-shaped thickenings are basic for the family, that complete baseplates characterize the core tribes of Mimosaceae, and that palmate baseplates characterize the clade containing the Mellittieae (Manning & Stirton 1994).

Asteraceae

Some confusion exists in the terminology of thickening patterns so far observed in the Asteraceae (Vincent & Getliffe 1988). Dormer (1962) reported three patterns of thickening within a single anther in the family, and these observations were subsequently confirmed by Nordenstam (1976, 1978) and Wetter (1983). In the 'polarized' pattern, bars of thickening occur on the periclinal walls; in the 'radial' pattern a baseplate is developed on the inner periclinal wall and the anticlinal walls are described as ribbed; and in the 'transitional' pattern the cell walls are more or less uniformly thickened. There is a distinct sequence in the progression of the endothecial patterns within an anther. The 'transitional' pattern occurs along the connective (and therefore should be ignored in comparative phylogenetic studies for reasons already stated), the 'polarized' occurs near the stomium, and the 'radial' occurs between the other two in the central part of the theca. In *Senecio* (Vincent & Getliffe 1988) the 'transitional' pattern is absent and the sequence is 'radial' to 'polarized' moving centripetally along the anther. The 'radial' cells *sensu* Vincent & Getliffe are uniformly thickened on both periclinal walls but the inner anticlinal wall has bars of thickening.

This asymmetric thickening is apparently different from the symmetrically thickened 'radial' type *sensu* Dormer, and is unique in flowering plants. Vincent & Getliffe (1988) speculate that this pattern may be peculiar to the Senecioneae, and certainly characteristic of *Senecio*, and accordingly use it as a derived character to exclude certain species from the genus. However, this asymmetric thickening pattern has been illustrated in Dormer (1962) and Wetter (1983) as their 'radial' type without any observation on its deviation from the typical 'radial' type as described by them. Their reports on the occurrence of the 'radial' type in other tribes may thus conceal the more widespread distribution of this asymmetric pattern.

Final comment

In all the families in which patterns of endothecial thickenings have proved phylogenetically informative, the informative distinction has been that between helical, baseplate and U-shaped types. Less extreme variations, including some basal anastomosis between adjacent U-shapes or even more extensive anastomosis resulting in a reticulum, seem to be less significant and by inference more prone to convergence.

Furthermore, the endothecial data have been most useful as corroborative data to be tested against existing phylogenetic hypotheses. Because endothecial patterns are highly convergent, any particular pattern is informative only in relation to those of all other representatives in the clade. This means that *a priori* polarization of the character states is uncertain, and endothecial data are almost impossible to use as a primary character in assessing relationships. The broader the group under investigation the worse the problems become, and endothecial data seem to be best employed below the level of the family.

Endothecial data have been markedly congruent with existing phylogenetic hypotheses in some families, and have provided significant extra synapomorphies supporting the monophyly of certain clades (Iridaceae, Fabaceae). Endothecial pattern may or may not vary within a family, and while some families are characterized by a single type and pattern of thickening, others

exhibit a range of thickening types. Certain genera may be more or less distinctive in subtle aspects of endothecial pattern, and endothecial thickenings are usually conservative at the generic level. Each family must be assessed individually, and if variation in endothecial patterns is evident, then the entire family must be thoroughly investigated before the data can be expected to yield useful information.

ACKNOWLEDGEMENTS

This paper is dedicated to the late Professor A.R.A. Noel. My thanks to the two referees and Mr A.T. de Villiers for comments on the manuscript.

LITERATURE CITED

Arora, K. & B. Tiagi. 1977. The role of endothecium in the identification of umbellifers. Curr. Sci. 46: 531.

Chatin, G.A. 1870. De l'Anthere. Baillière et fils, Paris.

Cronquist, A. 1981. An Integrated System of Classification of Flowering Plants. Columbia University Press, New York.

Dahlgren, R.M.T. 1980. A revised system of classification of the angiosperms. Bot. J. Linn. Soc. 80: 91–124.

Dahlgren, R.M.T. & H.T. Clifford. 1982. The Monocotyledons: A Comparative Study. Academic Press, London.

Dahlgren, R.M.T., H.T. Clifford & P.F. Yeo. 1985. The Families of Monocotyledons. Springer-Verlag, Berlin.

De Fossard, R.A. 1969. Development and histochemistry of the endothecium in the anthers of in vitro grown *Chenopodium rubrum* L. Bot. Gaz. 130: 10–22.

Dormer, K.J. 1962. The fibrous layer in anthers of Compositae. New Phytol. 61: 150–153.

Esau, K. 1977. Anatomy of Seed Plants, 2nd ed. Wiley New York.

French, J.C. 1985a. Patterns of endothecial wall thickenings in Araceae: subfamilies Calloideae, Lasioideae and Philodendroideae. Bot. Gaz. 146: 521–533.

French, J.C. 1985b. Patterns of endothecial wall thickenings in Araceae: subfamilies Pothoideae and Monsteroideae. Amer. J. Bot. 72: 472–486.

French, J.C. 1986. Patterns of endothecial wall thickenings in Araceae: subfamilies Colocasioideae, Aroideae and Pistioideae. Bot. Gaz. 147: 166–179.

Freudenstein, J.V. 1991. A systematic study of endothecial

thickenings in the Orchidaceae. Amer. J. Bot. 78: 766–781.

Gerenday, A. & J.C. French. 1988. Endothecial thickenings in anthers of porate monocotyledons. Amer. J. Bot. 75: 22–25.

Goldblatt, P. 1990. Phylogeny and classification of Iridaceae. Ann. Missouri Bot. Gard. 77: 607–627.

Kenrick, J. & R.B. Knox. 1979. Pollen development and cytochemistry in some Australian species of Acacia. Aust. J. Bot. 27: 413–427.

Kuhn, E. 1908. Über den Wechsel der Zelltypen im Endothecium der Angiospermen. Buchdruckerei Gebr. Leeman, Zurich.

Le Clerc Du Sablon, M. 1885. Recherches sur la structure et la dehischence des anthères. Annales des sciences naturelles botanique et biologie végétale ser. 7,1: 97–134.

Manning, J.C. & P. Goldblatt. 1990. Endothecium in Iridaceae and its systematic implications. Amer. J. Bot. 77: 527–532.

Manning, J.C. & H.P. Linder. 1990. Cladistic analysis of patterns of endothecial thickenings in the Poales/Restionales. Amer. J. Bot. 77: 196–210.

Manning, J.C. & C.H. Stirton. 1994. Endothecial thickenings and phylogeny of the Leguminosae. pp. 141–163 in Advances in Legume Systematics, Part 4 (edited by I.K. Ferguson & S. Tucker). Royal Botanic Gardens, Kew.

Meyen, F.J.F. 1828. Anatomisch-physiologische Untersuchungen über den Inhalt der Pflanzen-Zellen. Berlin.

Mohl, H. 1830. Ueber die fibrosen Zellen der Antheren. Flora 44: 697–740.

Noel, A.R.A. 1983. The endothecium – a neglected criterion in taxonomy and phylogeny? Bothalia 14: 833–838.

Nordenstam, B. 1976. Re-classification of Chyrsanthemum L. in South Africa. Bot. Not. 129: 137–165.

Nordenstam, B. 1978. Taxonomic studies in the tribe Senecioneae (Compositae). Opera Bot. 44: 3–83.

Page, Y.M. 1980. A Survey of the Endothecium. Unpublished Honours Thesis, University of Natal, Pietermaritzburg.

Purkinje, J. 1830. De cellulis antherarum fibrosis nec non de granorum pollinarium formis. Commentatio Phytotomica, Vratislaviae: Gruesonil.

Queckenberg, M. 1988. Endothecium: strukturen und ihre systematische Relevanz am Beispiel der Centrospermen. Unpublished thesis, Rheinischen Friedrich-Wilhelms Universitat, Bonn.

Simpson, M.G. 1990. Phylogeny and classification of the Haemodoraceae. Ann. Missouri Bot. Gard. 77: 722–784.

Untawale, A.G. & R.K. Bhasin. 1973. On endothecial thickenings in some monocotyledonous families. Current Science 42: 398–400.

Venturelli, M. & F. Bouman. 1988. Development of ovule and seed in Rapateaceae. Bot. J. Linn. Soc. 97: 267–294.

Vincent, P.L.D. & F.M. Getliffe. 1988. The endothecium in Senecio (Asteraceae). Bot. J. Linn. Soc. 97: 63–71.

Wetter, M.A. 1983. Micromorphological characters and generic delimitation of some new world Senecioneae (Asteraceae). Brittonia 35: 1–22.

Whatley, J.M. 1982. Fine structure of the endothecium and developing xylem in Phaseolus vulgaris. New Phytol. 91: 561–570.

APPENDIX 1.   Endothecial thickening types recorded in angiosperm families

(Numbers in parentheses following familial names indicate numbers of genera/species examined.)

LILIOPSIDA

Probably basically U-shaped, with repeated development of helices and baseplates.

Liliiflorae

Probably basically U-shaped, but with repeated development of helices, and less commonly of baseplates.

Dioscoreales   U-shaped, sometimes with basal anastomosis, or columnar.

Dioscoreaceae (1/3)   U-shaped.

Taccaceae (1/1)   Basal anastomosis, presumably correlated with radial elongation of the cells.

Smilacaceae (1/1)   Columnar, probably correlated with the radially elongate cell shape.

Asparagales   Characterized by U-shaped thinckenings, but with scattered occurrence of various derived types (basal anastomosis, reticulate) and particularly helical, which seems to occur in most families along with the more usual U-shaped patterns. Dracaenaceae is well characterized by derived patterns, and Tecophilaeaceae seem particularly variable.

Philesiaceae (1/1)   U-shaped.

Asparagaceae (1/3)   U-shaped.

Dracaenaceae (2/2)   U-shaped, massive with some basal anastomosis. This pattern is characteristic of the family, and similarly hypertrophied thickenings are known elsewhere only in Poaceae (*Zea*, *Coix*) and Caesalpiniaceae (*Cassia*).

Agavaceae (1/1)   U-shaped with some basal anastomosis.

Hypoxidaceae (3/3)   U-shaped and helical.

Tecophilaeaceae (2/2)   Helical and perforate baseplate. The perforate baseplate with some bars of thickening continuous across the outer periclinal wall is peculiar to *Walleria*. The family seems to possess a variety of derived patterns.

Hemerocallidaceae (1/1)   U-shaped.

Asphodelaceae (3/4)   U-shaped and helical.

Anthericaceae (3/4)   U-shaped (some development of reticulum) and helical.

Hyacinthaceae (5/5)   U-shaped and helical.

Alliaceae (3/4)   U-shaped with some basal anastomosis in species with isodiametric cells.

Amaryllidaceae (6/7)   U-shaped, some columnar and helical.

Liliales   Basically U-shaped, sometimes with basal anastomosis, but with repeated occurrence of the helical type. Thickenings rarely absent. Colchicaceae possess a variety of derived thickening patterns.

Alstroemeriaceae (1/1)   Basal anastomosis.

Colchicaceae (3/3)   Basal anastomosis and reticulate.

Liliaceae (1/1)   Basal anastomosis.

Iridaceae (28/45)   U-shaped, sometimes with basal anastomosis or annuli, baseplate and helical, sometimes double. It is most parsimonious to regard the U-shaped thickening as plesiomorphic, with the baseplate and helical types derived.

Burmanniales   Absent.

Burmanniaceae (1/1)   Absent.

Orchidales   The basal condition appears to be U-shaped, with the thickenings sometimes annular. Subsequent developments include basal anastomosis, reduction in the number of bars to 1 or 2, and loss of thickening on the periclinal walls leading to disorganized columnar thickenings.

Apostasiaceae (2/2)   U-shaped or annular.

Cypripediaceae (3/5)   U-shaped and columnar.

Orchidaceae (25/26)   U-shaped with or without basal anastomosis, annular or pseudannular, or columnar.

Ariflorae
Probably basically helical.

Arales   Variations and disruptions of a helical-annular pattern.

  Araceae (57/282)   Helical-annular, U-shaped sometimes with basal anastomosis, columnar and absent. This disruption of typical patterns and basal anastomosis is probably due to the radially elongated nature of the cells.

Alismatiflorae
U-shaped, helical and baseplate (rarely absent).

Alismatales   U-shaped, helical and baseplate.

  Aponogetonaceae (1/1)   U-shaped with some basal anastomosis. The degree of anastomosis is variable and presumably promoted by the isodiametric nature of the cells.

  Alismataceae (4/4)   Helical with some branching or tympanate baseplate with some thickenings continuous across outer periclinal wall. This type of baseplate is apparently diagnostic for some genera in Alismataceae, and is known elsewhere only in Pontederiaceae and rarely in Iridaceae (*Patersonia*). Its distribution in the Alismatiflorae should be further investigated.

  Hydrocharitaceae (1/1)   Helical with some branching.

  Najadales   Helical, U-shaped and absent.

  Juncaginaceae (1/1)   Double helix. This is an uncommon pattern known elsewhere only in some Iridaceae, *Hopkinsia* (Restionaceae) and some Solanaceae.

  Potamogetonaceae (2/2)   U-shaped and absent.

Bromeliiflorae
U-shaped, baseplate and helical (including annuli and pseudannuli).

Velloziales   Helical and annuli.

  Velloziaceae (2/2)   Helical and annuli. The latter are probably derived from the former as evinced by the co-occurrence of psuedannuli with the annuli.

  Bromeliales   U-shaped.

  Bromeliaceae (1/1)   U-shaped.

  Haemodorales   U-shaped and baseplate.

  Haemodoraceae (4/4)   U-shaped and baseplate, the baseplate sometimes disrupted into discrete U-shapes. The presence of a baseplate is probably of phylogenetic significance and the family should be thoroughly surveyed, particularly in the light of the findings by Simpson (1990) that Pontederiaceae is the sister group to Haemodoraceae.

  Pontederiales   U-shaped and baseplate.

  Pontederiaceae (2/2)   U-shaped and baseplate. The baseplate pattern present in *Eichhornia* is known elsewhere only in Alismataceae and Iridaceae.

  Typhales   Helical.

  Typhaceae (1/1)   Helical.

Zingiberiflorae

U-shaped and helical. It is not clear whether the subclass is basically helical or U-shaped, and more extensive sampling is necessary.

Strelitziaceae (1/1) U-shaped.

Zingiberaceae (1/2) Annuli or pseudannuli, possibly derived from helical pattern.

Costaceae (1/1) Columnar, possibly derived from U-shaped pattern, cells isodiametric.

Cannaceae (1/1) Helical, few, cells isodiametric.

Marantaceae (1/1) U-shaped, few with some basal anastomosis, cells isodiametric.

Commeliniflorae

U-shaped, helical and baseplates. Some orders or families are characterized by helical or basal thickenings, while in others there is a mixture of U-shapes and helices or U-shapes and baseplates. It appears that both helices and baseplates have been derived a number of times from U-shapes, which may be basic for the subclass.

Commelinales U-shaped, helical and baseplate.
  Commelinaceae (6/6) U-shaped and helical.
  Xyridaceae (1/1) U-shaped.
  Rapateaceae (9/12) Helical.
  Eriocaulaceae (1/1) Baseplate.
Cyperales U-shaped and helical.
  Juncaceae (1/1) U-shaped.
Cyperaceae (11/18) Helical. Fusiform cells ranged in regular rows.
Poales Baseplates seem to be basic in some families, sometimes disrupted into U-shapes, but helices are characteristic of others.
  Flagellariaceaeae (1/1) Baseplate, sometimes disrupted into U-shapes.
  Joinvilleaceae (1/1) U-shaped with branched basal strips and peripheral thickening.
  Poaceae (38/43) Baseplate, sometimes disrupted into U-shapes and annuli. It is not possible to polarize this transformation series, but existing cladistic topologies favor the baseplate as basic for the family.
  Ecdeiocoleaceae (1/1) Peripheral baseplate. Highly characteristic.
  Anarthriaceae (1/1) Helical but disrupted into annuli and pseudannuli with some basal anastomosis.
  Restionaceae (14/20) Helical with variations including disruption into annuli and pseudannuli, and doubling in *Hopkinsia*.

Centrolepidaceae (1/1) U-shaped, some annuli and pseudannuli.

Areciflorae

U-shaped.

Arecales U-shaped
  Arecaceae (1/1) U-shaped.

MAGNOLIOPSIDA

Magnoliidae

U-shaped and baseplate.

Magnoliales U-shaped.
  Winteraceae (1/1) U-shaped.
Piperales Baseplate.
  Piperaceae (1/1) Baseplate.
Aristolochiales Baseplate.
  Aristolochiaceae (1/1) Baseplate.
Nymphaeales U-shaped and baseplate.
  Nymphaeaceae (1/1) U-shaped and baseplate.
Ranunculales Baseplate.
  Ranunculaceae (2/2) Baseplate.
Papaverales U-shaped.
  Papaveraceae (1/1) U-shaped.

Caryophyllidae

U-shaped, baseplates, and some helices.

Caryophyllales Baseplate, U-shaped and helical.
  Phytolaccaceae (11/7) U-shaped, baseplate and helical.
  Achatocarpaceae (2/2) Helix.
  Nyctaginaceae (1/1) Baseplate.
  Aizoaceae (43/81) U-shaped, baseplate, helical, some columnar.
  Didieriaceae (1/1) U-shaped.
  Cactaceae (60/97) U-shaped, baseplate, helical, some columnar.
  Chenopodiaceae (13/25) U-shaped, some helical.
  Amaranthaceae (6/8) U-shaped, baseplate, helical.
  Portulaccaceae (6/13) U-shaped, baseplate.
  Basellaceae (4/5) U-shaped, baseplate.
  Molluginaceae (6/7) U-shaped, baseplate.
  Caryophyllaceae (28/84) U-shaped, baseplate, helical.
Polygonales Baseplate.
  Polygonaceae (1/1) Baseplate.
Plumbaginales Baseplate.
  Plumbaginaceae (2/2) Baseplate.

Dilleniidae
Baseplate and U-shaped.

Theales Baseplate.
 Ochnaceae (1/1) Baseplate.
Malvales U-shaped.
 Malvaceae (4/6) U-shaped.
Nepenthales Baseplate.
 Droseraceae (1/1) Baseplate.
Violales U-shaped and ?baseplate.
 Passifloraceae (1/1) U-shaped.
 Begoniaceae (1/1) U-shaped/baseplate.
Capparales Baseplate.
 Capparaceae (1/1) Baseplate.

Rosidae

Rosales U-shaped.
 Connaraceae (2/2) U-shaped.
 Bruniaceae (1/1) U-shaped.
 Crassulaceae (2/3) U-shaped.
 Rosaceae (6/6) U-shaped, some continuation across outer periclinal wall resulting in pseudannuli and annuli.
 Chrysobalancaeae (1/1) U-shaped, some anastomosis.
Fabales U-shaped and baseplate. It appears that the U-shaped type with some basal anastomosis is basic for the family and that both complete and palmate baseplates have been independently derived.
 Caesalpiniaceae (7/7) U-shaped, more or less anastomosing on inner periclinal wall, sometimes forming baseplate.
 Mimosaceae (5/5) Baseplate, rarely somewhat disrupted.
 Fabaceae (92/93) U-shaped with some basal anastomosis and palmate baseplate.
Haloragales U-shaped.
 Haloragaceae (1/1) U-shaped with some basal anastomosis.
 Gunneraceae (1/1) U-shaped.
Myrtales U-shaped and baseplate (rarely helical).
 Lythraceae (3/3) U-shaped with some basal anastomosis and helical.
 Thymelaeaceae (1/1) U-shaped and baseplate.
 Myrtaceae (2/2) U-shaped with some basal anastomosis.
 Combretaceae (1/1) Baseplate.
Rhizophorales U-shaped.
 Rhizophoraceae (1/1) U-shaped.
Celastrales U-shaped and baseplate.

Celastraceae (1/1) U-shaped/baseplate.
 Icacinaceae (1/1) Baseplate.
Euphorbiales U-shaped and baseplate.
 Euphorbiaceae (3/3) U-shaped and baseplate.
Linales U-shaped.
 Erythroxylaceae (1/1) U-shaped.
Polygalales Variously developed baseplates, sometimes disrupted into discrete annuli/U-shapes.
 Polygalaceae (12/14) Tympanate or palmate baseplate and basal anastomosis, sometimes disrupted into annuli/U-shapes.
 Krameriaceae (1/2) Basal anastomosis or annuli.
Sapindales U-shaped with basal anastomosis probably basic and baseplate derived.
 Sapindaceae (8/9) U-shaped with basal anastomosis and baseplate.
 Burseraceae (1/1) Baseplate.
 Zygophyllaceae (2/2) U-shaped with basal anastomosis.
Geraniales Variously developed baseplate, sometimes disrupted into U-shapes (and helical).
 Oxalidaceae (1/2) U-shaped/baseplate and helical.
 Geraniaceae (1/1) U-shaped/baseplate.
Apiales Baseplate.
 Araliaceae (1/1) Baseplate.
 Apiaceae (18/19) Baseplate.

Asteridae

Gentianales U-shaped and baseplate.
 Gentianaceae (1/1) U-shaped.
 Apocynaceae (2/2) U-shaped and baseplate.
Solanales Helical.
 Solanaceae (3/3) Helical.
Lamiales U-shaped and baseplate.
 Verbenaceae (5/5) U-shaped and baseplate.
 Lamiaceae (8/11) Baseplate, sometimes disrupted.
Callitrichales Peripheral baseplate.
 Callitrichaceae (1/1) Peripheral baseplate.
Plantaginales U-shaped.
 Plantaginaceae (1/1) U-shaped.
Scrophulariales U-shaped and baseplate (and helical).
 Buddlejaceae (1/2) U-shaped.
 Scrophulariaceae (6/6) U-shaped and baseplate.
 Gesneriaceae (1/1) U-shaped.
 Acanthaceae (2/2) Baseplate (absent in *Thunbergia*).
 Pedaliaceae (1/1) Helical.
 Bignoniaceae (1/1) Baseplate.
Campanulales Helical.
 Campanulaceae (1/1) Helical.

Rubiales   U-shaped and baseplate.
  Rubiaceae (5/5)   U-shaped and baseplate.
Dipsacales   Uncertain U-shaped/baseplate/helical.
  Valerianaceae (1/1)   U-shaped/baseplate.
  Dipsacaceae (1/2)   U-shaped/helical.
Asterales   U-shaped and baseplate.
  Asteraceae (8/9)   U-shaped and baseplate.

APPENDIX 2.   Endothecial patterns in the monocotyledons (Vouchers, unless otherwise indicated, are deposited in NU.)

LILIIFLORAE

Dioscoreales

Dioscoreaceae

  Dioscorea brownii Schinz   U-shaped (Hilliard & Burtt 7702).
  Dioscorea rupicola Kunth.   U-shaped (Hilliard & Burtt 5758).
  Dioscorea sylvatica Ecklon   U-shaped (Schelpe 933).
Taccaceae
  Tacca leontopetaloides (L.) O. Kuntze   U-shaped basal anastomosis (Cowie s.n.).
Smilacaceae
  Smilax kraussiana Meisn.   U-shaped, columnar (NU 5107).

Asparagales

Philesiaceae
  Behnia reticulata (Thunb.) Didrichs   U-shaped (*ex hort* NBG).
Asparagaceae
  Protasparagus africanus (Lam.)   U-shaped (Strey 7036).
  Protasparagus falcatus (L.)   U-shaped (Noel 2798).
  Protasparagus virgatus (Bak.)   U-shaped.
Dracaenaceae
  Dracaena hookeriana K. Koch   U-shaped, massive (Ward 4815).
  Sansevieria guineense (L.) Willd.   U-shaped, massive with some basal anastomosis.
Agavaceae
  Yucca gloriosa L.   U-shaped, few bars, some basal anastomosis (*ex hort*).
Hypoxidaxeae
  Empodium monophyllum (Nel) B.L. Burtt   Helical

(Rennie 79). *Hypoxis rooperi* S. Moore   U-shaped (Noel 2714).
  Rhodohypoxis baurii (Bak.) Nel   U-shaped (Hilliard & Burtt 8685).
Tecophilaeaceae
  Cyanella lutea L.   Helical with branching (Sim 1933).
  Walleria mackenzii Kirk.   Baseplate, perforate (Simpson 108).
Hemerocallidaceae
  Hemerocallis fulva (L.) L.   U-shaped (*ex hort*).
Asphodelaceae
  Bulbine frutescens (L.) Willd.   Helical (*ex hort*).
  Aloe arborescens Mill.   U-shaped (*ex hort* NU).
  Aloe kraussii Bak.   U-shaped (Noel 2726).
  Kniphofia laxiflora Kunth   U-shaped (Strey 9700).
Anthericaceae
  Anthericum cooperi Bak.   U-shaped.
  Chlorophytum comosum (Thunb.) Jacques   U-shaped (Stirton 234).
  Caesia contorta (L. f.) Dur & Schinz   Helical (Hilliard 3999).
Hyacinthaceae
  Drimiopsis maculata Lindl.   U-shaped (Frankish 40).
  Eucomis bicolor Bak.   Helical (Schelpe 1264).
  Ledebouria revoluta (L. f.) Jessop   U-shaped (Noel 2720).
  Ornithogalum longibracteatum Jacq.   U-shaped (Trauseld 1022).
  Scilla nervosa (Burch.) Jessop   U-shaped (Bews 3303).
Alliaceae
  Agapanthus campanulatus Leighton   U-shaped (Morris 226).
  Nothoscordum inodorum (Ait.) Nicholson   U-shaped (*ex hort*).
  Tulbaghia ludwigiana Harv.   U-shaped (Wright 1561).
  Tulbaghia simmleri Beauv.   U-shaped with some basal anastomosis (*ex hort*).
Amaryllidaceae
  Apodolirion buchananii Bak.   Helical (Phelan s.n.).
  Boophane disticha Harv.   U-shaped, some columnar (Pienaar 37).
  Cyrtanthus contractus N. E. Br.   U-shaped, some columnar (Commins 495).
  Cyrtanthus galpinii Bak.   U-shaped (Ward 2234).
  Haemanthus katherinae Bak.   U-shaped (Gordon–Grey 1365).
  Scadoxus multiflorus (Mart.)   U-shaped.

Burmanniales

Burmanniaceae
  Burmannia madagascariensis Mart.   Absent (Ward 7891).

Liliales

Alstroemeriaceae
Alstroemeria psittacina Lehm. U-shaped with some basal anastomosis.

Colchicaceae
Androcymbium striatum Hochst. U-shaped with some basal anastomosis (Martin s.n.).
Sandersonia aurantiaca Hook. Loosely reticulate (Smith s.n.).
Wurmbea alatior Nordenstam U-shaped with some basal anastomosis (Skead 101).

Liliaceae
Lilium regale Wilson U-shaped with basal anastomosis (Morris 240).

Iridaceae See Manning & Goldblatt (1990. U-shaped (including some basal anastomosis) and helical (presumed apomorphic).

Orchidales

See Freudenstein (1991). Basically U-shaped or annular (reduced to 1–2 Us or annuli in most Orchidoideae), with basal anastomosis in some taxa and reduction to columns in others.

ARIFLORAE

Arales

Araceae See French (1985a, b, 1986). Helical, with disruption into annuli, incomplete gyres or columns, rarely basal anastomosis or thickenings absent.

ALISMATIFLORAE

Alismatales

Aponogetonaceae
Aponogeton junceus Schlecht. U-shaped with some basal anastomosis (Edwards 2384).

Alismataceae
Alisma gramineum K.C. Gmel. Helical with some branching (Charette 2622).
Burnatia enneandra Micheli Baseplate tympanate, ribs continuous (Pooley 1579).
Limnophyton obtusifolium (L.) Miq. Baseplate tympanate, ribs continuous (Pooley 1575).
Sagittaria sagittifolia L. Baseplate tympanate, ribs continuous (Untawale & Bhasin 1973).

Hydrocharitaceae
Ottelia exserta (Ridley) Dandy Helical with some branching (Ward 7801).

Najadales

Juncaginaceae
Triglochin bulbosa L. Helical, double (Ward 6795).

Potamogetonaceae
Potamogeton crispus L. U-shaped (Ward 2667).
Ruppia spiralis Dum. Absent (Ward 8222).

BROMELIIFLORAE

Velloziales

Velloziaceae
Vellozia retinervis Bak. Helical, with annuli (Ross 1476).
Xerophyta viscosa Bak. Helical (Wingel 1).

Bromeliales

Bromeliaceae
Dyckia brevifolia Bak. U-shaped (*ex hort*).

Haemodorales

Haemodoraceae
Anigozanthus viridus Endl. Annular (*ex hort*).
Barberetta aurea Harv. Baseplate, derived from U-shaped (Hilliard 3476).
Lanaria lanata (L.) Druce U-shaped (Lewis 5334 NBG).
Wachendorfia paniculata L. Baseplate, derived from U-shaped.

Pontederiales

Pontederiaceae
Eichhornia crassipes (Mart.) Solms Baseplate, tympanate (Enseleni s.n.).
Pontederi cordata L. U-shaped.

Typhales

Typhaceae
Typha latifolia L. Helical (McDonald 215).

ZINGIBERIFLORAE

Zingiberales

Strelitziaceae
  Strelitzia nicolai R. & K.   U-shaped (Moll 2352).
Zingiberaceae
  Hedychium sp.   Annuli (*ex hort*).
Costaceae
  Stromanthe ?sanguinea Sond.   Columnar (*ex hort*).
Cannaceae
  Canna indica L.   Helical (Wood 38).
Marantaceae
  Calathea wallisii Regel   Basal anastomosis (Noel 2651).

COMMELINIFLORAE

Commelinales

Commelinaceae
  Commelina africana L.   Helical (Geekie 58).
  Cyanotis speciosa (L. f.) Hassk.   U-shaped (Geekie 56).
  Dichorisandra thyrsifolia Mihan. f.   Helical (Noel 2660).
  Floscopa glomerata (Schultes & Schultes f.) Hassk.   Helical (NU 5304).
  Tradescantia virginiana L.   U-shaped (*ex hort*).
  Zebrina pendula Schnizl.   U-shaped (Noel 2665).
Xyridaceae
  Xyris natalensis Nills.   U-shaped (Noel 2747).
Rapateaceae   See Venturelli & Bouman (1988). Helical, some branching.

Eriocaulaceae   See Manning & Linder (1990). Baseplate.
  Eriocaulon lanceolatus Miq.   Baseplate (Untawale & Bhasin 1973).

Cyperales

Juncaceae
  Prionium serratum (L. f.) Drege   U-shaped, massive (Strey 7764).
Cyperaceae   See Manning & Linder (1990). Helical.

Poales

Flagellariaceae   See Manning & Linder (1990). Baseplate.
Joinvilleaceae   See Manning & Linder (1990). U-shaped, some anastomosis.
Poaceae   See Manning & Linder (1990). Baseplate and annuli.
Ecdeiocoleaceae   See Manning & Linder (1990). Baseplate.
Anarthriaceae   See Manning & Linder (1990). Helical, disrupted.
Restionaceae   See Manning & Linder (1990). Helical.
Centrolepidaceae   See Manning & Linder (1990). U-shaped.

ARECIFLORAE

Arecales

Arecaceae
  Hyphaene natalensis Kuntze   U-shaped.

8 · The calcium oxalate package or so-called resorption tissue in some angiosperm anthers

WILLIAM G. D'ARCY, RICHARD C. KEATING AND
STEPHEN L. BUCHMANN

INTRODUCTION

In several plant families, crystalline calcium oxalate is accumulated in anthers and taken up along with the pollen by pollinators. In some plant groups (Solanaceae, Ericaceae) it is in the form of druses or crystal sand, while in others (Liliaceae, Balsaminaceae, Lemnaceae) it occurs as raphides. The crystals are found in a special structure commonly located beneath the epidermis at the stomium and running lengthwise around the anther (Figs. 1, 2). It is usually a row of cells that fill with fine calcium oxalate crystals. At maturity the cells bounding its sides break down to leave an oxalate package (OP) – a mass of crystals or raphide clusters. In poricidal anthers, which require sonication (buzz-pollination) by insects, the crystals are mixed with the pollen, while in anthers that open lengthwise, it is placed where it can be brushed onto the pollinator touching the anther.

In this paper we present (1) our own histological survey of the OP structure, mainly of the Solanaceae; (2) a review of the OP in other families based mainly on literature accounts; and (3) a discussion of possible functional roles for the OP structure and its contents.

Calcium oxalate crystals are widespread in plant and animal cells, and their occurrence in plants (except anthers) was surveyed by Küster (1956), Horner & Wagner (1980) and Smith (1982). Speculation continues as to their role in the plant's existence. Crystals of calcium oxalate are evident in many parts of the anther and have been discussed by many workers. In this paper we examine only those in the oxalate package, which we describe below. The crystal-bearing OP is best known to us in the Solanaceae and Ericaceae. The similarities in these cases suggest either a common function or a common origin, but we can show that it arose independently in different families. Previous workers have put forth various roles for the OP in the anther, which we discuss later. We deduce that it serves to enhance pollinator attraction through visual or chemical stimuli or both, and that it sustains an unreported kind of plant-pollinator interaction.

Perhaps the first note of this calcium oxalate-accumulating tissue in anthers was by Hegelmaier (1868: 107, fig. 6; 1871: fig. 9) in the Lemnaceae. It was later recorded in the Ericaceae (Artopoeus 1903: 320; Matthews & Knox 1926; Copeland 1943). Namikawa (1919) described it in anthers of the Solanaceae and identified its contents as calcium oxalate. In its best known configuration – hypodermal at the stomium – the tissue has been referred to as *Resorptionsgewebe* (resorption tissue) (Artopoeus 1903: 320; Puri 1970: 6) and 'disjunctive tissue' (Matthews and Knox 1926: 241). In Solanaceae it has been called a *Trennungsgewebe* (separation tissue) (Namikawa (1919: 385), 'resorption tissue' 'crystalliferous parenchyma' (Oryol & Zhakova 1977: 1723), 'stomium' (Horner & Wagner 1980: 1357), 'intersporangial septum' (Bonner & Dickinson 1989: 99), and 'hypodermal stomium' (Horner & Wagner 1992: 531). In some Ericaceae, (see below), tissue very like an OP has been referred to as 'granular pouches' (Palser 1951: 453; de Villamil 1980: 45). Chandler (1912: 53), who first noted the OP in the Theophrastaceae, called to it an 'open space, comparatively large, which is filled with acicular crystals in crystal dust,' and later Pohl (1932: 481) called this a *Kristallraume* (crystal cavity). In the Liliaceae, Balsaminaceae and Palmae, where the crystals are in the form of acicular raphides, Pohl (1941: 81) called the pollen-crystal mixture *Raphidenpollen* (raphides-pollen).

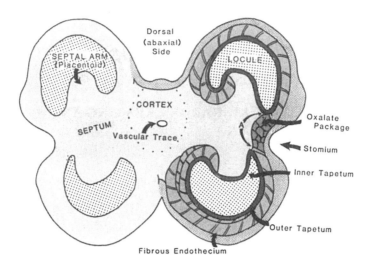

Fig. 1. Longitudinal diagram of a solanaceous anther showing the position of the oxalate package (A).

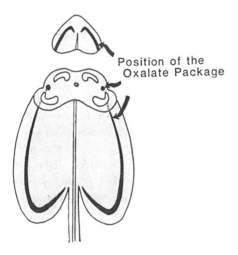

Solanoid Anther

Fig. 2. Cross-sectional diagram showing the major features of an anther in the Solanaceae.

For a general statement of anther morphology and its parts, see D'Arcy (this volume). In this paper we follow Hufford and Endress (1989: 303) where the term 'stomium is used in reference to the collective structural attributes that change over the course of anther ontogeny to give rise to the dehiscence region.' The stomium is often marked by a stomial furrow or 'lateral depression along the middle line of the thecae' (Venkatesh 1956: 170). In the Solanaceae and members of the Tubiflorae aggregation of families, the septum (Fig. 2) is evaginated, and the developing locules are C-shaped as seen in cross-section. These evaginations, which we refer to as septal arms, were termed placentoids by Chatin (1870: 2, 45). 'They have nothing in common with placenta' (Goebel 1905: 599). In most other flowers, the locules are round.

In many (or most) anthers, cells in the stomium beneath the epidermis break down or are resorbed just before dehiscence. This step towards confluence of the locules usually involves cells that seem to be unspecialized. The region was referred to as an *Öffnungsgewebe*, 'opening tissue' (OT) by Ziegler (1925: 405). When writing about the Melastomataceae, he equated the distal, small cells at the outer end of the septum with a *Trennungsgewebe*, or 'separation tissue.' In the Ericaceae, such cells were noted for their disappearance by collapsing (Matthews & Knox 1926: 246) and called 'collapse tissue' by Copeland (1943: 544; 1947: 84). He indicated it arose 'from subsidence and collapse of the thin plate of tissue the result of a pull from below, brought about by contraction of cells forming a part of the septum between the pollen sacs in each lobe.'

NEW INFORMATION

Methods

A wide taxonomic representation of the Solanaceae and some non-solanaceous taxa (see Appendix) was studied in serial cross-section. Most preparations for sectioning were made from living plants, but in some cases, dried herbarium material was used. When possible, each item was viewed at four stages of development, approximately pre-meiotic, meiotic, pre-anthesis, and post-anthesis. In addition to our own preparations, many slides were provided by Joseph Armstrong and Fredrick Utech. We also looked at living flowers of many species, particularly Solanaceae. Sometimes preparations did not show all developmental events but yielded considerable information.

Specimens from living plants were stored 48 hours or more in FAA. Dried material was first subjected to a variety of restorative techniques. All accessions, both our own and Armstrong's, were then cut into serial sections of 8–10 μm using a variety of standard methods. Staining used safranin and fast green, bismarck brown and methyl violet, and other stains. The staining was valuable in differentiating a variety of tissues, although the OP and fibrous endothecium could be clearly seen under polarized light and did not require staining.

To take up pollen and associated crystals, a microscope slide moistened with diluted glycerine was gently touched to the pollen-bearing surfaces of a series of recently dehisced anthers (Table 1). With a tuning fork, discussed below, we extracted pollen and crystalline material from poricidal anthers onto moistened slides.

The prepared slides were viewed with normal and polarized light at various degrees of extinction using stereo and compound microscopes at ×30 to ×100 magnifications. In addition to black and white photographs, some of which appear with this paper, color transparencies using tungsten-balanced Ektachrome 160 film and tungsten microscope illumination were prepared for review of observations.

Flowers of *Impatiens*, *Jaquinia nemophila*, *Deherania smaragadina*, *Typha angustifolia* and other taxa were examined using hand sections and dry mounts under a stereo dissecting microscope. Some samples were tested with lipid stains. These methods disclosed raphide bundles and other features.

Table 1. *Material that delivered calcium oxalate and pollen together when touched with a moistened microscope slide*

ARACEAE
Anthurium pallidiflorum Engler SOURCE: *Croat 50686* (MO) [heavy] – Ecuador
Anthurium sp. SOURCE: *Croat 53705* (MO) [light] – Brazil

BASAMINACEAE
Impatiens basamifer L. SOURCE: *D'Arcy 18270* (MO) cultivated, commercial source, St Louis, Missouri

ONAGRACEAE
Ludwigia pedunculatum (Griseb.) Gomez SOURCE: *D'Arcy 18269* (MO) cultivated Missouri Botanical Garden
Ludwigia grandiflora (Michx.) Greuter & Burdet SOURCE: *D'Arcy 18268* (MO) cultivated Missouri Botanical Garden 'from Cuba'

SOLANACEAE
Solanum americanum Mill. SOURCE: *D'Arcy 17705* (MO) – Comoro Is.
Solanum donianum Walp. SOURCE: *Pickett, s.n.* (MO) – Mexico
Solanum richardii Dunal SOURCE: *D'Arcy 17151* (MO) – Comoro Is.
Solanum sisymbrifolium Lam. SOURCE: *D'Arcy 17738* (MO) – Argentina
Capsicum annuum L. SOURCE: *D'Arcy 18271* (MO) commercial source – Mexico
Capsicum pubescens R. & P. SOURCE: *D'Arcy 18272* (MO) – South America
Cyphomandra betacea L. SOURCE: *D'Arcy 18273* (MO) – South America [very light]

THEOPHRASTACEAE
Jacquinia macrocarpa Cav. SOURCE: Progeny of D'Arcy 5236 (MO) – Panama
Jacquinia nemophila Pitt. SOURCE: *Progeny of D'Arcy 13695 (MO) – Panama*

Information is given as Taxon, Voucher – Geographical Distribution.

To determine contents of the staminodes of *Schizanthus*, crystal material from the larger staminodes of about 10 flowers (20 staminodes) was teased into watch glasses and reagents added under a stereo microscope.

Water, acetic acid, sodium hypochlorite, hydrochloric acid (2–5%) and sulfuric acid (2–10%) were added individually to each sample, and the watch glasses were re-examined after about 1 hour. In another test, material from the staminodes was picked up with a tungsten needle moistened in water and flamed.

In a tentative attempt to learn whether the calcium oxalate might be visible under ultraviolet light, recently opened anthers were examined in an otherwise dark room with a long wave UV 'black' light purchased from a hobby shop. The crystalline material could, indeed, be seen in this way. However, because we attributed other functionality to the material, this avenue of study was not pursued further.

In preliminary feeding studies (to be published by Buchmann, Schmidt, Li–Shen & D'Arcy) honeybees were presented with pollen adulterated with up to 2.5% dry weight of commercially obtained calcium oxalate. Controlled studies are part of the experimental design to be reported later.

Histological observations: Solanaceae

An oxalate package is found in most solanaceous anthers and develops in a similar way (Figs. 3–7). Typically, it appears in the early development of the anther, well before meiosis (Fig. 3), as a rib or band of 6–12 cells wide just inside the epidermis, extending the entire length of the anther at the stomium (Fig. 7). In serial sections, these early cells are densely staining and lack visible vacuoles, suggesting intensified metabolic activity. The cells of this single layer often divide periclinally into two or three cells (Fig. 4) and may give the tissue a triangular appearance. At about the time of meiosis, the cells entirely fill with numerous small calcium oxalate druses and become birefringent. Later, when pollen is mature and the tapetum senescent, cell walls in the OP degenerate to leave a mass of calcium oxalate crystals that does not enlarge further (Fig. 5). At about this time the septum and its arms undergo dieback. In anthers of small diameter, the septal arms may disappear completely, while in large anthers they persist through anthesis. As the septum withdraws from the side of the anther, an opening is formed. This opening, or stomial pouch (Fig. 5), is bounded inward by the end of the remaining septum, on the sides by the crushed remains of the tap-

etum in the two locules, and on the outer side by the epidermis. The OP is loosely enclosed in the pouch defined by these boundaries.

As shown in Figs. 8 and 9, the pouch or opening around the crystals is much larger than the crystalline mass itself, which by now consists of a lengthwise-running rib of calcium oxalate. This material may adhere to the epidermis of the stomium or other parts of the

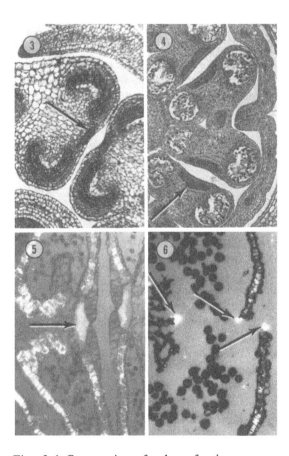

Figs. 3–6. Cross-sections of anthers of various Solanaceae at different developmental stages showing the oxalate package. Arrows indicate the oxalate package or remnants of it. Fig. 3. *Solanum americanum* before meiosis. Fig. 4. *Capsicum pubescens* at the stage when calcification begins. Fig. 5. *Graboswkia boerhaavifolia* just before dehiscence. (Light partially polarized.) Fig. 6. *Solanum lycopersicoides* just after dehiscence. (Light partially polarized.)

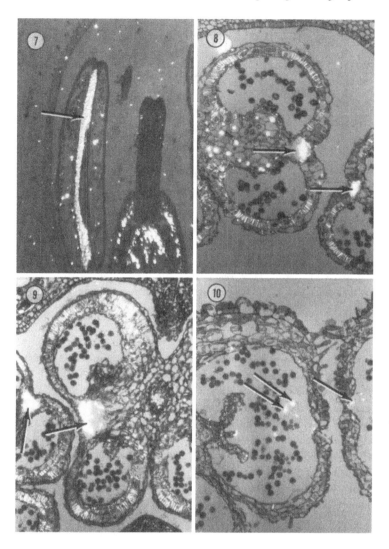

Figs. 7–10. Fig. 7. *Brachistus nelsonii* (Solanaceae), longitudinal section of anther showing the oxalate package extending along the entire length. (Light partially polarized.) Figs. 8–10. Cross-sections of anthers of Solanaceae showing various positions of crystals of the oxalate package at anther maturity, just before or just after dehiscence. (Light partially polarized.) Fig. 8. *Exodeconus miersii* showing pouch and loose crystals. Fig. 9. *Capsicum annuum* showing pouch and loose crystals. Fig. 10. *Lycopersicon esculentum* showing pouch and loose crystals.

pouch area. The next event is disruption of the two tapetal strands bounding the stomium chamber.

This results in confluence of the two locules of each theca into one locule, and the whole anther becomes two locular (Fig. 10). Finally, in longitudinally dehiscent anthers, the epidermis breaks, freeing the locules to open and expose the pollen (Fig. 6). This sequence was outlined by Keijzer (1987), but in the Liliaceae that he studied the locules are round, and there is no OP. In his examples, cells break down at the distal end of the septum at about the point where an OP might be expected. But the 'opening tissue' (OT) in his species

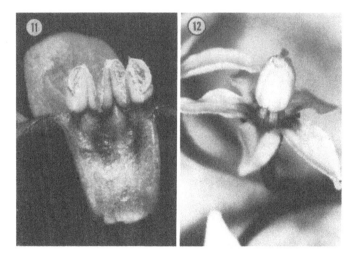

Figs. 11, 12. Fig. 11. Dissected flower of *Witheringia correae* with three anthers partially opened. White calcium oxalate crystals outline the openings. Fig. 12. *Solanum americanum* flower showing anthers with terminal pores forming a column.

shows no differentiation from other parenchymatous cells in the septum. *Schwenckia*, a solanaceous genus that lacks a crystalline OP, has a well-developed pouch area.

After dehiscence, granules from the OP remain in the stomial chamber or adhere to either the end of the septum or the edge of the longitudinal slit (Fig. 6). Sometimes they can be seen with the naked eye as a bright white line along the line of dehiscence. (Fig. 11). In anthers with poricidal dehiscence (Fig. 12), the OP is sited in the anther in much the same way.

Contents

Material from the OP in staminodes of *Schizanthus* remained intact with water, acetic acid and sodium hypochlorite but dissolved in hydrochloric acid, and in sulfuric acid. According to Yasue (1969), this is diagnostic for calcium oxalate. In the flame test, the orange flame characteristic of calcium (Lide 1993: sect. 4–6) was seen.

In most preparations, the crystalline material in the OP appears similar throughout. In some cases the crystals appeared much larger and coarser than in others, and in a few cases less defined and less birefringent. This suggests qualitative differences in the crystalline content of the OP that other techniques might elucidate. In the

Appendix, taxa are scored subjectively as having fine, granular and coarse OP.

Size

Obvious differences were seen in the size of the OP in various taxa, but because developmental-stage differences affecting size were only loosely approximate, size measurements only represent relative sizes. The estimates shown in Table 2 and Fig. 13 give only a general indication of relative size and are not appropriate for careful numerical comparison.

Size of the OP is not correlated with anther diameter. It is best developed, or largest, in the capsicoid genera *Acnistus*, *Capsicum*, *Tubocapsicum*, *Witheringia*, and in species of *Cyphomandra*. In at least one case, *Capsicum pubescens* (Fig. 4), the OP runs more than one third of the width of the anther. In general the OP is smallest, only two to four cells wide, in species of *Lycopersicon* and *Salpiglossis* (both longitudinally dehiscent), and in some species of *Solanum* subgenus *Leptostemonum* (dehiscent by terminal pores). In the genus *Capsicum*, the OP varies greatly in different species; *Capsicum annuum* and *C. baccatum* have a medium size or small OP, and *C. pubescens* and *C. chacoense* have large ones.

Table 2. *Size of oxalate package in selected species of the Solanaceae*

| Taxon | Oxalate package at the stage of: | |
	PMC	Pollen
Capsicum annuum	36	104
Capsicum lanceolatum	27	54
Capsicum pubescens	108	180
Chamaesaracha sordida	41	95
Cyphomandra betacea	36	90
Kalmia latifolia	n/a	95
Physalis alkekengi	45	81
Solanum antillarum	27	36
Solanum capsicoides	36	54

n/a, not available. Measurements in micrometers.

Cellular orientation

The OP is almost always abruptly demarcated from adjacent tissues, but in *Deprea*, crystals in the OP resembled those in the septum. In many Solanaceae the cells of the OP are aligned to appear as a palisade perpendicular to the epidermis (Fig. 14), and in other cases, they form a triangular configuration. In fertile anthers of *Solanum mahoriensis* (Fig. 15), and *Lycopersicon*, the OP is very small and might be regarded as transitional, either rudimentary or incipient.

Staminodes

Anther-bearing staminodes were examined for two species each of *Browallia* and *Schizanthus*. Those of *Browallia* had no pollen, no endothecium, and no OP. In *Schiz-*

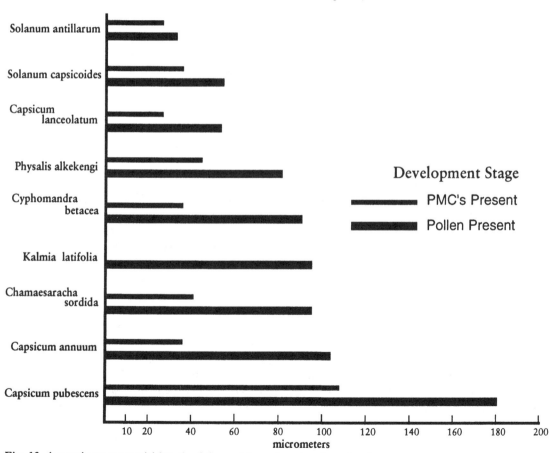

Fig. 13. Approximate tangential length of the oxalate package in selected anthers.

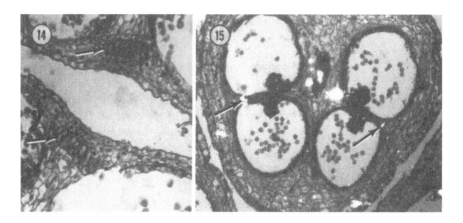

Figs. 14, 15. Unusual subhypodermal position of the oxalate package in two taxa of Solanaceae; cross-sections under partly polarized light. Fig. 14. *Cyphomandra*. The oxalate package is subhypodermal and arranged in a palisade formation. If viewed under polarized light, this preparation would display presence of crystals in the oxalate package. Fig. 15. *Solanum mahoriensis*. The oxalate package (at arrows) is deep within the anther tissue at the end of the septum and greatly reduced.

anthus (Fig. 15, 17) there are two fertile stamens and three sterile staminodes. The anther portion of the staminodes is much smaller than that of the fertile stamens and fills entirely with crystalline material as in the OPs of fertile anthers (Fig. 17). At anthesis, they dehisce to form small knobs of calcium oxalate (Fig. 16). These knobs presumably play a role in pollinator attraction or selection. In *Brunfelsia densiflora*, the two reduced anthers bear pollen, an OP, and an endothecium, and hence are fertile anthers, although their smaller size is suggestive of infertile staminodes.

Exceptions

In four genera – *Cestrum, Schwenckia, Nicandra* and *Nierembergia* – and in one or two species of *Solanum*, no OP was found, and it is believed to be absent.

In *Cestrum* (two taxa examined), at an early developmental stage, cells where an OP might be expected stain densely, but the region undergoes no further change, and no crystals are formed; locules become confluent without forming a stomial chamber, and the anther dehisces lengthwise as in many other plant families.

In *Nicandra* we saw no crystal formation, no stomial pouch formed, no OP. This monotypic genus was the

only one seen, in any family, where there is endothecial thickening in the septum distal to the septal arms. The configuration of the stomium allows no place for a structure such as an OP (Fig. 18).

In *Schwenckia* (two taxa examined) there is no sign of crystal formation, but a conspicuous and well-defined stomial pouch is formed.

In *Nierembergia*, although preparation quality was unsatisfactory, we saw no evidence of an oxalate package or pouch, and dehiscence may involve an opening tissue only. Species in this genus attract oil-collecting bees (Cocucci 1991), and perhaps this pollination mode is related to the absence of an OP.

Another example of absence of an OP is the anthers of pistillate flowers of *Solanum polygamum*. These anthers look like normal anthers, but pollen is completely lacking, and there is no sign of an OP.

The OP is usually strictly hypodermal, abutting the epidermis in the stomium, but in some cases, all with poricidal dehiscence, it is several cells below the anther epidermis. In most species of *Cyphomandra*, the OP is several cells below the epidermis, but in *C. hartwegii* it may be variable, ranging from hypodermal to two cells in from the epidermis within a single anther. In *Solanum allophyllum*, formerly placed in *Cyphomandra*, our prep-

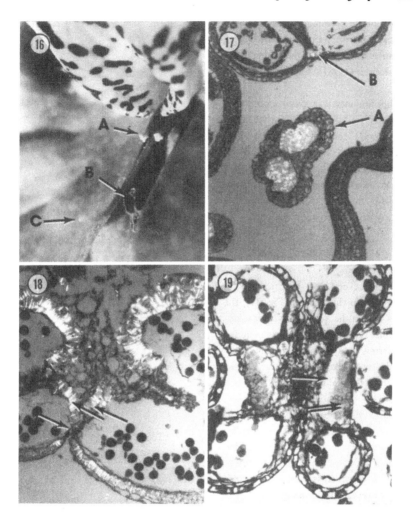

Figs. 16–19. Fig. 16. *Schizanthus pinnatus*. Photograph of the floral throat showing two staminodes with knobs of calcium oxalate (arrow A), fertile stamen (arrow B), and style (arrow C). Fig. 17. *Schizanthus pinnatus*. Cross-section of flower showing staminode. The photograph shows one staminode occupied by crystals of the oxalate package and parts of two fertile stamens showing the oxalate package (arrow). Fig. 18. *Nicandra physalodes*. Cross-section showing the absence of a site for an oxalate package (double arrow) and endothecial-type thickenings in cells of the septum distal to the septal arms (arrows). Fig. 19. *Kalmia latifolia*. Cross-section showing the pouch (arrow) and crystalline oxalate package (double arrow).

arations did not reveal crystals, but the pouch area was subhypodermal. A subhypodermal OP is also common in *Solanum* subgenus *Leptostemonum* (Fig. 15) and in some species of *Lycianthes*. In the perhaps intermediate case of *S. dulcamara*, most of the OP was manifestly subhypodermal, but at the point of stomium rupture at the pore,

it was hypodermal. In *Solanum madagascariense*, the OP was subhypodermal in origin, but the overlying layer was thin and soon crushed and hardly recognizable, and the OP appeared to be hypodermal.

One result of a subhypodermal placement of the OP is that at points of dehiscence (in the pore area), the stomium site that disrupts to permit pollen egress may be several layers thick. This points to different hygroscopic cellular movements for opening the anther from those outlined by Keijzer (1987) for longitudinally dehiscent anthers.

In a few cases, slide mounts that seemed otherwise well prepared and showed well-developed pouches in the OP region did not provide conclusive evidence as to presence or absence of an OP. Notable among these are *Nieremberia liniariaefolia*, *Datura stramonium*, *Solanum allophyllum*, *Solanum donianum*, *Withania somnifera*. These taxa, included in the table of materials (Appendix), are worthy of future reinvestigation.

Histological results: Ericaceae

Only *Kalmia latifolia* of several taxa of Ericaceae we examined showed an OP. In this species (Fig. 19), the OP greatly resembles that in the Solanaceae in development, topology and final appearance. Its size is included in Table 1. The epidermis does not have crystals in any part, and soon after the crystal-bearing cells of the hypodermal area break down, the radial walls of the pouch (tapetum) collapse. The tangential wall (epidermis) remains as a thin, hardly visible barrier to the exterior. As in many Solanaceae, a conspicuous pouch formed by pull-back of the septum then surrounds the crystalline OP. Fibrous-thickened endothecium cells are present only in the pore area. As seen in cross-section the locules are round in the Ericaceae, not C-shaped as in the Solanaceae.

Histological observations: other families

Slides were made of anthers of other families (Appendix), but none of our preparations revealed anything resembling an OP.

Pollen dusting observations

Moistened microscope slides gently dabbed or dusted with pollen from anthers of living plants and viewed under polarized light showed abundant druses or raphides mixed with the pollen in a number of taxa (Table 2, Figs. 20–22). This was found both in anthers with longitudinal and poricidal dehiscence.

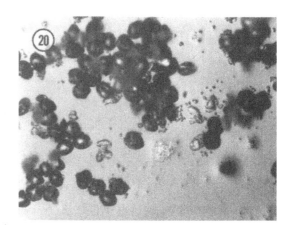

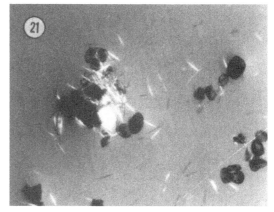

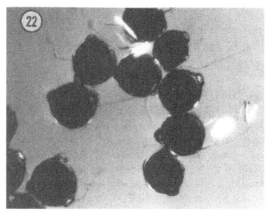

Figs. 20–22. Pollen and crystals obtained by gentle contact of microscope slides with recently opened anthers under partly polarized light. Fig. 20. *Capsicum annuum* with druses. Fig. 21. *Impatiens balsamifer* with raphides. Fig. 22. *Ludwigia peduncularis* with raphides and also viscin strands (dark threads) which are characteristic in this genus.

Table 3. *Occurrence of an oxalate package in the Solanaceae in a traditional systematic context (after D'Arcy 1991)*

Taxon	Present	Absent	Not examined or equivocal[a]
CESTROIDEAE			
ANTHOCERCIDEAE	*Anthocercis, Anthotroche, Crenidium, Cyphanthera, Duboisia, Grammosolen, Symonanthus*		
CESTREAE		*Cestrum*	*Metternichia, Pantacantha, Sessea, Vestia*
NICOTIANEAE	*Nicotiana, Petunia*	*Nierembergia*	*Benthamiella, Bouchetia, Calibrachoa, Combera, Fabiana, Latua*
SCHWENCKIEAE		*Schwenckia*	*Melananthus, Protoschwenkia*
SALPIGLOSSIDAE	*Browallia, Brunfelsia, Salpiglossis, Schizanthus, Streptosolen*		*Hunzikeria, Leptoglossis, Plowmania, Reyesia*
NOLANOIDEAE	*Alona, Nolana*		
SOLANOIDEAE			
DATUREAE			*Datura, Brugmansia*
HYOSCYAMEAE			*Anisodus, Atropanthe, Hyoscyamus, Physochlaina, Prezewalskia, Scopolia*
JABOROSEAE,	*Salpichroa*		*Jaborosa*
JUANULLOEAE	*Hawkesiophyton*		*Dyssochroma, Ectozoma, Juanulloa, Markea, Merinthopodium, Schultesianthus*
LYCIEAE	*Grabowskia, Lycium*		*Phrodus*
NICANDRAE		*Nicandra*	
SOLANDREAE	*Solandra*		*Trianaea*
SOLANEAE	*Acnistus, Brachistus, Capsicum, Chamaesaracha, Cuatresia, Cyphomandra, Deprea, Discopodium, Dunalia, Exodeconus, Iochroma, Jaltomata, Lycianthes, Lycopersicon, Physalis, Solanum[b], Tubocapsicum, Vassobia, Withania, Witheringia*		*Athenaea, Atropa, Aureliana, Dunalia, Hebecladus, Leucophysalis, Mandragora, Margaranthus, Mellissia, Nothocestrum, Oryctes, Pauia, Quincula, Saracha, Triguera*

[a] Equivocal.
[b] Present in most cases examined; missing in *S. antillarum, S. polygamum.*

SYSTEMATIC DISTRIBUTION OF THE OXALATE PACKAGE

Distribution in the Solanaceae

The distribution of the OP in the Solanaceae is indicated in Table 3, where its occurrence is shown in a traditional conspectus of the family (D'Arcy 1991). It can be seen that at least one species of each tribe in the family has been assessed. Each genus investigated and demonstrated as lacking an OP is found in a different tribe and the two major subfamilies are represented. So far as is known, each tribe either has an OP or does not, which argues for a certain taxonomic basis for its occurrence,

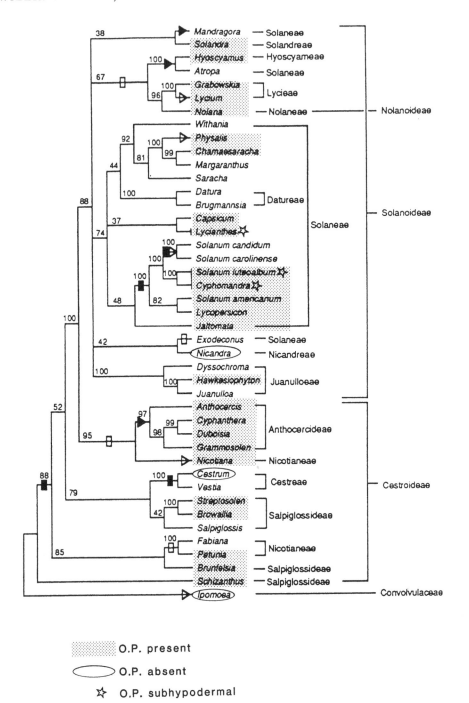

Fig. 23. Distribution of the oxalate package (OP) using the Solanaceae *rbc*L phylogeny of the family proposed by Olmstead & Palmer (1992).

or conversely, for the validity of the traditional classification.

Amplifying or supporting our findings are the records of OP in 10 taxa of Solanaceae by Namikawa (1919); *Capsicum annuum, Nicotiana alata* var. *grandiflora, Petunia violacea, Physalis alkekengi, Schizanthus pinnatus, Solanum dulcamara, S. melongena, S. nigrum, S. tuberosum,* and the report of antibody reaction in OP in *Capsicum annuum, Petunia hybrida* and *Lycopersicon esculentum* by Trull *et al.* (1991: 14). Horner & Wagner (1980) reported it in *Capsicum annuum* and *Lycopersicum esculentum.*

As a way of assessing the distribution of the OP in Solanaceae, Fig. 23 shows its occurrence on the phylogenetic scheme of Olmstead and Palmer (1992), which is based on chloroplast DNA (*rbcL*) sequences. The OP is present in most of the genera that they studied, including *Schizanthus*, which is basal in their Solanaceae tree. As the OP was not found in the outgroup, *Convolulus*, nor elsewhere in the Asterideae clade (Appendix), the structure probably arose within the Solanaceae clade (Solanaceae or its immediate ancestors). The absence of the OP in two taxa on different branches suggests that it has been lost independently in these cases.

Taxonomic Conclusions: Solanaceae

Until details of the adaptive role of the OP are known and interpretation of its structural features can be made, taxonomic conclusions will be few. We do note the following:

In *Nicandra*, which forms a monotypic tribe in some conspecti of the family (Wettstein 1892), there is endothecium-like thickening in the septum distal to the septal arms, and there is no site where an OP might develop. This argues for a substantial taxonomic distance from other members of the family.

In most taxa of Solanaceae, the OP is hypodermal, abutting directly on at least a few cells of the epidermis, although in some cases the outer layers are two or more cells thick. A notable exception to this is the subhypodermal position in *Cyphomandra*, many species of *Solanum* subgenus *Leptostemonum*, and some species of *Lycianthes*. While there is little in the appearance of these plants to suggest a special affinity, the chloroplast DNA analysis of Olmstead and Palmer does place *Cyphomandra* and subgenus *Leptostemonum* as closest neighbors on the Solanaceae tree.

Distribution in other families

Positive reports

Ericaceae. An OP (resorption tissue) has been reported in many taxa of Ericaceae. Our own slides, the photos of de Villamil (1980), and other reports, show an OP in the Ericaceae like that in the Solanaceae. In the Solanaceae, the OP always occurs at the stomium; crystalline tissue (OP) at this site in the ericaceous anther has usually been called resorption tissue, a synonym for the oxalate package (OP). Similar tissues, referred to as 'granular pouches,' occur in the stamens of some Ericaceae. Situated away from the stomium, even on the filament (*Gaultheria*), these pouches open and void their calcium oxalate crystals. de Villamil suggested – and her photographs bear her out – that these granular pouches are the same phenomenon as the OP but with a different topology. However, B. Palser (personal communication) had the impression that the contents of these pouches are gone before opening of the flower.

In some Ericaceae, the OP tissue has been described as 'involving the epidermis except at the extreme lower end where it is hypodermal (Copeland 1943: 557; de Villamil 1980).' In *Erica hirtiflora*, the adjacent epidermis is reported as becoming crystalliferous and also containing other 'granular contents' (Matthews & Taylor 1926: 241). We did not note crystalline epidermis in any Solanaceae, nor in *Kalmia*, where this condition was reported by Copeland (1943: 544). The latter author, Copeland (1943: 544), referred to a 'fluid mass' at the site of the OP in some Ericaceae, a condition we have not verified.

The distribution of OP and granular pouches in the Ericaceae is shown in Table 4, a traditional conspectus of the family (Stevens 1971). As in the Solanaceae, suprageneric aggregates tend either to have or not to have an OP. An exception is *Vaccinium*, where only one species (*V. albicans* Sleum: de Villamil 1980) is known to have it, but it is lacking in others that have been tested.

Oxalate package distribution is related to the chloroplast DNA (*rbcL*) arrangement of Ericales by Kron & Chase (1993) in Fig. 24. It is present in most taxa of the Ericaceae but is not found in the basal taxon, *Enkianthus*, or in the outgroup. Thus, the OP probably arose independently within the Ericaceae. Whether its loss in *Epacris, Dracophyllum* (these often recognized as

Table 4. *The oxalate package in the Ericales reported as resorption tissue[a] or as granular pouches (taxonomic arrangement modified from Stevens 1971)*

Taxon[b]	Oxalate package present	References
ERICACEAE		
Rhododendroideae		
Bejarieae		
Befaria	In stomium	Copeland (1943)
Rhodoreae	None found	Copeland (1943)
Cladothamneae		
Cladothamnus	In stomium	Copeland (1943)
Elliottia	In stomium	Copeland (1943); de Villamil (1980)
Epigaeae		
Phyllodoceae		
Bryanthus	In stomium	Copeland (1943)
Kalmia	In stomium	Copeland (1943); de Villamil (1980); Matthews & Knox (1926); this study
Ledothamnus	In stomium	Copeland (1943)
Phyllodoce	In stomium	Copeland (1943); Matthews & Knox (1926)
Rhodothamnus	In stomium	Copeland (1943)
Daboecieae		
Daboecia	In stomium	Copeland (1943); Matthews & Knox (1926)
Diplarcheae		
Diplarche	In stomium	Copeland (1943
Ericoideae		
Ericeae (Salaxideae)		
Bruckenthalia	In stomium	Artopoeus (1903)
Erica carnea	None found	de Villamil (1980)
Erica – many species	In stomium	Matthews & Knox (1926); Matthews & Taylor (1926)
Calluneae		
Calluna	Granular pouches throughout	de Villamil (1980); Artopoeus (1903)
Vaccinioideae		
Arbuteae		
Arbutus	In stomium	Matthews & Knox (1926)
Arctostaphylos	In stomium	Matthews & Knox (1926)
Enkiantheae		
Enkianthus	None found	Copeland (1943); de Villamil (1980)
Cassiopeae		
Cassiope	In stomium	B. Palser (personal communication 1994)
Harrimanella	Not found	B. Palser (personal communication 1994)
Andromedeae		
Gaultheria	Granular pouches	de Villamil (1980); Palser (1951); Matthews & Knox (1926)
Lyonia	Granular pouches	de Villamil (1980)
Leucothoe	Granular pouches	de Villamil (1980)
Oxydendrum	Granular pouches	de Villamil (1980)
Pieris	Granular pouches	de Villamil (1980)

Table 4. (*cont.*)

Taxon[b]	Oxalate package present	References
Vaccineae		
Agapetes (*Pentapterygium*)	None found	Matthews & Knox (1926)
Vaccinium albicans	In stomium	de Villamil (1980)
V. stamineum	None found	de Villamil (1980); this study
Vaccinium spp.	None found	Matthews & Taylor (1926)
Pyroloideae		
Chimaphila, Ramischia	In stomium	Copeland (1947)
Moneses, Pyrola	None found	Copeland (1947)
Monotropoideae		
Wittsteinioideae		
CLETHRACEAE	None found	Kavaljian (1952: 400)
EPACRIDACEAE		
Eleven genera	None found	Paterson (1961)
Sprengelia incarnata	In stomium	Paterson (1961)

[a] Reports of resorption tissue are noted here as being in the stomium. Taxa in this conspectus which have not been investigated have been left blank.

[b] Also *Orthaea, Thibaudia, Psammisia, Mycerinus, Lateropora, Notopora, Arbutus, Arctostaphylos* with pouches reported by Luteyn in de Villamil (1980: 100).

Epacridaceae), *Ceratiola, Rhododendron*, and species of *Vaccinium* was shared or independent is not known. Thus, absence of an OP may be considered plesiomorphic in the Ericales and its presence a synapomorphy of those Ericaceae more advanced than *Enkianthus*.

Within the advanced Ericaceae, it is plesiomorphic; its absences in this family (and in these Epacridaceae) are reversions that may be considered a derived state or synapomorphy. One member of the Epacridaceae not studied by Kron & Chase, *Sprengelia*, is reported to have an OP (Paterson 1961: 265).

Finally, the Solanaceae and Ericaceae are quite distant from one another on the chloroplast DNA-generated tree of the Asterideae of Olmstead & Palmer (1992). This suggests that the OP arose independently in the Solanaceae and Ericaceae, as is probably the case in other families where it has been reported.

Pollination by bees is widespread in the Ericaceae, e.g., *Calluna, Erica, Phyllodoce, Vaccinium* (Haslerud

1974); *Kalmia* (Rathcke & Real 1993); *Zenobia* (Dorr 1981); *Cavendishia* (Luteyn 1983: 41). Other pollinators are also reported, flies and thrips in *Calluna* (Faegri & van der Pijl 1979: 174), hummingbirds in *Cavendishia* (Luteyn 1983: 41).

Theophrastaceae. Chandler (1912) reported an open space between the pollen sacs 'filled with acicular crystals in crystal dust' in *Deherainia smargadina*, and her illustration is much like ours of *Kalmia* (Ericaceae) except that the OP seems to be subhypodermal. Pohl (1932) noted an OP, which he termed a crystal chamber (*Kristallraume*), containing various crystals, druses, needle-crystals and crystal sand in anthers of the same species. Chandler believed the species to be self-sterile, and Pohl thought that flesh-seeking flies were attracted to the corolla and glands on it.

This family is sometimes placed in the ericad alliance as one of its basal members (Kron & Chase 1993).

Table 5. *Families of seed plants known to have an oxalate package*

As druses, crystals, etc.	
Epacridaceae	Few species
Ericaceae	Many species, missing in few
Solanaceae	Most species, missing in few
Theophrastaceae	Some species, few investigated
Tiliaceae	Some species, few investigated
As raphids	
Araceae	Some *Anthurium*
Balsaminaceae	Some species
Bromeliaceae	Many species
Lemnaceae	Most species
Liliaceae	Some species, missing in most
Onagraceae	*Epilobium, Ludwigia*
Sparganiaceae	*Sparganum ramosum*
Uncertain reports or unknown type	
Amaryllidaceae	
Melastomataceae	
Ochnaceae	
On the petal	
Orchidaceae	*Stelis* sp., absent elsewhere

Balsaminaceae. Maurizio (1953: 524, 528) discussed and illustrated membrane-bound raphide bundles in honey-bee loads from *Impatiens*. We saw raphides, mostly dissociated, on moistened slides from *Impatiens* (Fig. 21). Kato *et al.* (1989) noted insects from 11 families in four orders visiting seven species of *Impatiens* in a lowland site in West Sumatra but did not identify primary pollinators (Hotta 1989: 5).

The nature and possible role of the OP in *Impatiens* was discussed in Vogel & Cucucci (1988).

Bromeliaceae. Kugler (1942) reported *Raphidenpollen*, pollen mixed with raphides located in an OP, in several genera. He illustrated anthers, and classified them on the orientation of the raphides in the OPs.

Araceae. Hegelmaier (1871: 648) noted an OP in *Atherurus* (*Pinellia*) and *Arum*. Hegelmaier also (1871: 649), citing van Tieghem (1866: fig. 10, error for fig. 9), reported mixing of crystals with pollen in *Richardia africana* (*Zantedeschia aethiopica* (L.) Sprengel, *fide* Traub 1948: 12). Reviewing this report Kugler (1942) concluded that the crystals are so placed in the anther that they cannot reach the pollen. However, without testing of its pollen dispersal, the report for an effective OP here cannot be discarded. Hegelmaier (1871: 646) and Pohl (1941: 85) reported raphides mixing with the pollen in *Calla palustris*. Hegelmaier (1871: fig. 9) showed an OP much like that in other families. Although moistened slides touched to living spadices of many species of *Anthurium* and *Spathiphyllum* at the Missouri Botanical Garden collected pollen but no calcium oxalate, abundant raphides were mixed with the pollen of *Anthurium pallidiflorum* and another species, and lighter amounts were found in a third *Anthurium* species. These species have dry pollen, unlike the moist or clumped pollen in most Araceae. When the flat side of a vibrating tuning fork tine was touched to the spadix, clouds of pollen and raphides were ejected from all sides of the region touched.

Lemnaceae. Hegelmaier (1868: 107, fig. 6) reported abundant small raphides mixed with the pollen in *Lemna trisulca* L. and provided an illustration showing the OP. Later (1871) he illustrated a cross-section of *Spirodela polyrrhiza* (L.) Schleiden showing clearly a well-defined OP in the same position as in our *Kalmia*.

Landolt (1986: 97) reported that 'raphide cells are present in the anther walls of *Spirodela* and *Lemna*,' and (p. 106) 'The line of dehiscence of the anther locules can be observed externally as a border line between two rows of cells at the place where the septa between the two locules touch the wall. At that point, just below the wall, there is a double row of raphide cells (Fig. 2.49) in *Spirodela* and *Lemna*. From here, the opening of the anthers takes place.' He also noted that 'the oxalate crystals of the raphide cells get mixed with the pollen grains when

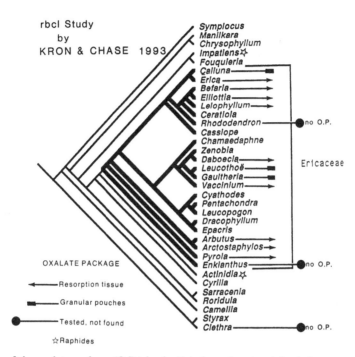

OXALATE PACKAGE

&

ERICAD PHYLOGENY

Fig. 24. Distribution of the oxalate package (O.P.) in the Ericales using the *rbc*L phylogeny proposed by Kron & Chase (1993). Thick lines indicate the Ericaceae, *sensu strictu*.

the anthers open (protection against feeding by arthropods?).'

Arecaceae. Pohl (1941: 87) cited an earlier report (Fickendey & Blommendaal 1929) of crystal needles in the anthers of the oilpalm, *Elaeis guineensis* Jacq., which is pollinated almost exclusively by weevils, thrips and beetles (Hartley 1988: 78).

Liliaceae. Pohl (1941) reported finding a raphide-bearing OP in a suite of Liliaceae genera, *Paradisia, Bulbinella, Bulbine, Anthericum, Chlorophytum, Esheandia* (subfam. *Asphodeloideae*), *Ornithogalum, Hyacinthus* (subfamily Scilloideae), and its absence in a number of other lili-aceous groups. It was absent in our views of *Scoliopus*

and *Stenanthium*. Pohl noted that all occurrences in the Liliaceae are in nectar-bearing flowers, which are other-wise uncommon in the family, and these are also bee-pollinated (P. Bernhardt, personal communication).

Onagraceae. Maurizio (1953: 528) reported raphides in honeybee pollen loads from *Epilobium angustifolium* L. We saw raphides, both in bundles and dissociated, in microscope slides touched with recently opened anthers of *Ludwigia peduncularis* (Fig. 22) and *L. grandiflora* (both section *Oligospermum*).

Tiliaceae. Demianowicz (1963) reported a druse-bearing OP in anthers of *Tilia*, sited as in many Solanaceae and in *Kalmia* (Ericaceae). He also showed that the druses are so abundant in honey made from *Tilia* nectar that they may be used to identify *Tilia* honey. In prepared microscope slides distributed by Triarch (George H. Conant, Ripon, Wisconsin, slide no. 17–18, set 14, no. 23), labeled *Tilia americana* bud, a well-developed OP much like that in many Solanaceae is clearly displayed. In this slide the OP appears to be hypodermal, and the druses are slightly larger than in most Solanaceae. In *Sparmannia*, also Tiliaceae, Demianowicz found a well-defined OP in the anthers. In this case, the druses were not deposited in the honey, seemingly being fixed in their cells and not presented for delivery with the pollen.

Equivocal and erroneous reports

Typhaceae. Kugler (1942: 388 and 1970: 13) interpreted a report by Kronfeld (1887) to mean raphides among the pollen. However, Kronfeld related how he moistened anthers from herbarium specimens of seven *Typha* species and dissected them with a needle before making observations. He said the raphides he found came from the anther wall, and he did not imply that they are mixed with the pollen. We took anthers that were open but pollen-bearing from herbarium specimens of *Typha angustifolia* L. (*Smith 2485*, Missouri (MO)) and teased out the pollen without disturbing the tissue of the anther itself. In this pollen, and in several kilograms of pollen of two other species gathered for another study (Schmidt *et al.* 1989), we found no crystalline material so we conclude that *Typha* anthers do not have an OP. Thus the finding by Schmidt *et al.* (1989) that *Typha* pollen is distasteful to and avoided by honeybees is probably not because of raphides.

Sparganiaceae. Pohl (1941: 85) tentatively reported raphides among the pollen grains of *Sparganium ramosum* Huds., suggesting that this case should be reviewed.

Leguminosae. An unspecialized OT was reported as resorption tissue by Matthews & Maclachlan (1930: 114) in *Cassia* (Leguminosae) and expanded upon by Venkatesh (1956: 170) in a series of morphological studies of the Caesalpiniaceae. Their illustrations do not show the degree of organization characteristic of an OP, and the cells are not indicated as different from surrounding cells. Neither Matthews & MacLachlan nor Venkatesh noted granularity or crystal formation, which is usually readily apparent even without light polarization. In our one preparation of *Cassia*, we found no sign of an OP, but we did note an occluded and dense area of epidermis corresponding to the granular stomial plug of Venkatesh (1957: 258).

Clethraceae. Kavaljian (1952: 400) found an opening tissue of very small but otherwise normal cells in the region of dehiscence in three species of *Clethra*.

Other families. Matthews & Maclachlan (1930: 114) made passing reference to resorption tissue in *Tibouchina*, Melastomataceae, and Eyde & Teeri (1967: 174) noted an unspecialized OT in *Rhexia*. Ziegler (1925) studied many taxa of Melastomataceae and reported only unspecialized opening tissue. Matthews & Maclachlan also noted resorption tissue in *Ochna*, Ochnaceae, and *Galanthus* (Amaryllidaceae), reports which we have not reinvestigated.

In serial sections, evidence of an OP is often inconspicuous, and an observer not seeking it might be unaware of its presence, especially if the anthers have dehisced or they are not viewed with polarized light. Thus, an OP may be present in many taxa not noted here. Capability of viewing serial sections under polarized light was not a common feature of microscopes until the 1970s.

NATURE AND FUNCTION OF THE OXALATE PACKAGE

Structural nature of the oxalate package: not idioblastic

Idioblasts, according to Esau (1965: 29), are 'cells markedly differing from other constituents of the same tissue in form, structure or contents'. This can refer to cells with druses, raphides, sclereids or other inclusions in leaves, stems, roots, etc., that are scattered among a parenchymatous ground tissue. In the oxalate package, however, the entire tissue comprises the same kind of cell, all filled with calcium oxalate crystals or oxalate raphide bundles. Because the package has a definite multicellular organization, sequence of development, placement, and form, it cannot be considered idioblastic; it is a specialized simple tissue. We call it an oxalate package (OP), but when its function is known a better term may be introduced.

The OP structures in the Ericaceae and Tiliaceae are so like those in the Solanaceae that they seem to be homologous, but in some plant groups the OP appears at different sites within the anther, and the histological origins may differ. In most druse-bearing examples, the OP is situated at the outer end of the septum and is perhaps derived from the connective, but in raphide-bearing examples (Liliaceae, Bromeliaceae) it is commonly at the inner end of the septum.

Content and ultrastructure of the oxalate package

Chemical tests by many workers (Solanaceae: Namikawa 1919; Bonner & Dickinson 1989; Trull *et al.* 1991: 14; Horner & Wagner 1980; Ericaceae: de Villamil 1980: 48; Theophrastaceae: Pohl 1932: 484; Lemnaceae: Al–Rais *et al.* 1971: 1217, etc.) and our own testing in *Schizanthus* have shown that the crystalline material in the OP is calcium oxalate.

Some plants have never been reported to have crystalline calcium oxalate of any kind, while in others the OP is only one representative of a wide array of shapes, topologies and amounts to be found. Crystalline calcium oxalate commonly occurs in two different forms, both often idioblastic: either as druses or crystal sand, or as raphides. In parts of a few taxa, e.g. *Spirodela* (Landolt 1986: 61), both forms are reported.

As druses or crystal sand, the calcium oxalate is mostly in the dihydrate form – calcium oxalate 2.25 H$_2$O or weddelite (Küster 1956: 496; Horner & Wagner 1992: 532; Landolt & Kandeler 1987: 43) – in various shapes (habits), and the cells where it occurs are often dead after formation (Frey–Wyssling 1981; Horner & Wagner 1980: 1359). However, Frey–Wyssling (1981: 141) noted so-called crystalline sand which consists of a dense accumulation of very small monohydrate grains (e.g. characteristic for the Solanaceae), which does not accord well with other statements that have been made. Horner & Wagner (1980: 1358) noted that in the OP in *Capsicum annuum* druses form around multiple nucleation sites in a single cell.

As raphides, the calcium oxalate occurs as stacks of minute, sharp needles in the monohydrate (whewellite) form (Al–Rais *et al.* 1971: 1217). Some plants form raphide-like crystals, actually acicular druses, which are called styloids or pseudo-raphides (Fahn 1990: 24). These are distinguishable from raphides by blunt or rounded ends and other features. Raphides occur as many-crystalled bundles or clusters (Arnott & Pautard 1965: 619) in membrane-bounded living cells. Kugler (1942: 393) recognized four types of raphide-containing OP according to the orientation of the raphides with respect to the axis of the anther.

Although the OP crystals are known to be calcium oxalate, there is little indication of its purity. Our studies found varying appearance of OP crystals in different taxa, suggesting differences in chemical composition. In plant tissues, small quantities of manganese, potassium, sulfur, chlorine, strontium, or barium can be added to such crystals or be substituted for part of the calcium (Landolt & Kandeler 1987: 43; Al–Rais *et al.* 1971: 1217; Franceschi 1985: 25; Rasmussen & Smith 1961: 100).

The matrix in which raphide bundles are embedded is of unknown content, but it apparently varies in different species (compare Perera *et al.* 1990: 1068 and Tang & Sakai 1983: 154). Raphides are credited with producing a sharp taste in plants used as human foods, such as taro (dasheen) (Black 1918; Tang & Sakai 1983) and kiwifruit (Perera *et al.* 1990), but in some Araceae which contain raphides, the acridity is caused by other substances within the raphide sacs, independent of any irritation the raphides may induce (Tang & Sakai 1983: 159).

The OP has been studied at the ultrastructure level, notably by Horner & Wagner (1980, 1992). At our suggestion, because we thought the OP might have an olfactory role, L.T. Durkee (personal communication) carried out a limited series of ultrastructure investigations on anthers of two Solanaceae species. She used *Vassobia breviflora* and *Witheringa solanacea*, which have large OPs. She found nothing suggestive of a scent-making apparatus, no proliferation of smooth or rough endoplasmic reticulum, no unusual amounts of mitochondria, nor any evidence of particular organelles. We also tried to stain anthers with Neutral Red, Sudan III and Sudan IV, but these stains gave no indication of neutral lipids (Lillie 1977) which might be associated with scent systems.

Trull *et al.* (1991) studied the OP of *Nicotiana tabacum* using a monoclonal antibody technique. This identified an antigen (epitope) specifically associated with the crystal-containing cells and not associated with the calcium oxalate crystals themselves nor with crystals (of other kinds) in the septum. They reasoned that the antigen must be associated with the synthesis or maintenance of the OP crystals. The antigen failed to react with the non-solanaceous plants, *Begonia* sp., *Zebrina pendula*, *Pilea* sp., and *Ginkgo biloba*, none of which is known to have an OP in the anthers.

Cost of the oxalate package

We assume that the OP is a required structure for all species where it has been found. Thus, conditions preventing its development would lead to suppressed flowering, especially in cases where the OP is large (Fig. 4). This puts constraints on the kinds of substrate where such plants can grow and reproduce that are not faced by other plants. The calcium cost of an OP in flowers is in addition to other calcium needs of plant tissues.

Plants of the Solanaceae are characteristically calciphilic, being most diverse and abundant on calcareous soils. We have observed in the greenhouse that many solanaceous plants grown under a generalized fertilizer regime seem healthy but flower poorly. These same plants flower well after addition of pieces of natural limestone to their pots. Also, many taxa of Solanaceae, especially those in refugial sites or of relictual status, are found only on notably calcareous substrates. The calcium requirement for an OP may be a factor limiting the natural distribution of many solanaceous taxa. In light of this, Solanaceae restricted to calcareous substrates might be expected to have the largest OP, but this is not the case (see statement on size under 'Histological observations' above).

In *Schwenckia*, a well-developed anther pouch is suggestive of an OP but no crystals were seen. The OP may have been lost under selective pressure in the region of occurrence, eastern South American lowlands, where calcium is at extremely low levels. Landolt (1981) concluded that in North Carolina, the Lemnaceae are geographically restricted in part because of low minerals in soils and waters, and Lüönd (1983) came to the same conclusion for northern Italy. This may in part reflect a special calcium requirement for OP formation.

Contrastingly, the Ericaceae, which commonly have calcium-bearing OPs, are seemingly calciphobic in their substrate tendencies. Plants of this family grow mainly in acidic sites where calcium limitation is to be expected. However, although calcium content of roots, leaves, and shoots is loosely correlated with calcium content of the soil, at least some ericaceous plants can take up calcium even where substrate calcium or the soil pH is unusually low (Marrs & Bannister 1978). The distribution of these plants on acidic substrate is probably not because of an avoidance of calcium: both blueberry and cranberry (*Vaccinium* spp.) crops benefit from addition of calcareous fertilizers as long as the soil pH is kept low (Eck 1988, 1990).

Mycorrhizal symbionts are associated with some Solanaceae and most Ericaceae (Harley & Harley 1987). In Solanaceae, widespread fungal species form mycorrhiza of the 'vescicular-arbuscular' type, which may aid in nutrient uptake (Pond *et al.* 1984). In Ericaceae, host-restricted fungi form 'arbutoid' or 'ericoid' mycorrhiza, which are known to improve nutrient uptake from the soil. 'Although this partnership is expensive in terms of energy, it may have permitted the Ericaceae to colonize poor, acidic soils' (Malloch *et al.* 1980: 1224). Enzymatic activity by the endophyte gives the plant access to soil nutrients (iron, nitrogen, phosphorus) that might not otherwise be available (Leake *et al.* 1990). Many endophytes (mycorrhizal associates) of 'calcifugic' Ericaceae are more enzyme-active at low pH values. This pH sensitivity by the mycorrhizal associate may be the major factor limiting Ericaceae to acid soils (Leake *et al.* 1990).

Thus, Solanaceae and Ericaceae differ in their ability to grow on substrates contrasting in pH, which is often correlated with calcium content. The Solanaceae are often restricted to relatively high pH substrates from which they can take up calcium directly. The Ericaceae embrace fungal intermediaries that permit, or require, exploitation of low pH substrates, and they are nevertheless able to take up enough calcium from such acidic soils (Marrs & Bannister 1978). In either case, calcium requirements for the OP may reduce flowering in situations where calcium is substrate-limited.

Additional genetic coding and energy cost of the formation and disappearance of dedicated cells, multiple druse centers, and calcium oxalate transport may be significant costs imposed by an OP. The presence of almost any unusual feature in a plant tissue must exact a cost of lessened structural integrity.

Possible functional role of the oxalate package

Theories of the function of the OP must consider that OP-bearing anthers are found in plants living under a wide range of environments – soils, climate, and latitude – displaying a wide variety of life forms – aquatics, herbs, trees, etc. – and facing a wide range of biological interactions such as kinds of pollinators, pathogens, and herbivores. Many longitudinally dehiscent and poricidal anther forms have OPs. It is possible, and even likely, that the OP occurs as a relictual structure in some taxa for which it is no longer adaptive, and it may have different selective value (function) in different taxa with quite similar OP structures. Any theory must also consider why the OP is found only in a few plant groups.

Thus, the occurrence of an OP in a limited range of plant families, and its widespread occurrence in those that do have it, suggest that in groups where the right conditions prevail, it is strongly selected for. The taxa where it is known are far from the basal taxa of angiosperms in almost any evolutionary scheme, and they are widely distributed on branches of evolutionary trees. Therefore, its presence must be viewed as an apomorphy: it must have arisen independently more than once.

Mechanical roles

Anther dehiscence

Most workers, both earlier (Artopoeus 1903: 320; Namikawa, 1919, etc.) and recent (Weberling 1989: 98; Horner & Wagner 1980: 1359), have assumed that the OP participates in anther dehiscence. The term 'resorption tissue' refers to a presumed disappearance of the material, leading to weakness and collapse of overlying tissue, which results in dehiscence (Palser & Murty 1967: 257). In many taxa, both porose and longitudinally

dehiscent, the epidermis becomes thinner before dehiscence, and in some cases (*Kalmia*) it becomes vanishingly thin, much as the anther's middle layers and tapetum, which are often referred to as 'crushed.' Horner & Wagner did not actually show that the OP relates to anther dehiscence, as was attributed to them by Mauseth (1988: 396).

In our preparations of Solanaceae, the crystalline material tends to remain coherent, but it does not consistently adhere to any particular areas of the pouch boundary nor appear to have any evident influence on its decay or rupture (Figs. 8–10).

In *Erica hirtiflora*, the crystalliferous epidermis and underlying layers break down to form the pore and 'there is no mechanical rupture' (Matthews & Taylor 1926: 241). The process of anther opening appears to differ, as in this account it is breakdown of the crystalliferous tissue itself that permits access to the outside air. One wonders, as did de Villamil (1980: 105), if the epidermis, being crushed at the point of opening, was too thin for these workers to recognize. Palser & Murty (1967) cite an earlier statement that dehiscence in *Erica* is 'by an apical pore in each half [of the anther] which forms by dissolution of resorption tissue (Artopoeus 1903; Matthews & Knox 1926),' which is a way of viewing the OP as participating in the dehiscence process. Some species of *Erica*, however, effect dehiscence by collapse tissue (de Villamil 1980: 105), what we have regarded as an undifferentiated opening tissue. de Villamil (p. 109) noted that in *Kalmia* 'after dissolution of the resorption tissue, the epidermis is still intact and must be broken before the cleft is open.' A helpful exploration of the opening of anthers by degeneration of localized areas of anther wall was presented by Bonner & Dickinson (1989).

We cannot show that the OP does not sometimes assist in effecting dehiscence, but the presence or absence of calcium oxalate crystals in necrotic tissue in the stomium does not seem related to the mechanics of effecting dehiscence. Hence, we do not think that dehiscence is the reason for its presence. As a synapomorphy in angiosperm anther structure, the OP should have been selected through conditions unique to the plants that have it. Nothing in the nature of the dehiscence mechanisms observed or recorded for these plants suggests evolution for dehiscence as a selecting arena: other plants dehisce, often in quite similar ways, without a crystalline OP.

Poricidal pollen delivery

Several surveys of poricidal anthers (Harris 1905; Buchmann 1978, 1983) have shown they are not uncommon, occurring in 5–6% or perhaps even 10% of flowering plants. Most of these cases are adapted to vibratile pollination, although in a few cases, e.g., Araceae, Viscaceae, Rafflesiaceae (Endress & Stumpf 1990: 231), pollen is presented to pollinators by extrusion.

In poricidal anthers, insects vibrate the anther to cause ejection of the pollen. This can be readily tested by touching a ripe anther with the tip of a C note tuning fork (512 Hz, or cycles per second), which mimics the bees' usual range of 400–600 vibrations. Dynamics of the pollen chamber required for this were examined by Buchmann & Hurley (1978), who likened the anther to the sonicating column of an organ pipe. In the poricidal Solanaceae and Ericaceae, dieback of the septum not only leads to confluence of the locules but provides empty space within which the pollen can move as it is vibrated toward the porose opening.

A function of the OP could be to act with septum withdrawal to provide space required for vibratile pollen delivery. However, it is unclear why a crystal-bearing area would be preferred for this over the unspecialized opening tissue (OT) noted above. We found that anthers of *Exacum affine* eject pollen in clouds when touched by the tuning fork. Histology showed retraction of the septum, but there was no sign of an OP.

Cranberries and blueberries have terminal-pored anthers that are probably mainly buzz-pollinated. In observations of *Vaccinium angustifolium* and *V. corymbosum* – native blueberries visited by bumblebees – flowers of the first species were hardly visited by honeybees and the second species not at all in a natural habitat in Ontario (Vander Kloet 1976). However these plants are evidently not restricted to this mode of pollen delivery, for commercial cranberry yields can be markedly increased by installation of honeybees (Eck 1990: 219), which do not buzz the flowers they visit (Buchmann 1983). The flowers and anthers hang down, and when a visitor struggles within the flower to collect nectar, it must brush against the anthers and probably induces dropping

of pollen from the pores. This bypasses the sonication for which these anthers are so well adapted.

Thus we think primary selection of the OP for a mechanical function in the anther is unlikely.

Physiological roles: metabolic sump or sink?

Calcium oxalate is commonly considered an end-point in various metabolic processes. Bornkamm (1965) outlined an oxidation pathway: glycolic acid → oxalic acid + Ca → calcium oxalate, which is moved by oxalic acid oxidase. Under genetic and to some extent environmental controls, plants can mediate the type, amount, and placement of crystals formed by adjusting components of this or perhaps other pathways. Thus Loetsch & Kinzel (1971) could say that *Spirodela* and *Lemna* are dependent on calcium to neutralize accumulations of oxalic acid. An intermediate in many plant vital processes, oxalic acid is toxic to many biological systems (Frey-Wyssling 1981: 141; Smith 1982). Precipitating it as calcium oxalate may be a way of removing excess oxalate from vital areas of the plant. Contrastingly, 'Many plants in their natural habitat are forced to take up more Ca than needed for the fundamental processes of life [and] this excess of Ca can be a burden under some circumstances' (Kinzel 1989: 105). By putting this precipitate in an ephemeral organ such as the flower, the oxalate and calcium can be removed not only from plant tissues but from the plant itself. But this does not address why the calcium oxalate should be structured into an OP sited adjacent to the pollen or what determines its precipitation as a dihydride druse or as a monohydrate raphide.

As to the last question, the ericad *rcbL* phylogeny of Kron & Chase (1993), shown in Fig. 24, is suggestive. On branches ancestral to the Ericaceae, monohydride raphides are formed, while in the Ericaceae clade dihydride druses are formed. This may signal that in this group, the dihydride druses are derived from a monohydride raphide ancestry.

Calcium is thought to have low mobility in plants (Baker 1983: 15). Franceschi (1987: 42) found that in *Lemna minor* roots, calcium oxalate formation can be slowly reversed. Calcium has been shown (Smith 1982: 258) to be removed from many tissues in the course of later development. Tomlinson (1961, 1969, 1982) stated that, for several monocotyledonous groups, raphides are most often found in young tissues. The sequestration of calcium in the anther septum (Horner & Wagner 1992) might be a calcium sink or reserve for later use. Most or all such crystals disappear before dehiscence, and calcium oxalate crystals formed in the young tapetum of some legumes are resorbed before anther maturity (Buss & Lersten 1972). However, crystals in the OP persist. The anther is a short-lived terminal structure which is soon discarded, and any reserve placed there is not long available to the plant.

The calcium ion is involved in many plant functions, and a physiological function for the OP cannot be ruled out. However, localization of calcium oxalate in an oxalate package in the anther to fulfill metabolic needs seems unlikely.

Biotic roles

Discouragement of herbivores (including pollen robbers)

Oxalates have long been suspected of inhibiting herbivory in plant tissues (Smith 1982: 258). The OP might check pollen plundering in several ways: by mixing with pollen and making it distasteful, by brushing against visitors and warding them off, by puncturing the gut of their offspring and killing them, or by making the anther tissue itself distasteful.

Pollen and seed are the most nutritious potential food sources made during the life of a plant, and avoidance of their loss to herbivores is an important selective pressure. There is unexpectedly little specialization by arthropods to pollen as a food source. Few, if any, arthropods other than Collembola, which produce an exinase enzyme (Scott & Stojanovich 1963), can digest pollen walls. Gisin (1948: 489) noted that pollen is the principal nourishment of epigeic Collembola, but that monophagy is unknown in Collembola. Any impact of collembolan feeding on floral evolution cannot be gauged with present knowledge. Honeybees make use of pollen by an elegant system of extracting its contents through its pores, sometimes after breaking it, and storing it as it passes through several gut chambers. The remnant exine is discharged in flight. Most feeders utilize the pollen they encounter by chewing. Among exceptions are oedemerid beetles, which are obligate pollen feeders with specialized intes-

tines. Some insects, nevertheless, do seek pollen for food. Beetle species feed on pollen of Zamiaceae, *Papaver*, Cucurbitaceae, Liliaceae (*Victoria*), palms (*Asterogyne*), and *Phlox* (Stanley & Linskens 1974: 109–10). Of these, some Liliaceae and palms have a raphide type of OP, but the *Phlox* species we examined had no sign of one. Syrphids (Haslerud 1974: 212), mosquitoes and some Lepidoptera (Gilbert 1972; DeVries 1979) also seek and feed on pollen. We have often encountered thrips inside the poricidal anthers of greenhouse *Solanum* plants where they presumably feed on pollen and other matter.

The most notable pollen feeding specialization is by bees – an interaction where plants trade a necessary service, pollination, for the food staple they present. Except for bees and small wasps of the family Masaridae, which must provision their nests, relatively few animals specifically seek out and collect pollen. Most pollen ingestion by animals is incidental to other feeding, as by the Colorado potato beetle and the pig. Perhaps most pollen-feeding is carried out by small animals such as arthropods or slugs. Although some birds, especially hummingbirds, bats or other mammals sometimes ingest pollen, no such cases are known to present a hazard to the life cycle of a flowering plant.

The world's most numerous crawling insects are ants, which, partly because of their small esophagus and disuse of solid foods, seldom, if ever specifically seek pollen from anthers. Nevertheless, it is not clear why they do not use pollen in their broods, as the fourth instar larvae can ingest particles larger than most pollen grains (Glancey et al. 1981: 396, 399, 401).

Whether calcium oxalate crystals or raphides can actually deter anther or pollen ingestion is not known, and a correlation between pollen seekers and plants with an oxalate package is nonexistent or unconvincing. Honeybees presented with calcium-oxalate-adulterated pollen ate the mixture without apparent adverse effects. And the crystalline material does not seem to be ideally placed if biotic repellence is the driving factor. The possible protective advantage of an OP seems so negligible that other selective forces must have led to this tissue.

Pollination enhancement

When poricidal anthers are vibrated by the bee or fly, the druses from the OP are mixed with the pollen as it is ejected. In most longitudinally dehiscent anthers, the oxalate druses or raphide-sacs are situated at the edges of the pollen mass (Fig. 19) so that the visitor must brush against them while gleaning pollen from the anther or gathering nectar.

In *Schizanthus* (Solanaceae), three of the five stamens are reduced to staminodes. These display balls of pure exposed OP crystals atop peg-like filaments that guard the nectar-bearing corolla throat. Two of these oppose the two fertile anthers (Fig. 16), and the third is held in the same plane. The bees (Arroyo *et al.* 1982: 86) visiting *Schizanthus* for nectar are dusted beneath with pollen and calcium oxalate from the two explosively dehiscent anthers and also with calcium oxalate from the two staminodes. In this genus, some of the calcium oxalate is delivered separately from the pollen to a different part of the bee from that which receives the pollen.

The orchid, *Stelis*, is another case where calcium oxalate is presented separately from the pollen (Chase & Peacor 1987). In this genus, nectar is found on the surface of the lip and petals, but in one species it is substituted by calcium oxalate raphides. Pollination, known in the genus only in other species, is by fungus gnats and root gnats. These authors suggest that the raphides are related to pollination.

Our dusting and dabbing of pollen onto glycerine-coated slides revealed calcium oxalate crystals mixed with pollen of *Capsicum, Ludwigia, Impatiens*, etc. (Figs. 20–24), and druses and raphides have been found in honeybee loads (Maurizio 1953: 128) or honey (Demianowicz 1963). This shows that bees regularly pick up crystals from OPs along with pollen and take them to the nest, if not from flower to flower.

While the contents of the OP are commonly distributed along with the pollen, in some cases they remain bound in the anther as in *Sparmannia* (Demianowicz (1963) and do not move with the pollen. *Cyphomandra* is a curious case. Species in this genus that we examined histologically have a large and typically developed OP but the species we were able to test with moistened slides discharged few crystals along with the pollen. Reports (Soares *et al.* 1989; Gracie 1993; Sazima *et al.* 1993) indicate that other species in the genus are visited by male, scent-gathering bees, which are not known to vibrate anthers or collect pollen. Further histological observations will be required to establish that these congeners

of OP plants do have an OP. Living plants must be observed to determine whether the OP crystals of these anthers eject with the pollen. And study in nature will be needed to identify a sonicating pollinator in addition to the non-sonicating male bees. If the OPs are present but not delivering crystals with the pollen, they are probably now nonaptive.

Presence of an OP and delivery of its contents with the pollen is not uncommon. Rough estimates of numbers of known species suggest that this kind of calcium oxalate–pollen delivery may occur in 1–2% of seed plants. What, if anything, the insect visitors do with the calcium oxalate, or if they even perceive that they have it, is a subject for future investigation.

It might be argued that the staminodes in *Schizanthus* and the crystals of *Stelis*, with their quantities of calcium oxalate positioned in the paths of flying visitors, deter frontal approaches by herbivores or direct pollinator visitation from other angles. However, as a deterrent they would likely deter as many acceptable pollinators as unacceptable visitors, and the hypothesis is implausible.

The calcium oxalate package might also enhance pollination by providing a visual signal. The crystalline material was visible when illuminated by an ultraviolet 'black' light, and it may also be visible to insects in ways not apparent to the human eye. This was suggested by Chase & Peacor (1987) for the raphides found on the surface of the *Stelis* petal. The calcium oxalate material may be both seen and removed by insect visitors, but because it is taken up, any visual role is likely only a part of its pollination enhancement function.

Destiny of the crystalline calcium oxalate materials

What happens to the crystal material when it is taken by an insect is unknown. When it is brought into the hive or nest, several fates can be envisioned.

When labeled calcium oxalate was fed in capsules to women, calcium was taken up in the blood stream (Heaney & Weaver 1989), showing that in humans at least, calcium oxalate can be a source of dietary calcium. The acidity of the human stomach probably aids in dissociation of the oxalate salt for assimilation; whether insects have such capability is not known. The abundant calcium oxalate druses reported in honey from *Tilia americana* suggests the bees do not dissociate or ingest the crystals. However, impurities in the plant-provided calcium oxalate may be of dietary value to the insects.

The addition of a basic cation and an often toxic anion in calcium oxalate to material within the nest may in some way permit the bees to adjust the pH of the nest or reduce fungal attack or parasite conditions.

Pollen of some plants is known to be scented (Vogel 1962; Percival 1965: 38, 41; Buchmann 1983: 88), and D'Arcy *et al.* (1990) have shown that many Solanaceae, including some that are bird-pollinated, have scented anthers. In his review of scent secretion in anthers. Vogel used the term druse (*Duftdrüsen*) to mean lipid-producing, osmophoric elements, and not the calcium oxalate containing elements discussed in this paper. H.T. Horner Jr has suggested (personal communication 1991) that calcium oxalate crystals in the septum may break down along with the septum to produce substances that could coat the pollen with scent. The role of the OP in such a pollen-coating system seems unlikely, as even with withdrawal of the septum and confluence of the locules, the calcium oxalate usually remains localized and does not approach the majority of pollen grains situated away from it. Even in buzzed anthers, the OP is probably not co-mingled with the pollen until the moment of buzzing.

Although L.T. Durkee (personal communication) could not demonstrate a scent-making role for the OP, anecdotal evidence suggests that the OP crystals may have a flavor or scent interesting to insects. When staminodes of *Schizanthus* were placed in ant pathways, the ants at first examined them for several seconds, even picking them up and carrying them a short distance. Soon, however, the ants lost all interest in the staminodes and crawled over them without interruption as they moved to and fro along the pathway. This suggests that the staminodes contained some scent or flavor unaccustomed and then uninteresting to the ants.

Enhancement of pollen performance

Little is known about features that may enhance performance of the pollen once it arrives on an appropriate stigma. The subject was recently reviewed by Stephenson *et al.* (1992). It is possible that the OP substances carried with pollen and at times probably deposited with it on the stigmatic surface can act to improve the success of pollen in actually fertilizing an ovule and developing a seed. It seems likely, however, that if this is the adaptive function driving selection of the OP, a more efficient and more widespread development of such a system would be found in nature.

GENERAL CONCLUSIONS

In the Solanaceae, a calcium oxalate druse-accumulating package occurs in anthers of most taxa. In the Ericaceae, also, the OP occurs widely, although it has not been reported in some tribes. Based on comparison with *rbc*L analyses, the oxalate package has arisen independently in these two families, and in independent events it has been lost in some cases. An OP also occurs in some other families, sometimes with the accumulation of membrane-bound raphides instead of druses.

We can say with confidence that crystals of the OP are taken up by insects along with pollen from anthers or staminodes in a significant number of plant groups. There are two major kinds of calcium oxalate material: loose, dead druses or crystal sand, and bound and sometimes membrane-enclosed raphides.

Previous suggestions that the OP serves mechanical or physiological functions or serves to reduce herbivory are discarded. Instead, a pollinator reward function is most likely.

Plants with an OP are pollinated by a wide variety of agents. Hymenoptera (bees) are seemingly the most frequent pollinators of such plants, having been reported in Solanaceae, Ericaceae, and other groups known to have an OP, but Lepidoptera (hawkmoths, butterflies, etc.) and Diptera (flies, mosquitoes, syrphids) are also frequently noted as pollinating OP-bearing plants. While we think that bees are the main pollinators to which the OP is adapted, we have not established this except to show that bees do indeed acquire calcium oxalate with pollen and take it to the nest. Presumably other pollinators receive the crystalline material from OP-bearing taxa too. The principal pollinators of *Stelis* are reported (Chase & Peacor 1987) to be flies, and Pohl (1932) thought *Deherania* also to be pollinated by flies. In Solanaceae, birds (*Iochroma*) and bats (*Markea*) are also pollinators of plants with a well-developed OP structure. The benefit of a calcium oxalate addition to pollen for nectar-collecting birds cannot be envisioned, and in these cases the OP may be a nonaptation (Gould & Vrba 1982) – a previously adaptive feature now no longer adaptive. Both *Cestrum* and *Iochroma* have groups of species with tubular red flowers pollinated by birds. In *Cestrum*, loss of the OP is all but complete, while in *Iochroma* there is still a well-developed and conspicuous OP. This might argue that the OP has been nonaptive in *Cestrum* for much longer than in *Iochroma*, which is another way of saying that *Iochroma* has only recently shifted to dependence on birds for pollination. The unusually small OP in the vibratile anthers of *Solanum* sect. *Leptostemonum* and *Lycopersicon* suggests that in these groups it may now be nonaptive and undergoing reduction.

This analysis suggests that calcium oxalate and perhaps the unknown impurities in the OP should be considered as another kind of reward, along with nectar, pollen, oils, etc., offered by flowers to encourage pollination. Is it significant that *Sprengelia*, the only member of the Epacridaceae reported by Paterson (1961) to have an OP, is also the only genus in the group to lack a nectary?

APPENDIX   Material used for histological preparations

Information is given in the form taxon, voucher, –geographical distribution, notes: (size, location, texture).

ACANTHACEAE
Ruellia humilis (Nees) Lindau SOURCE: *Keating 19vii89* (MO) –USA, Illinois NOTES: no indication of OP.
Thunbergia erecta Anders. SOURCE: commercial source – Asia NOTES: no indication of OP.

APOCYNACEAE
Catharanthus roseus (L.) G. Don SOURCE commercial source –Madagascar NOTES: no indication of OP.

ASCLEPIACACEAE
Calystegia sepium (L.) R. Brown. SOURCE: *D'Arcy & Keating 17763* (MO) –USA, Missouri NOTES: no indication of OP.

BIGNONIACEAE
Jacaranda jasminoides (Thunb.) Sandw. SOURCE: *Gentry 58768* (MO) –Brazil NOTES: no indication of OP.
Paulownia tomentosa (Thunb.) Steud. SOURCE: *Keating*

20ix90 (MO) –USA, Illinois NOTES: no indication of ºOP.

BORAGINACEAE
Cordia nervosa Lam. SOURCE: *Hahn 3970* (US) – Guyana NOTES: no indication of OP. *Cordia sinensis* Lam. SOURCE: *Miller 3734* (MO) –Madagascar NOTES: no indication of OP.
Ehretia saligna R. Br. SOURCE: *Willing 166b* (MO) –Australia NOTES: no indication of OP.

CALYCANTHACEAE
Calycanthus floridus L. SOURCE: *D'Arcy 17780* (MO) –USA NOTES: no indication of OP.

CAMPANULACEAE
Campanula floridana A. Gray SOURCE: *D'Arcy 17811* (MO) –USA, Florida NOTES: no indication of OP.

CONVOLVULACEAE
Convolvulus tricolor L. SOURCE: *D'arcy 17763* (MO) –commercial source NOTES: no indication of OP.

ERICACEAE
Cavendishia guatemalensis Loes. SOURCE: *D'Arcy 17932* (MO) –Honduras NOTES: no indication of OP.
Kalmia latifolia L. SOURCE: *D'Arcy 17779* (MO) –USA NOTES: medium, hypodermal, granular.
Vaccinium staminium L. SOURCE: *D'Arcy 17733* (MO) – USA, Missouri NOTES: no indication of OP. *Vaccinium* sp. SOURCE: *D'Arcy 17889* (MO) –USA, Missouri NOTES: no indication of OP.

GENTIANACEAE
Exacum affine Regel SOURCE: *D'Arcy 17737* (MO) –Socotra NOTES: no indication of OP.
Sabbatia sp. SOURCE: *D'Arcy 17735* (MO) –USA, North Carolina NOTES: no indication of OP.

GESNERIACEAE
Gloxinia sylvatica (Kunth) Wiehler SOURCE: *Miller 5600* (MO) *MBG–894271* –South America NOTES: no indication of OP.
Kohleria tubiflora (Cav.) Hanst. SOURCE: (MO) *MBG–750977* NOTES: no indication of OP.

GOODENIACEAE
Scaevola hookeri F. Muell. SOURCE: *MacDougal 5015* (MO) –Australia, Tasmania NOTES: no indication of OP.

LAURACEAE
Nectandra amazonicum Nees SOURCE: *C. Bosque 34* (MO) – Venezuela NOTES: no indication of OP.

LEGUMINOSAE

Cassia sp. SOURCE: *D'Arcy 17988* (MO) NOTES: no indication of OP, but an unusual opaqueness in the epidermis at the stomium.

Chamaechrista fasciculata (Michx.) Rydb. SOURCE: *D'Arcy 17736* (MO) NOTES: no indication of OP.

LILIACEAE

Scoliopus bigelovii Torr. SOURCE: *Utech H 5\7* –USA, California NOTES: no indication of OP.

Stenanthium occidentale A. Gray SOURCE: *Utech H3* –USA, California NOTES: no indication of OP.

MAGNOLIACEAE

Magnolia stellata Maxim. SOURCE: *Keating 2224* (MO) – Japan NOTES: no indication of OP.

POLEMONIACEAE

Polemonium reptans L. SOURCE: *Bush 956* (MO) –USA, Missouri NOTES: no indication of OP. *Polemonium reptans* L. SOURCE: *Plank 1020* (MO) –USA, Missouri NOTES: no indication of OP.

SOLANACEAE

Acnistus arborescens (L.) Schldl. SOURCE: *Gentry & Burger 2722* (MO) –Costa Rica NOTES: large, hypodermal, granular.

Alona coelestis Lindley SOURCE: *Zoellner 9628* (MO) –Chile NOTES: medium, hypodermal, granular.

Anthocercis genistoides Miers SOURCE: *Haegi 1783* (NSW) – Australia NOTES: small, hypodermal. *Anthocercis littorea* Labill. SOURCE: *Haegi 1994* (NSW) –Australia NOTES: small–medium, hypodermal, granular. *Anthocercis viscosa* R. Br. SOURCE: *Haegi 1049* (ADW) –Australia NOTES: medium, hypodermal. *Anthocercis viscosa* subsp. *caudata* Haegi SOURCE: *Haegi 1237* (ADW) – Australia NOTES: small–medium, hypodermal, granular.

Anthotroche pannosa Endl. SOURCE: *Haegi 1825* (NSW) – Australia NOTES: medium, hypodermal, granular.

Brachistus nelsonii (Fern.) D'Arcy et al. SOURCE: *Breedlove 41727* –Mexico NOTES: hypodermal.

Browallia americana L. SOURCE: *Tyson 7125* (MO) –Panama NOTES: medium, hypodermal.

Browallia speciosa Hook. f. SOURCE: *D'Arcy 18204* (MO) – South America NOTES: medium, hypodermal, coarse.

Brunfelsia densifolia Krug & Urban SOURCE: *Plowman 3365* (ECON) –Puerto Rico NOTES: small. *Brunfelsia pauciflora* (Cham. & Schlecht.) Benth. '*Macrantha*' SOURCE: *Plowman 2739* (ECON, MO) – Brazil NOTES: medium, hypodermal, coarse and granular.

Capsicum annuum L. SOURCE: Commercial source –Mexico NOTES: medium, hypodermal. *Capsicum baccatum* L. SOURCE: *D'Arcy 17715* (MO) –Bolivia NOTES: large, hypodermal, coarse. *Capsicum chacoense* Hunziker SOURCE: *GRIN-26049* – Bolivia NOTES: large, hypodermal, coarse. *Capsicum lanceolatum* (Greenm.) Morton & Standley SOURCE: *Taylor & Breedlove 41082* (E) –Mexico NOTES: medium–large, hypodermal. *Capsicum pubescens* R. & P. SOURCE: *Eshbaugh E-1529* (MU) – Brazil NOTES: very large, hypodermal. *Capsicum sinense* Jacq. SOURCE: BG Univ. Nijmegen – Caribbean NOTES: large, hypodermal, granular.

Cestrum aurantiacum Lindley SOURCE: *D'Arcy 17715* (MO) – Guatemala NOTES: initials only, no OP. *Cestrum elegans* Schldl. SOURCE: *D'Arcy 18274* (MO) – Mexico NOTES: no OP.

Chamaesaracha nana A. Gray SOURCE: *Chuang & Chuang 6820* (ISU) –USA, California NOTES: medium, hypodermal, granular. *Chamaesaracha sordida* A. Gray SOURCE: *Miller 4740* (MO) –USA, Texas NOTES: medium, hypodermal, granular.

Crenidium spinescens Haegi SOURCE: *Haegi 2001* (NSW) – Australia NOTES: medium, hypodermal.

Cuatresia exiguiflora (D'Arcy) Hunziker SOURCE: *D'Arcy 16414* (MO) –Panama NOTES: medium, hypodermal, granular. *Cuatresia riparia* H.B.K. SOURCE: *D'Arcy 16356* (MO) –Panama NOTES: medium, hypodermal.

Cyphanthera albicans (Cunn.) Miers SOURCE: *Haegi 1378* (ADW) –Australia NOTES: hypodermal, granular. *Cyphanthera racemosa* (F. Muell.) Haegi SOURCE: *Haegi 1936* (NSW) –Australia NOTES: hypodermal, granular.

Cyphomandra betacea (Cav.) Sendt. SOURCE: *Bohs 2274* (GH), *2275* (VT) –Ecuador NOTES: large, subhypodermal. *Cyphomandra corymbiflora* Sendtner SOURCE: *Bohs (Pringle) 2343* (GH, VT) –Brazil NOTES: medium, subhypodermal (3 or 4 cells). *Cyphomandra diploconos* (Mart.) Sendt. SOURCE: *Bohs 2335* (VT) NOTES: large, partly subhypodermal (2 cells). *Cyphomandra diversifolia* (Dunal) Bitter SOURCE: *Vico 7662* (MO); *Benitez 3672* (MY) –Venezuela NOTES: large, hypodermal or subhypodermal, granular. *Cyphomandra hartwegii* (Miers) Dunal SOURCE: *D'Arcy 3949* (MO); *Mori & Kallunki 1826* (MO) –Panama NOTES: large, hypodermal or subhypodermal. *Cyphomandra hartwegii* subsp. *ramosa* Bohs SOURCE: *Bohs 1644* (GH) –Colombia NOTES: large, hypodermal or subhypodermal.

Datura stramonium L. SOURCE: *Keating s.n.* (MO) –USA, Illinois NOTES: hypodermal pouch, druses not seen.

Deprea orinocensis (H.B.K.) Raf. SOURCE: *D'Arcy 10670*

(MO) –Panama NOTES: medium, OP resembles crystals in septum.

Discopodium penninervium Hochst. SOURCE: *Lewalle 2316* (MO) –Burundi NOTES: medium.

Duboisia hopwoodii (F. Muell.) F. Muell. SOURCE: *Maconochie 1886* (MO); *Haegi 2169* (NSW) –Australia NOTES: medium–large, hypodermal. *Duboisia myoporoides* R. Br. SOURCE: *D'Arcy 17805* (MO) –Australia NOTES: medium–large, hypodermal.

Dunalia lycioides Miers SOURCE: *Sagastegui et al.* 8101 (MO) –Peru NOTES: HYPODERMAL.

Exodeconus miersii (Hook. f.) D'Arcy SOURCE: *D'Arcy 17809* (MO) –Ecuador NOTES: large, hypodermal.

Grabowskia boerhaavifolia (L. f.) Schld. SOURCE: *D'Arcy 17709* (MO) –Argentina NOTES: large, hypodermal.

Grammosolen dixonii (F. Muell. & Tate) Haegi SOURCE: *Haegi 676* (ADW) –Australia NOTES: hypodermal.

Hawkesiophytum ulei (Dammer) Hunziker SOURCE: *Tyson et al.* 4726 (MO)–Panama NOTES: medium–large, hypodermal.

Hyoscyamus albus L. SOURCE: *D'Arcy 17776* (MO) –Europe NOTES: medium, hypodermal. *Hyoscyamus aureus* L. SOURCE: *D'Arcy 17774* (MO) –Lebanon NOTES: unusual epidermis at stomium. *Hyoscyamus desertorum* (Ascher. & Boiss.) Tackholm SOURCE: BG Univ. Denmark –Eastern Mediterranean NOTES: very small, hypodermal.

Iochroma grandiflora Bentham SOURCE: *D'Arcy 17710* (MO); *Plowman 4594* (GH) –Peru NOTES: large, hypodermal. *Iochroma* sp. SOURCE: cultivated ISU –Peru NOTES: hypodermal.

Jaltomata procumbens (Cav.) Gentry SOURCE: *Haber 1562* (MO); *Diaz 622* (MO) –Mexico, Costa Rica NOTES: medium, hypodermal. *Jaltomata grandiflora* (Greenm.) D'Arcy *et al.* SOURCE: *D'Arcy 17749* (MO) –Mexico NOTES: medium, hypodermal.

Lycianthes lycioides (L.) Hassler SOURCE: *D'Arcy 17713* (MO) –Colombia NOTES: medium, subhypodermal. *Lycianthes maxonii* Standl. SOURCE: *D'Arcy 15543* (MO) –Panama NOTES: pouch subhypodermal, druses not seen. *Lycianthes ocellata* (J.D. Smith) Morton & Standley SOURCE: *Breedlove 35135* (MO) –Mexico NOTES: large, hypodermal, granular. *Lycianthes rantonei* (L.) Carriere SOURCE: *Symon 48404* (ADW) – South America NOTES: medium, subhypodermal. *Lycianthes repens* (Spr.) Bitter SOURCE: *D'Arcy 17815* (MO) –Venezuela NOTES: medium, hypodermal and subhypodermal. *Lycianthes* sp. SOURCE: *D'Arcy 2375* (MO) –Panama NOTES: subhypodermal (3 cells).

Lycium acutifolium Dunal SOURCE: *D'Arcy 15387* (MO) – Madagascar NOTES: hypodermal, granular. *Lycium brevipes* Benth. SOURCE: *MBG-822310 (Lievens 5667)* (MO) –Mexico NOTES: small, medium, hypodermal. *Lycium barbarum* L. SOURCE: *Chase 17089* (ISU) – USA, Illinois NOTES: medium, hypodermal.

Lycopersicon chmielewskii Rick et al. SOURCE: *D'Arcy 17748* (MO) –Peru NOTES: small. *Lycopersicon chilense* Dunal SOURCE: *Ferreyra 12021* (MO) –Peru NOTES: druses not seen. *Lycopersicon esculentum* Miller SOURCE: D'Arcy commercial source –Mexico NOTES: small, hypodermal. *Lycopersicon hirsutum* H.B.K. SOURCE: *Hutchison 4428* (MO) –Peru NOTES: small. *Lycopersicon hirsutum* H.B.K. var. *glabratum'* SOURCE: *D'Arcy 17757* (MO) –Peru NOTES: small, hypodermal. *Lycopersicon pimpinellifolium* (Jusl.) Miller SOURCE: *Ferreyra 11805* (MO); *D'Arcy 17755* (MO) –Peru NOTES: small, hypodermal.

Nicandra physalodes (L.) Gaertner SOURCE: *D'Arcy 17784* (MO) –Chile NOTES: special.

Nicotiana attenuata S. Wats. SOURCE: *Chuang & Heckard 6444-B* (ISU) –USA, Nevada NOTES: medium, hypodermal. *Nicotiana glauca* (L.) A. Graham SOURCE: *French 8* (ISU) –Argentina NOTES: hypodermal. *Nicotiana longiflora* Cav. SOURCE: BG Univ. Denmark – Argentina NOTES: small, hypodermal, septum collapse. *Nicotiana rustica* L. SOURCE: BG Geneva–America NOTES: hypodermal. *Nicotiana silvestris* Speg. & Comes SOURCE: *D'Arcy 17746* (MO) –Argentina NOTES: large, hypodermal. *Nicotiana trigonophylla* Dunal SOURCE: *Miller 5157* (MO) –USA, Texas NOTES: medium, hypodermal.

Nierembergia linariaefolia A. Graham SOURCE: *Renvoize 3012* (MO) –Chile NOTES: no OP.

Nolana napiformis Philippi SOURCE: *D'Arcy 17773* (MO) NOTES: small–medium, hypodermal, coarse. *Nolana paradoxa* Lindl. SOURCE: BG Kew NOTES: hypodermal, granular. *Nolana prostrata* L. f. SOURCE: BG Univ. Nijmegen –Peru NOTES: small, subhypodermal or hypodermal.

Petunia hybrida (J.D. Hooker) Vilmorin SOURCE: commercial source NOTES: medium, hypodermal. *Petunia nyctaginflora* Juss. SOURCE: BG Univ. Nijmegen– Argentina NOTES: small–medium, hypodermal, granular.

Physalis alkekengi L. SOURCE: D'Arcy 17707 (MO) –Europe NOTES: medium–large, hypodermal, subhypodermal (1 cell). *Physalis hederaefolia* A. Gray SOURCE: *Chuang & Heckard 6724* (ISU) –USA, Illinois NOTES: medium, hypodermal, granular, hardly birefringent.

Salpichroa origanifolia (Lam.) Baillon SOURCE: *Symon 51426* (ADW); *D'Arcy 17712* (MO) –USA, Florida, South America NOTES: small, hypodermal, granular.

Salpiglossis sinuata R. & P. SOURCE: *D'Arcy 18199* (MO) – Chile NOTES: hypodermal, poorly organized, little calcification.

Schizanthus pinnatus R. & P. SOURCE: *D'Arcy 18224* (MO) – Chile NOTES: small–medium, hypodermal, coarse, also in staminodes. *Schizanthus retusus* Graham SOURCE: BG Univ. Nijmegen –Chile NOTES: small, hypodermal, also in staminodes.

Schwenckia lateriflora Vahl SOURCE: *Benitez 3901* (MY) – Venezuela NOTES: good pouch, hypodermal, no OP. *Schwenckia trujilloi* Benitez SOURCE: *Benitez 31189* (MY) –Venezuela NOTES: good pouch, hypodermal, no OP.

Solandra brachycalyx Kuntze SOURCE: *Gentry 2957* (MO) – Costa Rica NOTES: large, hypodermal. *Solandra longiflora* Tussac SOURCE: *D'Arcy 18221* (MO) –Cuba NOTES: medium–large, hypodermal.

Solanum allophyllum Miers (Miers) Dunal SOURCE: *Moreno 16658* (MO) –Nicaragua NOTES: pouch subhypodermal, druses not seen. *Solanum americanum* Miller SOURCE: *D'Arcy 17705, 17706* (MO) –USA, Comoro Islands NOTES: small, hypodermal, coarse. *Solanum andigena* Juzepczuk & Bukazov SOURCE: *D'Arcy 17711* (MO) –Colombia NOTES: medium, hypodermal, coarse. *Solanum antillarum* O.E. Schulz SOURCE: *D'Arcy 5243* (MO) –Panama NOTES: small–medium, hypodermal. *Solanum bummeliaefolium* Dunal SOURCE: *Leandri & Saboureau 4076* (P) –Madagascar NOTES: very small, subhypodermal.

Solanum capsicoides All. SOURCE: *D'Arcy 17729* (*GRIN-390818*) (MO) –Argentina NOTES: small, subhypodermal (1 cell). *Solanum crinitum* Lamarck SOURCE: *Plowman 9196* (MO) –Brazil NOTES: medium, hypodermal, granular. *Solanum croatii* D'Arcy & Keating SOURCE: *B. Descoings 1446* (MO) – Madagascar NOTES: small–medium, subhypodermal, coarse. *Solanum donianum* Walp. SOURCE: *Pickett s.n.* (MO) –Mexico NOTES: medium, pouch subhypodermal (1 cell), druses not seen. *Solanum dulcamara* L. SOURCE: *Armstrong s.n.* (ISU) –USA NOTES: small, hypodermal –Chiapas NOTES: large, subhypodermal, druses only at pore, OP big at pore.

Solanum erythracanthum Dunal SOURCE: *D'Arcy 15206* (MO) –Madagascar NOTES: small, subhypodermal (three or four cells). *Solanum guineense* L. SOURCE: *Loubser 3807* (MO) –South Africa NOTES: small, pouch hypodermal, druses not seen. *Solanum hazenii*

Britton SOURCE: *D'Arcy 5226* (MO) –Panama NOTES: medium, hypodermal and partly subhypodermal. *Solanum heineanum* D'Arcy & Keating SOURCE: *D'Arcy 15456* (MO) –Madagascar NOTES: small, subhypodermal (three or four cells), coarse and granular. *Solanum humblotii* Damm. SOURCE: *Rakoto 6481* (MO); *Schatz 1713* (MO) –Madagascar NOTES: hypodermal, druses not seen.

Solanum incanum L. SOURCE: Phillipson 2805 (MO) –Madagascar NOTES: small, subhypodermal, coarse. *Solanum juglandifolium* Dunal SOURCE: *Asplund 7317* (LL) – Ecuador NOTES: small, hypodermal. *Solanum laciniatum* Aiton SOURCE: *Gleason & Bloomfield 117* (MO) – New Zealand NOTES: small–medium, hypodermal. *Solanum luteoalbum* Persoon SOURCE: *Bohs 2337* (GH) –Peru NOTES: medium, subhypodermal (two or three cells at pore). *Solanum lycopersicoides* Dunal SOURCE: *Rick 8619639* –USA NOTES: small, hypodermal. *Solanum madagascariense* Dunal SOURCE: *Decary 5068* (P) –Madagascar NOTES: medium, hypodermal NOTES: medium, hypodermal, granular.

Solanum mahoriensis D'Arcy & Rakotozafy SOURCE: *D'Arcy 15487a* (MO) –Madagascar NOTES: small, subhypodermal (six or seven cells), greatly reduced. *Solanum mauritianum* Scopoli SOURCE: *Schatz & Miller 2381* (MO); *Croat 28471* (MO) –Madagascar NOTES: small, hypodermal. *Solanum nigrum* L. SOURCE: *Armstrong 554* (ISU) –USA, Illinois NOTES: small, hypodermal. *Solanum ochranthum* Dunal SOURCE: *Holm–Nielsen et al. 6533* (AAU) –Ecuador NOTES: medium–large, hypodermal, granular. *Solanum pensile* R. & P. SOURCE: *Neill et al. 9176* (MO) –Ecuador NOTES: large, hypodermal, granular.

Solanum persicaefolium Dunal SOURCE: *D'Arcy 5008* (MO) – Puerto Rico NOTES: very small, hypodermal. *Solanum polygamum* Vahl SOURCE: *D'Arcy 2093* (MO) –Virgin Islands NOTES: staminate flower, no pollen, no OP. *Solanum pyracanthos* Lam. SOURCE: *D'Arcy & Rakotozafy 15332, 15334* (MO) –Madagascar NOTES: very small–medium, hypodermal and subhypodermal. *Solanum richardii* Dunal SOURCE: *D'Arcy 17515* (MO) – Comores NOTES: small, subhypodermal proximally, granular. *Solanum rugosum* Dunal SOURCE: *Benitez et al. 3614* (MY) –Venezuela NOTES: large, hypodermal.

Solanum simile F. Muell. SOURCE: *Adelaide BG SR 4538* (AD) NOTES: small, hypodermal, granular. *Solanum sisymbriifolium* Lamarck SOURCE: *Symon 81487* (ADW) –South America NOTES: small, subhypodermal at apex, hypodermal proximally. *Solanum sitiens* I. M. Johnston SOURCE: *D'Arcy 17786* (MO) –Chile

NOTES: small, hypodermal, coarse. *Solanum sublobatum Roemer & Schultes* SOURCE: BG Univ. Coimbra – Argentina NOTES: small, hypodermal. *Solanum tolearaea* D'Arcy & Rakotozafy SOURCE: *D'Arcy & Rakotozafy 15460* (MO); *D'Arcy 17771* –Madagascar NOTES: small, subhypodermal, reduced.

Solanum tomentosum L. SOURCE: BG Univ. Nijmegen – Africa NOTES: small, pouch hypodermal, druses not seen. *Solanum trizygum* Bitter SOURCE: *Tyson 7144* (MO) –Panama NOTES: pouch hypodermal, druses not seen.

Streptosolen jamesonii Miers SOURCE: *D'Arcy 17781* (MO) – Peru NOTES: small–medium, hypodermal, granular.

Symonanthus aromaticus (st.) Haegi SOURCE: *Haegi 1820* (NSW) –Australia NOTES: granular.

Triguera osbeckii (L.) Willk. SOURCE: *Miller et al. 833* (MO) –Africa NOTES: small, hypodermal.

Tubocapsicum anomalum (Franch. & Sav.) Makino SOURCE: *Chen 231* (MO) –Taiwan NOTES: medium, hypodermal, granular.

Vassobia breviflora (Sendt.) Hunz. SOURCE: *D'Arcy & Hunziker 13949* (MO) –Argentina NOTES: large, hypodermal, granular.

Withania somnifera (L.) Dunal SOURCE: *Symon 55142* (ADW); *D'Arcy 17745* (MO); *DeWilde 4072* (MO) – Mediterranean NOTES: small, druses not seen.

Witheringia correana D'Arcy SOURCE: *D'Arcy 16415* (MO) –Panama NOTES: very large, hypodermal. *Witheringia solanacea* L'Her. SOURCE: *D'Arcy 5358* (MO), 16399 (MO) –Panama NOTES:large, hypodermal, granular.

VERBENACEAE

Lantana camara L. SOURCE: *D'Arcy 17769*, commercial source –Caribbean NOTES: no indication of OP.

ACKNOWLEDGMENT

Many people contributed to this study in major ways. Those making slides or other preparations: Joseph E. Armstrong, Illinois State University, Normal, IL; Lenore T. Durkee, Grinnell College, Grinnell, IA; Suzanne Eder, Southern Illinois University, Edwardsville, IL; Barbara Palser, University of Massachusetts, Amherst, MA; Fredrick H. Utech, Carnegie Museum of Natural History, Philadelphia, PA.

Providors of plant material for study: W. Hardy Eshbaugh, Miami University, Oxford OH; United States Department of Agriculture, Germplasm Services Laboratory, Beltsville, MD; United States Department of Agriculture, Plant Introduction Station, Sturgeon Bay, WS.

Contributors of information and advice: Roger C. Anderson, Illinois State University; Mark W. Chase, Royal Botanic Gardens, Richmond, United Kingdom; Peter C. Hoch, Missouri Botanical Garden; Harry T. Horner, Iowa State University; Kathleen A. Kron, University of North Carolina; Charles D. Michener, University of Kansas; Richard G. Olmstead, University of Colorado, Boulder; Tatyana Shulkina, Missouri Botanical Garden; Richard Snyder, Michigan State University; James Trager, Missouri Botanical Garden; D.P. Wojcik, United States Department of Agriculture, MAVERL.

Gratitude is extended to all of the above and others who helped.

LITERATURE CITED

Al-Rais, A.H., A. Myers & L. Watson. 1971. The isolation and properties of oxalate crystals from plants. Ann. Bot. 35: 1213–1218.

Arnott, H.J. & F.G.E. Pautard. 1965. Development of raphid idioblasts in Lemna. Amer. J. Bot. 52: 618–619.

Arroyo, M.T.K, R. Primack & J. Armesto. 1982. Community studies in pollination ecology in the high temperate Andes of central Chile. I. Pollination mechanisms and altitudinal variation. Amer. J. Bot. 69: 82–87.

Artopoeus, A. 1903. Über den Bau und die Öffungsweise der Antheren und der Entwickelung der Samen der Erikaceen. Flora 92: 309–330.

Baker, D.A. 1983. Uptake of cations and their transport within the plant. pp. 3–20 in Metals and Micronutrients: Uptake and Utilization by Plants (edited by D.A. Robb & W.S. Pierpoint). Academic Press, New York.

Black, O.F. 1918. Calcium oxalate in the dasheen. Amer. J. Bot. 6: 447. nv.

Bonner, L.J. & H.G. Dickinson. 1989. Anther dehiscence in *Lycopersicon esculentum* Mill. New Phytol. 113: 97–115.

Bornkamm, R. 1965. Die Rolle des Oxalats im Stoffwechsel hoherer gruner Pflanzen. Untersuchungen an *Lemna minor* L. Flora A 156: 139–171.

Buchmann, S.L. 1978. Vibratile ('Buzz') pollination in angiosperms with poricidally dehiscent anthers. Ph.D. Thesis. University of California, Davis.

Buchmann, S.L. 1983. Buzz pollination in angiosperms. pp. 73–113 in Handbook of Experimental Pollination

Biology (edited by C.E. Jones & R.J. Little). Van Nostrand Reinhold, New York.

Buchmann, S.L. & J.P. Hurley. 1978. A biophysical model for buzz pollination in angiosperms. J. Theor. Biol. 72: 639–657.

Buss, P.A., Jr & N.R. Lersten. 1972. Crystals in tapetal cells of the Leguminosae. Bot. J. Linn. Soc. 65: 81–85.

Chandler, B. 1912. Deherainia smaragdina Dcne. Notes R. Bot. Gard. Edinb. 5: 49–56.

Chase, M.W. & C.R. Peacor. 1987. Crystals of calcium oxalate hydrate on the perianth of Stelis Sw. Lindleyana 2: 91–94.

Chatin, 1870. De l'Anthère. Baillière et fils, Paris.

Cocucci, A.A. 1991. Pollination Biology of Nierembergia (Solanaceae). Pl. Syst. Evol. 174: 17–35.

Copeland, H.F. 1943. A study, anatomical and taxonomic, of the genera of Rhododendroideae. Amer. Midl. Nat. 30: 533–625.

Copeland, H.F. 1947. Observations on the structure and classification of the Pyroloeae. Madroño 102: 65–102.

D'Arcy, W.G. 1991. The Solanaceae since Birmingham, 1976, with a review of its biogeography. pp. 75–137 in Solanaceae 3: Taxonomy–Chemistry–Evolution (edited by J.G. Hawkes, R. Lester, M. Nee & N. Estrada). Royal Botanical Gardens, Richmond, U.K.

D'Arcy, W.G., N.S. D'Arcy & R.C. Keating. 1990. Scented anthers in the Solanaceae. Rhodora 92: 50–52.

de Villamil, P.H. 1980. Stamens in the Ericaceae: a developmental study. Ph.D. Thesis Rutgers University.

DeVries, P.J. 1979. Pollen-feeding rainforest Parides and Battus butterflies in Costa Rica. Biotropica 11: 237–238.

Demianowicz, Z. 1963. Sur l'origine des macles d'oxalate de calcium contenues dans les miels de tilleul. Ann. Abeille 6: 249–255.

Dorr, L.J. 1981. The pollination ecology of Zenobia (Ericaceae). Amer. J. Bot. 68: 1325–1332.

Eck, P. 1988. Blueberry Science. Rutgers University Press, New Brunswick.

Eck, P. 1990. The American Cranberry. Rutgers University Press, New Brunswick.

Endress, P.K. & S. Stumpf. 1990. Non-tetrasporangiate stamens in the angiosperms: structure, systematic distribution and evolutionary aspects. Bot. Jahrb. Syst. 112: 193–240.

Esau, K. 1965. Plant Anatomy, 2nd ed. Wiley, New York.

Eyde, R.H. & J.A. Teeri. 1967. Floral anatomy of Rhexia virginica (Melastomataceae). Rhodora 69: 163–178.

Faegri K. & L. van der Pijl. 1979. The Principles of Polli-

nation Ecology, 3rd revised ed. Pergamon Press, Oxford.

Fahn, A. 1990. Plant Anatomy, 4th ed. Pergamon Press, Oxford.

Fickendey, E. & H.N. Blommendaal. 1929. Ölpalme. Bangerts Ausland-Bücherei No. 35; Reihe Wohltmann-Bücher Bd. 7: Hamburg & Leipzig. nv.

Franceschi, V.R. 1985. Pathways for formation of oxalate in Lemna minor. Pl. Physiol. 77(4 Suppl.): 25.

Franceschi, V.R. 1987. Calcium oxalate formation in Lemna minor is a rapid and reversible process. Pl. Physiol. 83(4 Suppl.): 42.

Frey-Wyssling, A. 1981. Crystallography of the two hydrates of crystalline calcium oxalate in plants. Amer. J. Bot. 68: 130–141.

Gilbert, L.E. 1972. Pollen feeding and reproductive biology of Heliconius butterflies. Proc. Natl. Acad. Sci. USA 69: 1403–1407.

Gisin, H. 1948. Etudes écologiques sur les collemboles épigés. Mitt. Schweizer. Entomol. Ges. 4: 485–515.

Glancey, B.M., R.K. Vander Meer, A. Glover, C.S. Lofgren & S.B. Vinson. 1981. Filtration of microparticles from liquids ingested by the red imported fire ant Solenopsis invicta Buren. Insects Sociaux 28: 395–401.

Goebel, K. 1905. Organography of plants. II. Trans., I.b. Balbfour. [Clarendon Press, Oxford] repr. Hafner, New York, 1969, pp. 599–603.

Gould, S.J. & E.S. Vrba. 1982. Exaptation – a missing term in the science of form. Paleobiology 8: 4–5.

Gracie, C. 1993. Pollination of Cyphomandra endopogon var. endopogon (Solanaceae) by Eufriesea spp. (Euglossini) in French Guiana. Brittonia 45: 39–46.

Harley, J.R. & E.L. Harley. 1987. A check-list of mycorrhiza in the British flora. New Phytol. 105(2 Suppl.): 1–102.

Harris, J.A. 1905. The dehiscence of anthers by apical pores. Missouri Bot. Gard. Ann. Rep. 16: 167–257.

Hartley, C.W.S. 1988. The Oil Palm, 3rd ed. Tropical Agriculture Series. Longman, London.

Haslerud, H.D. 1974. Pollination of some Ericaceae in Norway. Norw. J. Bot. 21: 211–216.

Heaney, R.P. & C.M. Weaver. 1989. Oxalate: effect on calcium absorbability. Amer. J. Clin. Nutr. 50: 830–832.

Hegelmaier, F. 1868. Die Lemnaceen. Verlag von Wilhelm Engelmann, Leipzig.

Hegelmaier, F. 1871. Ueber die Fructifikationsheile von Spirodela. Bot. Zeit. 29(38): 622–629, (39): 646–656.

Horner, H.T., Jr & B.L. Wagner. 1980. The association of druse crystals with the developing stomium of Capsicum annuum (sweet pepper) (Solanaceae) anthers. Amer. J. Bot. 67: 1347–1360.

Horner, H.T., Jr. & B.L. Wagner. 1992. Association of four different calcium crystals in the anther of connective tissue and hypodermal stomium of *Capsicum annuum* (Solanaceae) during microsporogenesis. Amer. J. Bot. 79: 531–541.

Hotta, M. 1989. Biological problems in West Malesian tropics: remarks for the 1987–1988 Sumatra research. pp. 1–10 in Diversity and Plant–Animal Interaction in Equatorial Rain Forests (edited by M. Hotta). Sumatra Nature Study (Botany), Kagoshima University, Kagoshima.

Hufford, L.D. & P.K. Endress. 1989. The diversity of anther structures and the dehiscence patterns among Hamamelididae. Bot. J. Linn. Soc. 99: 301–346.

Kato, M., T. Ichino, M. Hotta, I. Abbas & H. Okada. 1989. Flower visitors of 32 plant species in West Sumatra. pp. 15–31 in Diversity and Plant–Animal Interaction in Equatorial Rain Forests (edited by M. Hotta). Sumatra Nature Study (Botany), Kagoshima University, Kagoshima.

Kavaljian, L.G. 1952. The floral morphology of *Clethra alternifolia* with some notes on *C. acuminata* and *C. arborea*. Bot. Gaz. 113: 392–413.

Keijzer, C.J. 1987. The processes of anther dehiscence and pollen dispersal. I. The opening mechanism of longitudinally dehiscing anthers. New Phytol. 105: 487–498.

Kinzel, H. 1989. Calcium in the vacuoles and cell walls of plant tissue. Flora 182: 99–125.

Kron, K.A. & M.W. Chase. 1993. Systematics of the Ericaceae, Empetraceae, Epacridaceae and related taxa based upon *rbc*L sequence data. Ann. Missouri Bot. Gard. 80: 735–741.

Kronfeld, M. 1887. Über Raphiden bei *Typha*. Bot. Centralbl. 30: 154–156.

Kugler, H. 1942. 'Raphidenpollen' bei Bromeliaceen. Ber. Dsch. Bot. Ges. 60: 388–393.

Kugler, H. 1970. Blutenökologie. Gustav Fischer Verlag, Jena.

Küster, E. 1956. Oxalatkristalle. pp. 492–506. in Die Pflanzenzelle, 3rd ed. Gustav Fischer Verlag, Jena.

Landolt, E. 1981. Distribution pattern of the family Lemnaceae in North Carolina. Veroeff. Geobot. Inst. ETHS Rübel, Zurich 77: 112–148.

Landolt, E. 1986. The family Lemnaceae – a monographic study. Veroeff. Geobot. Inst. ETHS Rübel, Zurich 71: 1–566.

Landolt, E. & R. Kandeler. 1987. The family Lemnaceae – a monographic study, Vol. 2. Veroeff. Geobot. Inst. ETHS Rübel, Zurich 95: 1–638.

Leake, J.R., G. Shaw & D.J. Read. 1990. Proteinase activity in mycrorrhizal fungi. I. The effect of extracellular pH on the production and activity of proteinase by ericoid endophytes from soils of contrasted pH. New Phytol. 115: 243–250.

Lide, D.R. 1993. CRC Handbook of Chemistry and Physics, 74th ed. CRC Press, Boca Raton.

Lillie, R.D. 1977. H.J. Conn's Biological Stains, 9th ed. Williams & Wilkins, Baltimore.

Loetsch, B. & H. Kinzel. 1971. Zum Calciumbedarf von Oxalatpflanzen. Biochem. Physiol. Pfl. 162: 209–219.

Lüönd, A. 1983. Das Wachstum von Wasserlinsen (Lemnaceae) in Abhängigkeit des Nährstoffangebots, insbesondere Phosphor und Stickstoff. Veroeff. Geobot. Inst. ETHS Rübel, Zurich 80: 1–116.

Luteyn, J. 1983. Ericaceae. I. Cavendishia. Fl. Neotropica 35: 1–289.

Malloch, D.W., K.A. Priozynski & P.H. Raven. 1980. Ecological and evolutionary significance of mycorrhizal symbioses in vascular plants (a review). Proc. Natl. Acad. Sci. USA 77: 2113–2118.

Marrs, R.H. & P. Bannister. 1978. The adaptation of *Calluna vulgaris* (L.) Hull to contrasting soil types. New Phytol. 81: 753–761.

Matthews, J.R. & E.M. Knox. 1926. The comparative morphology of the stamen in the Ericaceae. Trans. Bot. Soc. Edinb. 29: 243–281.

Matthews, J.R. & C.M. Maclachlan 1930. The structure of certain poricidal anthers. Trans. Bot. Soc. Edinb. 30: 104–122.

Matthews, J.R. & G. Taylor. 1926. The structure of the stamen in *Erica hirtiflora*. Trans. Bot. Soc. Edinb. 29: 233–242.

Maurizio, A. 1953. Weitere Untersuchungen an Pollenhöschen. Beih. Schweiz. Bienen-Zeit. 20: 485–556.

Mauseth, J.D. 1988. Plant Anatomy. Benjamin/Cummings, Menlo Park.

Namikawa, I. 1919. Ueber das Oeffnen der Antheren bei einigen Solanaceen. Bot. Mag. Tokyo 33: 385–396.

Olmstead, R.G. & J.D. Palmer. 1992. A chloroplast DNA phylogeny of the Solanaceae: subfamilial relationships and character evolution. Ann. Missouri Bot. Gard. 79: 346–360.

Oryol, L.I. & M.A. Zhakova. 1977. The mechanism of anther dehiscence of tomato *Lycopersicon esculentum* Mill. (Solanaceae). Bot. Zhurn. 62: 1720–1730.

Palser, B. 1951. Studies of floral morphology in the Ericales. I. Organography and vascular anatomy in the Andromedeae. Bot. Gaz. 112: 447–485.

Palser, B. & Y.S. Murty. 1967. Studies of floral morphology in the Ericales. VIII. Organography and vascular anatomy in *Erica*. Bull. Torrey Bot. Club 94: 243–320.

arbuscular mycorrhizal fungi collected from saline soils. Mycologia 76: 74–84.

Paterson, B.R. 1961. Studies of floral morphology in the Epacridaceae. Bot. Gaz. 122: 259–279.

Percival, M.S. 1965. Floral Biology. Pergamon Press, London.

Perera, C.O., I.C. Hallett, T. Nguyen & J.C. Charles. 1990. Calcium oxalate crystals: the irritant factor in kiwifruit. J. Food Sci. 55: 1066–1069, 1080.

Pohl, F. 1932. Über sich Öffnende Kristallräume in den Antheren von *Deherainia smaragdina*. Jahrb. Wiss. Bot. 75: 481–493.

Pohl, F. 1941. Über Raphiden pollen und seine blütenëkologische Bedeutung. Öst. Bot. Zeit. 90: 81–96.

Pond, E.C., J.A. Menge & W.M. Jarrell. 1984. Improved growth of tomato in salinized soil by vesicular-arbuscular mycorrhizal fungi collected from saline soils. Mycologia 76: 74–84.

Puri, V. 1970. Anther sacs and pollen grains: some aspects of their structure and function. J. Palynol. 6: 1–17.

Rathcke, B. & L. Real. 1993. Autogamy and inbreeding depression in mountain laurel, *Kalmia latifolia* (Ericaceae). Amer. J. Bot. 80: 143–146.

Sazima, M., S. Vogel, A. Cocucci & G. Hausner. 1993. The perfume flowers of *Cyphomandra* (Solanaceae): pollination by euglossine bees, bellow mechanism, osmophores, and volatiles. Pl. Syst. Evol. 187: 51–88.

Schmidt, J.O., S.L. Buchmann & M. Glaiim. 1989. The nutritional value of *Typha latifolia* pollen for bees. J. Apic. Res. 28: 155–165.

Scott, H.G. & C.J. Stojanovich. 1963. Digestion of juniper pollen by collembola. Florida Entomol. 46: 189–191.

Smith, D.L. 1982. Calcium oxalate and carbonate deposits in plant cells. pp. 253–261 in The Role of Calcium in Biological Systems (edited by L.J. Anghileri & A.M. Tuffet-Anghileri). CRC Press, Boca Raton.

Soares, A.A., L.A. Campos, M.F. Vieira & G.A. Melo. 1989. Relaçoaes entre *Euglossa* (*Euglossella*) *mandibularis* Friese, 1899 (Hymenoptera, Apidae, Euglossini) e *Cyphomandra calycina* (Solanaceae). Ciencia e Cultura 41: 903–905.

Stanley, R.G. & H.F. Linskens. 1974. Pollen: Biology, Biochemistry, Management. Springer-Verlag, New York.

Stephenson, A.G., T. Lau, T. Quesada & J.A. Winsor. 1992. Factors that affect pollen performance. pp. 119–136 in Ecology and Evolution of Plant Reproduction (edited by R. Wyatt). Chapman & Hall, New York.

Stevens, P.F. 1971. A classification of the Ericaceae: subfamilies and tribes. Bot. J. Linn. Soc. 64: 1–53.

Tang, C. & W.S. Sakai. 1983. Acridity of taro and related plants. pp. 148–163 in Taro: A Review of *Colocasia*

esculenta and its Potentials (edited by Jaw-Kai Wang). University of Hawaii Press, Honolulu.

Tomlinson, P.B. 1961. Palmae. pp. 1–453 in Anatomy of the Monocotyledons, Vol. 2 (edited by C.R. Metcalfe). Clarendon Press, Oxford.

Tomlinson, P.B. 1969. Commelinales-Zingiberales. pp. 1–446 in Anatomy of the Monocotyledons, Vol. 3 (edited by C.R. Metcalfe). Clarendon Press, Oxford.

Tomlinson, P.B. 1982. Helobiae (Alismatidae). pp. 1–522 in Anatomy of the Monocotyledons, Vol. 3 (edited by C.R. Metcalfe). Clarendon Press, Oxford.

Traub, H.P. 1948. The genus *Zantedeschia*. Pl. Life 4: 9–32.

Trull, M.C. B.L. Holoway, W.E. Friedman & R.L. Malmberg. 1991. Developmentally regulated antigen associated with calcium crystals in tobacco anthers. Planta 186: 13–16.

van Tieghem, 1866. Recherches sur la structure des Aroïdées. Ann. Sci. Nat. Sèr. 5, 6: 72–211.

van Tieghem, 1867. Recherches sur la Structure des Aroïdées, pp. 1–140. E. Martinet, Paris.

Vander Kloet, S.P. 1976. Nomenclature, taxonomy, and biosystematics of *Vaccinium* section *Cyanococcus* (the blueberries) in North America. I. Natural barriers to gene exchange between *Vaccinium angustifolium* Ait. and *Vaccinium corymbosum* L. Rhodora 78: 503–515.

Venkatesh, C.S. 1956. The form, structure and special modes of dehiscence in anthers of *Cassia* I – subgenus *Fistula*. Phytomorphology 6: 168–176.

Venkatesh, C.S. 1957. The form, structure and special modes of dehiscence in anthers of *Cassia* II – subgenus *Senna*. Phytomorphology 7: 253–73.

Vogel, S. 1962. Duftdrüsen im Dienste der Bestäubung. Abh. Math. -Naturwiss. Kl. Akad. Wiss. Lit., Mainz. 1962(10): 602–763.

Vogel, S. & A. Cucucci. 1988. Pollen threads in *Impatiens*: their nature and function. *Beitr. Biol. Plantzen* 63: 271–87.

Weberling, F. 1989. Morphology of Flowers and Inflorescences. Cambridge University Press, Cambridge, (English translation of Weberling, 1981. Morphologie der Blüten und der Blütenstände. Eugen Ulmer, Stuttgart.)

Wettstein, R. von. 1892. Über die Systematik der Solanaceen. Verh. Zool.-Bot. Ges. Wien 42 (Sitzber.): 29.

Yasue, T. 1969. Histochemical identification of calcium oxalate. Acta Histochem. Cytochem. 2: 83–95.

Ziegler, A. 1925. Beiträge zur Kenntnis des Androeciums und der Samenentwicklung einiger Melastomaceen. Bot. Arch. 9:398–467.

9 · Anther adaptation in animal pollination

PETER BERNHARDT

Despite the past 150 years' intensive field and laboratory work on pollination biology, the literature on the roles of anthers in animal-pollinated flowers remains depauperate for three, overlapping reasons. First, the pollination of any angiosperm taxon is usually treated as a pattern or problem in which the whole is not broken down into the sum of its floral parts. Reviews dealing specifically with the anther's role in zoophily are almost nonexistent.

Second, when pollination biology has addressed the individual role of floral organs in the evolution of zoophily, the authors have emphasized the adaptive function of perianth segments over that of the androecium. This bias is well reflected in the classic work on the subject by Faegri & van der Pijl (1979) in which the floral forms related to pollination syndromes were classified according to the form and/or presence of the perianth. It was only when the perianth was absent, or reduced to whorls of scales, that the androecia were considered sufficiently prominent to refer to the whole blossom as a brush flower. Leppik (1977), in contrast, attempted to incorporate aspects of staminal presentation into his treatment of floral symmetry but this has not met with general acceptance. Consequently, anther biology has long been dominated by developmental botanists, histologists and pollen physiologists (see this volume) and not by pollination ecologists.

Third, since the majority of angiosperms and several gymnosperms (see review by Norstog 1987) appear to be pollinated by animals any attempt to classify the ecology of anthers remains a staggering task due to the sheer number of species of seed plants. In contrast, the literature reviewing the mechanisms of anthers belonging to the minority of flowers and strobila pollinated by air and/or water currents has received far more comprehen-sive treatments (Knox 1979; McConchie 1982; Dafni 1993). Consequently, this contribution is necessarily incomplete and will reveal gaps in the subject instead of presenting a cohesive picture of the subject. If a full review of the pollination literature were combed for references to anther adaptations it would mean poring over hundreds of unrelated papers and probably end in a review the size of this volume. Therefore, this chapter has had to be limited to examples in a few, well-studied taxa. These examples may not serve ultimately as model systems for all zoophilous angiosperms in the future.

ADAPTIVE SIGNIFICANCE OF ANTIIER DEHISCENCE

In many angiosperms the timing of anther dehiscence occurs independently of either the opening of the perianth or stigmatic receptivity. Staggered timing of pollen release (dichogamy) should increase the reproductive fitness of outbreeding species as it promotes the establishment of a self-isolation barrier. In the best known examples, protandry and protogyny (dichogamy) lower the chances of self-pollination in a bisexual flower or an inflorescence with synchronous flowering. A review of relictual angiosperms currently supports the view that the ancestral condition in the evolution of bisexual flowers is protogyny (Bernhardt & Thien 1987). However, dichogamy cannot be interpreted as an adaptation confined only to zoophilous flowers (Knox 1979; McConchie 1982).

The timing of anther dehiscence, in relation to the opening of the perianth, may relate directly to the exploitation of the animal pollinators in some plant families. This is particulary dramatic in genera of Old World

Loranthaceae. In these passerine-pollinated flowers anther dehiscence occurs commonly before the petals separate completely. For a period of hours or days, the corolla enters a 'Chinese lantern' phase (Bernhardt 1983) in which the anthers have dehisced, nectar is secreted at the base of the floral tube and the bird's probing bill finds access to the nectar only through the open slits of the corolla (Johri & Bhatnagar 1972). The petals are under such intense internal pressure that the entrance of the bird's bill causes the corolla to spring open explosively so that the bird's head is dusted with pollen or is actually struck by detachable anthers (Evans 1895; Gill & Wolf 1975; Kuijt 1969). An explosive or sudden release of pollen is found in other families, including many genera of papilionoid legumes (Free 1970), the Ericaceae (e.g., *Kalmia*: Reader 1977) and up to three genera of Proteaceae (Wrigley & Fagg 1989). Therefore, it is usually expected that rapid pollen release in many flowers depends on modifications of the anthers in conjunction with one or more perianth segments and/or the style (e.g., *Conospermum*: Wrigley & Fagg 1989).

Kearns & Inouye (1993) reviewed the relationship between anther dehiscence and the quantity of pollen released. The time of day when anthers dehisce varies greatly both between different species and within individual flowers belonging to the same species (Percival 1955). The staggered release of pollen by anthers in the same androecium may be far more common in nature than a synchronous release of pollen and should result in pollen dispersed to greater numbers of pollinators and to more receptive stigmas over time. Kearns & Inouye argue that staggering the release of pollen in a flower may reduce the loss of pollen viability as climatic conditions change. This remains theoretical and does not take into account the fact that many flowers produce porose-porate anthers in which pollen is never exposed to the atmosphere outside the thecae until it is removed by insect pollinators 'buzzing' anthers (Buchmann 1983; and see below).

THE ROLE OF NONFUNCTIONAL ANTHERS BEFORE ANTHESIS

The role anthers play as sources of floral attractants and rewards is greatly limited in some taxa. In these species the anthers have *no* function *after* the flower bud opens.

These 'nonfunctional' anthers are not staminodes since true staminodes often contribute directly to the pollination syndrome. For example, in some zoophilous flowers staminodes direct the movements of the pollinators and/or produce floral fragrances and/or edible rewards (Endress 1986; Gottsberger 1977; Weberling 1989). Nonfunctional anthers are fertile anthers that dehisce prior to the opening of the perianth but do not present or contain viable pollen at anthesis. Nonfunctional anthers are dry, empty and are either concealed by the perianth or abcise prior to anthesis. This secondary presentation of pollen has evolved independently in many angiosperms orders but occurs more frequently in dicotyledons and has been treated as a rare modification in monocotyledons (e.g., *Calathea*: Kennedy 1978).

In flowers with nonfunctional anthers a second floral organ retains and exposes the pollen at anthesis. For example, in *Acrotriche serrulata* (Epacridaceae; McConchie *et al.* 1986) petal combs are the pollen presenters. More commonly, a portion of the stigma-style complex is modified for pollen presentation as in some Rubiaceae (see Manning 1990), most Proteaceae and many species within the Asteraceae, Campanulaceae and Goodeniales (Knox & Freiderich 1974; Faegri & van der Pijl 1979; Burns–Balogh & Bernhardt 1985). Within these taxa pollen presentation may be further enhanced by the development of an expanded, adhesive disc (e.g., the protostigmas of the Proteaceae) or a hairy, cuplike indusium as in the Goodeniaceae and Brunoniaceae *sensu stricto*.

Although the evolution of secondary pollen presentation appears to be unique to zoophilous flowers, the mere form or presence of a pollen presenter in a flower is insufficient to predict either the major pollinators or to characterize the pollination syndrome. Both the petal combs of *Acrotriche* and most subfamilies of Australian Proteaceae appear to depend largely on vertebrates, especially passerines, loriid parrots and nonflying marsupials, for pollen dispersal (Hopper & Burbidge 1982; Turner 1982; McConchie *et al.* 1986; Paton 1986). In contrast, the pollination ecology of the goodenialean and campanulalean families with pollen presenters remains understudied, but the floral syndromes seem more likely to encourage insect vectors (Faegri & van der Pijl 1979).

ANTHERS AS PRIMARY ATTRACTANTS

It is often presumed that anthers are natural sources of visual cues contributing to the overall attractiveness of the entire blossom. However, there is comparatively little direct evidence to determine whether all pollinators can still see the anthers when they are a distance from the flower or even when they are in the process of probing floral bases for nectar. It has usually been assumed, for example, that most flower beetles 'see their world through their sense of smell.' Only recent work by Dafni *et al.* (1990) shows that scarabs in the genus *Amphicoma* discriminate between floral organs by sight, not scent.

In a flower with radial symmetry the exerted anthers of many species commonly offer color patterns that contrast with the perianth. Anther and perianth colors in the same flower often contrast in spectra visible and invisible to the human eye (Barth 1985; Menzel 1990; Lunau 1992). In the flowers of *Zygadenus nuttalii* the tepals, basal nectaries and ovaries strongly reflect ultraviolet (UV) light, while the anthers and filaments do not (Figs. 1, 2). To the *Dialictus* bees or dipterans the androecium of *Z. nuttallii* may show up like a bright asterisk against a darker background. Barth suggests that UV reflections are especially common towards the center of the flower so that the anthers may reflect UV patterns overlapping with the bases of their perianths (see *Potentilla, Swertia, Bryonia* and *Vinca* in Barth 1985). In the zygomorphic *Orchis laxiflora*, the column's solitary anther relects an UV pattern at much the same intensity as does the sculpted center of the flower's labellum.

The same pattern, in which the anthers have the same color as the center of the perianth segments, also occurs in spectra visible to humans. Many red, bowl-shaped flowers of the Mediterranean region show a striking convergence in color patterns. These flowers are pollinated primarily by scarab bettles in the genus *Amphicoma*. Experimentation shows that the beetles approach model flowers based on visual not olfactory cues. Red appears to be the preferred color of attaction. However, there is a particularly striking convergence in the *Amphicoma*-pollinated species *Papaver rhoeas* (Papaveraceae), *Anemone coronaria* (Ranunculaceae), *Ranunculus asiaticus* (Ranunculaceae) and *Tulipa aegenensis* (Liliaceae *sensu lato*). The centers of these flowers are uniformly dark and often a glossy blackish color to the human eye. In *P. rhoeas, T. aegenensis* and *A. coronaria* the 'blackening' of whole stamens overlaps with the dark bases of the perianth segments (Dafni *et al.* 1990). Even though a number of *Tulipa* spp. have relatively light colored

Figs. 1, 2. Fig. 1. Flowers of *Zygadenus nuttalii* under white light. The androecium is a whitish-cream color to the human eye. Fig. 2. The same flowers of *Z. nuttalii* under ultraviolet (UV) using a Corning filter at 350 nm between the flowers and the camera lens (Kodak 2415 film at 30 seconds for optimum exposure). (Photograph by W. Stark.

anthers they extrude and shed a sooty, black pollen upon dehiscence. Further observations by Gack (1979) suggest that when color patterns at the base of the perianth overlap with those of the anthers this makes the androecium more attractive to pollen collectors, as the pattern artificially inflates the actual size of the anther cluster. This requires experimental evidence to be confirmed.

Within monocotyledons there is increasing evidence that ornamentation/pigmentation of the apices of the filaments may enhance the attractiveness of narrow, or short, or dull colored anthers. The filament apices may be swollen, wrinkled-glossy, and/or ornamented with papillae or trichomes (Fig. 3) and may be a bright white or yellow-orange and/or blue-purple. This trend was first noted by Vogel (1978) who treated the gaudiness of the filaments as examples of evolutionary shifts from reward to deception in pollen flowers. That is, Vogel interpreted these colorful appendages, papillae and tri-

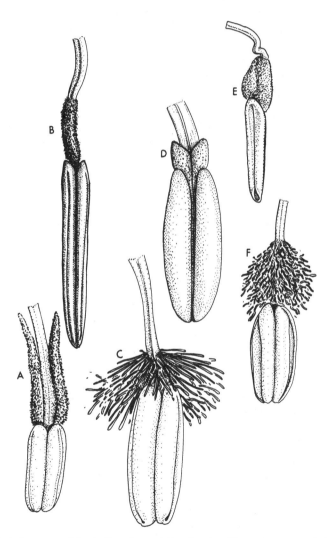

Fig. 3. Stamens of selected genera of Australian Asparagales showing filaments ornamented at their apices. A. *Dichopogon capillipes*. B. *Stypandra caespitosa*. C. *Bulbine bulbosa*. D. *Dichopogon strictus*. E. *Dianella caerulea* var. *assera*. F. *Arthropodium milleflorum*.

chomes on the filaments as 'false pollen' luring foragers that would otherwise ignore the depauperate rewards in the true anthers (pseudanthery). This interpretation was accepted, in part, by Gack (1979) and Faden (1992).

Two additional observations of floral pigmentation made by the author are mentioned here in the hope of stimulating further research on anther functions. First, the deep-purple anthers and filament tips of many night-blooming flowers should be considered for future experimentation. As potential pollinators approach the nocturnal flowers of *Capparis* and some amaryllids (e.g., *Crinum*) do the darkened anthers offer a three-dimensional, nectar guide against the white background of the broad, perianth lobes? Second, does the contrasting pigmentation of inedible staminodes, stigmas, column hoods (e.g., *Thelymitra*) and perianth calli (e.g., *Caladenia*, *Glossodia*) reflect deceptive coloration encouraging pseudanthery in those flowers pollinated primarily by pollen eating/pollen collecting insects (Dafni & Bernhardt 1989)?

Anthers are known sources of floral scents (D'Arcy *et al.* 1990) but the biochemical analysis of anther scents remains in a state of infancy as most analyses use whole flowers and not the centers of scent production (osmophores). Pollen is often scented in bee-pollinated flowers with pollenkitt acting as a matrix for volatiles of lower molecular weights (Dobson 1988). Scented pollen may be an important adaptation in flowers pollinated by solitary bees. Dobson (1987) showed that adult bees are more likely to visit the flowers of particular species when they are raised on the pollen of those species as larvae. In fact, the adult bee recognizes and is attracted to the pollen scent from extracts of the pollenkitt dabbed onto filter paper.

The anther may produce scents exogenously or endogenously. Stamens have not been classically associated with large, glandular osmophores as have been the inflorescences of some aroids or the perianths of some orchids and asclepiads (Vogel 1990). The androecia of the Commelinaceae and Phormiaceae may be an exception as floral odors may be released by distinctive, trichome tufts (Faden 1992) or papillose structures known as strumae (Bernhardt 1995). Osmophores on the staminal filaments may occur in some lilioid genera (Fig. 3 and see below). In most angiosperms with scented anthers, though, epidermal cells releasing volatile oils are usually

not recognizable to the naked eye (Endress 1986) and may require advanced histological techniques to discern in depth (Sazima *et al.* 1993). Neutral Red tests, while crude, indicate that the anther walls may be sites of scent production. For example, in the inflorescences of the protogynous species of *Acacia* the number of scent components and the intensity of the floral scent increases over the floral lifespan as the stigmatic receptive/stamens unexpanded (female phase) period wanes and the stigma nonreceptive/stamens expanded (male phase) begins (Holman & Bernhardt, work in progress). This correlates positively with the increasing absorption of Neutral Red by the anthers over the floral lifespan (Bernhardt 1989).

Dobson *et al.* (1990) have found that the scent produced by the artificially emptied anthers of *Rosa rugosa* still differs from the scents produced by both the pollen and the other floral organs. Different organs on the same rose blossom do not always produce the same volatiles and when they do such compounds may be secreted in entirely different proportions. The anthers of *R. rugosa* appear to be the only organs to secrete the terpenoids, neryl acetate and citronellyl acetate. In contrast to the petals, sepals, gynoecium and pollen the anthers of *R. rugosa* release a far higher proportion of the benzenoid compound, eugenol, than of nine other known volatiles.

The production of floral terpenes and other volatiles collected by male bees in the tribe Euglossini (Apidae) is more typically associated with the perianths of orchids, gesneriads and the inflorescences of some aroids. A recent, and most important, exception to this rule has been found in the solanaceous genus, *Cyphomandra*. Male bees repeatedly collect volatiles produced by the anther connectives in at least four species within this genus (Gracie 1993; Sazima *et al.* 1993). In contrast to most other flowers with tubular anthers opening by means of terminal pores the anthers of these *Cyphomandra* spp. do not release pollen as a result of direct, anther vibration by the floral forager. Instead, the anther walls are modified to form a 'bellows-like mechanism' in which pollen is literally blown out of the terminal pores as the anther wall caves in at the slightest touch of the perfume-collecting bee (Sazima *et al.* 1993).

Volatile oils may also be produced by the genus *Pleurothyrium* (Lauraceae). The anthers in this genus are non-secretory but the androecial glands or staminodia of some species are so large they surround and enclose fertile stamens (van der Werff 1993). These glands produce a spicy odor reminiscent of the flowers of those epidendroid orchids pollinated by male, euglossine bees (Dodson & van der Werff, unpublished data). However, there is no morphological evidence of a bellows-like mechanism in the anthers of these Lauraceae.

ANTHERS PRODUCING REWARDS

Pollen is regarded as the primary reward produced by most anthers in those flowers pollinated by animals. Knox (1979) was among the first to note that the cytoplasm of pollen grains belonging to animal-pollinated flowers was higher in digestible amino acids and lipids than in flowers pollinated by air or water currents, suggesting coevolution between the digestive physiology of pollinators and resource partitioning of many angiosperms. There are important exceptions to this interrelationship, though. Since pollen grains released as large, interconnected pollinia (Asclepiadaceae, Orchidaceae) cannot be removed and consumed easily by the vector or its offspring (e.g., *Diuris*: Beardsell *et al.* 1986), it is probable that the nutrient level of orchid or asclepiad pollinia will be lower than the pollen of flowers released separately as monads or polyads.

The insect superfamily, Apoidea (bees), remains the dominant pollinator of angiosperms based on both diversity (over 20 000 species; Armstrong 1979), their pandemic distribution (Michener 1979) and the pollen-foraging habits of the females as adults. Virtually all species of Apoidea feed on some pollen as adults but, more important, their larvae must be fed exclusively on pollen with the exception of a few tropical taxa that scavenge alternately on carrion and feces (Roubik 1989). In contrast, although some pollinating bats, loriid parrots, flies and beetles require pollen as a dietary supplement as adults they do not feed their young *directly* on pollen.

The anthers of some genera in the Actinidiaceae, Cesalpinaceae, Commelinaceae, Dilleniaceae, Lecythidaceae, Melastomataceae, Myrtaceae, Solanaceae and Tiliaceae appear to be structurally and cytologically dimorphic, reflecting their dependency on female bees as pollinators (Percival 1965; Bowers 1975; Schmid 1978; Anderson 1979; Prance & Mori 1979; Haber & Bawa 1984; Faden 1992; Cane 1993). That is, some anthers produce pollen grains in which the exine retains the usual germinal pores or slits and is always nucleate. These are the viable, pollen grains. A second set of anthers produces grains in which exine development is arrested, germinal pores are usually absent and the cell contents may be anucleate. These 'feeder' grains are often produced in great quantities, and may be larger or smaller than the viable grains, but feeder pollen is not known to germinate. Taxa offering dimorphic pollen rarely produce floral nectar so pollen is the *only* edible reward offered to female bees foraging for their offspring.

The arrangement and proportion of anthers manufacturing viable pollen versus those containing feeder pollen depends on the expression of floral sexuality. In the bisexual flowers of some Caesalpiniaceae, Commelinaceae, Lecythidaceae and Solanaceae dimorphic pollen is usually produced by two different anther forms so the androecium is heterantherous (Percival 1965; Prance & Mori 1979; Gottsberger & Silberbauer-Gottsberger 1988; Faden 1992). For example, in *Solanum rostratum* the large anthers produce viable pollen while the small anthers offer feeder pollen (Bowers, 1975). The dimorphic anthers of *Commelina coelestris* segregate into anthers producing viable grains and anthers that release a milky fluid (Faegri & van der Pijl 1971).

Second, dimorphic pollen is also offered by some plants that are cryptically dioecious (Cane 1993). The flowers of female plants bear only anthers that offer feeder pollen, while the male plants produce all the viable pollen (Schmid 1978; Anderson 1979; Haber & Bawa 1984). The flowers of male plants usually contain more stamens than do the flowers of female plants and the anthers of a male flower are usually larger than those of a female (Cane 1993). The relative nutritional value of viable versus feeder remains inconsistent in the absence of biochemical evidence. Haber & Bawa (1984) insist that feeder pollen is an important food of bees pollinating *Saurauia* while Schmid (1978) described the inaperturate pollen of *Actinidia* as a false reward. However, the polyphyletic evolution of feeder pollen in angiosperms provides some indirect evidence that pollinating bees, in particular, may exert selective pressures on both pollen and anther evolution.

Therefore since certain pollinators will come to a flower that produces only pollen, and since all bee larvae are raised on an obligate diet of pollen, the selective foraging of some mature animals appears to have had an important selective force on floral evolution in general. This has most probably encouraged the evolution of flowers that offer pollen as their *only* edible reward. These 'pollen flowers' do not secrete nectar and do not offer any other rewards unless the reward is located on the anther (e.g., volatile oils or food bodies) yet they remain animal-pollinated, are usually outcrossing and occur repeatedly throughout the major orders of angiosperms. Vogel's (1978) attempt to classify variation in these pollen flowers based on stamen number, stamen form and mode of pollen dehiscence continues to have considerable merit. He derived three overlapping types that currently require some additional expansion following continued research over the past 15 years.

1. The *Magnolia* type is produced primarily by relictual dicots (Magnoliales and Nymphaeales). The flower is multi-staminate (polyandrous) and the number of stamens in a flower is often indeterminant. The stamens are short and do not usually subdivide into true anthers and filaments. Their pollen is extruded or shed at dehiscence and may collect on the floral receptacle or on the tepals. The primary pollinators (beetles, thrips, small flies, micropterigid moths, and short-tongued bees) tend to have rather short mouthparts. However, these mouthparts are usually modified for the active collection and consumption of pollen (e.g., Syrphidae in the Diptera; Boganiidae, Catereitidae, Chrysomelidae, Nitidulidae etc. in the Coleoptera) although their legs tend to lack specific modifications for the manipulation of anthers or the extraction of grains from thecal slits or pores (Crowson 1981; Bernhardt & Thien 1987; Samuelson 1994).

2. The *Papaver* type is particularly well distributed throughout the dicotyledons. The androecium is also multistaminate and stamen numbers are often indeterminant but the stamens are elongated and subdivided into true filaments and boxlike anthers. The primary pollinators tend to have both longer mouthparts and legs adapted for the removal of pollen from anthers. Most bee families and some fly families (Syrphidae) are the primary pollinators of these flowers, and they harvest pollen from anthers opening via slits by scraping, brushing or applying thoracic vibration.

3. The *Solanum* type has evolved in both monocotyledons and dicotyledons. The androecium has comparatively fewer stamens (oligandrous) than either of the preceding types. The anthers tend to be physically large and are often equal to or longer than the truncated filaments. Anthers tend to have terminal, porose-porate openings and removal of the pollen is usually restricted to thoracic vibration by certain bee genera (e.g., *Bombus*, *Lasioglossum*, *Xylocopa*, etc.), and the syrphid fly genus, *Volucella* (see Buchmann *et al.* 1978; Buchmann 1983). Evolution within anthers of this flower-type has become so specialized that the androecium may become heterantherous (e.g., neotropical Cassiinae: Gottsberger & Silberbauer-Gottsberger 1988 and see above) with filaments of different lengths and a combination of both staminodes, fertile anthers and feeder anthers of different shapes and performing different functions.

The recent work on *Cyphomandra* (see above) strongly suggests that anthers with a bellows-like mechanism are derived directly from the standard, *Solanum*-type flower. For this to occur pollen-collecting female bees must have been replaced by males collecting scents. Both *Solanum*-type and bellows-type flowers are found in the genus, *Cyphomandra*. Furthermore, the derivation of bellows-type anthers from vibratile ancestors may have evolved independently in the Acanthaceae and Melastomataceae (Sazima *et al.* 1993).

While pollen has been proposed as the first, edible reward in the bisexual flowers of the protoangiosperms (van der Pijl 1978), it is not the preferred food of many pollinators. Most flower-visiting birds, Lepidoptera and certain families of flies (e.g., Nemestrinidae: Goldblatt & Bernhardt 1990) have not been observed to feed on pollen and appear to lack appropriate mouthparts and gut modifications for pollen ingestion and digestion. The ability to collect and digest the pollen may vary widely in closely related animals associated with floral foraging.

For example, beetles have been classically associated with the active consumption of both pollen and such solid structures as staminodes and starch bodies in relictual magnolioids (Table 1; Bernhardt & Thien 1987). However, pollen grains often pass through the tracts of some flower-foraging beetles without significant digestion of their cytoplasmic contents (Samuelson

Table 1. *Location of rewards on the organs of the androecium and gynoecium of relictual dicotyledons*

	Organ			
Taxon	Anther	Filament	Staminode	Carpel
Annonaceae				
Guatteria	EO	EO	—	—
Calycanthaceae				
Calycanthus (1)	FB	—	FB	—
Eupomatiaceae				
Eupomatia (1)	—	—	EO	—
Lauraceae				
Cinnamomum (1)	—	N	N	—
Laurus (1)	—	N	N	—
Persea (2)	—	N	N	—
Pleurothyrium (21)	—	N?	VO?	—
Magnoliaceae				
Magnolia (2)	—	—	—	SS?
Talauma (1)	EO	—	EO	SS?
Myristicaceae				
Myristica (1)	—	—	—	SS
Nelumbonaceae				
Nelumbo (2)	FB?	—	—	SS?
Nymphaeaceae				
Victoria (1)	—	—	—	FB
Trochodendron (1)	—	—	—	N
Winteraceae				
Drimys (1)	—	—	—	SE
Pseudowintera (2)	—	—	—	SE

EO, edible organ; FB, food body; N, nectar; SS, stigmatic secretions (protonectar); VO, volatile odors. Numbers in parentheses refer to *known* number of species examined.

Sources: Prance & Arias (1975), Gottsberger (1977), Schneider & Buchanan (1980), Weberling (1989), Chaw (1992), Lloyd & Wells (1992).

1994). *Dascillus* species have elongated hairy lobes on their maxillae and/or labia associated with nectar collection but lack both the three known modifications for pollen consumption in the Coleoptera (R.A. Crowson, personal communication; Crowson 1981). First, in most cases, beetles dependent on pollen as a food source have a molar on their mandibles that cracks or grinds the pollen wall exposing the contents of each grain directly to the digestive system *before* it enters the gut (Crowson 1981; Barth 1985; Samuelson 1994). Secondly, in some

Oedmeridae the pollen grains are stored in the beetle's crop. The crop also stores flower nectar so grains begin to hydrate and/or germinate *before* they enter the gut and are more likely to release their contents directly onto the lumen (Arnett 1968; Samuelson 1994). This second modification would seem far less useful in beetle taxa associated most consistently with nectarless flowers. Finally, Rickson *et al.* (1990) suggested that the scarab beetle, *Cyclocephala amazona*, swallowed the lignified trichomes on the inflorescences on the palm, *Bactris gasi-*

paes, to use them as a form of 'grit' to help crush or break open pollen walls.

Some bee taxa do not even visit those flowers that lack nectar but produce copious quantities of pollen. There is a recurrent trend towards oligolecty in most of the families that make up the superfamily Apoidea (Michener 1979). Oligolectic species forage only on a few, closely related angiosperm species to provision their larvae and the flowers they do visit offer pollen *with* nectar. In southeastern Australia, for example, mega-chilid species are dependent on the pollen of different legume genera offering both pollen and nectar (Armstrong 1979). However, megachilids are rarely observed or collected on the nectarless florets of the *Papaver*-type inflorescences of *Acacia* spp., although *Acacia* is the largest genus of legumes in Australia (Bernhardt 1989).

Therefore, although nectarless, pollen flowers occur repeatedly throughout the angiosperms (Buchmann 1983) species of animal-pollinated flowers that offer both pollen *and* other rewards (especially nectar) are far more comon. The topography of the nectary, or other reward-producing body, in relation to that of the androecium tends to be 'strategic', ensuring maximum contact between the pollinator and the dehiscent anthers. This mode of presentation can occur in virtually any flowers with filamentous stamens. In flowers with radial symmetry filaments keep the anthers propped up over or around the nectar resource. In zygomorphic flowers the nectary is most often located under the androecium or behind it as in flowers with nectar-containing spurs or pouches (Henslow 1893; Leppik 1977).

In the relictual magnolioids there is often no distinction between anther and filament or the flattened filament is relatively short in relation to anther dimensions. Therefore, there appears to be a trend within this group of plants to display edible rewards, other than pollen, *on* androecial structures especially staminodes and filaments proper (Table 1). This includes nectar glands, discrete food bodies and edible staminodes (Endress 1986; Weberling 1989).

Anthers bearing solid, edible bodies or glands that release an edible exudate, to be consumed separately or to be mixed with pollen, are not confined to the magnolioids. The anthers of *Acacia bidwilii* bear stalked bodies staining positively with Neutral Red. These bodies are plucked off and eaten by beetles (see Hawkeswood as cited by Bernhardt 1982). More detailed information on the food bodies of legumes was provided by Chaudhry & Vijayaraghavan (1992) who found that the globose head of the anther gland of *Prosopis juliflora* lyses to release a protein–carbohydrate exudate. In *Thryptomene calycina* (Myrtaceae) the secretion of the anther gland is hydrophobic; it mixes with the pollen to raise levels of edible lipids for pollen foraging insects (Beardsell *et al.* 1989). It should be noted that when edible rewards are located on dehiscent anthers the edible reward is a solid food body or the secretion is water-poor.

In fact, although pollen and nectar remain intrinsic rewards in the majority of bee, syrphid fly, bat and certain beetle-pollinated angiosperms there is a relative paucity of records of true nectar glands attached to anthers proper. Pollen release by the anther is dependent on the dehydration of anthers. Pollen hydrates rapidly upon contact with solutions of sugar and water (Linskens & Stanley 1974). The presence of a water-secreting gland on a dehiscent anther would seem maladaptive in most cases since the precocious hydration of a pollen grain *before* it reaches an appropriate stigma causes it to explode or lose viability within a hour or less.

Within this century anther nectaries have been identified in three families: Asclepiadaceae, Melastomataceae and Violaceae. As anatomical and ontogenetic studies have continued the number of genera with anther nectaries has actually declined. For example, when Fahn (1953) first addressed the topography of anther nectaries he listed the classical examples of *Asclepias* and *Viola*. He later dropped *Asclepias* (Fahn 1979). The position of the floral nectaries of *Asclepias* is now treated as on the receptacle *between* the filaments but as *Asclepias* filaments are very short the nectary or nectar-holding cup tends to be confused with the anther slit. The actual extension of the nectariferous tissue up the filament tube and onto the anther slit has been reported in some genera of the Stapelieae (Kunze 1990).

Locating the nectar glands on the anther connectives of *Viola* spp. is so old in the literature it remains the standard example of anther nectaries to this day (Weberling 1989). However, Beattie (1971, 1974) challenged this interpretation. Beattie interpreted the topography of *Viola* nectaries as on the highly reduced, staminal filaments and not on the anther connectives proper. Examination of fresh flowers of *Viola hederacea* and rehydrated flowers of *V. papilionacea* and *V pedata* by Bernhardt (personal observations) suggests that the nectaries actually develop on the stamens where the filament apex grades *into* the base of the anther sacs. In the species with the largest nectar glands (*V. pedata*) the base of each nectary extends well onto the anther connectives (Fig. 4).

Although the actual position of the nectar gland in *Viola* may be a moot point the fact remains that water-secreting glands appear to be extraordinarily close to dehiscent anthers in the Violaceae. Even so, pollen viability does not appear to be compromised in members of the Violaceae with zygomorphic flowers for two important reasons. First, the pollen grains of *Viola* spp., and at least one *Hybanthus* sp., are released by extended anther awns, not by longitudinal slits, so individual grains are further isolated from the staminal nectary. Secondly, the nectar is unable to contact the pollen as the nectar glands are always inserted up a corolla pouch or spur in the living flowers of *Viola* and *Hybanthus prunifolius* (Augspurger 1980).

That does not explain how pollen of the male flowers of *Melicytus* spp. (Violaceae) escape hydration by secretions of the staminal nectaries. The nectaries of *Melicytus* spp. are as long as the anthers but the corollas of these flowers are radially symmetrical. In *M. lanceolatus* anther awns appear to release pollen as in zygomorphic genera while staminal nectar drains off into a short tube formed by the bases of the petals so nectar accumulates much *below* pollen grains released by the anther awns (Fig. 5). Furthermore, the anthers of *M. lanceolatus* have fused together so nectar cannot reach the pollen by seeping between the anthers. In *M. ramiflorus* and *M. micranthus*, though, the anthers are free, the anther awns are so reduced that they are not involved directly in pollen release and the petals do not form a nectar-collecting tube. In these two species nectar accumulates in a depression at the tip of each columnar, staminal, nectary.

Illustrations by Powlesland (1984) indicate that in *M. ramiflorus* and *M. micranthus* each anther awn has formed a widened, scale-like hood *between* the dehiscent anther and its nectary (Fig. 6). This barrier is the only isolation between the exposed pollen in the slits of the introrsively dehiscent anthers and the nectar droplets produced by their dorsal glands. In the Violaceae, then, it would appear that the secretion of staminal or anther nectaries

are kept away from the dry, viable, pollen by a combination of morphological features; corolla symmetry, the formation of petal tubes or true spurs and the presence of nonabsorbent awns or scales that form a barrier between the nectary and the dehiscent anther.

Lastly, there is the report of nectar secreted by the curved, horn-like nectar gland on the anther of *Memecylon heyneanum* (Melastomataceae). The region of secretion is reported to be confined to the basal end of the gland (Subramanyam 1949). If this interpretation is correct then *Memecylon* is a great exception to the rule that anthers are protected from the secretions of their nectaries.

Documented reports of nectaries on the staminal filaments seem to be more common but are not always easier to verify as there have been few morphological and biochemical studies to confirm that these glands do, indeed, secrete a true nectar. Fahn (1979) lists genera in three families (Caryophyllaceae, Papilionaceae and Liliaceae sensu lato) and illustrates a fourth (Lauraceae) in which the staminal filaments of some genera bear nectaries. This is a partial list to which we may add the Oxalidaceae (see review in Bernhardt 1990b). In these cases nectaries are often basal to the filament or located midway along the filament, especially when the filament is highly elongated.

Is it possible that some genera of Ericaceae have nectaries on their anthers as ericoid anthers often bear paired, gland-like structures known as spurs? Spurs are usually associated with anthers that also bear tubular awns. Therefore, do the anthers of some Ericaceae ever function with a pollen release/nectar secretion system as described above in *Viola*? Ontogenetic studies by de Villamil (1980) showed great variability in spur morphology from bristles to fan shapes but the prospects for secreting nectar remain unpromising. The spur cells contain granular tissue that does not appear to develop into nectar-secreting cells. Field observations by Cane *et al.* (1985), Dorr (1981), Haselrud (1974) and Reader (1977) found that when insects visit the flowers of *Chamaedaphne*, *Erica*, *Gaultheria*, *Kalmia*, *Ledum*, *Vaccinium* and *Zenobia* for nectar they probe the base of the floral tube for nectar but not the anthers.

While it is true that Daumann (1970) provided a more complete treatment of the classification of nectaries on stamens this survey must be approached with considerable caution. It is obvious that part of Daumann's review often identifies nectar glands as distinctive sculptures on staminal epidermis but provides no observations or evidence of secretions by such sculptures. Consequently, staminal filaments with swollen, often papillose-hairy apices were classified as nectaries. In contrast to this blanket classification of filament ornamentation Bernhardt (1995 and personal observation) has examined the ornamented filaments within some genera of the Asparagales (*sensu* Dahlgren & Clifford 1982). Nectar secretions were not found to be produced by the filament sculptures of *Arthropodium*, *Bulbine*, *Dianella*, *Dichopogon*, *Echeandia*, *Stypandra* or *Tricoryne*. These genera have nectarless flowers with modes of floral presentation suggestive of buzz-pollination (Buchmann 1983; Fig. 3) and limited references in the literature tend to confirm that they show the pollination ecology of *Solanum*-type flowers (Bernhardt & Montalvo 1979; Bernhardt & Burns-Balogh 1986; Bernhardt 1995).

ANTHER PRESENTATION AND POLLINATION SYNDROMES

Anther phenology, attractants and rewards cannot define modes of anther presentation alone. Other factors have influenced the way in which the anthers may manipulate a prospective pollinator as the animal probes or investigates the floral interior. These factors include the number of anthers within the androecium, the mode of anther dehiscence and the role of pollen as a reward for the pollinator. All three factors must be considered jointly when examining the presentation of anthers as it relates to the pollination syndrome.

There are probably no more than two basic, overlapping modes of anther presentation. These two modes of presentation are determined by the foraging behavior and digestive system of the animal versus modifications to the adaptive morphology and physiology of the flower. The two modes may be summarized as passive pollen collection versus active pollen collection. The floral characters of both modes are reviewed in Table 2.

Passive pollen collection

The passive pollen collection mode is undoubtedly more common in nature. The evolution of this mode correlates

positively with both the evolution of radial and bilateral symmetry in perianths and the elongation of staminal filaments. In this mode pollen is *not* the primary reward to be collected by the pollinator. Pollen is either a secondary reward in this mode or it is ignored by the vector. To locate and forage upon the primary reward (nectar, liquid oils, food bodies) the forager must pass through, on, or under the dehiscent anthers. These anthers swab or brush the forager with pollen as it searches for rewards usually confined to the base of open, salver bowl-shaped flowers (e.g., *Leptospermum flavescens*: Hawkeswood 1987a; *Krameria*: Simpson 1989; *Nierembergia*: Cocucci 1991) or partially to completely concealed within modified pouches, tubes, spurs or hoods (e.g., *Aquilegia*: Macior 1966; *Linaria Melampyrum*, *Mentha*, *Salvia*: Faegri & van der Pijl 1979; *Lecythis poiteaui*: Mori & Prance 1990; *Calceolaria*: Molau 1988).

The actual depth at which the reward is concealed correlates with the length of mouthparts and foraging habits of the pollinator. For example, the nectar cups of the flowers of some shrubby Myrtaceae tend to be extremely shallow (e.g., *Baeckea*, *Eucalyptus*, *Leptospermum*) and photographic evidence shows clearly that some nectar-drinking beetles must literally push through, or climb over, the incurving whorl of stamens as if they were attempting to make their way through the bars of a cage (Hawkeswood 1987a, b, 1990). In contrast, the anthers of *Nivenia* (Iridaceae) dust the head and mouthparts of the nemestrinid flies and anthophorid bees that feed on nectar while hovering and inserting their long proboscés down the elongated floral tube (Goldblatt & Bernhardt 1990).

Since pollen is not actively sought after in this passive mode, the flowers are more likely to show a trend towards zygomorphy of the perianth. In the majority of bilabiate, gullet or flag flowers (*sensu* Faegri & van der Pijl 1979) the anthers are concealed under the perianth (e.g., *Pedicularis*: Macior 1982), within a keel petal (e.g., the papilionoid legumes), or hidden under the crests of a central, expanded style (e.g., Iridaceae: Goldblatt *et al.* 1989).

A second variation in zygomorphic flowers showing the passive collection mode is observed commonly in the Lamiaceae. The elongated stamens extend beyond the perianth lips and form a curved brush which the pollinator contacts upon landing on the lower lip or when it

Table 2. *Floral characteristics of the two modes of anther presentation*

Floral character of pollination syndrome	Passive mode	Active mode
Perianth		
Flowers nodding with reflexed-recurved perianth exposing anthers prominently	±	+
Perianth tubular and/or spurred	+	±
Zygomorphic perianth common often concealing anthers	+	−
Whole stamens or staminal filaments		
Androecium polyandrous	+	+
Stamens fasciculate	−	+
Filaments restrict access to edible rewards	+	−
Filaments of tubular–zygomorphic flowers much exserted and curved, compressing anthers into a brush or swab	+	−
Filament apices highly ornamented/pigmented	−	+
Filament apices articulate, anthers usually peltate	+	−
Anthers		
Anthers porose/porate	−	+
Anthers bearing eliaphor	−	+
Anthers extrusive, shedding pollen copiously	±	+
Anthers explosive	+	−
Thecae much inflated	−	+
Gynoecium		
Stigma positioned outside anther cone or tuft	±	−
Adnate to stamens forming gynostemia or column	+	±
Stigma-style becomes pollen presenter	+	±
Pollen		
Pollen dimorphic (viable and feeder grains in the same androecium)	−	+
Pollen embedded in mucilage	+	−
Pollinia attached directly/indirectly to viscidia or corpuscula	+	±
Pollen forms true pollinia	+	±
Attractants and rewards		
Anthers are common osmophores	?	+
Nectary or eliaphor present	+	−
Pseudantherous calli, staminodes, stigma, etc., present	−	+
Extra-floral rewards present	+	−

inserts its bill down the floral tube while hovering (Spira 1980). The length and angle of the stamens in *Trichostema* shows a positive correlation with the physical dimensions and foraging habits of the dominant pollinators. Furthermore, the length of the stamens shows a positive correlation with the degree of outbreeding within *Trichostema* spp. In some of the Lecythidaceae,

on the other hand, it is not the perianth but the androecium itself that forms a cap-like androphore that must be pushed up by the bee or bat to reach the reward concealed in the staminodal roof or hood while fertile anthers dust the animal with pollen concealed in a basal disc (Prance & Mori 1979; Mori & Prance 1990).

Since the pollinators may actually ignore the anthers,

additional structural and physiological modifications to the anthers may have evolved to facilitate heavier pollen deposition on an otherwise, unresponsive forager. These modifications occur at one or two levels. First there are those modifications that increase the levels of swabbing the pollinators body with pollen or literally shovelling pollen onto the vector as it forages for interior rewards. The anthers of *Oxalis* (Bernhardt 1990b) and *Passiflora*, for example, are fixed to articulated filament tips (MacDougal 1983) that turn or roll the dehiscent surfaces of the anthers (Masters 1871) over the probing bodies of their pollinators. This is observed in the photographs of hummingbird pollination of *Passiflora vitifolia* (Bernhardt 1993). The concave, spade-like awns of the anthers of *Viola* spp. and *Hybanthus* (Beattie 1971, 1974; Augspurger 1980) fill with pollen grains that will be dumped on a nectar-seeking insect when it triggers the anthers with its head or probosces. Perhaps the most specialized mechanism for bringing the anthers to the body of the pollinator is seen in the irritable gynostemium of most Stylidiaceae. As the insect draws nectar from the floral tube it contacts a series of sensitive, perianth hairs stimulating the gynostemia which swings downward in a predictable arc usually striking the insect and depositing an imprint of pollen on the exoskeleton before reseting itself (Erickson 1958). Second, there is a trend towards fixing discrete clumps of pollen to the pollinator's body with some sort of adhesive. Pollenkitt is found coating the exines of most animal-pollinated grains whether they belong to the passive or the active mode. As pollenkitt is a lipophilous substance, its presence on exines is not expected to influence negatively rates of ingestion by pollen foragers (Thien *et al.* 1985; Bernhardt & Thien 1987).

In contrast, true deposits of carbohydrate (or carbohydrate and lipid) mucilages are produced less commonly by anthers. These secretions form a much stronger adhesive as they dry upon contact with the air so pollinators are less likely to scrape off and swallow the pollen. Therefore, carbohydrate-based glues seem more likely to be found in the passive pollen collection mode. These glues may be manufactured exogenously by trichomes on the anthers as in some Cucurbitaceae (Vogel 1981) and the basal cells of *Calliandra* (Hernandez 1986). Recently, though, de Frey *et al.* (1992) have found that the anthers of *Tylosema esculentum* manufacture a glue

endogenously. This mucilage is produced by internal, anther connective tissue and is released into the thecae so that, at dehiscence, the pollen grains emerge already embedded in a sticky matrix. Stigmatic secretions may also be a source of mucilage for pollen in some magnolioids.

Gottsberger (1977) has found that beetles foraging in *Talauma* become sticky following direct contact with stigmatic secretions increasing pollen adherence. *Talauma* is a nectarless flower, however, and the beetles gnaw on most of the floral organs. Gottsberger suggested that when beetles gnaw on carpels they contacted the stigmatic secretions indirectly. Recent work by Lloyd & Wells (1992) suggests that the stigmatic secretions of some magnolioids attract floral foragers as some insects ingest them as a source of proto-nectar. If stigmatic secretions are a regular reward for the pollinators of magnolioid flowers the status of such flowers as 'pollen' flowers *sensu* Vogel (1978) will have to change for many genera (Table 1).

The production of true pollinia has required the evolution of gynostemia in the Asclepiadaceae and Orchidaceae. There is a shared trend in both families for a mucilaginous liquid and/or cellular adherent to be manufactured by the gynoecium instead of by the androecium. The broad variation in the construction of pollinaria is particularly well studied in the Orchidaceae and remains a basic taxonomic character (Dressler 1981). The subfamily Cypripedioideae is the only subfamily of orchids in which pollinia adhere directly to the pollinator without a viscidium manufactured by the stigma. In all other orchids with pollinia the pollen masses must be fixed to the pollinator with some adhesive produced by the stigma. In some cases this adhesive is a viscous liquid produced by the undifferentiated stigma lobes. In other cases it is a shoe-like, cellular structure produced by the rostellar lobe of the stigma. The pollinia may be united directly to the viscidium or indirectly via a series of anther-derived caudicles and/or rostellum-derived stipes (in Burns-Balogh & Funk 1986).

Although caudicles and stipes perform much the same function it is interesting to note that caudicles, produced only by anthers, are regarded as more primitive structures than the stipes derived from the rostellar lobe of the stigma (Dressler 1981). A few orchid genera produce *both* caudicles and stipes which interconnect within the same pollinarium (e.g., *Dipodium*: Bernhardt & Burns-Balogh 1983).

Historically, the floral architecture of the passive pollen collection mode was well understood by the floral biologists of the last century who first attempted to catalogue the pollination systems of the angiosperms and relate their origins to natural selection. In particular, Henslow (1893) must have had a passive mode in mind when he addressed correlations of floral nectaries with pollination and wrote 'It is hardly worth while giving other cases to prove the universal rule, that the position of the honey and its gland is always where it is most accessible; and the position of the anthers is, at the same time, just where they will be most conveniently struck by the insect.'

Active pollen collection

In contrast, anthers provide the only rewards in the active pollen collection mode. The pollinator seeks out the anthers in the flower for its primary reward. This reward is almost always pollen, although pollen may be offered with food bodies (e.g., *Calycanthus*, Table 1) or with volatiles (e.g., *Cyphomandra*: Sazima *et al.* 1993). The primary pollinators of these flowers do not probe other floral organs for rewards. Therefore, flowers following the active pollen collection mode have a more limited range of prospective pollinators than those adopting the syndrome of the first model. Pollinators of the active pollen collection mode may be pollinated by some beetles, certain flies (especially syrphids) and female bees.

Flowers adopting the active pollen collection mode are more likely to show restricted perianth forms. Since anthers offer the primary rewards they are not concealed under perianth segments and the mode is less common in bilabiate blossoms. Species with multistaminate flowers tend to display their anthers in simple bowl or cup-shaped perianths or the perianth segments are reduced to scales or short tubes and the stamens form clustered brushes in which florets may become compacted into dense inflorescences; e.g., some mimosoid legumes (Bernhardt 1989, 1990a). The Lecythidaceae, once again, are important exceptions to the rules. In the nectarless flowers of *Couroupita* spp. and *Lecythis alba*, for example, the androecium forms a hooded androphore. Bees must push up the hood of stamens to collect feeder pollen from the anthers comprising the inner surface of the hood (Prance & Mori 1979).

Polyandry (many stamens in the androecium), however, is *not* a character unique to the active pollen collection mode. Many families contain genera that have polyandrous and nectariferous flowers in which stamens form several whorls or a continuous spiral (e.g., Cactaceae, Cistaceae, Lecythidaceae, Mimosaceae, Myrtaceae, Ranunculaceae, etc.; and see Fahn 1979; Pellmyr 1985). If insect pollinators show a regular and active foraging pattern by seeking stigmatic secretions as a primary reward (Gottsberger & Silberbauer–Gottesberger 1988; Lloyd & Wells 1992) the blossoms of many magnolioids can no longer be classified as examples of the active pollen collection mode (Table 1).

Oligandry (few stamens in the androecium) is not a character unique to this mode either. However, the floral presentation of oligandrous taxa with active pollen collection modes often show a broad convergence throughout the angiosperms. Since the anthers contain the only reward, the flowers tend to nod on their pedicels, and have recurved-reflexed perianth segments further exposing the androecium. The anthers are usually porose/porately dehiscent and are often as long, or longer, than their filaments (Vogel 1978; Buchmann 1983). These anthers tend to be clustered and often form a tuft, cone, or stiff tube around the style. Early work by Dobson (1987), Bernhardt (1989) and D'Arcy *et al.* (1990) suggests the anthers may be scented. In the petalloid monocots filaments in the oligandrous models may be swollen, highly pigmented and/or ornamented.

We may then interpret some filament ornamentations as anther guides or side advertisements for the anthers as they appear to increase the surface area and visibility of the few, clustered stamens (Bernhardt & Montalvo 1979; Faegri 1986; Bernhardt 1995). Faden (1992) has suggested that filament ornaments in the Commelinaceae may serve as a holdfast for the pollen-extracting insect but in such lilioid genera as *Echeandia* (Bernhardt & Montalvo 1979), *Dianella* (Bernhardt 1995) and other lilioid monocots (Bernhardt & Burns–Balogh 1986) the female bees seem unable to grasp the ornamented filaments with the claws on their feet while shaking pollen from the anther apices.

The active pollen collection mode may seem to be less

variable than the passive mode since a plant adopting the active mode is pollinated by a more limited spectrum of vectors. Certainly the morphological apparatus increasing the anthers' capacity to swab, shovel or dump pollen onto a forager is unlikely to evolve frequently in the active collection mode, based on a review of available literature (Table 2). On the other hand, modifications in the active mode may simply be more subtle, and have not yet been recognized as specific adaptations in this mode.

For example, while perianth zygomorphy is uncommon in the active collection mode androecium zygomorphy has evolved independently in such unrelated families as the Dilleniaceae (*Hibbertia*: Stebbins & Hoogland 1976), Caesalpinaceae (Cassineae: Gottsberger & Silberbauer-Gottsberger 1988), Commelinaceae (Faden 1992) and some Lecythidaceae (*Couroupita*, *Lecythis alba*: Prance 1976; Mori & Prance 1990). In the Cassineae, Commelinaceae and *Couroupita* spp. androecial zygomorphy appears to correlate with either the production of dimorphic pollens isolated in different anthers and/or the evolution of pollen mimectic staminodes (see below). In *Hibbertia* (Bernhardt 1984, 1986) there is still no direct evidence that the zygomorphic androecium produces different pollen in different anthers or that staminodes deceive pollen vectors.

Another modification, which may be unique to the active pollen collection mode, has been the convergent evolution of a petal or sepal tube that surrounds the androecia of flowers with many, longitudinally dehiscent anthers. The encircling organ functions like one large, compound anther as it opens by means of a terminal pore and pollen falls through this opening when the tube is vibrated by bees, as in *Anemopsis* (Pellmyr 1985), *Ternstroemia* and possibly in some Ericaceae (Bittrich *et al*. 1993). Evolution of the petal tube appears to have converted a polyandrous *Papaver*-type flower into a 'falsely' oligandrous, *Solanum*-type. In *Ternstroemia* each anther connective is so elongated that it forms an elongated stalk at the tip of each anther. Bittrich *et al*. (1993) suggests that these elongated connectives function like tuning forks transmitting the insect-applied vibrations from the petal tube to the dehiscent anthers.

Descriptions of angiosperms with an active pollen collection mode are found in Sprengel (1793) and Delpino (1868–1875). However, the function of floral attractants, anther presentation and rewards in these flowers has been less well understood, historically, than the functional architecture of angiosperms with a passive pollen collection mode. There are two possible explanations for inconsistencies in the study of the active mode. First, as noted above, the active mode is less common in nature than the passive mode and may have often gone unrecognized. Second, some early floral biologists seem to have refused to believe that insects such as female bees and syrphid flies would actively seek out nectarless flowers and then forage for pollen exclusively.

This bias may have begun with Sprengel (1793) who coined the term 'pollen flower' to refer to blossoms that contained no nectaries, offered pollen as their only reward and were insect-pollinated. However, he insisted this was a form of pollination-by-deceit. Sprengel believed that nectar-gathering insects were attracted to nectarless flowers initially because nectarless flowers produced scents and color patterns similar to those of nectariferous flowers. Pollen was transferred to the deceived forager when it contacted the anthers while probing for nonexistent nectar or while foraging for pollen prior to searching for nectar. The general belief that all insects tend to avoid flowers that are rich in pollen but nectarless has led to two misinterpretations that persisted into our century. First, nonsecretory sculptures on the anthers or filaments of nectarless flowers have been described repeatedly as nectaries (Daumann 1970; see above) or some fieldworkers insist that *Solanum*-type flowers conceal nectar within 'deep staminal tubes' formed by the cone of porose anthers (e.g., *Dodecatheon*: Robertson 1895; Macior 1964). Second, the color patterns on some flowers showing the active pollen collection mode (e.g., *Solanum dulcamara*) have been interpreted as translucent or false nectar guides (Macior 1964). In fact, the terminology used by pollination biologists still does not subdivide the visual cues of different flowers into true nectar guides versus pollen or anther guides.

Stigma versus anther presentation

A unique feature of the active pollen collection mode is that anther presentation accommodates cross-pollination by 'luring' the vector into contacting the receptive stigmas while the animal is still laden with pollen from another genotype. This would tend to account for so many hooked or curved styles in buzz-pollinated flowers (see illustrations in Buchmann 1983; Gottsberger & Silberbauer-Gottsberger 1988) that are poised within, or directly behind, the tuft or cone of anthers. The pollinator cannot harvest pollen from the anthers without colliding with, or backing into, the receptive stigma. In the passive collection mode, by contrast, the stigma is usually just one more brush or swab mechanism the pollinator must push past, or under, to reach the reward in the floral spur, pouch or basal receptacle. Therefore, enantiostyly appears to occur more commonly in flowers showing the active collection mode (Bowers 1975) while heterostyly is expressed in more than 22 angiosperm families in which the flowers of at least one morph must offer nectar (Ornduff 1975; Beach & Bawa 1980).

Comparatively minor modifications to the anthers may strongly influence the final pollination syndrome of a flower. These modifications include the actual number of fertile anthers in the androecium and the way in which the anthers dehisce (Figs. 7, 8). The variation of anther characters in *Hibbertia sensu stricto* makes a useful model to understand the adaptive radiation of pollination systems within a genus offering active pollen collection flowers.

Although there are earlier reports of beetle pollination dominating *Hibbertia* (Armstrong 1979; Keighery 1975), and the Dilleniaceae in general (Gottsberger 1977), more recent fieldwork suggests the dependence on bee pollinators (Bernhardt 1984, 1986; Hawkeswood 1989; Keighery 1991) in southeastern Australia, Western Australia and Madagascar (Bernhardt & Thien 1987). There may also be some secondary pollination by syrphid flies (e.g., *H. dentata*; Bernhardt, work in progress).

Floral symmetry within *Hibbertia*, as mentioned above, is determined by the presence and organization of staminal bundles and not by perianth morphology. In fact, taxonomists have used androecium symmetry as another primary character to subdivide *Hibbertia* into several sections (see Stebbins & Hoogland 1976). Androecium symmetry alone cannot be used to predict the full spectrum of pollinators for each *Hibbertia* species. For example, *Hibbertia stricta* (section Pleurandra) has a zygomorphic androecium consisting of a dense but single fascicle of stamens. In contrast, *H. fasciculata* (section *Hibbertia*) has a single, rotate whorl of separately initiated stamens yet both species are pollinated by the

same genera of halictid bees in Victoria, Australia (Bernhardt 1984, 1986).

Therefore, divergent anther presentation has not relegated the two *Hibbertia* spp. to two different sets of pollinators. When the *same* bee genus visits the flowers of different *Hibbertia* species with different androecial symmetries pollen is deposited on different parts of the bees' bodies and the bees contact receptive stigmas of the respective flowers at different angles. Variations in androecium symmetry within *Hibbertia* may increase interspecific isolation despite shared pollinators (Bernhardt 1984, 1986).

Within section Hibbertia anther–stigma presentation, coupled with the sheer number of stamens and the mode of anther dehiscence, determines the comparative effectiveness of a wide range of pollen-foraging insects as pollinators. *Hibbertia dentata* has an androecium of approximately 100 stamens distributed in several whorls. The three styles are so crowded by the inwardly pressing stamens that their three receptive stigmas appear suspended within a dense 'rug' of longitudinally dehiscent anthers (Figs. 7, 9). Small- to medium-sized bees (4–7 mm; e.g., *Homalictus* and *Hylaeus*) and large syrphid flies (9 mm; e.g., *Melangyna*) pollinate these flowers by merely crawling over or colliding with the receptive stigmas while removing pollen from adjacent anthers. The naturalized honeybee (*Apis mellifera*) is considerably larger (13 mm) but can also pollinate *H. dentata* as thoracic vibration is not required to harvest pollen from longitudinal slits (Bernhardt, work in progress).

The androecium of *Hibbertia scandens* (section Hibbertia), is also multiwhorled but contains over 200 fertile anthers requiring a floral receptacle twice the circumference of *H. dentata* (Fig. 5). Bees most likely to contact the five stigmas of *H. scandens*, while foraging for pollen, were a minimum of 7 mm (e.g., *Exoneura*; Anthoporidae) to 22 mm long (e.g., *Amegilla* and *Lestis*; Anthophoridae). Unlike *H. dentata*, the anthers of *H. scandens* are terminally porose and have inflated thecae (Fig. 8) as in most members of the genus. Syrphid flies may visit the anthers but are unable to remove viable pollen. The primary pollinators of *H. scandens* represent three families of bees (Anthophoridae, Colletidae and Halictidae). Each bee family contains at least one genus that extracts pollen by applying some degree of thoracic vibration (Bernhardt, in progress).

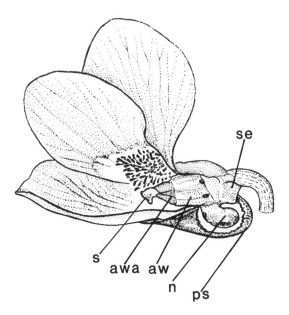

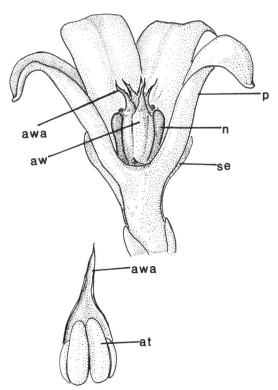

Fig. 4. *Hybanthus–Viola* type presentation of androecial nectaries using a longitudinal section of a flower of *Viola soraria*. Note that dehiscent anthers are not exposed to the secretory nectaries due to the enclosing barriers of the awns. aw, awn; awa, awn appendage; n, nectary; ps, petal spur; se, sepal; s, stigma.

Fig. 5. Male flower of *Melicytus lanceolatus* with petal removed. at, anther sac; p, petals (all other abbreviations as in Fig. 4).

Other *Hibbertia* spp. in section Hibbertia (*H. diffusa, H. fasciculata and H. saligna*) have a completely different mode of anther presentation although the androecium still shows radial symmetry. Androecia in these species bear less than 40 fertile anthers in their single whorl. The anthers are porose and push inwards forming a central tuft over the three ovaries (Fig. 10). The styles of the gynoecia are bent and push between the staminal filaments until the styles extend *outside* the anther tuft. In these flowers the receptive stigmas form an equilateral triangle outside the circumference of the tuft of anthers. Bees land on the anther tuft and vibrate the anthers by spinning around in a circle (Figs. 11, 12). The vast majority of halictid and anthophorid pollinators bear patches of scopal hairs on both their hindlegs and on the undersurfaces of their abdomens (see Bernhardt 1986,

1989). As the bee spins around to remove pollen from the anther tuft her hindlegs and hairy abdomen contact the receptive stigmas transferring pollen collected from other flowers (Bernhardt 1986, work in progress).

This mode of pollination seems to have become increasingly specialized in *H. conspicua* (section Candollea) as the stamens are now fasciculate and several anthers may emerge from the apices of one, elongated trunk resembling a candelabra (Stebbins & Hoogland 1976). The anthophorid bee is capable of grasping up to three anthers on one trunk at a time as it spins around the androecium harvesting pollen while its abdomen contacts the three, receptive stigmas (Kieghery 1991).

However, the extension and arching of styles or stylodia outside a central tuft of cluster of anthers is *not* a mechanism exclusive to buzz-pollinated flowers. Fur-

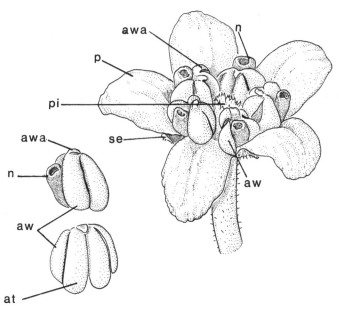

Fig. 6. Male flower of *Melicytus micranthus*. pi, pistilode (all other abbreviations as in Figs. 4, 5).

Figs. 7, 8. Fig. 7. Scanning electron micrograph of the anthers of *Hibbertia dentata* (section *Hibbertia*) showing broad, longitudinal dehiscence. Fig. 8. Scanning electron micrograph of the anthers of *Hibbertia scandens* (section *Hibbertia*) showing terminal (porose) dehiscence.

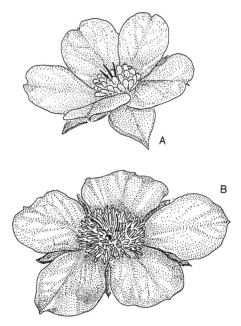

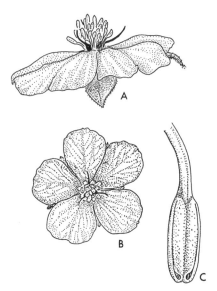

Fig. 9. Anther and stigma presentation (carpels blackened) in the flowers of *Hibbertia* spp. (section *Hibbertia*) in which the androecium consists of more than 50 stamens. A. *H. dentata* with three carpels and fewer than 100 stamens. B. *H. scandens* with five carpels and more than 200 stamens.

Fig. 10. Anther and stigma presentation (carpels blackened) in the flowers of *Hibbertia* spp. (section *Hibbertia*) in which the androecium consists of fewer than 40 stamens. A. *H. saligna* with three carpels and fewer than 40 stamens. B. *H. diffusa* (lateral view) with three carpels and fewer than 20 stamens. C. Stamen of *H. diffusa* showing porose anther approximately as long as its filament.

Fig. 11. Pollination of *H. scandens* by a female halictid (*Nomia* sp.). Position of stigmas is indicated by arrows.

Fig. 12. A worker *Apis mellifera* clasps the anther tuft of *H. saligna*. Stigma (indicated by black arrow) contacts the bee's abdomen while it revolves around the anther tuft.

thermore, it is not a form of anther/stigma organization unique to the active pollination mode. Much the same organization is found in the short-styled morph of *Oxalis* spp. (Bernhardt 1990b) in which the bee contacts both dehiscent anthers and stigmas while probing for nectar, not for pollen.

PSEUDANTHERY

Pseudanthery will be broadly defined here as a structure or organ on a nectarless flower that mimics a dehiscent anther. In some rare cases this organ may actually slough a pseudopollen made of loose epidermal cells or some other tissues (e.g., the labellum of *Gastrodia* spp. and other orchid genera; Dressler 1981; Jones 1981). Pseudanthery appears to have evolved independently in families of monocotyledons and dicotyledons. Organs and/or sculptures interpreted as false anthers include true staminodes, filament sculptures, perianth appendages, perianth color patterns and modified stigmas or stylodia (Gack 1979; Bernhardt & Burns-Balogh 1983; Buchmann 1983; Faegri 1986; Dafni & Bernhardt 1989).

The convergent evolution of pseudanthery becomes much easier to understand when compared to the broad distribution of the active pollen collection mode in different habitats. Since pseudantherous flowers mimic flowers exhibiting the active mode these mimectic flowers are most likely to be pollinated by the same insect groups that pollinate the *Papaver*-type and *Solanum*-type flowers (*sensu* Vogel 1978), especially syrphid flies and female bees.

Pseudanthery may be based on automimicry due to dichogamy or dicliny. For mimesis to succeed it is usually expected that the mimic occurs at a far lower density than its model. For example, in the nectarless and protogynous *Acacia* spp. of southern Australia inflorescences in the early, female phase exhibit the same colors and scents as in the later male phase but they have not extended their stamens yet. The female phase takes up only about 20% of the lifespan of the inflorescence. Therefore, the proportion of male (pollen-rewarding) phase inflorescences on an *Acacia* is usually much higher than the number of female (pollen-concealed) phase inflorescences on the same branch (Bernhardt 1989).

Both Vogel (1978) and Faegri (1986) suggested that the swollen, pigmented and papillose apices of the stami-

nal filaments of some buzz-pollinated flowers were actually indicative of a trend towards pseudanthery. Although these ornamented filaments were connected to fertile anthers containing edible pollen, both Vogel and Faegri continued to argue that convergent trends towards expanded, colorful filaments made small, dull-colored anthers look bigger. Filament sculptures then mimicked an anther bursting with pollen. However, field observations on the flowers of *Arthropodium*, *Dianella*, *Dichopogon*, *Echeandia* and *Stypandra* fail to show, so far, that either pollen foragers or pollen thieves mistake the filament ornaments for true pollen or attempt to scrape them off (Bernhardt & Montalvo 1979; Bernhardt & Burns–Balogh 1983; Bernhardt 1995, and personal observations).

However, within the nectarless, bisexual flowers of the Commelinaceae insects may be attracted to the large, brightly pigmented staminodes and only brush against the smaller, fertile anthers (Faden 1992). Gack (1979) suggested that the yellow, ball-shaped lobes of the stigmas of some *Orobanche* spp. are also anther mimics (Barth 1985). Within unisexual flowers the best-known, most obvious examples are the stylodia and stigmas of *Begonia* spp. Once again, though, male flowers (pollen sources) tend to outnumber females (pollen mimics) (Bernhardt & Montalvo 1979; Agren & Schemske 1991).

Pseudanthery, then, becomes important as an evolutionary option within a purely ecological context and this may explain its convergent evolution within both monocot and dicot lineages. Since the active pollen collection mode offers no nectar the pollinators of such species must visit a wider range of co-blooming species to acquire simple sugars for immediate energy. The absence of nectar may not impede the pollen foraging activities of some Coleoptera (with pollen-cracking molars) and Diptera but it is a major problem of female bees which are really collecting pollen primarily for their offspring. Long-term research has shown that female bees that are the major foragers on the active collection mode are almost always polylectic foragers (*sensu* Michener 1979). That is, the bees visit many different plants in bloom regardless of angiosperm phylogeny. A single bee may visit four to six different species in the course of a single foraging bout but she typically bears pollen from both nectarless and nectariferous flowers (Bernhardt & Montalvo 1979; Bernhardt; 1984, 1986,

1989, 1990b). It is these visits to the nectariferous, or passive mode, flowers that provide an immediate source of chemical energy.

Apis spp. are exceptions to this rule. Although *Apis* spp. visit the flowers of many angiosperms, and seem preadapted to unfamiliar honey floras when naturalized on other continents, individual workers often visit only one species of nectarless flower during a foraging bout. *Apis* spp. are the most eusocial of the Apoidea (Michener 1974). Workers exchange and share nectar reserves or the bee may leave the colony with sufficient energy reserves obtained from the common store. Furthermore, although *Apis mellifera* has been commonly recorded while collecting pollen from nectarless flowers it is not often associated with successful foraging on flowers that have porate-porose anthers requiring thoracic vibration (Buchmann 1983).

Pseudanthery should then be more likely to appear in habitats or regions where active pollen collection modes are relatively common but *Apis* spp. are not endemic. This may explain why the few literature reviews have described the mode so often in the mediterranean to temperate south of the Australian continent (Dafni & Bernhardt 1989) and from the montane-lowland regions of Mesoamerica (Bernhardt & Montalvo 1979; Agren & Schemske 1991). This may be based, in part, on both phytogeographic and phylogenetic constraints as the importance of pseudanthery varies greatly from family to family. For example, pseudanthery is particularly common within two subfamilies of the Orchidaceae *sensu stricto*. (Orchidoideae and Epidendroideae; *sensu* Dressler, 1993). These two families show a high degree of diversity through southern Australia and Mesoamerica, respectively, and the genus *Apis* is not endemic to either region (Michener 1979; Roubik 1989).

However, pseudanthery must still be treated as only a minor aspect of floral evolution within the Orchidaceae. The labellum appendages encouraging both pseudanthery and, the far more common, pseudocopulatory syndromes in orchidoid tribes have a common, ontogenetic origin (Dafni & Bernhardt 1989). In fact, more orchid taxa appear to mimic the passive pollination mode. So many orchid species exhibit well-pigmented nectar guides, leading only to empty pouches or spurs, that the family is used as a common example of the polyphyletic evolution of pollination-by-deceit (van der Pijl & Dodson 1966; Dressler 1981).

What appears to be an extensive trend towards pseudanthery in the terrestrial Orchidaceae of southern Australia depends on pigmented sculptures ornamenting the labellum *or* the column hood (e.g., Thelymitreae; *sensu* Burns-Balogh & Bernhardt 1985) which exploits the foraging behavior of some sawflies, syrphid flies and female bees. Pseudanthery in southern Australia has been documented in some species in the genera *Caladenia*, *Gastrodia* and *Thelymitra* (Dafni & Bernhardt 1989). The genera *Eriochilus* (Erickson 1965), *Epiblema*, *Elythranthera* and *Glossodia* also show morphological evidence of pseudanthery although no ecological studies have been published to date (Jones 1981; Dafni & Bernhardt 1989). Pseudanthery is not restricted to orchidoid genera of the southern hemisphere, though, and was first described in North America by Thien & Marcks (1972) for *Arethusa*, *Calopogon* and *Pogonia*. The genus *Apis* is not native to North America either (Michener 1974) and the major bee pollinators of pseudantherous genera in North America and southern Australia appear to be genera of polylectic foragers especially *Bombus* (Apidae), *Exoneura* (Anthophoridae), *Nomia* and *Lasioglossum* (Halictidae).

In the pseudantherous orchids pollinated by female bees the flowers exploit what Buchmann (1983) referred to as the carryover phenomenon. That is, a polylectic, female bee that has applied thoracic vibration to a series of buzz-pollinated flowers with porate or porose anthers is most likely to carry this mode of foraging to the next taxon she visits regardless of the flower's actual mode of anther dehiscence. For example, *Lasioglossum* spp. (Halictidae) attempt to vibrate the trichome brushes on the column hood of *Thelymitra nuda* after they have foraged on co-blooming lilioids that have nectarless flowers with porose-porate anthers (Bernhardt & Burns-Balogh 1986).

EVOLUTIONARY TRENDS IN ANTHER PRESENTATION MODES

How flexible have the two anther presentation modes been in the evolution of zoophily within angiosperms? When the adaptive radiation of a lineage is considered do these two models of presentation represent reversible trends in floral evolution? The answer varies with lineage.

Obviously, the active and passive modes of presentation intergrade within the Ericaceae. The passive mode is observed within *Zenobia* (Dorr 1981) and *Erica* (Haselrud 1974). In *Kalmia* (Reader 1977) the triggering of the sudden release of pollen from anthers freed from their petal pockets would appear to be a far more specialized example of the passive mode paralleling the evolution of the explosive release of pollen in the Loranthaceae and Proteaceae. In contrast, the flowers of *Vaccinium* spp. produce nectar but also have porate anthers. Although bees actively probe for nectar in *V. stamineum* they collect pollen by thoracic vibration indicative of the active mode. This species appears to be one of the few taxa in which mechanisms of active and passive modes overlap.

In some lineages differences between active and passive pollen collection syndromes seem to have involved little more than suppressing development of the nectary and changing visual and/or scent cues. The adaptive radiation of pollination systems within the Ranunculaceae that have actinomorphic flowers is exemplified by most *Ranunculus* spp. having yellow-white corollas and anthers (to the human eye) with nectar glands at the base of the petals. These nectar glands are absent in *R. asiaticus* and the petals of this species are red-orange while the anthers are blackish (Dafni *et al.* 1990). So far, it is the only member of the genus known to be pollinated almost exclusively by pollen-eating scarabs in the genus *Amphicoma*, an obvious but atypical shift towards the active pollen collection mode.

Within the polyandrous flowers of the Cimicifugae (Ranunculaceae), *Actaea* spp. have nectarless, strongly scented (citrus–rose-like) blossoms and are pollinated by pollen-eating beetles and some syrphid flies. The members of the allied genus, *Cimicifuga*, also have polyandrous flowers but they are usually nectariferous and either weakly scented or the scents are based on compounds distinct from those of *Actaea*. In contrast, *Cimicifuga* spp. are pollinated by nectar-foraging bumblebees and the fritillary butterfly, *Argynnis paphia*. Within this tribe of Ranunculaceae the nectarless, rotate and polyandrous flower of *Anemopsis* is sometimes regarded as a floral form ancestral to both *Actaea* and *Cimicifuga* (Pellmyr 1985). The deep purple, nodding flowers of *Anemopsis* have a modified sepal tube containing over 100 stamens restricting pollination to buzz-pollinating *Bombus* spp. (Pellmyr 1985).

Anther presentation models in the Ranunculaceae undoubtedly shift with symmetry. It is usually presumed

that genera of Ranunculaceae with zygomorphic flowers (e.g., *Aconitum* and *Delphinium*) were derived from a radially symmetrical ancestor (e.g., see review in Barth 1985). This would suggest a labile trend from ancestral taxa with rotate flowers, expressing either the passive *or* active mode of anther presentation, to plants in which each petal is deeply spurred (e.g., *Aquilegia*) *or* the perianth is zygomorphic and the passive mode dominates (e.g., *Aconitum, Delphinium*).

However, the derivation of flowers with zygomorphic perianths and passive modes of anther presentation from ancestors with actinomorphic perianths and active modes of anther presentation does not appear to be a universal constant within the angiosperms. Powlesland (1984), for example, has interpreted the unisexual, radially symmetrical flowers of *Melicytus* in New Zealand as floral forms derived from continental ancestors with zygomorphic flowers similar to modern *Viola* or *Hybanthus* species. Anther presentation in *Melicytus* continues to express the passive mode despite an apparent reduction in the size of anther awns and the absence of petal spurs.

Flowers that have a bizarre, zygomorphic perianth and a buzz-pollination syndrome may have had nectariferous ancestors with either an irregular or tubular perianth containing nectar at its base. Within the floral evolution of *Pedicularis* (Scrophulariaceae) the nectarless, 'elephant nose,' buzz-pollinated species appear to be most closely allied to those *Pedicularis* species with nectariferous flowers and bilabiate (snapdragon type), tubular corollas based on comparative illustrations by Macior (1982). Nectar-producing species tend to flower in the spring and are pollinated by *Bombus* queens searching for nectar before these reproductively mature queens raise their first broods of neuter workers. The nectarless, buzz-pollinated species of summer are pollinated primarily by *Bombus* workers (Macior 1984). Examination of the flowers of buzz-pollinated species of *Pedicularis* suggests that the symmetry of the original, almost radially symmetrical flower has been bilaterally exaggerated and diversified.

Anther presentation in the neotropical Lecythidaceae suggests that the zygomorphic, nectariferous flower is derived from an actinomorphic, nectarless ancestor but changes in floral symmetry are based on the androphore not on the perianth. Genera with polyandrous, radially symmetrical flowers are nectarless and pollinated by large, female bees indicative of the active pollen collection mode (Prance 1976; Prance & Mori 1979). With the evolution of the androphore, though, there appears to be a positive correlation towards the addition of nectar glands to the bilateral hood as it has evolved increasingly elaborate shapes and coils as seen by comparing nectarless *Couroupita* against nectarless-nectariferous *Lecythis* spp. and with nectariferous *Eschweilera* (Mori & Prance 1990).

Evolution of both the passive pollen collection mode and true pseudanthery in the Orchidaceae may be among the most diversified in the angiosperm families. Both nectariferous and pseudantherous genera may occur in the same subfamily or tribe (e.g. Bletiinae, *sensu* Dressler 1981). Anther morphology and presentation in the diandrous genus *Apostasia* suggests buzz-pollination (Burns-Balogh & Bernhardt 1985; Burns-Balogh & Funk 1986). Considering the phylogeny of orchid subfamilies (Dressler 1993), this would indicate that floral evolution in the family showed a trend towards the active pollen collection mode, but this trend has not diversified to the present day. However, there is still no evidence of a true, active pollination collection mode within the monandrous flowers that dominate the family Orchidaceae. Indeed, within such tribes as Geoblasteae and the Thelymitreae (*sensu* Burns-Balogh & Funk 1986) pseudantherous orchids appear to share a far closer alliance with both pseudocopulatory and nectariferous taxa (Burns-Balogh & Bernhardt 1985; Dafni & Bernhardt 1989).

The evolution of the passive mode has involved the infrequent but recurrent presence of an extra-floral reward especially where inflorescences consist of many, compact, brush-like florets devoid of nectaries. This has involved the evolution of edible bracts to reward birds or rodents (e.g., *Freycinetia*: Cox 1983; *Protea*: Wiens & Rourke 1978). In *Boerlagiodendron* birds are reputed to search the inflorescence for pseudofruits treading upon the true flowers and carrying away the pollen, atypically, on their feet (see review by Faegri & van der Pijl 1979). Extrafloral nectar and/or resin may serve as common rewards for insect pollinators of the inflorescences (pseudanthia) of some neotropical Araceae and Euphorbiaceae (Simpson & Neff 1981).

Is it likely that a passive mode, dependent on extrafloral rewards to lure vectors, will be derived from an active mode ancestor? This seems to have been the evolutionary pathway leading to the remarkable form of bird-pollination in *Acacia terminalis* (Knox *et al.* 1985). This Australian species may make a useful model system within evolutionary ecology as both active and passive modes still intergrade when different populations are compared (Kenrick *et al.* 1987).

Floral nectaries appear to be common to *Acacia* spp. (Bernhardt 1982) outside Australia. In most *Acacia* spp. in Australia an active pollen collection mode occurs in which insects must scrape up the semi-extruded polyads from the horizontal slits on the anther's surfaces without recourse to floral nectar (*Papaver*-type; *sensu* Vogel 1978 and see above). Successful pollination of most Australian *Acacia* spp., then, is limited largely to pollen-eating beetles, syrphid flies and polylectic bees (Bernhardt *et al.* 1984; Bernhardt 1989).

In contrast, nectar-feeding passerines pollinating the autumn–winter flowering *A. terminalis* do not consume pollen. These birds forage for nectar on large, red nectar glands confined to the leaf petioles of montane populations of *A. terminalis*. These glands secrete large quantities of nectar while the synfloresences of *A. terminalis* bloom in the leaf axils. The dehiscent anthers of compressed florets powder the birds' heads and bills with polyads while they collect nectar from the leaf glands. However, insect-pollination is still common in lowland populations as autumnal bees and flies continue to forage on the individual inflorescences for the same polyads.

The size of the leaf glands and the nectar volume and chemistry of *A. terminalis* appear to be distinct from the foliar glands of other Australian *Acacia* spp. that secrete nectar to attract wasp and ant guards (Bernhardt 1982, 1987). In this partial shift from the active to the passive mode of anther presentation the pollination system appears to have evolved along a line of least possible resistance. That is, red, tubular perianths or elongated, staminal tubes with basal nectaries are the characters most commonly associated with bird-pollinated mimosoids (Arroyo 1981). In *A. terminalis*, though, the shift towards bird-pollination has emphasized the modification of an extra-floral gland over the actual flowers. The florets of *A. terminalis* remain reduced, yellowish and devoid of both elongated floral tubes and nectar glands so they are not distinguishable immediately from any insect-pollinated *Acacia* spp. native to southeastern Australia.

CONCLUSIONS

Anthers contribute both directly and indirectly to the pollination syndromes of animal-pollinated species. The timing of anther dehiscence appears to be more than a mechanism restricting self-pollination. Variation in the timing of anther dehiscence may also protect pollen from environmental damage over the floral lifespan, encourage the explosive release of pollen onto the forager, and anther dehiscence in the bud may be part of an ontogenetic process in the establishment of a trend towards the secondary presentation of pollen in some angiosperm families.

Anthers may function as sources of primary attractants and primary rewards. As sources of primary attractants, anther colors often contrast with perianth colors (sometimes along the ultraviolet end of the spectrum) and anthers release biochemically distinct odors that may also contrast with perianth scents. Pollen remains the primary reward offered by anthers and many taxa offer pollen as their only edible reward. Anatomical and morphological variation of flowers that encourage those animals that will forage exclusively for pollen includes the mode of anther dehiscence, the number of anthers in the androecia and the mode of floral presentation. The anthers and/or staminal filaments of some taxa may also offer volatile oils, food bodies and nectar. However, the location of true nectar glands on or near the anthers occurs infrequently in nature. When it does occur at all morphological modifications separate nectar droplets from viable pollen.

Despite the domination of animal-pollination throughout the angiosperms there are really only two modes of anther presentation: the passive pollen collection mode and the active pollen collection mode. The passive mode is more diverse with the primary pollinator(s) either ignoring the pollen entirely or treating it as a secondary reward. Flowers adopting the passive mode show the greatest number of modifications to the stamens effecting maximum pollen deposition on the forager. In the active mode the pollinator(s) requires pollen, or materials found only on the anther (e.g., volatile oils) as its primary reward. Flowers encouraging the active pollen collection mode are usually nectarless and their anthers may show a trend towards porose dehiscence and heteranthery. Flowers showing this active mode are pollinated either by some beetles, flies, female bees and, rarely, male bees in the tribe Euglossini (e.g., *Cyphomandra*). Both the comparative effectiveness of these insects and the presentation of receptive stigmas within flowers with active modes varies according to the mode of anther dehiscence, the number of anthers in each flower and the length and angle of the styles.

Pseudanthery is interpreted as a polyphyletic syndrome dependent on the diversity and density of plants showing the active mode. Pseudanthery may have been derived repeatedly from ancestors that had an active mode of anther presentation. Comparisons of angiosperm lineages suggest that active and passive modes may represent reversible trends requiring relatively few changes to anther morphology and biochemistry.

LITERATURE CITED

Agren, J. & D.W. Schemske. 1991. Pollination by deceit in a neotropical monoecious herb, *Begonia involucrata*. Biotropica 23: 235–241.

Anderson, G.J. 1979. Dioecious *Solanum* species of hermaphroditic origin is an example of broad convergence. Nature 282: 836–838.

Armstrong, J. 1979. Biotic pollination mechanisms in the Australia flora – A review. N.Z. J. Bot. 17: 467–508.

Arnett, R.H. 1968, Pollen feeding by Oedemeridae (Coleoptera). Bull. Entomol. Soc. Amer. 14: 184, 204 (abstract).

Arroyo, M.T.K. 1981. Breeding systems and pollination biology in the Leguminosae. pp. 723–769 in Advances in Legume Systematics, Part 2, (edited by R.M. Polhill, & P.H. Raven). Royal Botanic Gardens, Kew.

Augspurger, C.K. 1980. Mass-flowering of a tropical shrub (*Hybanthus prunifolius*): influence on pollinator attraction and movement. Evolution 34: 475–488.

Barth, F.G. 1985. Insects and Flowers: The Biology of Partnership. Princeton University Press, Princeton.

Beach, J.H. & K.S. Bawa. 1980. Role of pollinators in the evolution of dioecy from distyly. Evolution 34: 1138–1142.

Beardsell, D.V., M.A. Clements, J.F. Hutchinson & E.G. Williams. 1989. Pollination of *Diuris maculata* R. Br. (Orchidaceae) by floral mimicry of the native legumes, *Daviesia* spp. and *Pultenaea scabra* R. Br. Aust. J. Bot. 34: 165–173.

Beattie, A.J. 1971. Pollination mechanisms in *Viola*. New Phytol. 70: 343–360.

Beattie, A.J. 1974. Floral evolution in *Viola*. Ann. Missouri Bot. Gard. 61: 781–793.

Bernhardt, P. 1982. Insect pollination of Australian *Acacia*. pp. 84–101 in Pollination 82 (edited by E.G. Williams, R.B. Knox, J.H. Gilbert & P. Bernhardt). University of Melbourne Press, Australia.

Bernhardt, P. 1983. The floral biology of *Amyema* in southeastern Australia. pp. 87–100 in The Biology of Mistletoes (edited by D.M. Calder & P. Bernhardt). Academic Press, London.

Bernhardt, P. 1984. The pollination biology of *Hibbertia stricta* (Dilleniaceae). Pl. Syst. Evol. 147: 267–277.

Bernhardt, P. 1986. Bee-pollination in *Hibbertia fasciculata* (Dilleniaceae). Pl. Syst. Evol. 152: 231–241.

Bernhardt, P. 1987. A comparison of the diverity, density, and foraging behavior of bees and wasps on Australian *Acacia*. Ann. Missouri Bot. Gard. 74: 42–50.

Bernhardt, P. 1989. The floral ecology of Australian *Acacia*. pp. 263–282 in Advances In Legume Biology (edited by C.H. Stirton & J.L. Zarucchi) Monographs in Systematic Botany from the Missouri Botanical Gardens 29, St. Louis.

Bernhardt, P. 1990a Anthecology of *Schrankia nuttalii* (Mimoscaceae) on the tallgrass prairie. Pl. Syst. Evol. 170: 247–255.

Bernhardt, P. 199b. Pollination ecology of *Oxalis violacea* (Oxalidaceae). Pl. Syst. Evol. 170: 147–155.

Bernhardt, P. 1993. Natural Affairs: A Botanist Looks at Attachments Between Plants and People. Villard Press, New York.

Bernhardt, P. 1995 Buzz-pollination of *Dianella caerulea* var. *assera* (Phormiaceae). Cunninghamia. 4:9–20.

Bernhardt, P. & P. Burns-Balogh. 1983. Pollination and pollinarium of *Dipodium punctatum* (Sm.) R. Br. Victorian Nat. 100: 197–199.

Bernhardt, P. & P. Burns-Balogh. 1986. Floral mimesis of *Thelymitra nuda* (Orchidaceae). Pl. Syst. Evol. 151: 97–101.

Bernhardt, P.J. Kenrick & R.B. Knox. 1984. Pollination biology and the breeding system of *Acacia retinodes* (Leguminoseae: Mimosoidae). Ann. Missouri Bot. Gard. 72: 135–142.

Bernhardt, P. & E.A. Montalvo. 1979. The pollination ecology of *Echeandia macrocarpa* (Liliaceae). Brittonia 31: 64–71.

Bernhardt, P. & L.B. Thien. 1987. Self-isolation and insect pollination in the primitive angiosperms: new evaluations of older hypotheses. Pl. Syst. Evol. 156: 159–176.

Bittrich, V., M.C.E. Amaral & G.A.R. Melo. 1993. Polli-
nation biology of *Ternstroemia laevigata* and *T. dentata* (Theaceae). Pl. Syst. Evol. 185: 1–6.

Bowers, K.A.W. 1975. The pollination ecology of *Solanum rostratum*. Amer. J. Bot. 62: 633–638.

Buchmann, S.L. 1983. Buzz pollination in angiosperms. pp. 73–113 in Handbook of Experimental Pollination (edited by C.E. Jones & R.J. Little). Van Nostrand Reinhold, New York.

Buchmann, S.L., C.A. Jones & L.J. Colin. 1978. Vibratile pollination of *Solanum douglasii* and *S. xanthi* (Solanaceae) in southern California. Wassman J. Biol. 35: 1–25.

Burns-Balogh, P. & P. Bernhardt. 1985. Evolutionary trends in the androecium of the Orchidaceae. Pl. Syst. Evol. 149: 481–485.

Burns-Balogh, P. & V. Funk. 1986. A Phylogenetic Analysis of the Orchidaceae. Smithsonian Contributions to Botany 61. Smithsonian Institution Press, Washington, D.C.

Cane, J.H. 1993. Reproductive role of sterile pollen in cryptically dioecious species of flowering plants. Curr. Sci. 65: 223–225.

Cane, J.H., G.C. Eickwort, F.R. Wesley & J. Spielholz. 1985. Pollination of *Vaccinium stamineum* (Ericaceae: Vaccinoideae). Amer. J. Bot. 72: 135–142.

Chaudhry, B. & M.R. Vijayaraghavan. 1992. Structure and function of the anther gland in *Prosopis juliflora* (Leguminosae, Mimosoideae): a histochemical analysis. Phyton 32: 1–7.

Chaw, S. 1992. Pollination, breeding syndromes, and systematics of *Trochodendron aralioides* Sieb. & Zucc. (Trochodendraceae), a relictual species in Eastern Asia. pp. 63–77 in Phytogeography and Botanical Inventory of Taiwan (edited by C. Peng). Monograph Series No. 12, Institute of Botany, Academia Sinica. Taipei.

Cocucci, A.A. 1991. Pollination biology of *Nierembergia* (Solanaceae). Pl. Syst. Evol. 174: 17–36.

Cox, P.A. 1983. Extinction of the Hawaiian avifauna resulted in a change of pollinators for the ieie, *Freycinetia arborea*. Oikos 41: 195–199.

Crowson, R.A. 1981. The Biology of The Coleoptera. Academic Press, London.

Dafni, A. 1993. Pollination Ecology: a Practical Approach. Oxford University Press. Oxford.

Dafni, A. & P. Bernhardt. 1989. Pollination of terrestrial orchids of southern Australia and the Mediterranean region: systematic, ecological and evolutionary implications pp. 193–252. in Evolutionary Biology, Vol 24. (edited by M.K. Hecht, B. Wallace & R.J. Macintyre). Plenum Press, New York.

Dafni, A., P. Bernhardt, A. Shmida, Y. Ivri, S. Greenbaum, C. O'Toole & L. Losito. 1990. Red bowl-shaped flowers: convergence for beetle pollination in the Mediterranean region. Isr. J. Bot. 39: 81–92.

Dahlgren, R. & T. Clifford. 1982. The Monocotyledons: A Comparative Study. Academic Press, London.

D'Arcy, W.G., N.S. D'Arcy & R.C. Keating. 1990. Scented anthers in the Solanaceae. Rhodora 92: 50–53.

Daumann, E. 1970. Das Bluttennektarium der Monocotyledone unter besonderer Berucksichtigung seiner systematischen und phylogenetischen Bedeutung. Feddes Repert. 80: 463–590.

De Frey, H.M., L.A. Coetzer & P.J. Robbertse. 1992. A unique anther-mucilage in the pollination biology of *Tylosema esculentum*. Sex Pl. Reprod. 5: 298–303.

Delpino, G. 1868–1875. Ulteriori oservazioni sulla dicogamis nel regno vegetale. I (1868, 1869), II (1870, 1875). Estratto Dagli Atti Soc. Ital. Sci. Nat. Milano 11, 12.

De Villamil, P.H. 1980. Stamens in the Ericaceae: a developmental study. Ph.D. Thesis, Rutgers University, New Brunswick.

Dobson, H.E.M. 1987. Role of flower and pollen aromas in host-plant recognition by solitary bees. Oecologia 72: 618–623.

Dobson, H.E.M. 1988. Survey of pollen and pollenkitt lipids – chemical cues to flower visitors? Amer. J. Bot. 75: 170–182.

Dobson, H.E.M., E.M. Dobson, G. Bergstrom & I. Groth. 1990. Differences in fragrance chemistry between flower parts of *Rosa rugosa* Thunb. (Rosaceae). Isr. J. Bot. 39: 143–156.

Dorr, L.J. 1981. The pollination ecology of *Zenobia* (Ericaceae). Amer. J. Bot. 68: 1325–1332.

Dressler, R.L. 1981. The Orchids: Natural History and Classification. Harvard University Press, Cambridge, Mass.

Dressler, R.L. 1993. Phylogeny and Classification of the Orchid Family. Dioscorides Press, Portland, Oregon.

Endress, P.K. 1986. Reproductive structures and phylogenetic significance of extant primitive angiosperms. Pl. Syst. Evol. 152: 1–28.

Erickson, R. 1958. Triggerplants. Paterson, Brokensha, Perth, Western Australia.

Erickson, R. 1965. Orchids of The West, 2nd ed. Paterson Brokensha, Perth, Western Australia.

Evans, M.S. 1895. The fertilization of *Loranthus kraussianus* and *Loranthus dregei*. Nature 51: 235–236.

Faden, R.B. 1992. Floral attraction and floral hairs in the Commelinaceae. Ann. Missouri Bot. Gard. 89: 46–52.

Faegri, K. 1986. The solanoid flower. Trans. Bot. Soc. Edinb. 150th Anniversary Suppl: 51–59

Faegri, K. & L. van der Pijl. 1979. The Principles of Pollination Ecology, 3rd ed. Pergamon Press, New York.

Fahn, A. 1953. The topography of the nectary in the flower and its phylogenetical trend. Phytomorphology 3: 424–426.

Fahn, A. 1979. Secretory Tissues in Plants. Academic Press, London.

Free, J.B. 1970. Insect Pollination of Crops. Academic Press, London.

Gack, C. 1979. Zur Ausbildung, Evolution und Bedeutung von Staubgefassimitationen bei Bluten als Signale für die Bestauber. Dissertation. Albert-Ludwigs-Universität, Freiburg.

Gill, F.B. & L.L. Wolf. 1975. Foraging strategies of East African sunbirds at mistletoe flowers. Amer. Nat. 109: 491–510.

Goldblatt, P. & P. Bernhardt. 1990. Pollination biology of *Nivenia* (Iridaceae) and the presence of heterostylous self-incompatibility. Isr. J. Bot. 39: 93–111.

Goldblatt, P., P. Bernhardt & J. Manning. 1989. Observations on the pollination biology of two species of *Moraea* (Iridaceae). Pl. Syst. Evol. 170: 247–255.

Gottsberger, G. 1977. Some aspects of beetle pollination in the evolution of flowering plants. Pl. Syst. Evol. (Suppl. 1): 211–226.

Gottsberger, G. & I. Silberbauer-Gottsberger. 1988. Evolution of flower structures and pollination in Neotropical Cassinae (Caesalpiniaceae) species. Phyton 28: 293–320.

Gracie, C. 1993. Pollination of *Cyphomandra endopogon* var. *endopogon* (Solanaceae) by *Eufriesea* spp. (Euglossini) in French Guinea. Brittonia 45: 39–46.

Haber, W.A. & K.S. Bawa 1984. Evolution in *Saurauia* (Dilleniaceae). Ann. Missouri Bot. Gard. 71: 289–293.

Haselrud, H.D. 1974. Pollination of some Ericaceae in Norway. Norw. J. Bot. 21: 211–216.

Hawkeswood, T.J. 1987a. Pollination of *Leptospermum flavescens* SM. (Myrtaceae) by beetles (Coleoptera) in the Blue Mountains, New South Wales, Australia. G. Ital. Entomol. 3: 261–169.

Hawkeswood, T.J. 1987b. Notes on some Coleoptera from *Baeckea stenophylla* F. Muell. (Myrtaceae) in New South Wales, Australia. G. Ital. Entomol. 3: 285–290.

Hawkeswood, T.J. 1989. Notes on *Diphucephala affinis* (Coleoptera, Scarabaeidae) associated with flowers of *Hibbertia* and *Acacia* in Western Australia. Pl. Syst. Evol. 68: 1–5.

Hawkeswood, T.J. 1990. Insect pollination of *Bursaria spi-*

nosa (Pittosporaceae) in the Armidale area, New South Wales, Australia. G. Ital. Entomol. 5: 67–87.

Henslow, G. 1893. The Origin of Floral Structures through Insect and other Agencies. Kegan Paul, Trench, Trubner & Co., London.

Hernandez, H.M. 1986. *Zapoteca*: a new genus of neotropical Mimosoideae. Ann. Missouri Bot. Gard. 73: 755–763.

Hopper, S.D. & A.A. Burbidge, 1982. Feeding behaviour of birds and mammals on flowers of *Banksia grandis* and *Eucalyptus angulosa*. pp. 67–75 in Pollination and Evolution (edited by J.A. Armstrong, J.M. Powell & A.J. Richards), Royal Botanic Garden, Sydney.

Johri, B.M. & S.P. Bhatnagar. 1972. Loranthaceae. Coun. Sci. Ind. Sci. Calcutta, Bot. Monogr. No. 8.

Jones, D.L. 1981. The pollination of selected Australian orchids. Proc. Orchid Symp. 13th Int. Bot. Congr. 1981: 40–43.

Kearns, C.A. & Inouye, C.A. 1993. Techniques for Pollination Biologists. University Press of Colorado, Boulder.

Keighery, G. 1975. Pollination of *Hibbertia hypericoides* (Dilleniaceae) and its evolutionary significance. J. Nat. Hist. 9: 681–684.

Keighery, G. 1991. Pollination of *Hibbertia conspicua* (Dilleniaceae). W. Aust. Nat. 18: 163–164.

Kennedy, H. 1978. Systematics and pollination of the closed-flowered species of *Calathea* (Marantaceae). Univ. Calif. Berkeley Publ. Bot. 71: 1–90.

Kenrick, J., P. Bernhardt, R. Marginson, G. Beresford, R.B. Knox, I. Baker & H.G. Baker. 1987. Pollination-related characteristics in the mimosoid legume *Acacia terminalis* (Leguminosae). Pl. Syst. Evol. 157: 49–62.

Knox, R.N. 1979. Pollen and Allergy. Studies in Biology 107. Edward Arnold, London.

Knox, R.N. & E. Friederich. 1974. Tetrad pollen grain development and sterility in *Leschenaultia formosa* (Goodeniaceae). New Phytol. 73: 251–258.

Knox, R.N., J. Kenrick, P. Bernhardt, R. Marginson, G. Beresford, I. Baker & H.G. Baker. 1985. Extrafloral nectaries as adaptations for bird pollination in *Acacia terminalis*. Amer. J. Bot. 72: 1185–1196.

Kuijt, J. 1969. The Biology of Parasitic Flowering Plants. University of California Press, Berkeley.

Kunze, H. 1990. Structure and function of asclepiad pollination. Pl. Syst. Evol. 176: 227–253.

Leppik, E.E. 1977. Floral Evolution in Relation to Pollination Ecology. Today and Tomorrow's Printers and Publishers, New Delhi.

Linskens, H.F. & R.G. Stanley. 1974. Pollen: Biology, Biochemistry, Management. Springer-Verlag, New York.

Lloyd, D.G. & M.S. Wells. 1992. Reproductive biology of a primitive angiosperm, *Pseudowintera colorata* (Winteraceae), and the evolution of pollination systems in the Anthophyta. Pl. Syst. Evol. 181: 77–95.

Lunau, K. 1992. A new interpretation of flower guide colouration: absorption of ultraviolet light enhances colour saturation. Pl. Syst. Evol. 183: 51–65.

MacDougal, J. 1983. Revision of *Passiflora* L. Section Pseudodysosmia (Harms) Killip emend. J. MacDougal, The Hooked Trichome Groups (Passifloraceae). Ph.D. Thesis. Duke University. North Carolina.

Macior, L.W. 1964. An experimental study of the floral ecology of *Dodecatheon meadia*. Amer. J. Bot. 61: 96–108.

Macior, L.W. 1966. Foraging behavior of *Bombus* (Hymenoptera: Apidae) in relation to *Aquilegia* pollination. Amer J. Bot 53: 302–309.

Macior, L.W. 1982. Plant community and pollinator dynamics in the evolution of pollination mechanisms in *Pedicularis* (Scrophulariaceae). pp. 29–45 in Pollination and Evolution (edited by J.A. Armstrong, J.M. Powell & A.J. Richards). Royal Botanic Garden, Sydney, Australia.

Macior, J.M. 1984. Behavioral coadaptation of *Bombus* pollinators and *Pedicularis* flowers. Vème Symposium International sur la Pollinisation (INRA Publ. Eds.). Colloques de INRA 21: 257–261.

Manning, S.D. 1990. *Pavetta*, subgenus *Baconia* (Rubiaceae) in Cameroon. (Ph.D.) Thesis. St Louis University, St Louis.

Masters, M.T. 1871. Contributions to the natural history of the Passifloraceae. Trans. Linn. Soc. Lond. 27: 593–645.

McConchie, C.A. 1982. The diversity of hydrophilous pollination in monocotyledons. pp. 148–165 in Pollination 82 (edited by E.G. Williams, R.B. Knox, J.H. Gilbert & P. Bernhardt). University of Melbourne Press, Melbourne.

McConchie, C.A., T. Hough, M.B. Singh, R.B. Knox, 1986. Pollen combs in the geoflorous heath *Acrotriche serrulata* (Epacridaceae). Ann. Bot. 57: 155–164.

Menzel, R. 1990. Color vision in flower visiting insects. Institut für Neurobiologie der Freien Universität Berlin, Germany.

Michener, C.D. 1974. The Social Behavior of the Bees: A Comparative Study. The Belknap Press, Harvard University, Cambridge, Mass.

Michener, C.D. 1979. Biogeography of the bees. Ann. Missouri Bot. Gard. 66: 277–347.

Molau, U. 1988. Scrophulariaceae. I. Calceolarieae. Flora Neotropica Monogr. 47: 1–326. New York Botanical Garden, New York.

Mori, S.A. & G.T. Prance. 1990. Lecythidaceae. II. The zygomorphic-flowered New World Genera. Flora Neotropica. New York Botanical Garden, New York.

Nilsson, L.A., L. Jonsson, L. Rason & E. Randrianjohany. 1986. The pollination of *Cymbidiella flabellata* (Orchidaceae) in Madagascar: a system operated by sphecid wasps. Norw. J. Bot. 6: 411–422.

Norstog, K. 1987. Cycads and the origin of insect pollination. Amer. Sci. 75: 270–279.

Ornduff, R. 1975. Complementary roles of halictids and syrphids in the pollination of *Jepsonia heterandra* (Saxifragaceae). Evolution 29: 371–373.

Paton, D.C. 1986. Honeyeaters and their plants in southeastern Australia. pp. 9–19 in The Dynamic Partnership: Birds and Plants in Southern Australia (edited by H.A. Ford & D.C. Paton Handbook of the Flora and Fauna of South Australia, Government Printer, South Australia.

Pellmyr, O. 1985. Pollination adaptations in the Cimicifugae and the evolutionary origin of pollinator-plant mutualism. Acta Universitatis upsaliensis: Comprehensive Summaries of Uppsala Dissertations from the Faculty of Science 2: 1–34.

Percival, M.S. 1955. The presentation of pollen in certain angiosperms and its collection by *Apis mellifera*. New Phytol. 54: 353–368.

Percival, M.S. 1965. Floral Biology. Pergamon Press, Oxford.

Powlesland, M.H. 1984. Reproductive biology of three species of *Melicytus* (Violaceae) in New Zealand. N.Z.J. Bot. 22: 81–94.

Prance, G.T. 1976. The pollination and androphore structure of some Amazonian Lecythidaceae. Biotropica 8: 235–241.

Prance, G.T. & J.R. Arias. 1975. A study of the floral biology of *Victoria amazonica* (Poepp.) Sowerby (Nymphaeaceae). Acta Amazonica 5: 109–139.

Prance, G.T. & S.A. Mori. 1979. Lecythidaceae. I. Flora Neotropica. New York Botanical Garden, New York.

Reader, R.J. 1977. Bog ericad flowers: self-compatibility and relative attractiveness to bees. Can. J. Bot. 55: 2279–2287.

Rickson, F.R., M. Cresti & J.H. Beach. 1990. Plant cells which aid in pollen digestion within a beetle's gut. Oecologia 82: 424–426.

Robertson, C. 1895. Flowers and insects: XIII. Bot. Gaz. 20: 104–110.

Roubik, D.W. 1989. Ecology and Natural History of Tropical Bees. Cambridge University Press, Cambridge.

Samuelson, G.A. 1994. Pollen consumption and digestion by leaf beetles. pp. 179–183 in Novel Aspects of the Biology of Chrysomelidae (edited by P.H. Jolivet, M.L. Cox & E. Pettipierre), Kluwer, Dordrecht.

Sazima, M., S. Vogel, A. Cocucci & G. Hausner. 1993. The perfume flowers of *Cyphomandra* (Solanaceae): pollination by euglossine bees, bellows mechanism, osmophores, and volatiles. Pl. Syst. Evol. 187: 51–88.

Schmid, R. 1978. Reproductive anatomy of *Actinidia chinensis* (Actinidiaceae). Bot. Jahrb. Syst. 100: 149–195.

Schneider, E.L. & J.D. Buchanan. 1980. Morphological studies of the Nymphaeaceae. XI. The floral biology of *Nelumbo pentapetala*. Amer. J. Bot. 67: 182–193.

Simpson, B.B. 1989. Krameriaceae. Flora Neotropica, Monograph 49, pp. 1–108. New York Botanical Garden, New York.

Simpson, B.B. & J.L. Neff 1981. Floral rewards: alternatives to pollen and nectar. Ann. Missouri Bot. Gard. 68: 301–322.

Spira, T.P. 1980. Floral parameters, breeding system and pollinator type in *Trichostema* (Labiatae). Amer. J. Bot. 67: 278–284.

Sprengel, C.K. 1793. Das entdeckte Geheimniss der Natur im Bau und in der Befruchtung der Blumen. Vieweg, Berlin.

Stebbins, G.L. & R.D. Hoogland. 1976. Species diversity in ecology and evolution in a primitive angiosperm genus: *Hibbertia* (Dilleniaceae). Pl. Syst. Evol. 125: 139–154.

Subramanyam, K. 1949. On the nectary in the stamens of *Memecylon heyneanum* Benth. Curr. Sci. 18: 415–416.

Thien, L.B., P. Bernhardt, G.W. Gibbs, O. Pellmyr, G. Bergstrom, I. Groth & G. McPherson. 1985. The pollination of *Zygogynum* (Winteraceae) by a moth *Sabatinca* (Micropterigidae): an ancient association? Science 227: 540–543.

Thien, L.B. & B.G. Marcks, 1972. The floral biology of *Arethusa bulbosa, Calopogon tuberosus* and *Pogonia ophioglossoides* (Orchidaceae). Can. J. Bot. 50: 2319–2325.

Turner, V. 1982. Marsupials as pollinators in Australia. pp. 55–66 in Pollination and Evolution (edited by J.A. Armstrong, J.M. Powell & A.J. Richards). Royal Botanic Garden, Sydney.

van der Pijl, L. 1978. Reproductive integration and sexual disharmony in floral functions. pp. 79–88 in The Pollination of Flowers by Insects (edited by A.J. Richards). Linnean Society Symposium Series No. 6. Academic Press, London.

van der Pijl, L. & C.H. Dodson. 1966. Orchid Flowers:

Their Pollination and Evolution. University of Miami Press, Coral Gables, Florida.

van der Werff, H. 1993. A revision of the genus *Pleurothyrium* (Lauraceae). Ann. Missouri Bot. Gard. 80: 39–118.

Vogel, S. 1978. Evolutionary shifts from reward to deception in pollen flowers. pp. 89–96 in The Pollination of Flowers By Insects (edited by A.J. Richards). Linnean Society Symposium Series, Number 6. Linnean Society of London, Academic Press, London.

Vogel, S. 1981. Die klebstoffhaare an den antheren von *Cyclanthera pedata*. Pl. Syst. Evol. 137: 291–316.

Vogel, S. 1990. The Role of Scent Glands in Pollination. Smithsonian Institution Library, Amerind Publishing Co., New Delhi.

Weberling, F. 1989. Morphology of Flowers and Inflorescences. Cambridge University Press, Cambridge.

Wiens, D. & J.P. Rourke. 1978. Rodent pollination and the South African *Protea* spp. Nature 276: 71–73.

Wrigley, J.W. & M. Fagg, 1989. Banksias, Waratahs and Grevilleas: and all other plants in the Australaian Proteaceae Family. Collins, Sydney.

10 · Anther differentiation in the Asclepiadaceae – Asclepiadeae: form and function

SIGRID LIEDE

INTRODUCTION

The intricate pollination system of the Asclepiadaceae has fascinated both floral morphologists and pollination biologists for a long time (Brown 1833; Corry 1884), and some of its features have attracted recent attention (Schill & Jäkel 1978; Kunze 1982). In this family, androecium and gynoecium are joined into a highly specialized structure, the gynostegium. The gynoecial part is modified from the usual pistil in other families, consisting of two almost free carpels, each with its own style, but a common, large stylar head. The five receptive stigmatic surfaces are hidden behind the guide rails on the basal part of the stylar head, while its visible part is sterile (Kunze 1982). The characters of the gynoecial part are remarkably uniform throughout the family. The androecial part consists of five highly specialized stamens, which vary considerably across the family. A typical stamen consists of a filament and the anther. The anther comprises anther sacs containing the pollinia, the connective appendage, often called anther appendage, and two more or less well differentiated anther wings (Fig. 1). Two pollinia from adjacent anthers, linked by an acelluar mass, the translator apparatus, are called the pollinarium. In the Asclepiadoideae, the translator apparatus consists of the central corpusculum and two lateral caudicles (translator arms), to which the pollinia are attached (see Kunze 1993 for details). Anther wings of adjacent anthers together form a functional unit, the guide rail, designed to trap an insect body part (leg, proboscis, or only a stiff hair) and to guide it towards the corpusculum topping the guide rail. The insect can free itself only by removing the pollinarium. If the insect is too weak to remove the pollinarium, it remains stuck in

the guide rail and dies. Large flowers, such as those of *Oxypetalum* R. Br. and *Mitostigma* Decne., can be deadly traps even for quite large insects (butterflies, moths, flies). The intricacy of asclepiad flower structure is further increased by the presence of a corona. It is formed by appendages of either the stamens or the corolla, or both. Its basic structure is often used to delimit genera, while differences in shape and color constitute the characters most frequently used for species delimitation. These androecial structures yield the most

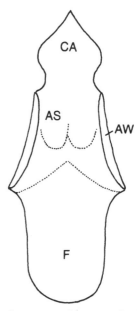

Fig. 1. General structure of a stamen in the Asclepiadaceae – Asclepiadoideae. AS, anther sacs; AW, anther wings; CA, connective appendage; F, filament part.

important characters for higher level classification. In fact, the delimitation of subfamilies and tribes relies almost entirely on androecial characters (Bruyns & Forster 1991). A critical systematic evaluation of these characters is presented here.

Among the structures taking part in the flower–insect interaction, the corona (Bookman 1981; Liede & Kunze 1993) and the pollinarium (El-Gazzar *et al.* 1974; Schill & Jäkel 1978; Kunze 1993) have received some detailed attention. The differentiations of the gynostegium, especially of the anther and the anther wings, are much less known. This is surprising as these elements play a crucial part in trapping the insect by a body part and guiding this body part in such a way that it can be extricated only by removing the pollinarium. Schill & Jäkel (1978) provide a good comparative overview of pollinaria structure throughout the family, even though they do not include the orientation of pollinia in their study, and Kunze (1982) discusses the ontogenetic differentiation of the anther. Anther differentiations are particularly pronounced in the tribe Asclepiadeae. As an example, the wide variety of anther structures in the large genus *Cynanchum* L. will be discussed. The circumscription of this difficult genus is not critical in the present context, as almost all variation is encountered in species unanimously placed in *Cynanchum*, while members of *Ditassa* R. Br., *Metastelma* R. Br. and *Vincetoxicum* Moench are much more uniform among themselves in these characters.

MATERIALS AND METHODS

This study forms part of a long-term project in Asclepiadaceae, which focuses on the subtribe Cynanchinae, tribe Asclepiadeae, Asclepiadoideae. Spirit material has been collected on several trips throughout southern and eastern Africa, Madagascar, the United States, Mexico, Argentina, Chile and Bolivia. A total of more than 3000 herbarium specimens from BM, BOL, EA, K, MEXU, MO, NBG, P, PRE, SAM and STEU, representing more than 600 species and all genera in the Asclepiadeae, have supplemented my own collections.

Cladistic analysis was conducted using the widely used computer program PAUP version 3.1 (Swofford 1991). The family Apocynaceae, widely recognized as representing the sister group of the Asclepiadaceae (e.g.,

Fallen 1986), was used as outgroup. The ingroup is understood to comprise the six tribes of Asclepiadaceae delimitated by Bruyns & Forster (1991), the Periploceae (only member of the subfamily Periplocoideae), the Secamoneae (only member of the subfamily Secamonoideae), the Marsdenieae, Stapelieae, Gonolobeae and Asclepiadeae (all subfamily Asclepiadoideae). For a list of genera included into these tribes see Liede & Albers (1994). In addition, the newly delimited tribe Fockeeae (Kunze *et al.* 1994) has been included in the analysis. The Fockeeae, comprising the genera *Fockea* and *Cibirhiza*, have been segregated from the Marsdenieae because of their highly complex corona structure, their large, sac-like connective appendages, and differences in the translator apparatus (see below).

RESULTS

Androecial structures and tribal classification of the Asclepiadaceae

If the androecial characters of the seven tribes of the Asclepiadaceae are reviewed using the Apocynaceae as outgroup, several lines of structural progression can be followed. In the Periplocoideae, pollen tetrads are formed, which are presented by a spoon-shaped translator.

Both the aggregation of pollen grains and the formation of a translator are apomorphic in comparison with the Apocynaceae. In the Secamoneae four pollinia per anther are formed, which are joined by a small and undifferentiated translator. The pollinia are in an apical position, the translator in a basal one. The formation of pollinia can be regarded as apomorphic compared with the formation of pollen tetrads.

In the Fockeeae only the two ventral pollen sacs are fertile, as is the case in all Asclepiadoideae (Kunze 1982); this is certainly an apomorphic condition. The construction of the translator, attachment of the pollinia and orientation of translator and pollinia, however, resemble closely what is found in the Secamoneae (Kunze 1993). In the Marsdenieae, a typical Asclepiadoideae pollinium with a well-developed translator apparatus and distinct caudicles (apomorphy) is formed. The 'erect' orientation of pollinia, historically thought to be advanced (e.g., Schumann 1895; Volk 1949), however, has to be considered plesiomorphic in the light of the orientation of the pollinia in the hitherto discussed more basal taxa. The Stapelieae are characterized by possession of a well-developed pellucid germination pore of the pollinium in an endolateral position (apomorphy). In this tribe, the erect, primitive orientation of pollinia is combined with a highly developed caudicle structure. The caudicle structure enables a change in pollinium orientation after the pollinarium is removed from the gynostegium, facilitating its deposition in the guide rail of the next flower (e.g., Kunze 1991). The Gonolobeae are characterized by a horizontal orientation of the pollinia, and a sterile pollinium tip (apomorphy). The Asclepiadeae, finally,

are characterized by a hanging orientation of the pollinia (apomorphy).

An exhaustive search has been performed for the data matrix in Appendix 1. Characters 1 and 5 were coded as ordered, employing the assumption that a transition from pollen grains to pollinia leads via pollen tetrads and that a change from erect to hanging pollinia (and back) leads via horizontally oriented ones. Unweighted analysis results in nine trees (L = 14, Consistency Index (CI) = 0.857, Retention Index (RI) = 0.833); the strict consensus tree shows the topology depicted in Fig. 2A. This tree shows that the new tribe Fockeeae clearly represents a member of the subfamily Asclepiadoideae. It also sup-

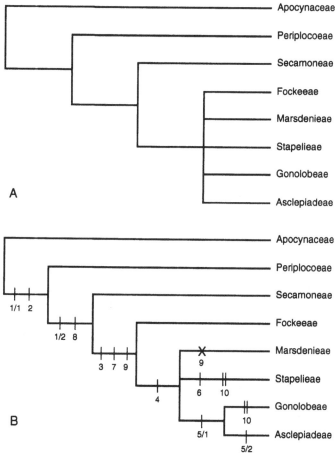

Fig. 2. Cladogram of the tribes of the Asclepiadaceae. A. Consensus tree derived from unweighted analysis. B. Consensus tree derived after assigning double weight to characters 1–5 (For details, see text.)

ports the often-stated conviction that the three subfamilies form three well-delimited clades, with the Periplocoideae occupying the basal and the Asclepiadoideae the most derived position. The procedure of successive weighting of characters according to their best fit onto the tree results in weight 10 for all characters except four and five, which are assigned a weight of 0. As this does not at all reflect the importance of these characters for Asclepiadaceae phylogeny, the procedure has been abandoned as inadequate for the present data set. Instead, characters 1–5 have been assigned a weight of 2. To justify this action, it needs to be pointed out that these five characters represent major steps in asclepiad evolution and are expressed in every single member of the tribe concerned. The only exception known to me is *Karimbolea verrucosa* Descoings, which was shown to belong in the tribe Asclepiadeae (Liede *et al.* 1993), but possesses erect pollinia. An exhaustive search under these conditions results in three most parsimonious trees (L = 23, CI = 0.913, RI = 0.946), and the strict consensus is depicted in Fig. 2B. The Fockeeae now resume the basal position in the Asclepiadoideae, and the Gonolobeae and Asclepiadeae are identified as sister groups. The relationships between the Marsdenieae, the Stapelieae and the Asclepiadeae/Gonolobeae clade remain unresolved.

A serious problem of this analysis lies in the inapplicability of characters 4–6 in both the outgroup and the Periploceae. Platnick *et al.* (1991) have pointed out the difficulties of such characters, but do not suggest a way of dealing with them. In an attempt to clarify how much the above results were influenced by the inapplicability problem, characters 4–6 were recoded by adding a primitive state to the characters in question (Appendix two). The reasoning behind this action is that, e.g., formation of pollinia is a prerequisite for orienting them in any way. While this is certainly an oversimplification, it is interesting to note that the analysis of the artificially completed data set yields almost the same results as the original one. Unweighted analysis results in nine trees (L = 18, CI = 0.833, RI = 0.812), and the strict consensus tree shows the same topology as the tree depicted in Fig. 2A. An analysis with a weight of two assigned to characters 1–5 results in three trees (L = 29, CI = 0.897, RI = 0.870). The strict consensus tree shows the same topology as the tree in Fig. 2B. While these findings do nothing to increase or decrease confidence in the results

of the first analysis, they show that the inapplicability of characters 4–6 is not a serious problem in this particular data set.

Androecial variation in the Asclepiadeae: hexangular patterns

Of all tribes, the Asclepiadeae display by far the most varied structure of androecial parts (see also Kunze 1982). Probably this fact is correlated with the hanging orientation of the pollinia, but a strictly functional explanation is still lacking. Six gynostegial and androecial character transformations have been recognized for *Cynanchum sensu stricto* alone. These are shown in the analytical diagrams in Figs. 3–8. In the diagrams character states are arranged along the sides of a hexangle, depicting the shape of the gynostegium, the means of elevation of the gynostegium, presence or absence of a filament, anther shape, relative guide rail length and guide rail shape. The icons displayed were drawn from material shown in the scanning electron micrographs in Figs. 9–11. As shown at the top of the hexangle, the gynostegium can be cone-shaped (Figs. 3.1, 9A) or cylindrical (Figs. 3.2, 9E). The gynostegium can be sessile (Figs. 3.3, 9A), situated on a short, thick bulge (Fig. 3.4), or long-stipitate (Figs. 3.5, 9B), as shown along the upper right side of the hexangle. Some anthers lack a visible filament (Figs. 3–6, 9D), while others have one (Figs. 3.7, 9C), depicted along the lower right side of the hexangle. The anthers can display a variety of shapes (Figs. 3.8–3.11, 10), shown at the bottom of the hexangle. The guide rails can equal the anther in length (Figs. 3.13, 9A) or be longer (Figs. 3.12, 9E, F) or shorter (Figs. 3.14, 9C, D) than the anther, as depicted along the lower left side of the hexangle. Finally, as shown at the upper left side of the hexangles, the guide rails can be parallel to each other (Figs. 3.15, 11A), basally divergent (Figs. 3.16, 11B), form a basal opening (Figs. 3.17, 11C), protrude beyond the plane of the anther and form a 'mouth' with its basal margin (Figs. 3.18, 11D), or even be contorted (Figs. 3.19, 11E).

Outgroup comparison allows the identification of the primitive states of each of these characters. In general, outgroup selection is difficult in the Asclepiadaceae because of uncertainty about genus limits and the interrelationships between genera. However, within the vast

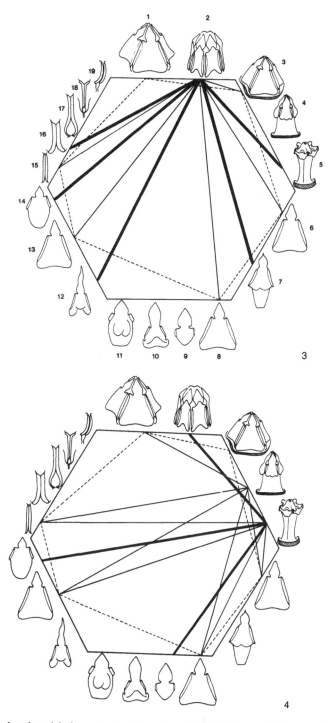

Figs. 3–8. Gynostegial and androecial characters in *Cynanchum*. Dashed lines, primitive combination; thin lines, derived state in combination with primitive state; thick lines, derived state in combination with derived state. Examples for 1–19: 1, *C. bojerianum*; 2, *C. baronii*; 3, *C. bojerianum*; 4, *C. altiscandens*; 5, *C. mahafalense*; 6, *C. bojerianum*; 7, *C. danguyanum*; 8, *C. bojerianum*; 9, *C. africanum*; 10, *C. madagascariense*; 11, *C. repandum*; 12, *C. fimbriatum*; 13, *C. bojerianum*; 14, *C. andringitrense*; 15, *C. bojerianum*; 16, *C. lenewtonii*; 17, *C. marnieranum*; 18, *C. decaryi*; 19, *C. eurychiton*.

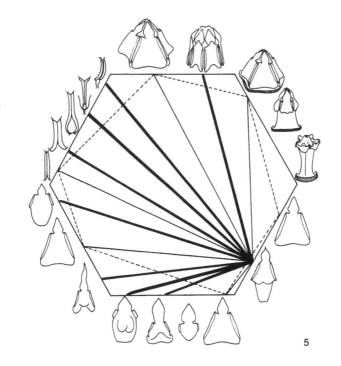

5

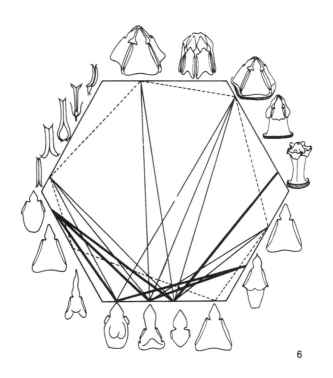

6

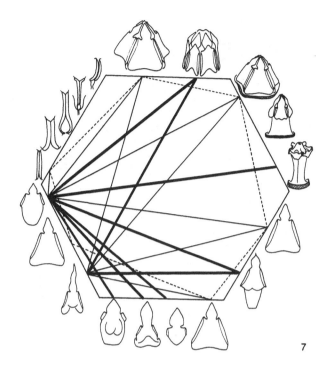

7

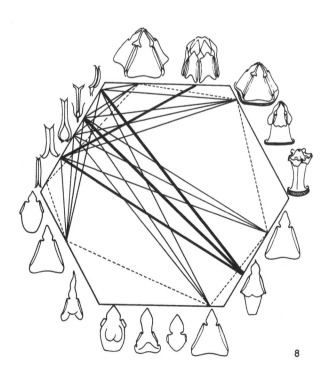

8

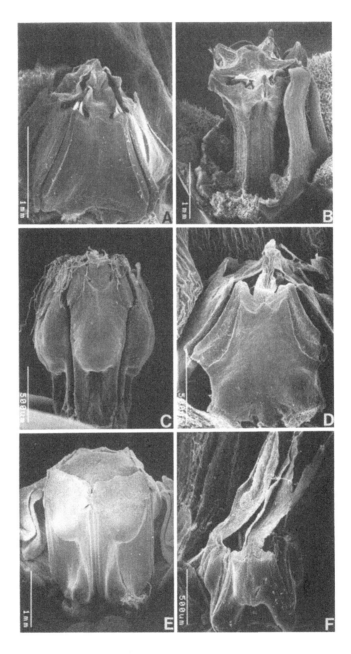

Fig. 9. Positioning of the lower guide rail entrance. A. Guide rail entrance not raised; other characters also in the primitive state (*C. bojerianum*, Peltier 1802). B. Stipe (*C. messeri*, Liede & Conrad 2721). C. Extended filament; note shortened guide rails (*C. andringitrense*, Bosser 7599). D. Shortened guide rails, forming a 'pseudostipe' (*C. leucanthum* subsp. *elongatum*, Humbert 25670). E. Elongated guide rails (*C. bisinuatum*, Hardy 2901), cylindrical gynostegium. F. Elongated guide rails, causing an 'arched' anther shape (*C. fimbriatum*, Decary 7316).

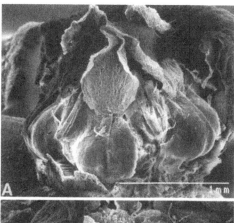

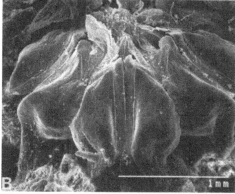

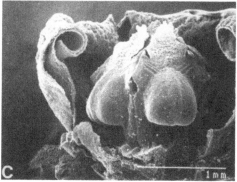

Fig. 10. Anther shapes in *Cynanchum*. A. Rounded (*C. bosseri*, Humbert 23686). B. Biconvex (*C. leucanthum* subsp. *leucanthum*, Perrier de la Bâthie 16877). C. Gibbose (*C. repandum*, Liede & Conrad 2867).

majority of genera in the tribe Asclepiadeae (e.g., *Asclepias* L., *Sarcostemma* R. Br., *Vincetoxicum* Wolf) as well as parts of the genus *Tylophora* R. Br. (hitherto placed in the Marsdenieae) little gynostegial and androecial variation is displayed. Most likely, in *Cynanchum*, a member of the Asclepiadeae, the original character states are the same as those displayed in these genera. Thus, a cone-shaped, sessile gynostegium, flat anthers without visible filament, guide rails equalling the anthers in length and parallel to each other and without any particular differentiation are presumably primitive states of the six characters analyzed. In Figs. 3–8 this primitive character state combination, which is found, for example, in *Cynanchum bojerianum* Choux, is represented by a dashed line.

In Figs. 3–8 derived states of each character transformation have been analyzed for their co-occurrence with primitive states of other characters (represented by thin lines), as well as for their co-occurrence with derived states of other characters (represented by thick lines).

Fig. 3 shows that a cylindrical gynostegium is frequently combined with features elevating the guide rail entrance, such as a stipitate gynostegium, anthers with a visible filament and shortened anthers. It does not occur in conjunction with specialized anther shapes or guide rail structures.

An elevation of the lower guide rail entrance can be accomplished by raising the whole gynostegium (Fig. 4), either by means of a short, thick bulge, as in *C. altiscandens* (even more pronounced in the Astephaninae *Goydera somaliense* Liede) or a long, slender stipe (Fig. 9B). Presence of a bulge, a rare condition both in *Cynanchum* and in the Asclepiadeae, is not combined with any other derived character state. A stipe, in contrast, is often combined with a cylindrical gynostegium, deltate anthers, and, more rarely, with shortened anther wings. A stipe is a fairly frequent condition in *Cynanchum* and is also known from other genera in the Asclepiadeae (*Amblystigma*, *Ditassa*, *Folotsia*, *Metastelma*, *Schizostephanus*). It seems to have evolved several times independently. Another way to raise the entrance of the guide rails is by elongation of the filament (Fig. 9C). This construction seems to be the precondition for differentiation of different guide rail shapes and also for variation in anther shape (Fig. 5). Anther shape is highly

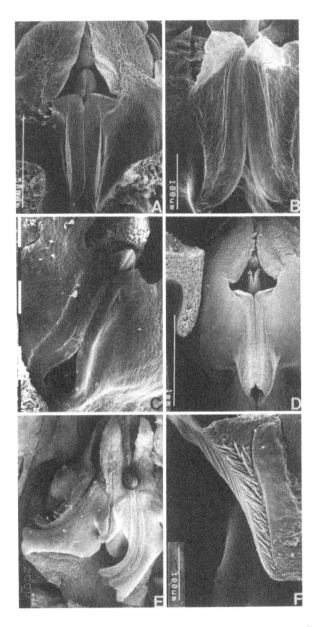

Fig. 11. Guide rail shapes in *Cynanchum*. A. Simplest form: two adjacent, straight, parallel anther wings in the same plane as the anther (*C. cucullatum*, Liede & Conrad 2868). B. Straight anther wings divergent towards the base, in the same plane as the anther (*C. gerrardii*, Rauh s.n.). C. two adjacent, straight, parallel anther wings, in the same plane as the anther, widened at the lower entrance (*C. marnieranum*, Rauh s.n.). D. Two adjacent, straight, parallel anther wings, projecting distally from the plane of the anther at the lower entrance (*C. decaryi*, Liede & Conrad 2691). E. Contorted guide rails (*C. eurychiton*, Cours 4512). F. Guide rail structure in *Cynanchum* (*C. lineare*, Liede & Conrad 2863).

variable within *Cynanchum* (Fig. 10). Fig. 6 shows that derived shapes are correlated with a stipitate gynostegium (only deltate anthers), shortened guide rails and an extended filament. Unusual anther shapes such as deltate, biconvex (Fig. 10B), or gibbose (Fig. 10C) are never correlated with guide rail differentiation.

The guide rail can equal the anther in length, exceed it in length or be shorter. A shortened guide rail (Fig. 9C, D) is the third means available to raise the guide rail entrance and this, in fact, occurs in combination with both other elevating mechanisms: the stipe and the extension of the filament (Fig. 7). It is also frequent in conjunction with special anther shapes. Elongated guide rails, in contrast, occur only with cylindrical gynostegia and extended filaments (Fig. 9E, F).

The most striking differentiation in *Cynanchum* anthers is the variability of guide rail structure (Fig. 11). The simplest form consists of two adjacent, straight, parallel anther wings, in the same plane as the anther (Fig. 11A). Progression occurs towards two adjacent, straight, anther wings in the same plane as the anther, divergent towards the base (Fig. 11B), or widened at the lower entrance, forming kind of a mouth (Fig. 11C). More complex are two adjacent, straight, parallel anther wings, projecting distally from the plane of the anther at the lower entrance and forming a distinct entrance together with the basal lateral margin of the anther (Fig. 11D, E). Even more unusual are two adjacent, tortuous anther wings (Fig. 11F).

Analysis (Fig. 8) shows that the sole precondition for even the most unusual guide rail shape is an extended filament. Internally, the construction of the guide rails is remarkably uniform throughout the genus (Fig. 12A). Proximal and distal horny ridges are formed. The distal ridge is adapted to guiding the trapped part of the insect body, the proximal one to guiding the pollinium towards the stigmatic surface (Kunze 1982). The space between the two ridges is normally filled with upwardly directed bristles, and the distal ridge can show marked striations. Both features most probably prevent the trapped insect body part from slipping out through the basal guide rail entrance without having removed the pollinarium. In other Asclepiadeae genera, derived guide rail structures are rare. Examples are members of the Malagasy genus *Folotsia*, the two species of *Schizostephanus*, some species of *Microloma* (Kunze 1991: fig. 41). In these three

genera, the basal 'mouth' formed by the guide rails is covered by papillae (Liede 1993: fig. 21b), the function of which is unknown. In the genus *Pentatropis* Wight & Arn. the guide rail follows the basal margin of the anther to the point where the staminal corona originates (Fig. 12C). Internally, its structure is also unusual, with long hairs (Fig. 12D) originating from the receptacle. In *Morrenia*, and in some members of the genus *Oxypetalum*, the distal ridge of the guide rail ends at about half of the anther height, forming an opening (Fig. 12B). The inner ridge turns outward at this point and continues along the whole length of the anther, to form a second, basal opening. Finally, differentiation of a proximal guide rail ridge is not a constant feature in the tribe. *Sarcostemma* and most of its relatives possess only one distal ridge. As one single guide rail ridge is the common condition in other Asclepiadaceae tribes (e.g., most Marsdenieae; Kunze 1982) the differentiations encountered in some members of the Asclepiadeae can be judged as apomorphies.

DISCUSSION

Androecial characters and tribal classification in the Asclepiadaceae

In the Asclepiadaceae, androecial characters are crucial for tribal delimitation. Even with occasional exceptions (one case discussed in Liede *et al.* 1993), these androecial features provide by far the most stable set of characters for tribal delimitation, and the tribes recognized here represent by and large natural groups. Although not included in the present analysis because of incomplete data, the connective appendage follows a similar tribal pattern. A small connective appendage is formed in the Secamoneae, Marsdenieae and Stapelieae, while the Asclepiadeae are characterized by large laminar connective appendages, which differ little between the genera. The Fockeeae take a special position with large, sac-like connective appendages.

The phylogenetic analysis of androecial data results in a hierarchical arrangement of the tribes involved. The Periploceae and Secamoneae, long assumed to be primitive taxa, always assume a basal position. The new tribe Fockeeae, despite its complex corona structure (Kunze *et al.* 1994), takes a basal position in the Asclepiadoideae, which are characterized by the possession of two pollinia

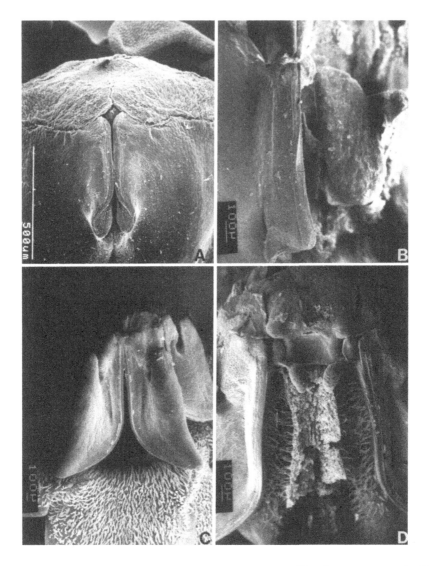

Fig. 12. Derived guide rail structures in other Asclepiadeae genera. A. Papillate lower entrance of guide rails (*Folotsia madagascariense*, Liede & Conrad 2695). B. Anther wing extending along the basal anther margin in *Pentatropis*. C. Guide rail structure in *Pentatropis* (*Pentatropis madagascariense*, Liede & Conrad 2749). D. Shortened outer guide rail ridge in *Morrenia* (Liede & Conrad 3009).

per anther. As a corollary, it is interesting to note that all 'primitive' taxa, the two subfamilies Periplocoideae and Secamoneae, as well as the Fockeeae, are restricted to the Old World. An Old World origin for the whole family thus seems plausible (see Good 1951). The dispersal of Asclepiadaceae into the New World needs further investigation.

The relationships between the four higher tribes of the Asclepiadaceae remain obscure. The Marsdenieae appear as a paraphyletic group; the relationships between the Marsdenieae, Stapelieae, and the Asclepiadeae/Gonolobeae are all but clear. Two key issues need further investigation. First, would a removal of parts of the genus *Tylophora* R. Br. with small, horizontal pollinia from the Marsdenieae leave the latter a monophyletic group? Rao & Kumari (1979) found a completely different position of the germination locus for *Tylophora indica* (N. Burmann) Merrill than for the other two members of the Marsdenieae studied. Unfortunately, *Tylophora* itself is a heterogeneous assemblage of species, so that this issue can only be addressed by a monograph on the large genus *Tylophora* itself. Second, the horizontal pollinia position in the Gonolobeae can be interpreted as a transitional stage from erect towards hanging pollinia (as it is done here), but this could also originate from a reorientation of hanging pollinia towards an erect position. In this case, recognition of the tribe Gonolobeae would leave the Asclepiadeae as a paraphyletic group. The detailed investigations required on the floral biology and ontogenetic development in the Gonolobeae are in progress elsewhere (H. Kunze personal communication).

How to trap an insect's leg

The gynostegium as a whole must offer resistance against the forces of the insect trying to free its caught body part. If insects that are too forceful work on the flowers, destruction of the flower is likely to occur (e.g., in *Cynanchum foetidum* visited by honeybees (Liede 1994). A conical arrangement of sessile anthers seems to be the most stable construction – and is certainly the one most frequently encountered. Cylindrical arrangement of the anthers is not only less stable, it also offers no hook to catch on to the insect. According to present knowledge, a cylindrical gynostegium occurs only in connection with pseudolobe-forming coronas, in which the staminal parts are pressed against the back of the anthers and the interstaminal parts form folds in front of the guide rails. In this corona type staminal and interstaminal parts are so highly fused – often overtopping the gynostegium – that only an insect's proboscis can penetrate, which is then guided into the guide rail entrance by the folding of the corona.

Several of the gynostegial and androecial specializations aim at raising the lower guide rail entrance. A stipe, which raises the whole gynostegium, seems to be well suited for the purpose, but diminished stability of the whole gynostegium is the price paid for this advantage. Slender stipes and rather large stylar heads are often combined with long caudicles, which probably facilitate

the removal of the pollinarium (e.g., *C. messeri*). Another frequent combination is a corona stabilizing the whole structure (e.g., *Folotsia madagascariense*). Other mechanisms to raise the lower guide rail entrance are shortening of the guide rails or extension of the filament.

Probably the most interesting correlation found in the present study is the strong correlation between an extended filament and derived guide rail shapes (Figs. 5, 8). Anthers with both an extended filament and derived guide rail shapes are particularly frequent in Madagascar.

The guide rail differentiations mentioned for other genera of the Asclepiadeae emphasize their importance in this tribe. Unfortunately, the internal differentiations, such as bristles and proximal guide rail ridges, are difficult to study in dried specimens, especially small-flowered ones; however, the present results indicate a possible systematic value at the genus and subtribe level.

CONCLUSIONS

The highly adaptive value of all characters directly linked to pollination – pollinia, guide rails and their arrangement, as well as the corona – poses a particularly difficult problem for the systematics of the Asclepiadaceae. Traditionally, tribal classification has been based on androecium characters, while corona characters were commonly used for subtribal and generic classification. Both sets of characters seem to be under almost identical adaptive pressure, but it is not yet understood why androecial characters seem to reflect natural relationships much better than corona characters. The need for different data sets to check on the hypotheses derived from these characters is evident. However, taxonomically usable morphological features are rare and often difficult to observe in other parts of the asclepiadaceous plant body.

The genus *Cynanchum* not only displays an extraordinary variability in corona structure, but also exhibits an unusual diversity in gynostegial and androecial characters. It is remarkable, though, that especially the guide rail differentiations occur exclusively in Malagasy species. Madagascar is particularly rich in *Cynanchum* species (*c.* 70 species, compared with *c.* 25 in all of mainland Africa). We can thus speculate that the high degree of androecial variability is linked directly with the species richness in *Cynanchum* in Madagascar. Nonetheless, the functional details of these structures are anything but clear and deserve more thorough investigation, especially in relation to possible coevolution with pollinating insects.

ACKNOWLEDGMENTS

The present study was supported by a habilitation grant of the Deutsche Forschungsgemeinschaft (LI 496–1 and 496–2). Thanks are due to the directors of the above-mentioned herbaria for long-term loans of specimens or permission to work on the premises, or both. Prof. F. Albers, Münster, provided the opportunity for SEM work. Dr H. Kunze, Minden, and Dr U. Meve, Münster, have contributed ideas, discussions and criticisms to drafts of this paper. I am grateful to the Ulmer Universitätsgesellschaft for financial aid to present this paper at the 15th International Congress of Botany and to Prof. F. Weberling, Ulm, for his generous support of my work. Figs. 3–8 have been drawn by G. Hintze, Ulm. W.G. D'Arcy helped in editing this paper, and an anonymous reviewer contributed useful comments.

LITERATURE CITED

Bookman, S.S. 1981. The floral morphology of *Asclepias speciosa* (Asclepiadaceae) in relation to pollination and a clarification in terminology for the genus. J. Bot. 68: 675–679.

Brown, R. 1833. On the organs and mode of fecundation in Orchideae and Asclepiadeae. Trans. Linn. Soc. London 16: 685–745.

Bruyns, P.V. & P.I. Forster. 1991. Recircumscription of the Stapelieae (Asclepiadaceae). Taxon 40: 381–391.

Corry, T.H. 1884. On the structure and development of the gynostegium and the mode of fertilization in *Asclepias cornuti*, Decaisne (*A. syriaca*, L.). Trans. Linn. Soc. London, Ser. 2, 2: 173–208.

El-Gazzar, A., M.K. Hamza & A.A. Badawi. 1974. Pollen morphology and taxonomy of Asclepiadaceae. Pollen Spores 16: 227–238.

Fallen, M.E. 1986. Floral structure in Apocynaceae: morphological, functional and evolutionary aspects. Bot. Jahrb. Syst. 106: 245–286.

Good, R. 1951. Atlas of the Asclepiadaceae. New Phytol. 51: 198–209.

Kunze, H. 1982. Morphogenese und Synorganisation des Bestäubungsapparates einiger Asclepiadaceen. Beitr. Biol. Pflanzen 56: 133–170.

Kunze, H. 1991. Structure and function in asclepiad pollination. Pl. Syst. Evol. 176: 227–253.

Kunze, H. 1993. Evolution of the translator in Periplocaceae and Asclepiadaceae, Pl. Syst. Evol. 185: 99–122.

Kunze, H., U. Meve & S. Liede. 1994. *Cibirhiza albersiana*, a new species of Asclepiadaceae, and establishment of tribe Fockeeae. Taxon 43: 367–376.

Liede, S. 1993. A revision of the genus *Cynanchum* in southern Africa. Bot. Jahrb. Syst. 114: 503–550.

Liede, S. 1994. Some observations on pollination in Mexican Asclepiadaceae. Madroño 41: 266–276.

Liede, S. & F. Albers. 1994. Tribal disposition of Asclepiadaceae genera. Taxon 43: 201–231.

Liede, S. & H. Kunze. 1993. A descriptive system for corona analysis in the Asclepiadaceae. Pl. Syst. Evol. 185: 275–284.

Liede, S., U. Meve & P.G. Mahlberg. 1993. On the position of the genus *Karimbolea* Descoings. Amer. J. Bot. 80: 215–221.

Platnick, N.I., C.E. Griswold & J.A. Coddington. 1991. On missing entries in cladistic analysis. Cladistics 7: 337–344.

Rao, O.M. & O.L. Kumari. 1979. Germination loci of pollinia and their taxonomical significance. Geobios 6: 163–165.

Schill, R. & U. Jäkel. 1978. Beitrag zur Kenntnis der Asclepiadaceen-Pollinarien. Trop. Subtrop. Pfl. 22: 53–170.

Schumann, K. 1895. Asclepiadaceae. pp. 189–305 in Die natürlichen Pflanzenfamilien, 4, 2 (edited by A. Engler & K. Prantl). Engelmann, Leipzig.

Swofford, D.L. 1991. PAUP: Phylogenetic Analysis Using Parsimony, version 3.1. Computer program distributed by the Illinois Natural History Survey., Champaign, Illinois.

Volk, O.H. 1949. Zur Kenntnis der Pollinien der Asclepiadaceen. Ber. Dtsch. Bot. Ges. 62: 68–72.

APPENDIX 1.   CHARACTERS AND DATA MATRIX FOR THE TRIBES OF THE ASCLEPIADACEAE

CHARACTERS

1. Pollen in grains (0) – Pollen in tetrads (1) – Pollen in pollinia (2)

2. Translator apparatus absent (0) – Translator apparatus present (1)

3. Four anther sacs fertile (0) – Two anther sacs fertile (1)

4. Translator arms indistinctly developed (0) – Translator arms distinctly developed (1)

5. Pollinia erect (0) – Pollinia horizontal (1) – Pollinia hanging (2)

6. Endolateral pellucid germination pore absent (0) – Endolateral pellucid germination pore present (1)

ADDITIONAL CORONA CHARACTERS

7. Corolline corona (Cc) present (0) – absent (1)

8. Free staminal corona (Cs) absent (0) – present (1)

9. Corona of connate staminal and interstaminal parts (C(is)) absent (0) – present (1)

10. Annular corona absent (0) – present (1)

DATA MATRIX

Apocynaceae	00----	0000
Periploceae	11----	0000
Secamoneae	210000	0100
Fockeeae	211000	1110
Marsdenieae	21110{01}	1100
Stapelieae	211102	1111
Gonolobeae	211111	1111
Asclepiadeae	211121	{01}110

APPENDIX 2.   CHARACTER LIST AND DATA MATRIX

Inapplicable characters of Appendix 1 have been replaced by adding an additional primitive state for characters 3–6.

CHARACTERS

1. Pollen in grains (0) – Pollen in tetrads (1) – Pollen in pollinia (2)

2. Translator absent (0) – Translator present (1)

3. Four anther sacs fertile (0) – Two anther sacs fertile (1)

4. No translator arms (0) – Translator arms indistinctly developed (1) – Translator arms distinctly developed (2)

5. No pollinia (0) – Pollinia erect (1) – Pollinia horizontal (2) – Pollinia hanging (3)

6. No pollinia (0) – Endolateral pellucid germination pore absent (1) – Endolateral pellucid germination pore present (2)

Additional corona characters as in Appendix 1.

DATA MATRIX

Apocynaceae	000000	0000
Periploceae	110000	0000
Secamoneae	210111	0100
Fockeeae	211111	1110
Marsdenieae	21121{12}	1100
Stapelieae	211212	1111
Gonolobeae	211221	1111
Asclepiadeae	211231	{01}110

11 • Stamen structure and development in legumes, with emphasis on poricidal stamens of caesalpinioid tribe Cassieae

SHIRLEY C. TUCKER

INTRODUCTION

Of all the floral organs, stamens are perhaps the least frequently considered in morphological and systematic studies. They are considered to be rather uniform throughout angiosperms, although papers by Schmid (1976) and Endress and colleagues show the wide diversity among stamens (Endress & Hufford 1989; Endress & Stumpf 1991; Hufford & Endress 1989). As part of a long-term comparative study of floral development in legumes (summarized in Tucker 1987, 1989, 1992b), this chapter will consider the range of diversity of stamens, and examine some aspects of the underlying developmental processes that bring about the diversity.

METHODS AND MATERIALS

Inflorescences and floral buds of various sizes were collected and immediately fixed in FAA (5 parts formalin, 5 parts acetic acid, 90 parts 70% ethyl alcohol). The material was transferred to 70% ethanol for storage. For dissection, buds were dehydrated to 95% ethanol to give them a crisp texture. Buds of various sizes were dissected, removing outer organs to reveal all developmental stages. The resultant pieces were further dehydrated through an ethanol–acetone series, critical point dried in a Denton DCP-1 apparatus, mounted on aluminum stubs with colloidal graphite or other products, and coated with gold–palladium in an Edwards S-150 sputter coater. The preparations were studied and micrographs taken with a Cambridge S-260 scanning electron microscope (SEM) at 25 kV.

Sections of buds were also made and studied. Buds of various sizes were dehydrated through a tertiary butyl alcohol series, embedded in paraffin, and sectioned at 7–8 μm. They were mounted and stained using safranin and alcian blue. A list of taxa studied and sources is given in the Appendix.

RESULTS

The parameters of stamen diversity are: shape and surface features (hairs, glands, stomata, etc.); size, fusion, loss or suppression of organs; filament features; connective features; androecial heteromorphy; and type of dehiscence. Exomorphic features that help to define particular taxonomic groups are of especial interest. There are internal features (endothecium, number of sporangia) that are of interest as well (see Endress & Stumpf 1991), but these will not be considered here.

The leguminous flower typically is pentamerous, with single whorls of sepals and petals, two whorls of stamens, and a single carpel, although exceptions to all of these are known. The two whorls of stamens are initiated successively, although at maturity all appear to be part of a single whorl of ten. The order within each whorl differs according to the taxon: simultaneous in taxa of subfamily Mimosoideae, and unidirectional in most taxa of subfamily Papilionoideae. In the third subfamily, Caesalpinioideae, order within each stamen whorl may be either helical or unidirectional.

Shape

Most leguminous stamens have a distinct anther and filament (Figs. 1–4), the former usually versatile. The anther is usually dorsifixed, although basifixed anthers also occur: *Aeschynomene katangense* (Fig. 37) and

Myroxylon balsamum (Fig. 3) among papilionoids, *Bauhinia malabarica* (Fig. 14), *Senna* spp. (*S.* × *floribunda*, Fig. 61), and other taxa in tribe Cassieae among caesalpinioids, as well as some mimosoids. In New World Psoraleeae the two whorls of anthers are said to be alternately basifixed and dorsifixed (Grimes 1990), although I have not been able to confirm it. All anthers in Leguminosae are tetrasporangiate. Dehiscence is usually introrse, although taxa of the *Sclerolobium* group of caesalpinioid Caesalpinieae have lateral dehiscence and taxa of caesalpinioid tribe Cassieae exhibit dehiscence by pores or short slits (Fig. 9). Stamens of the papilionoid tribe Sophoreae (Figs. 1–4) are shown as examples of relatively unspecialized leguminous stamens. The ten stamens are free, in an apparent single whorl, and with flared filament bases.

Developmentally, legume anthers are always basifixed at first (Figs. 33, 36, 38). In most taxa, the anthers later elongate basally and the connective expands abaxially so that the filament attachment becomes displaced to an adaxial position (Figs. 35, 37, 39). In others (Fig. 3), the basifixed position during development persists.

Anther features

The anther surface usually is smooth in most papilionoid taxa (Fig. 9, *Cassia javanica*) but is papillate or verrucose in *Ateleia herbert-smithii* (Fig. 4), *Baphia chrysophylla*, *Castanospermum australe*, *Cercis canadensis* (Fig. 7), *Crotalaria mucronata*, *Desmodium lineatum*, *Eysenhardtia texana*, and *Lupinus affinis*. The surface is verrucose with individual epidermal cells bulging outward in some mimosoids (*Albizia carbonaria*, Fig. 19; *Calliandra houstoniana*, Fig. 24; *Dichrostachys cinerea*, *Pararchidendron pruinosum*, Figs. 20, 21) and the caesalpinioids *Berlinia tomentella*, *Cercis canadensis*, *Chamaecrista nicitans*, *Detarium microcarpum* (Fig. 18), *Eurypetalum tessmannii*, *Gillettiodendron pierreanum*, *Haematoxylon campechianum* and *H. brasiletto*, *Oddoniodendron micranthum*, *Tessmannia africana*. The abaxial sporangia in *Gleditsia triacanthos*, *Petalostylis labicheoides* (Figs. 5, 6), and *Tamarindus indica* are larger than the adaxial ones at anthesis.

Trichomes are rare on the anther, but in species of *Bauhinia* (*B. binata*, *B. malabarica*, *B. thonningii*; Figs. 14, 15), there are abundant short, broad, uniseriate trichomes over the sporangial surfaces on one or both sides. These trichomes are also present on the petals, stamens,

carpel, stem and leaves in various species, and have been shown to be hollow (Tucker *et al.* 1984). Other taxa having trichomes on the anthers include *Moldenhauera*, *Harleyodendron*, *Leucaena*, and *Mucuna*.

Dehiscence

Most leguminous stamens dehisce by introrse or, less commonly, lateral slits. The slit may extend over the upper and lower shoulder in mimosoids such as *Albizia* and *Archidendron* (Endress & Stumpf 1991). The same authors say that in Caesalpinioideae and Papilionoideae, the dehiscence line extends only over the upper shoulder of each theca, except in *Genista tinctoria* and *Baptisia*, where the line extends also over the lower shoulder. The dehiscence line does, however, extend over both upper and lower shoulders in caesalpinioids *Berlinia tomentella*, *Brownea latifolia*, *Crudia*, *Cynometra ananta*, *Detarium microcarpum*, *Gilbertiodendron ogouense*, *Gillettiodendron pierreanum*, *Gleditsia triacanthos*, *Tamarindus indica*, and *Tessmannia africana*, and in papilionoids *Amorpha fruticosa*, *Apios americana*, *Calpurnea aurea*, *Castanospermum australe*, *Dioclea mucronata*, *Erythrina*, *Pisum sativum*, and *Wistaria* × *formosa*.

Poricidal dehiscence is found in taxa in *Chamaecrista* and *Senna* (Cassieae tribe of Caesalpinioideae), to be discussed later. Stamen form in these taxa is often considered diagnostic (Bentham 1871; Irwin 1964; Lasseigne 1979; Irwin & Barneby 1981, 1982).

Filament features

Filaments are usually thin, although the base may be larger in diameter (Figs. 1, 2, *Sophora*, *Castanospermum*). Filament length may vary between members of the two stamen whorls (Fig. 47). The surface is usually smooth but may have bulging cells (Figs. 20, 21, 24). Hairs are present on the filament base in many taxa of legumes such as caesalpinioids *Afzelia quanzensis*, *Bussea*, *Caesalpinia*, *Cercidium*, *Colvillea*, *Detarium microcarpum*, *Gymnocladus*, *Parkinsonia*, *Peltophorum*, *Sclerolobium*, *Stenodrepanum*, *Stuhlmannia*, and *Wagatea* (Hutchinson 1964), and papilionoids *Airyantha*, *Baphia chrysophylla* (also Solodoye 1985), *Lonchocarpus*, and *Sophora* (*S. davidii*, Fig. 16). Hairs are usually absent on mimosoid stamen filaments.

In heteromorphic androecia, stamen form varies within each flower. In three of the ten stamens of *Cassia javanica*, the filaments are dilated in one area (Fig. 17). These anthers have sigmoid-shaped filaments and are much larger than the other stamens. Dilated filaments are also found in *Lotus purpurea, L. scoparius* and *Trifolium ochroleucum*.

The stamen filaments of many mimosoids have two distinctive features: they are long and coiled in the bud (Figs. 19, 23, 25; *Albizia carbonaria, Calliandra houstoniana, Dichrostachys cinerea, Inga tonduzii, Mimosa pudica*), and anthers are caducous (*Neptunia pubescens, Dichrostachys cinerea*). A narrow isthmus is common at the top of the filament (Fig. 22) that enables the entire anther to abscise. The coiled state of stamen filaments is associated with packing numerous elongate filaments into the bud (Figs. 19, 25). Caesalpinioid flowers also show coiled or declinate filaments in bud (Fig. 18, *Detarium microcarpum*; *Hymenostegia floribunda, Gilbertiodendron ogouense, Intsia bijuga*, and *Pentaclethra*), but these filaments are thicker, with fewer coils than those in mimosoid taxa.

Connective features

A connective may be present (Fig. 3, *Myroxylon balsamum*), or absent, where the sporangia are abutting (Figs. 9, 10, *Cassia javanica, Chamaecrista nicitans; Sophora* species, *Tamarindus indica, Peltophorum, Haematoxylon*). Inflated or locally thickened connectives are found in species of *Cassia sensu stricto, Galega, Hedysarum, Lathyrus, Onobrychis, Pisum*, and *Vicia* (in part from Endress & Stumpf 1991). The tip of the connective may be unornamented (Fig. 37, *Aeschynomene; Ceratonia, Pisum*), or it may have a small knob (*Castanospermum, Afzelia quanzensis*), an acute protrusion (Fig. 12, *Indigofera hirsuta; Myroxylon balsamum, Robinia hispida*), a tuft of hairs (*Indigofera incarnata*, Fig. 11); or a gland (*Burkea africana, Dalea, Petalostemon, Neptunia pubescens*, Fig. 22; *Pentaclethra macrophylla* and *P. macroloba*, Fig. 13). Terminal connective glands are found in about 25 mimosoid genera in three tribes (Adenanthereae, Mimoseae pr. p., Parkieae: Elias 1981; Tucker 1988a). Function of the glands has not been investigated.

Size

The stamens of Mimosoideae are very small, while those of Caesalpinioideae and Papilionoideae vary greatly from small to large (Endress & Stumpf 1991). Anthers of *Medicago lupulina* are 0.3 mm long (Endress & Stumpf 1991); *Bauhinia petersiana* has anthers 2 cm long, with total stamen length 7 cm (personal observation). Bird-pollinated sophoroid taxa such as *Castanospermum australe* (Fig. 2), *Myroxylon balsamum* (Fig. 3; Tucker 1993), and *Erythrina* also have very large anthers. Stamens of *Erythrina arborescens* are 4.6 cm long. Within one genus, *Sophora* of the papilionoid tribe Sophoreae, anther length differs greatly between different species; the largest stamens (section Edwardsia) are associated with bird-pollination (Tucker 1994).

Fusion

The most common specialization of stamens in legumes is fusion of the filaments, found in most taxa of Papilionoideae (except tribe Sophoreae), in tribe Ingeae of Mimosoideae, and to a lesser extent in tribes Amherstieae and Detarieae of Caesalpinioideae (Cowan 1981). A slight amount of basal fusion is found among stamens of some Mimoseae and Acacieae.

Flowers of most papilionoids (excepting Sophoreae) generally show one of three types of androecial fusion: monadelphy, diadelphy, or pseudomonadelphy. In all, the stamens originate individually as free organs (Figs. 33, 36, 38). Later in development, there is zonal (intercalary) growth below the stamen filament bases which raises the stamens upward on a 'fused' cylindrical base. The difference between the first two types of fusion is the extent of zonal growth; in monadelphy (e.g., *Aeschynomene katangense*, Fig. 37), all ten stamens are raised on a continuous cylindrical base, while in diadelphy (e.g., *Clitoria fairchildiana*, Figs. 34, 35), the cylinder includes only nine stamens, leaving one free. In pseudomonadelphy (e.g., *Erythrina caffra*, Figs. 38–40), the androecium goes through a diadelphous stage, followed by surface fusion (cell-to-cell appression) of the tenth stamen to the free sides of the cylinder.

Of 23 papilionoid tribes showing a stamen cylinder, 16 show monadelphy, 20 show diadelphy, and 10 show

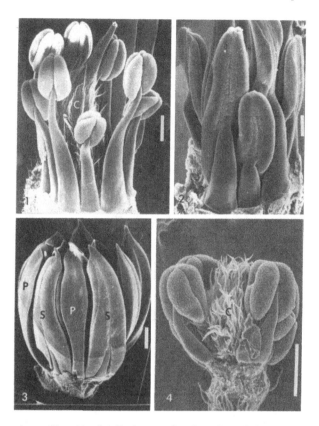

Figs. 1–4. Androecia of taxa in papilionoid tribe Sophoreae. Sepals and petals have been removed in all. Fig. 1. *Sophora secundiflora*. Free stamens of various heights with flared filament bases. Fig. 2. *Castanospermum australe*. Dorsifixed stamens; the inner whorl shorter than the outer. Fig. 3. *Myroxylon balsamum*. Basifixed stamens with an acutely pointed connective. Fig. 4. *Ateleia herbert-smithii*. Stamens with dorsifixed anthers having a verrucose surface texture. C, carpel; S, stamen. Scale bar represents 500 μm.

pseudomonadelphy. There is considerable overlap between diadelphy and pseudomonadelphy in the same tribe, but very little between monadelphy and pseudo-monadelphy. The key to this discrepancy is that diadel-phy is an intermediate stage to pseudomonadelphy. Developmental differences and taxonomic distribution of these three types of staminal fusion have been discussed previously (Tucker 1987, 1989). Other variations are reported in type of fusion among stamens, such as the fluctuation over time in flowers of New World Psoraleeae (Grimes 1990). The vexillary stamen is free at inception, then becomes fused above to the tube, and then, with age, becomes free again on one or both sides of the fila-

ment tube. This temporal and temporary fusion is likely due to surface fusion by interlocking epidermal cells, as described by Tucker (1984).

The term 'stemonozone' was coined by Lewis & Elias (1981) for the basal tube bearing petals and stamens in some taxa, particularly of mimosoids. It usually does not include the calyx or the gynoecium. The stemonozone forms after petal and stamen initiation, by intercalary growth below the bases of all petal and stamen primordia.

Flowers of the caesalpinioids *Bauhinia divaricata* (Fig. 32), *B. malabarica* (Fig. 29), *Haematoxylon* (two species), and *Tamarindus indica* show an androecial sheath basally,

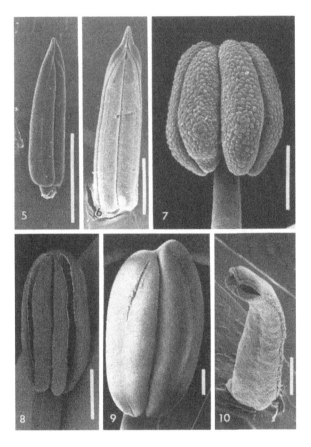

Figs. 5–10. Anther structure. Figs. 5, 6. *Petalostylis labicheoides*. Abaxial sporangia are longer than the adaxial pair. Fig. 7. *Cercis canadensis*. Verrucose surface of anthers. Fig. 8. *Baphia chrysophylla*. Sporangia have dehisced by longitudinal slits. Fig. 9. *Cassia javanica*. Dehiscence is by two slits, at top and bottom of the sporangia. Fig. 10. *Chamaecrista nicitans*. Porate anther. Scale bar represents 1 mm in Figs. 5, 6, 9; 500 μm in Figs. 7, 8, 10.

the result of zonal growth after initiation of stamen primordia as free structures.

Changes in organ number

Loss or suppression

Decrease in stamen number from the usual ten occurs in caesalpinioid tribes Amherstieae (*Cryptosepalum, Didelotia, Macrolobium, Pellegriniodendron*), Detarieae (*Augouardia, Hylodendron, Saraca declinata*, Fig. 26; *Stemonocoleus*), Caesalpinieae (*Acrocarpus*: Hutchinson 1964; *Arcoa, Dimorphandra, Gleditsia*: Tucker, 1991; *Mora, Tetrapterocarpon*), Cassieae (*Ceratonia, Dialium*, Fig. 28; *Dicorynia, Koompassia, Labichea, Martiodendron, Petalostylis, Storkiella* pr. p.), and Cercideae (*Bauhinia*, Figs. 29, 32; *Brenierea*). Among Mimosoideae, species of *Desmanthus* pr. p. and *Parkia* have only five stamens, plus 5–15 staminodia. Stamen number can decrease due to either of two developmental sequences: initiation followed by suppression, or absence of initiation (Tucker 1988c). Some stamens are suppressed after initiation in *Bauhinia divaricata* (Fig. 32; Tucker 1988b); all stamens are suppressed after initiation in functionally female flowers of *Ceratonia siliqua* (Tucker 1992a) and *Bauhinia malabarica* (Fig. 29; Tucker 1988b).

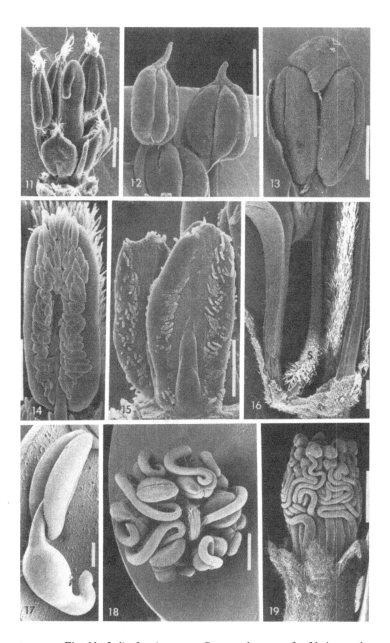

Figs. 11–19. Stamen structure. Fig. 11. *Indigofera incarnata*. Stamens have a tuft of hairs on the connective tip. Sepals and petals have been removed. Fig. 12. *Indigofera hirsuta*. The connective tip is narrow and attenuate. Fig. 13. *Pentaclethra macroloba*. The connective is a massive glandular structure. Figs. 14, 15. Hairs on anthers of *Bauhinia* species. Fig. 14. *Bauhinia malabarica*. Fig. 15. *Bauhinia binata*. Fig. 16. *Sophora davidii*. Stamen filament (S) is hairy toward base. Fig. 17. *Cassia javanica* with dilated filament. Fig. 18. *Detarium microcarpum*, polar view of flower with sepals and petals removed. Fig. 19. *Albizia carbonaria*. Side view of flower in bud, with petals removed. Scale bar represents 500 µm in Figs. 11–15; 1 mm in Figs. 16–19.

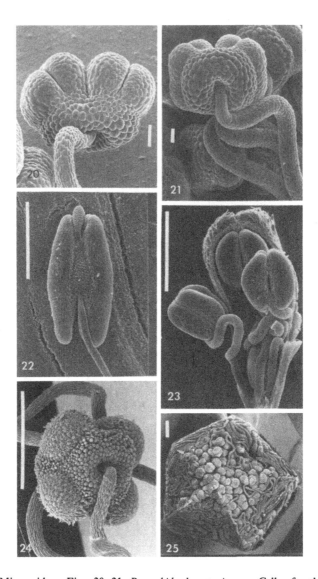

Figs. 20–25. Stamens of Mimosoideae. Figs. 20, 21. *Pararchidendron pruinosum*. Cells of anther and filament are bulging. Fig. 22. *Neptunia pubescens*. Functional stamen with gland on connective. Fig. 23. *Mimosa pudica*, with coiled filaments and anther with bulging cells. Figs. 24, 25. *Calliandra houstoniana*. Fig. 24. Anther with ornamented epidermal pattern. Fig. 25. Flower bud showing numerous stamens fused basally. Sepals and petals have been removed. Scale bar represents 50 μm in Figs. 20, 21; 500 μm in Figs. 22–25.

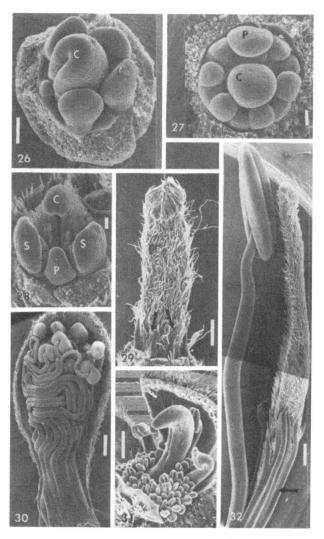

Figs. 26–32. Flowers with reduced or proliferated number of stamens. Sepals and petals have been removed in all. Fig. 26. *Saraca declinata*. Flower with only four stamens; a rudiment of the fifth member of the whorl is visible at left next to the carpel. The second whorl of five stamens does not form in this species. Fig. 27. *Ateleia herbert-smithii*. Flower bud (calyx removed) in which stamen primordia are initiated in erratic order, different from most other Papilionoideae. Fig. 28. *Dialium guineense*. Flower (calyx removed) with one petal, two stamens, and the carpel. No other organ primordia are initiated. Fig. 29. *Bauhinia malabarica*, carpellate flower with staminodia at base (arrows). Fig. 30. *Samanea saman*. Numerous stamens, with their filaments fused basally. Fig. 31. *Swartzia macrosperma* has three large stamens, then produces 60–70 small stamens. Fig. 32. *Bauhinia divaricata*. Staminodial tube at arrows. C, carpel; P, petal; S, stamen. Scale bar represents 50 μm in Figs. 26–28; 500 μm in Figs. 29–32.

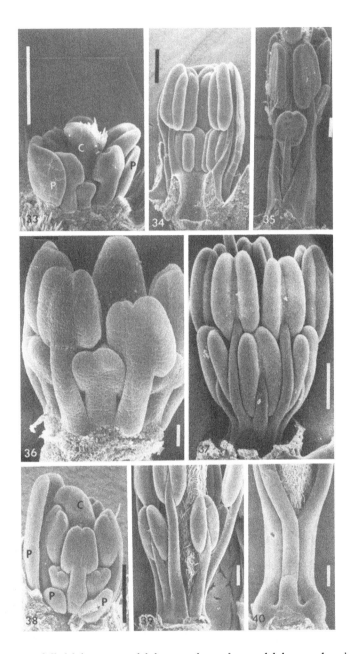

Figs. 33–40. Development of diadelphous, monadelphous, and pseudomonadelphous androecial tubes in papilionoids. Sepals and petals have been removed in all. Figs. 33–35. Development of diadelphy in *Clitoria fairchildiana*. Fig. 33. Free stamens at midstage of development. Fig. 34. Zonal growth has formed a short incomplete stamen tube including nine stamens. Fig. 35. Nine stamens are fused into a cylindrical partial tube, while the tenth is free. Figs. 36, 37. Development of monadelphy in *Aeschynomene katangense*. Fig. 36. Free stamens at midstage in development. Fig. 37. Zonal growth below the ten stamen bases has produced a continuous stamen tube. Figs. 38–40. Development of pseudomonadelphy in *Erythrina caffra*. Fig. 38. Free stamens at midstage in development. Fig. 39. Zonal growth below nine of the stamens has produced an intermediate diadelphous stage. Fig. 40. The nine-stamen tube has fused with the tenth stamen to form a continuous tube. C, carpel; P, petal. Scale bar represents 50 μm in Fig. 36; 500 μm in Figs. 33–35, 37–40.

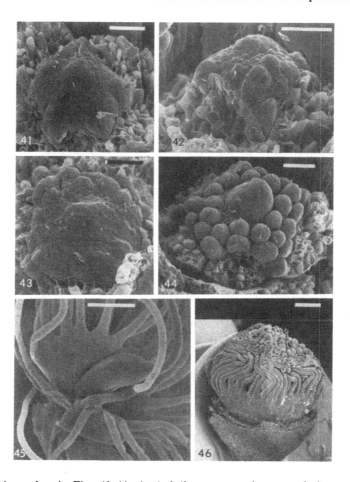

Figs. 41–46. Proliferating androecia. Figs. 41–44. *Acacia baileyana*, successive stages during stamen initiation. Sepals and petals have been removed. Fig. 41. Five equidistant stamen primordia are located around the young carpel. Fig. 42. Additional stamen primordia are being initiated around each of the five original primordia (side view). Fig. 43. Polar view of flower with stamens proliferating on all sides of the central carpel, after the five centers of initiation become confluent. Fig. 44. Oblique view of the flower after all stamen primordia have been initiated. Fig. 45. Androecium of *Zapoteca lambertiana* showing the fused basal part of the androecial tube. Fig. 46. *Calliandra houstoniana*, oblique polar view of flower showing the numerous stamens. Scale bar represents 25 µm in Figs. 41–44; 500 µm in Fig. 45; 1 mm in Fig. 46.

Some mimosoids have only four (e.g., *Mimosa pudica*) or five stamens (e.g., *Desmanthus illinoiensis*, *Neptunia gracilis* and other Australian species) rather than the usual ten, a decrease caused by the absence of initiation of the second whorl. In *N. pubescens* (Tucker 1988a), one of the three floral morphs, the neuter flower, lacks stamens altogether. Stamen primordia are initiated in this morph, but they develop as showy petalloid staminodia (Fig. 52).

Proliferation of organs

A stamen number greater than ten is found in the mimosoid tribes Acacieae and Ingeae (Figs. 25, 30, 45, 46), a few taxa in caesalpinioid tribes Caesalpinieae (*Campsiandra*), Cassieae (*Mendoravia*, *Storkiella* pr. p.), Detariae (*Brownea*, *Browneopsis*, *Colophospermum*, *Maniltoa*), and Amherstiae (*Englerodendron*, *Polystemonanthus*, *Pseudomacrolobium*), and in the anomalous tribe

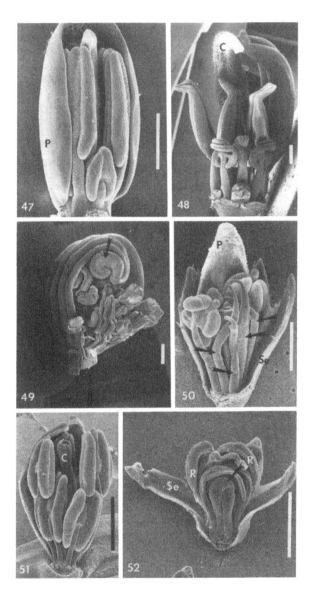

Figs. 47–52. Heterostameny in legumes. Sepals and petals have been removed except in Fig. 52. Fig. 47. *Genista tinctoria*, with two whorls of stamens of different size classes. Fig. 48. *Senna nicaraguensis* flower with large functional stamens at the rear, and small bizarrely shaped staminodia in the foreground. Fig. 49. *Swartzia simplex* with two sizes of stamens. Arrow indicates one of the three large stamens. Figs. 50–52. *Neptunia pubescens*. Fig. 50. Staminate flower with staminodia at arrows. Fig. 51. Hermaphrodite flower. Fig. 52. Neuter flower, with stamens modified as petalloid staminodia at arrows. C, carpel; P, petal; Se, sepal. Scale bar represents 500 μm in Figs. 47, 50–52; 1 mm in Figs. 48, 49.

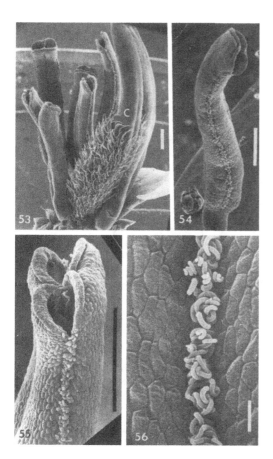

Figs. 53–56. Poricidal stamens of *Chamaecrista nicitans*. Fig. 53. Androecium and carpel (C). Sepals and petals have been removed. Fig. 54. Individual stamen with paired apical pores and lateral lines of hairs. Fig. 55. Tip of stamen with open pores. Fig. 56. Line of interlocking hairs along side of anther. Scale bar represents 500 μm in Figs. 53–55; 50 μm in Fig. 56.

Swartzieae (*Swartzia macrosperma*, Fig. 31) currently associated with Papilionoideae. The developmental processes by which stamen proliferation occurs in legumes vary greatly, indicating parallel evolution for this character. In *Acacia baileyana* (Figs. 41–44) the initiation of stamens after the first five is sectorial, with proliferation of new primordia around all sides of each of the first five stamens (Derstine & Tucker, 1991). In addition to the

sectorial pattern, Ramírez-Domenech (1989) reported the following additional patterns of stamen proliferation among mimosoids: lateral multistameny, acropetal multistameny, and helical sequence. Another pattern, a ring meristem, was described for *Acacia neriifolia* by Gemmeke (1982). Stamen proliferation in three species of *Acacia* was described by Newman (1933a, b, 1934, 1936) but the study was based on sections, so that the pattern of organ proliferation over the apical surface could not be determined. After initiation of the numerous stamens, zonal growth below their bases raises them up, resulting in an apparent fusion (Figs. 45, 46).

The five types of polyandry reported by Ronse de Craene and Smets (1987) among angiosperms include the helical, lateral and acropetal multistameny recorded by Ramírez-Domenech (1989), but not sectorial, in the sense of acropetal, lateral, and even basipetal proliferation from the first five points. One type of polyandry of Ronse de Craene and Smets, in which additional whorls are initiated after the first, does not occur among legumes.

Order of stamen initiation

Each whorl of stamens shows one of three patterns of inception: helical, simultaneously whorled, or unidirectional. Mimosoideae all have simultaneously whorled initiation of stamens, and Papilionoideae all have unidirectional order. Both patterns are represented among Caesalpinioideae. In *Ateleia* (Fig. 27; Tucker 1990) order of stamen initiation is erratic. Previously placed in the papilionoid tribe Sophoreae, several anomalous features of *Ateleia* floral development suggest that it is closer to Swartzieae, where it is now placed by Polhill (1994).

Androecial heteromorphy

In many legumes, members of the two stamen whorls differ somewhat, commonly in size, filament length (*Centrosema virginica*, *Galega orientalis*; *Genista tinctoria*, Fig. 47), anther shape, or time of dehiscence. Dimorphy is more evident in *Swartzia macrosperma* (Fig. 31) and *S. simplex* (Fig. 49) in which there are three large stamens and about 60–70 small ones. *Moldenhauera* (Caesalpinioideae) has one stamen much larger than the rest.

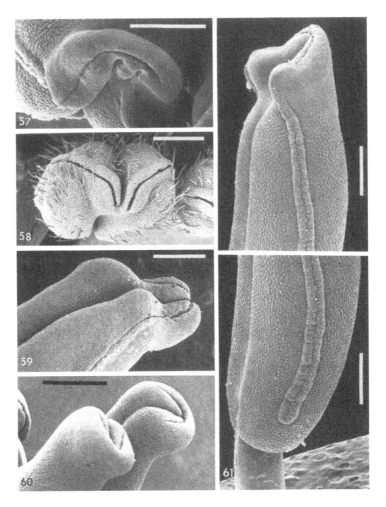

Figs. 57–61. Poricidal stamens of *Senna*. Figs. 57–60. Variation in the apical stamen pores of different *Senna* species. Fig. 57. *S.* × *floribunda*. Fig. 58. *S. bicapsularis*. Fig. 59. *S. nicaraguensis*. Fig. 60. *S. pendula*. Fig. 61. Individual stamen of *S.* × *floribunda*, with lateral ridge. Scale bar represents 500 μm.

Heterostameny occurs in the zygomorphic flowers of *Cassia sensu stricto* and *Senna* (Caesalpinioideae: Cassieae: Cassiinae). Differentiation occurs along the median sagittal plane of the flower with three or four stamen morphs present (Fig. 48, *Senna nicaraguensis*). Although the two genera are superficially alike in zygomorphy and dorsiventral distinctions among the stamens, the two differ in a way that suggests convergence rather than close relationships. The largest two stamens in *Senna* species are abaxial in the inner whorl, while the three large sta-

mens in the flowers of *Cassia* species are abaxial in the outer whorl. Some of the other stamens are modified, such as 'feeder' stamens.

Heterostameny occurs in a different form in *Neptunia pubescens* (Tucker 1988a), in which the inflorescences include three floral morphs along a vertical sequence: perfect in the upper half, male below, and neuter at the bottom. Normal stamens are present in perfect (Fig. 51) and male flowers (Fig. 50), but the number is diminished in male flowers. The remaining stamen primordia of male flowers differentiate as staminodia or transitional structures between staminodia and normal stamens (Fig. 51). In the neuter flower (Fig. 52) all stamen primordia differentiate as petalloid staminodia.

Staminodia, stamen homologues originating in stamen sites but not differentiating as stamens, are found in many caesalpinioid legume flowers, for example, in Amherstieae (*Amherstia nobilis*, *Anthonotha*, *Brachystegia*, *Cryptosepalum*, *Didelotia*, *Englerodendron*, *Gilbertiodendron ogouense*, *Humboldtia*, *Macrolobium*, *Paramacrolobium*, *Pellegriniodendron*, Hutchinson 1964; *Tamarindus indica*), Caesalpinieae (*Arcoa*, *Mora*, *Dimorphandra*), Cassieae (*Distemonanthus*, *Kalappia*, *Petalostylis*, *Zenia*); Cassiinae (*Cassia sensu stricto*, *Chamaecrista*; *Senna*, Fig. 48; Irwin & Barneby 1981, 1982), Cercideae (*Bauhinia*; Tucker 1988b), Detarieae (*Augouardia*, *Brachycylix*, *Elizabetha*, *Heterostemon*, *Leucostegane*, *Paloveopsis*, *Saraca*, *Sindora*) and in mimosoids (*Dichrostachys*; *Neptunia*, Fig. 50; *Pentaclethra*; Elias 1981).

Cassieae

Stamen specialization in Cassiinae of the caesalpinioid tribe Cassieae exhibits parallel evolution. Many of the taxa are buzz-pollinated by large bees that vibrate the flowers and dislodge the pollen out of the anther pores or slits (Hardin *et al.* 1972; Buchmann 1974, 1983; Thorp & Estes, 1975; Delgado *et al.* 1977; Fontanelle 1979; Dulberger 1981; Saradhi & Mohan Ram 1981; Gottsberger and Silberbauer-Gottsberger 1988; Wolfe & Estes 1991). Some species of *Chamaecrista* (Figs. 53–56) and *Senna* (Figs. 57–61) have evolved paired apical pores, the structure of which differs considerably among species (Figs. 57–60; Harris 1905; Venkatesh, 1956a,b, 1957; Lasseigne 1979). These pores are found only in the larger stamens of each flower, while smaller stamens in the flower are staminodial and lack pores (Fig. 48). The apical pores are specialized structures located on the upper shoulders. Anthers of *Cassia sensu stricto* are visibly different (Fig. 9) from those with pores, although their restricted slits may function in a similar way. These restricted slits may be at the top, the base, or in both positions on the anther (Fig. 9) depending on the species. In place of the lateral lines of dehiscence, *Chamaecrista* has rows of interlocked hairs (Fig. 56), while *Senna* (Fig. 61) has a prominent ridge.

DISCUSSION

The current work shows that minor characteristics of stamen structure, such as nature of the connective, anther surface texture, and shape of the filament, as seen with scanning electron microscopy, vary considerably within Leguminosae. At least some of these small differences have functional significance, as in the tripping mechanism that repositions stamens in some papilionoids (Faegri & van der Pijl, 1979).

Regardless of the variability in stamen structure, Kunze (1978) used developmental evidence to assert that angiosperm stamens of greatly differing mature form are homologous. He showed that they all begin as a bifacial primordium that develops by marginal growth. The marginal meristems split lengthwise, continue to grow, and eventually form microsporangia. Legumes show similar patterns of development.

Poricidal anthers

Flowers having poricidal anther dehiscence belong to 24 orders of angiosperms (Buchmann, 1983) including families Solanaceae and Leguminosae (tribe Cassieae). The activities of the bees in vibrating flowers are described by Harris (1905), Vogel (1978), and Buchmann (1983). Buzz-pollination of flowers of *Cassia* or *Chamaecrista* species is described by Buchmann (1974), Delgado *et al.* (1977), Dulberger (1981), Gottsberger & Silberbauer-Gottsberger (1988), Hardin *et al.* (1972) Salinas & Sanchez (1977), Saradhi & Mohan Ram (1981), Thorp & Estes (1975), Venkatesh (1956a,b, 1957) and Wolfe & Estes (1991).

Heteranthery and enantiostyly, commonly associated with buzz-pollination, are both found in some members of Cassiinae. Vogel (1978) calls these pollen flowers, which use deception as well as reward in the form of attractive fodder stamens. The heterantherous flowers have a few unusually large and showy anthers that contain copious pollen for dispersal, plus fodder stamens that act as food for the pollinators (Buchmann 1983). Vogel (1978) notes that the deception continues after pollen is completely removed, since the papery yellow pollen sac walls persist and continue to attract pollinators.

Closure of the lateral slits in poricidal anthers appears

to be partly mechanical in *Chamaecrista*, where there is a line of interlocking hairs along the position of the slit. Histological work in progress in our laboratory indicates that the hairs form quite late in development, and that no internal tissue resembling a stomium is present.

ACKNOWLEDGMENTS

This work was supported in part by National Science Foundation grant DEB9207671. The author thanks Andrew Douglas for his skilled technical assistance. She also thanks the following people and institutions for facilitating collections of material: J.E. Armstrong, F.J. Breteler, B. Browning, B. Jackes, D. Janzen, A. Lievens, G. Lewis, M. Wiemann, J. Pruski, and O.L. Stein; Royal Botanic Gardens, Kew; Fairchild Tropical Garden, Miami, Florida; Missouri Botanical Garden, St Louis; Foster Botanic Garden, Honolulu, Hawaii; Pacific Tropical Botanical Garden, Lawai, Kauai, Hawaii; National Botanic Garden, Harare, Zimbabwe. Joe Armstrong and James Grimes were helpful in suggesting revisions. The following gave permission to use previously published photographs: *American Journal of Botany, Missouri Botanical Garden Monographs in Systematic Botany*; Royal Botanic Garden, Kew, and J. Cramer.

APPENDIX. TAXA STUDIED AND SOURCES

CAESALPINIOIDEAE

Afzelia quanzensis Welw. National Botanic Garden, Harare, Zimbabwe, B. Browning

Amherstia nobilis Wall. Lyon Arboretum, Tucker 26706, and Foster Garden, Tucker 26711, Honolulu, Hawaii; Wahiawa Botanic Garden, Wahiawa, Oahu, Hawaii, Tucker 30455

Arcoa gonavensis Urb. Dominican Republic: Sierra de Baoruco; Prov. Barahona, T. Zanoni and R. Garcia 40990

Bauhinia binata Blanco Cultivated, Fairchild Tropical Garden, Miami, Florida, Tucker 25179

B. divaricata L. Cultivated, Fairchild Tropical Garden, Miami, Florida, Tucker 25458

B. malabarica Roxb. Cultivated, Fairchild Tropical Garden, Miami, Florida, Tucker 25181, 25184

B. petersiana C. Bolle in Peters Cultivated, National Botanic Garden, Harare, Zimbabwe, Browning 52

B. thonningii Schum. Cultivated, Fairchild Tropical Garden, Miami, Florida, Tucker 25186, 26915

Berlinia grandiflora (Vahl.) Hutch. & Dalz. A.J.M. Leeuwenberg 9405, Tucker 32306

B. tomentella Keay J.J. Bos 1942, Tucker 32308

Brownea latifolia Jacq. Lyon Arboretum, Honolulu, Hawaii, Tucker 26708, 26710

Campsiandra angustifolia Spr. ex Benth. Mitu, Vaupes, Colombia, J. Zarucchi 1860

Cassia javanica L. Cultivated, Fairchild Tropical Garden, Miami, Florida, Lievens and Gregory 2001; Manoa campus, Univ. Hawaii, Honolulu, Hawaii, Tucker 26702

Ceratonia siliqua L. Fairchild Tropical Garden, Miami, Florida, Tucker 25183; Univ. California Arboretum, Davis, California, Tucker 28894

Cercis canadensis L. Burden Plantation, Baton Rouge, Louisiana, Tucker 26975

Chamaecrista nicitans var. *glabrata* (Vogel) Irwin and Barneby Adventive, Pacific Tropical Botanic Garden, Lawei, Kauai, Hawaii, Tucker 26671

C. nictitans (L.) Irwin and Barneby Villahermosa, Tabasco, Mexico, Lievens and Gregory 2346

Crudia choussyana Standl. La Libertad, El Salvador, Hughes 1249

Cynometra ananta Hutch. & Dalz. J. de Koning 5060, Tucker 32310

Detarium microcarpum Guill. & Perr. P. de Wit 3148, Tucker 32312

Dialium guianense (Aubl.) Sandw. Sacha Biol. Reserve, Misahualli, Napo Prov., Ecuador, Palacios 1364

D. guineense Willd. National Botanic Garden, Harare, Zimbabwe, Browning and Mur 40

Dimorphandra gardneriana Tul. G. Lewis *et al.* 1349, Tucker 26958

Eurypetalum tessmannii Harms F.J. Breteler 11368, Tucker 32317

Gilbertiodendron ogouense (Pellegr.) J. Leon J.J. de Wilde 8431; Tucker 32318

Gillettiodendron pierreanum (Harms) J. Leon F.J. Breteler & J.J. de Wilde 6939 (Tucker 32319)

Gleditsia triacanthos L. Missouri Botanical Garden, St Louis, Missouri, Tucker 28270; Normal, Illinois, J.E. Armstrong 1107; Royal Botanic Gardens, Kew, Tucker 24735, 24759; Baton Rouge, Louisiana, A. Lievens, 1990

Haematoxylon campechianum L. Fairchild Tropical Garden, Miami, Florida, Tucker 25340

H. brasiletto Karsten Motagua Valley, Guatemala, Lewis & Hughes 1711

Hymenostegia floribunda (Benth.) Harms   J.J. de Wilde *et al.*, 9032, Tucker 32320

Intsia bijuga (Colebr.) Kuntze   Waimea Botanical Garden, Oahu, Hawaii, A. Bruneau 929; A.J.M. Leeuwenberg 13258, Tucker 32321

Maniltoa pseudoobscura   Cultivated, Normal, Illinois, J.E. Armstrong, Mar. 4, 1991

Oddóniodendron micranthum (Harms) Bak.   J.J. de Wilde 7668, Tucker 32325

Parkinsonia aculeata L.   Cultivated, Baton Rouge, Louisiana, Tucker 30580, 30715

Peltophorum adnatum Griseb.   Fairchild Tropical Garden, Miami, Florida, Tucker 25474

Petalostylis labicheoides R. Br.   Gammon Ranges, South Australia, Grimes 3208

Saraca declinata Miq.   Lyon Arboretum and Foster Garden, Honolulu, Hawaii, Tucker 26709, 26715; 30453

S. indica L.   Cultivated, Royal Botanic Gardens, Kew, Tucker 24727

Senna bicapsularis (L.) Roxb.   Fairchild Tropical Garden, Miami, Florida, Tucker 28871

S. × *floribunda* (Cav.) Irwin and Barneby   Royal Botanic Gardens, Kew, Tucker 24698, 24832

S. nicaraguensis (Benth.) Irwin and Barneby   Cultivated, Foster Garden, Honolulu, Hawaii, Tucker 26714

S. pendula (Willd.) Irwin and Barneby   Puente Chilapilla, Tabasco State, Mexico, Lievens & Gregory 2374

Tamarindus indica L.   Fairchild Tropical Garden, Miami, Florida, Tucker 25175; Mwanza Dist., Tanganyika, R. Tanner 333

Tessmannia africana Harms   J.J. de Wilde 7716; Tucker 32236

Wagatea spicata Dalz.   Fairchild Tropical Garden, Miami, Florida, Tucker 26982; Lyon Arboretum, Honolulu, Hawaii, Tucker 25163, 25343

PAPILIONOIDEAE

Aeschynomene katangense de Wilde   Tanganyika, Watermeyer 145

Amorpha fruticosa L.   Cultivated, Royal Botanic Gardens, Kew, Tucker 24742

Apios americana Medic.   Thompson's Creek, West Feliciana Parish, Louisiana, Tucker 26584

Ateleia herbert-smithii Pittier   Guanacaste Prov., Costa Rica, Janzen, 1984

Baphia chrysophylla Taub. in Engl.   F.J. Breteler 8607, Tucker 32304

B. racemosa (Hochst.) Walp.   Natal Herbarium Garden, Durban, South Africa, Stirton 8035A

Baptisia australis (L.) R. Br.   Cultivated, Royal Botanic Gardens, Kew, Tucker 24692

Burkea africana Hook.   Marandellas Rhodesia, Corby 2163; comm. O.L. Stein; Tucker 25196

Calpurnia aurea (Pam.) Benth.   M. Lavin 6198, Royal Botanic Gardens, Kew; Africa, Tucker 28455

Castanospermum australe A. Cunn. & C. Fraser   Atherton, Queensland, Australia, J.E. Armstrong 16–85/86, 17–85/86

Clitoria fairchildiana Howard   Fairchild Tropical Garden, Miami, Florida, J. Pruski 2902

Crotalaria mucronata Desv.   Nigeria, Meikle 960, Kew 11455

Desmodium lineatum DC.   Idlewild Research Station, Clinton, Louisiana, Tucker 25811

Dioclea aff. *ucayalina* Harms   Peru: Junin Dept, Chanchamayo Prov., D.N. Smith 5624

Erythrina arborescens Roxb.   Cultivated, Waimea, Oahu, Hawaii Tucker 26435

E. caffra Thunb.   Cultivated, Missouri Botanical Garden, St Louis, Missouri, Tucker 26485

Eysenhardtia texana Scheele   Travis Co., Texas, Lievens 1530

Galega orientalis L.   Cultivated, Royal Botanic Gardens, Kew, Tucker 24690

Genista tinctoria L.   Cultivated, Royal Botanic Gardens, Kew, Tucker 24715

Hedysarum flavescens Regel.   Turkestan source, cultivated, Royal Botanic Gardens, Kew, Tucker 24708

Indigofera hirsuta L.   Alachua Co., Florida, Lievens 4119

Indigofera incarnata (Willd.) Nakai   Cultivated, Baton Rouge, Louisiana, Tucker 25441

Lathyrus grandiflorus Sibth. & Sm.   Cultivated, Royal Botanic Gardens, Kew, Tucker 24685

Lonchocarpus violaceus H.B. & K.   Fairchild Tropical Garden, Miami, Florida, Tucker 25171

Lotus scoparius L.   Morro Bay, San Luis Obispo Co., California, Tucker 28780

L. purpureus Webb. in Hook. (*Tetragonolobus purpureus*)   Cultivated, Royal Botanic Gardens, Kew, Tucker 24777

Lupinus affinis Agardh   Cultivated, Royal Botanic Gardens, Kew, Tucker 24723

Mucuna sloanei Fawc. & Rendle   Fairchild Tropical Garden, Miami, Florida, Tucker 25461, 26456

Myroxylon balsamum v. *pereirae* Harms   Laguna, El Salvador, G. Lewis; Fairchild Tropical Garden, Miami, Florida, Tucker 26981

Onobrychis rotundifolia Desv.   Cultivated, Royal Botanic Gardens, Kew, Tucker 24712

Ononis spinosa L.   Cultivated, Royal Botanic Gardens, Kew, Tucker 24800a, 24814, 28476

Petalostemon purpureum (Vent.) Rydb.   Normal, Illinois, J.E. Armstrong, June 3, 1986

Pisum sativum L. var. 'Sugar Snap' Cultivated, Baton Rouge, Louisiana, Tucker 25621

Robinia hispida L.   Cultivated, St Paul, Minnesota, Tucker 24871

Sophora davidii (Franch.) Skeels   Royal Botanic Gardens, Kew, Tucker 26896

S. secundiflora Lag. ex DC.   Cultivated, Univ. California Arboretum, Davis, California, Tucker 28775; cultivated, Baton Rouge, Louisiana, Tucker 25444

Swartzia macrosperma Bertal.   Bomborza, Morona, Santiago Prov., Ecuador, D. Neil 7565

S. simplex (Swartz) Spreng.   Heredia Province, Costa Rica, Wiemann T24

Trifolium ochroleucum Huds.   Cultivated, Royal Botanic Gardens, Kew, Tucker 24703

Wistaria × *formosa* Rehd.   Cultivated, Baton Rouge, Louisiana, Tucker 25655, 26862

MIMOSOIDEAE

Acacia baileyana F. Muell.   Royal Botanic Gardens, Kew, Tucker 24700, 24836; Univ. California Arboretum, Davis, California, Tucker 28898

Albizia carbonaria Britt.   Braulio Carrillo National Park, San Jose Prov., Costa Rica, M. Wiemann T19

Calliandra houstoniana (Miller) Standl.   Dept Izabal, Guatemala, Hughes 1271

Desmanthus illinoiensis (Michx.) MacM.   Baton Rouge, Louisiana, Tucker 25680

Dichrostachys cinerea (L.) Willd.   Cultivated, Tucson, Arizona, R. Gilbertson, Oct. 1985

Inga tonduzii J.D. Smith   Estacion La Selva, Heredia Prov., Costa Rica, Wiemann J26

Leucaena leucocephala (Lam.) de Wit   Cultivated, Royal Botanic Gardens, Kew, Tucker 24697

Mimosa pudica L.   Cultivated, Royal Botanic Gardens, Kew, Tucker 24830

Neptunia gracilis Benth.   Townsville, Queensland, Australia, B. Jackes, Mar. 6, 1988

N. pubescens Benth. in Hook.   Fairchild Tropical Garden, Miami, Florida, Tucker 25624, 25857

Pararchidendron pruinosum (Benth.) Nielsen   Curtain Fig State Forest, Atherton Tableland, Queensland, Australia, J.E. Armstrong 10–85/86

Pentaclethra macroloba (Wild.) Ktze.   Heredia Province, Costa Rica, Wiemann T22

P. macrophylla Benth.   Costa Rica, J. Denslow, 1988

Samanea saman (Jacq.) Merr.   Fairchild Tropical Gardens, Miami, Florida, Ramirez–Domenech and Lievens 2006

Zapoteca lambertiana (G. Don.) H.M. Hernandez   Mexico, Hernandez H871, cultivated at Missouri Botanical Garden, Hernandez, 1985

LITERATURE CITED

Bentham, G. 1871. Revision of the genus *Cassia*. Trans. Linn. Soc. Lond. 27: 503–591.

Buchmann, S.L. 1974. Buzz pollination of *Cassia quiedonilla* (Leguminosae) by bees of the genera *Centris* and *Melipona*. Bull. S. Cal. Acad. Sci. 73: 171–173.

Buchmann, S.L. 1983. Buzz pollination in angiosperms. pp. 73–113 in Handbook of Experimental Pollination Biology (edited by C.E. Jones & R.J. Little). Scientific Academic Editions, New York.

Cowan, R.S. 1981. Caesalpinioideae. pp. 35–54 in Advances in Legume Systematics, part 1 (edited by R.M. Polhill & P.H. Raven). Royal Botanic Gardens, Kew.

Delgado, S., A.O. Sanchez & M.S. Sanchez. 1977. Biología floral del género *Cassia* en la region de los Tuxlas, Veracruz. Bol. Soc. Bot. Mexico 37: 5–52.

Derstine, K.S. & S.C. Tucker. 1991. Organ initiation and development of inflorescences and flowers of *Acacia baileyana*. Amer. J. Bot. 78: 816–832.

Dulberger, R. 1981. The floral biology of *Cassia didymobotrya* and *C. auriculata* (Caesalpiniaceae). Amer. J. Bot. 68: 1350–1360.

Elias, T.S. 1981. Mimosoideae. pp. 143–151 in Advances in Legume Systematics, part 1 (edited by R.M. Polhill & P.H. Raven). Royal Botanic Gardens, Kew.

Endress, P.K. & L.D. Hufford. 1989. The diversity of stamen structures and dehiscence patterns among Magnoliidae. Bot. J. Linn. Soc. 100: 45–85.

Endress, P.K. & S. Stumpf. 1991. The diversity of stamen structures in 'Lower' Rosidae (Rosales, Fabales, Proteales, Sapindales). Bot. J. Linn. Soc. 107: 217–293.

Faegri, K. & L. van der Pijl. 1979. The Principles of Pollination Biology, 3rd ed. Pergamon Press, New York.

Fontanelle, G.B. 1979. Contribuição ao estudo de biologia floral de *Cassia silvestris* Vell.: considerações anatômicas. Leandra 8–9: 49–84.

Gemmeke, V. 1982. Entwicklungsgeschichtliche Untersuchungen an Mimosaceen-Blüten. Bot. Jahrb. Syst. 103: 185–210.

Gottsberger, G. & I. Silberbauer-Gottsberger. 1988. Evolution of flower structures and pollination in neotropical

Cassiinae (Caesalpiniaceae) species. Phyton (Austria) 28: 293–320.

Grimes, J.W. 1990. A revision of the New World species of Psoraleeae (Leguminosae: Papilionoideae). Mem. N. Y. Bot. Gard. 61: 1–114.

Hardin, J.W., G. Doerksen, D. Herndon, M. Hobson & F. Thomas. 1972. Pollination ecology and floral biology of four weedy genera in southern Oklahoma. Southwest. Nat. 16: 403–412.

Harris, J.A. 1905. The dehiscence of anthers by apical pores. Ann. Rep. Missouri Bot. Gard. 16: 167–257.

Hufford, L.D. & P.K. Endress. 1989. The diversity of anther structures and dehiscence patterns among Hamamelididae. Bot. J. Linn. Soc. 99: 301–346.

Hutchinson, J. 1964. The Genera of Flowering Plants (Angiospermae). Vol. I. Dicotyledones. Oxford University Press, Oxford.

Irwin, H.S. 1964. Monographic studies in *Cassia* (Leguminosae: Caesalpinioideae). I. Section Xerocalyx. Mem. N. Y. Bot. Gard. 12: 1–114.

Irwin, H.S. & R. Barneby. 1981. Cassieae. pp. 97–106 in Advances in Legume Systematics 1 (edited by R.M. Polhill & P.H. Raven). Royal Botanic Gardens, Kew.

Irwin, H.S. & R. Barneby. 1982. The American Cassiinae: a synoptical revision of Leguminosae, tribe Cassieae, subtribe Cassiinae in the New World. Mem. N. Y. Bot. Gard. 35: 1–918.

Kunze, H. 1978. Typologie und Morphogenese des Angiospermen-Staubblattes. Beitr. Biol. Pfl. 54: 239–304.

Lasseigne, A. 1979. Studies in *Cassia* (Leguminosae: Caesalpinioideae). III. Anther morphology. Iselya 1: 141–160.

Lewis, G.P. & T.S. Elias. 1981. Tribe Mimoseae. pp. 155–168 in Advances in Legume Systematics, Part 1 (edited by R.H. Polhill & P.H. Raven). Royal Botanic Gardens, Kew.

Newman, I.V. 1933a. Studies in the Australian Acacias. I. General introduction. J. Linn. Soc. Bot., Lond. 49: 133–143.

Newman, I.V. 1933b. Studies in the Australian Acacias. II. The life history of *Acacia baileyana*. Part 1. Some ecological and vegetative features, spore production, and chromosome number. Bot. J. Linn. Soc. 49: 145–171.

Newman, I.V. 1934. Studies in the Australian Acacias. III. Supplementary observations on the habit, carpel, spore production, and chromosomes of *Acacia baileyana*. Proc. Linn. Soc. N. S. W. 59: 237–251.

Newman, I.V. 1936. Studies in the Australian Acacias. VI. The meristematic activity of the floral apex of *Acacia longifolia* and *A. suaveolens* as a histogenetic study of the ontogeny of the carpel. Proc. Linn. Soc. N. S. W. 61: 56–88.

Polhill, R.M. 1994. Classification of the Leguminosae. In Phytochemical dictionary of the Leguminosae Part 1, pp. 35–56, Chapman & Hall, New York.

Ramírez-Domenech, J.I. 1989. Floral ontogenetic studies in mimosoid legumes. Ph.D. dissertation, Louisiana State University, Baton Rouge.

Ronse de Craene, L.-P. & E. Smets. 1987. The distribution and systematic relevance of the androecial characters oligomery and polymery in the Magnoliophytina. Nord. J. Bot. 7: 239–253.

Salinas, A.O.D. & M.S. Sanchez. 1977. Biología floral del género *Cassia* en la region de los Tuxtlas, Veracruz. Bol. Soc. Bot. Mexico 37: 5–45.

Saradhi, P., & H.Y. Mohan Ram. 1981. Some aspects of floral biology of *Cassia fistula* (the Indian Laburnum), Part. 1. Curr. Sci. 50: 802–805.

Schmid, R. 1976. Filament histology and anther dehiscence. Bot. J. Linn. Soc. 73: 303–315.

Solodoye, M.O. 1985. A revision of *Baphia* (Leguminosae: Papilionoideae). Kew Bull. 40: 29–38.

Thorp, R.W. & J.R. Estes. 1975. Intrafloral behavior of bees on flowers of *Cassia fasciculata*. J. Kansas Entomol. Soc. 48: 175–184.

Tucker, S.C. 1984. Origin of symmetry in flowers. pp. 351–395 in Contemporary Problems in Plant Anatomy III (edited by R.A. White & W.C. Dickison). Academic Press, London.

Tucker, S.C. 1987. Floral initiation and development in legumes. pp. 183–239 in Advances in Legume Systematics, part 3 (edited by C.H. Stirton). Royal Botanic Gardens. Kew.

Tucker, S.C. 1988a. Heteromorphic flower development in *Neptunia pubescens*, a mimosoid legume. Amer. J. Bot. 75: 205–224.

Tucker, S.C. 1988b. Dioecy in *Bauhinia* resulting from organ suppression. Amer. J. Bot. 75: 1584–1597.

Tucker, S.C. 1988c. Loss versus suppression of floral organs. In Aspects of Floral Development (edited by P. Leins, S.C. Tucker & P.K. Endress). Gebrüder Borntraeger, Stuttgart.

Tucker, S.C. 1989. Evolutionary implications of floral ontogeny in legumes. In Advances in Legume Biology (edited by C.H. Stirton & J.L. Zarucchi). Monogr. Syst. Bot. Missouri Bot. Gard. 29: 59–75.

Tucker, S.C. 1990. Loss of floral organs in *Ateleia* (Leguminosae: Papilionoideae: Sophoreae). Amer. J. Bot. 77: 750–761.

Tucker, S.C. 1991. Helical floral organogenesis in *Gleditsia*, a primitive caesalpinioid legume. Amer. J. Bot. 78: 1130–1149.

Tucker, S.C. 1992a. The developmental basis for sexual expression in *Ceratonia siliqua* (Leguminosae: Caesalpinioideae: Cassieae). Amer. J. Bot. 79: 318–327.

Tucker, S.C. 1992b. The role of floral development in studies of legume evolution. Can. J. Bot. 70: 692–700.

Tucker, S.C. 1993. Floral ontogeny in Sophoreae (Leguminosae: Papilionoideae). I. *Myroxylon* (Myroxylon group) and *Castanospermum* (Angylocalyx group). Amer. J. Bot. 80: 65–75.

Tucker, S.C. 1994. Floral ontogeny in Sophoreae (Leguminosae: Papilionoideae). 2. *Sophora* (Sophora group). Amer. J. Bot. 81: 368–380.

Tucker, S.C., S.R. Rugenstein & K.S. Derstine. 1984. Inflated trichomes in flowers of *Bauhinia* (Leguminosae: Caesalpinioideae). Bot. J. Linn. Soc. 88: 291–301.

Venkatesh, C.S. 1956a. The form, structure and special modes of dehiscence in anthers of *Cassia*. I. Subgenus Fistula. Phytomorphology 6: 168–176.

Venkatesh, C.S. 1956b. The form, structure and special ways of dehiscence in anthers of *Cassia*. II. Subgenus Lasiorhegma. Phytomorphology 6: 272–277.

Venkatesh, C.S. 1957. The form, structure and special ways of dehiscence in anthers of *Cassia*. III. Subgenus *Senna*. Phytomorphology 7: 253–273.

Vogel, S. 1978. Evolutionary shifts from reward to deception in pollen flowers. pp. 89–96. in The Pollination of Flowers (edited by A.J. Richards). Academic Press, London.

Wolfe, A.D. & J. Estes. 1991. Pollination and the function of floral parts in *Chamaecrista fasciculata* (Fabaceae). Amer. J. Bot. 79: 314–317.

12 · Anther investigations: a review of methods

RICHARD C. KEATING

INTRODUCTION

Stamens are usually transitory organs, developing to maturity in a few days and presenting pollen over a period of a few hours. Events occurring before and after pollen presentation occasion overlapping studies of inflorescence structure, floral organography and histology, microgametophyte development, pollination biology, and palynology. Clearly, of the wide array of methods brought to the study of anthers, few apply to them alone. Advancing one's knowledge of these organs requires an integrated understanding of reproductive biology and some appreciation of the various approaches bearing on its study.

This overview of diverse techniques addresses the disciplines of morphology, histology, systematics, development, pollination biology, biochemistry, and molecular genetics. Each field has produced a larger literature than can be condensed into a short review. My present intention is to provide an entry into methodology for those considering research in the field. More accessible techniques are treated in the most detail. Investigators new to research in this area can find discussions of the morphological and life history contexts of stamens in Raven et al. (1992), Gifford & Foster (1989), Bold et al. (1987) or Bierhorst (1971).

Investigations of plant material begin with the acquisition or propagation of specimens. It bears re-emphasizing that voucher specimens should be prepared at the time of sampling, unless the material is being removed from herbarium sheets. Arrangements should be made to deposit vouchers in a recognized, curated herbarium where the application of a scientific name, or binomial, will allow organized additions to the science.

Techniques for collecting and preparing specimens are described in Kearns & Inouye (1993), Forman & Bridson (1989), Radford et al. (1974), and Fosberg & Sachet (1965). A comprehensive bibliography of collection and herbarium techniques can be found in Hicks & Hicks (1978).

Students of embryology have routinely included anther features in their data sets, but often in a peripheral way. Characters that should be investigated have been discussed by Davis (1966), Bhandari (1984), Palser (1975), and Herr (1984). As listed by Herr, these are: number of microsporangia; number of vertical rows of cells in each archesporium; type of wall development (basic, dicotyledonous, monocotyledonous, reduced); nature of endothecium (level of differentiation, number of layers); persistence of the epidermis; type of tapetum (glandular or amoeboid, number of nuclei per cell, ploidy level); delimitation of microspores (simultaneous or successive cytokinesis, by furrowing or cell plate formation); type of tetrad (tetrahedral, decussate, isobilateral, linear, T-shaped, rhomboidal); cohesion of microspores (separate, tetrads, or larger clusters). If more attention were given to the parent organ, such descriptions could include general anther configuration, dehiscence type and stomial histology, extent, distribution and frequency of crystals or other inclusions, and connective and vasculature patterns.

As with ovular characters, one may approach anthers by observing their structure, the focus of the first part of this review. With each species studied, the investigator must collect floral bud material of all sizes, making dissections to ascertain the appropriate stages of development for the phenomena being investigated. One must decide from among a variety of techniques the ones that

255

will yield the desired data most efficiently. To begin, I will review specimen preparation for light microscopy and electron microscopy. Less attention will be given alternative microscopical techniques, histochemistry and molecular techniques. Finally, I discuss field-based techniques for pollination studies and related pollen preparation methods.

LIGHT MICROSCOPY AND MICROTECHNIQUE

The compound optical microscope, with magnifications up to about 1000 diameters, has grown up with the biological sciences. Neither the instrument's evolution over the past 300 years nor its effective use is presently widely taught or understood. Most microscope manufacturers offer basic guides to elementary theory and practice. Those by Mollring (1981), Patzelt (1974) and Determann & Lepusch (1974) are especially good as are the general short texts by Culling (1974), Wilson (1976), Bradbury (1989), and Rawlins (1992).

Of central importance in light microscopy practice is achieving 'Koehler illumination', a type of critical illumination yielding the best performance that objective lenses can deliver. This involves centering and focusing the lamp filament image on the back focal plane of the objective lens, focusing the substage condenser to image the field diaphragm at the edge of the plane of the specimen, and closing the substage iris slightly so as to enhance contrast. Secondly, microscopists should know that empty magnification (enlargement of the image without revealing detail) begins between 500–1000 times the numerical aperture (NA) of the objective. This sets the limit of useful magnification of a photographic negative (Mollring 1981). Numerical aperture controls the resolving power (capacity of a lens to distinguish between two closely spaced details). As an example, the new × 63, 1.4 NA oil immersion lenses have a better working distance, larger field, and produce more useful negative enlargement than the older × 100, 1.25 NA oil lenses.

Photographic documentation (photomicrography) is an essential part of microscopic investigations and can be accomplished well using elaborate or simple camera setups. Various forms of digital imaging are becoming available but, as of this writing, conventional color and black and white (silver) films provide much better resolution of detail. They are also much more cost-effective, even including time and labor. Nonetheless, the obsolescence of films for this work is on the horizon.

A basic knowledge of photography transfers directly to photomicrography and can be gained from any elementary level manuals available at camera dealers. Excellent contemporary short texts covering microscopic and photomicrographic techniques are the well-illustrated works by Delly (1989), and Smith (1993). At photographic suppliers, ask for the technical data sheets printed by manufacturers for all films, papers, and most chemicals.

The most common and versatile camera format for photomicrography uses 35 mm film. Many color transparency, color negative, and silver negative films are widely available in this format and are easily processed. The camera is usually mounted on the microscope on a third ocular tube. A focusing telescope should be present and calibrated so that when the image is sharply focused inside the framing reticle (lines seen superimposed over the image), the film will record exactly the same composition. It is possible but less satisfactory to view the image directly through the viewing prism of a through-the-lens type camera. As an aid to focusing, use a focusing telescope eyepiece (Bertrand lens). First, focus it on a distant object then place it on top of the viewing prism to focus the image.

A light meter should be used to calibrate the exposure with the film speed. A test strip must be made for every film/developer combination. This involves photographing some typical specimens using a range of shutter speeds, keeping careful records of all variables for each exposure.

Microscope illumination for color films must usually be filtered for good balance and a clear white background. Either daylight- or tungsten-balanced film can be used with the appropriate compensating filters. Processing is usually best done by a color laboratory. Silver film and developer combinations vary widely but development can be easily accomplished by the microscopist. There is no reason to use the finest-grained silver films since the recording of detail is limited first by the numerical aperture of the objective. It is better to use a higher contrast film/developer combination since getting good contrast is usually the more serious problem

(Spencer 1982). A discussion of some of the best options is found in Smith (1993).

Microtechnique includes a variety of specimen preparation methods for examination with the compound optical microscope. Its contemporary practice finds many older procedures and formulas co-existing with recently introduced materials and methods. Many early procedures still produce beautiful and informative preparations and one should comb the older literature for such numerous half-forgotten contributions to the science. The journal *Stain Technology* is an excellent source. Bracegirdle (1978) has reviewed the history of the field.

Paraffin techniques

Preparation of permanent slides using paraffin embedment dates back to about 1880 when paraffin replaced spermaceti as an embedding medium (Malies 1959). The sectioning of paraffin-embedded specimens has been practised by nearly all anatomists in the twentieth century, and by many exclusively. It has been the source of much of the large literature of twentieth-century plant anatomy. The procedures involve killing and fixing plant parts in a liquid fixative (or restoring dried plant parts to resemble their original hydrated form), dehydrating in a graded series of solutions in preparation for embedment, then infiltration and embedment in a paraffin-based medium. Specimens are then sectioned serially using a rotary microtome. The resulting sections are affixed to glass slides, run through a staining sequence and covered with resin and a cover glass. From specimen to finished slide may require 2 weeks, often too time-consuming for single specimen preparation. However, most laboratories have specimens in all stages of production, which leads to operational efficiencies.

We can now view the 'modern' era to have begun in 1940 with the publication of *Plant Microtechnique* by D.A. Johansen, a text that has remained a standard reference for over 50 years. Many of the formulas and schedules described therein are still in general use. Better than many authors, Johansen goes into specimen handling in considerable detail. Since then, Foster (1949), Sass (1958), Venning (1958), Jensen (1962), Purvis *et al.* (1966), Berlyn & Miksche (1976), and Gahan (1984) have expanded this literature and updated methodology. Baker (1958) has provided an analysis of the theory

behind many of the staining methods and Gray (1954) has compiled a large number of useful techniques. The outstanding recent contribution is the compendium of O'Brien & McCully (1981), privately published and now out of print. Their work considered in detail the theory behind the use of various chemical treatments, in the tradition of Baker (1958).

The nature of stains (dyes of some authors) and their recommended uses is thoroughly discussed in the standard reference, *Conn's Biological Stains* (Lillie 1977). This volume is cross-indexed with *Staining Procedures used by the Biological Stain Commission* (Conn et al. 1960; Clark 1973). The two editions of *Staining Procedures* are sources of many literature citations for application of stain schedules to all kinds of biological material.

For general anatomical investigations and taxonomic surveys, slide production using paraffin techniques can be used to produce large numbers of informative microscope slides of flowers and anthers. For general histology, paraffin techniques are especially forgiving of field-collected specimens, stored in fixatives of dubious origin, and where sections cut at 5–15 μm will yield useful information. The work can be performed by students or professional technicians with equivalent results, with one caveat: O'Brien and McCully (1981) and I have noted the tendency for slide quality to 'run down hill' with time, that is, for the product to get progressively less acceptable unless regular quality control is exercised. When relying on technicians, daily monitoring of the work is essential.

As in any craft, learning microtechnique under the tutelage of an experienced worker is invaluable. While not inherently difficult to master for a person with good manual dexterity, there are at least hundreds of movements, decisions, and modified tools that are not intuitive to the novice. What follows is only a supplement to instruction or to extended narratives found in the standard references noted above. I will describe a few improvements that are not presently widely known.

Killing and fixing

Collected plant parts should be placed in a liquid fixative with the goal of tissue or cellular preservation in as life-like a condition as possible. Different approaches to liquid preservation, each with specific advantages, have been described by O'Brien & McCully (1981). Each type involves a compromise between a good 'fixation image' for one structure at the expense of a poor one for another. The investigator must select from among alternatives based on the nature of the problem. For faithful recording of histological patterns of plant tissues, and not especially for cytological details, the 'coagulating' formulas remain in wide use. Formalin, acetic acid, 50% ethanol (FAA) (5:5:90) or formalin, propionic acid, 50% ethanol (FPA) (5:5:90) are both common fixatives, as are the chromic acid/acetic acid/formalin (CRAF) formulas. They provide good images of lignified and cellulosic walls, and general identification of histological features. Liquid preserved specimens collected in remote locations in the tropics (see Tomlinson 1965) are most usually preserved in FAA as the ingredients are easily found worldwide. I have found no deterioration of general structure from storing plant parts for years in FAA or FPA, as long as metal jar caps are not used. However, J.M. Herr, Jr (personal communication) recommends transferring fixed plant material, after 24–48 hours, for storage in 70% ethanol.

As an ingredient in fixatives, formaldehyde has been recommended for abandonment as being unacceptably toxic (Forman & Bridson 1989). The replacement fixative, 70% ethanol and glycerol, recommended by Forman & Bridson has not been extensively evaluated for its fixation image in general plant histology. No good proposals have yet emerged for the replacement of such excellent rapid penetrating fixing agents as the various aldehydes. Formulas using them should be handled in well-ventilated areas and kept from contact with the skin. Safety brochures detailing the hazards of such reagents as aldehydes, and organic ring solvents, can be readily obtained from one's institutional hazardous wastes office, or from any chemical supplier.

Dissection

Liquid-fixed and stored material should be rinsed in water or 30% ethanol for safer handling and then trimmed with sharp blades into the size needed for microtoming. Pieces should be as small as feasible to show the desired structure. Dried, flattened material can sometimes be revived to lifelike shape using dilute solutions of detergent, Aerosol-OT, Contrad 70, or any of the various

clearing solutions (see Lersten 1967; Gardner 1975). A short time in clearing reagents, not to the point of clearing, is usually adequate for revival of the tissue. Dry plant tissues vary in their resistance to restoration so experimentation is necessary. Placing restored tissues in a fixative often improves their reactivity with stains.

Dehydration

Zirkle's tertiary butanol schedule (Johansen 1940) is entirely satisfactory for dehydration (removal of water) and clearing (removal of ethanol). The end solution in the series is pure tertiary butanol which is miscible in ethanol and is a paraffin solvent. In laboratories where butanol tends to freeze at ambient temperatures, I use limonene (Histoclear):butanol (1:4) to prevent freezing and increase the paraffin solvent properties of the final t-butanol soak. Sterling (1964) has had good results using Johansen's (1940) N-butanol method.

Infiltration and casting

The dehydrated specimens should be placed in lipless vials and covered with the end t-butanol solution. Paraplast (or other brand of paraffin-plastic) chips are added and the vials corked. The vials may be placed on a 43 °C warming tray for up to a week, with daily checking that the specimen is always immersed in the liquid of the butanol–paraffin slurry. After several days, the cork is removed and the vial placed in a 60 °C paraffin oven. The melted slurry is poured off and the final casting mixture is added. Let this mixture remain in the oven for 12–24 hours to drive off remaining alcohol. Some investigators use paraffin oil in the final mix with the chips, claiming that more thorough infiltration takes place (D. Stevenson, personal communication). This is not necessary for materials that infiltrate easily but can be helpful if the material is resistant to good infiltration. Drawing a vacuum while the specimen is immersed in melted paraffin will also help remove butanol from infiltration-resistent specimens. The final casting mixture should only be kept melted when needed so that the plastic polymers do not break down and destroy good sectioning properties. The use of small molds, one per specimen, assists the alignment process when specimens and Paraplast are being cast.

Sectioning

Any commercially available, well-lubricated rotary microtome will produce reliable sections. The specimens can be cut at 1–15 μm, depending on the purpose. Resharpenable, heavy steel blades provide the most reliable results, especially with difficult specimens. Having a good knife sharpener and practice with making good edges is central to success if steel blades are being used. Clean cuts with undamaged tissues cannot be made using blade edges obtained by hand stropping. Wide-edged, ultrasharp glass knives of the Ralph type (Bennett *et al.* 1976) can be useful although their use may lead to excessive compression at most thicknesses. Ralph knives can be cut on an expensive machine or by hand as detailed in Bennett *et al.* (1976).

Mounting ribbons

Rows of attached serial sections, called ribbons, are affixed to clean glass slides prior to staining. Haupt's adhesive (Johansen 1940) is reliable, but serum adhesive (Keating 1969) is better if specimens tend to fall off in the stain sequence. This formula calls for horse, swine or bovine serum : glycerine (1:1) where serum replaces gelatin in the original Haupt's recipe. Add 0.01% (v/v) sodium salicylate as a preservative and refrigerate between uses. On top of a thinly smeared layer of the adhesive on a clean slide, ribbons are floated on a meniscus of water. The formalin solution previously recommended is not necessary. Slides can be stored in boxes after drying on a 43°C warming tray.

Staining

Conn *et al.* (1960) and Clark (1973) and the microtechnique manuals previously cited describe numerous staining schedules. Lersten's (1986) Chlorazol black E schedule, applied to cleared leaves should also be quite useful for cleared floral parts. The dye is capable o. very delicate enhancement of cell walls. After trying n any stain formulas and procedures, I concluded that he Safranin-O/Fast Green-FCF staining schedule, usii g Johansen's (1940) stain formulas, provides the most luminous results on a variety of specimens. It is entirely satisfactory for paraffin sections, is easy to master, and

provides good photogenic preparations. The sample working schedule given below can be used for months for mass production of slides with infrequent changes of solutions. The stain recipes are stable and can be used indefinitely (at least a decade), needing only occasional added solvent.

I use 500 ml, wide mouth, screw cap jars with Teflon gasket inserts, and handmade, soldered copper stain racks as modified from Sass (1958: 56). Stainless steel racks designed to carry the slides vertically will also fit these containers. Numerous commercial 'stain jars' with loose fitting lids are available. These are not satisfactory and require replacing solutions constantly.

The most important recent development is that the toxic organic ring solvents, xylene and toluene, can be replaced in all procedures calling for them with an entirely safe solvent. Limonene (sold as Histoclear, from National Diagnostics) has a lower vapor pressure, is safe to breathe, and is more satisfactory in all respects.

In the following schedule, all solution times are as short as 2–5 minutes unless otherwise specified. The dyes are prepared according to Johansen (1940: 59, 62) except that methyl salicylate replaces clove oil.

Limonene I → limonene II → limonene/absolute ethanol → absolute ethanol → 95% ethanol → 70% ethanol → safranin, 2–24 hours → water rinse → 50% ethanol → 70% ethanol → 95% ethanol → absolute ethanol → fast green, ten seconds to one minute, by test → absolute ethanol [this step saves the solutions in the next step, and is replaced when it gets dark with stain] → differentiating solution (methyl salicylate/absolute ethanol/limonene 1:1:1) → absolute ethanol/limonene → limonene III → limonene IV → limonene V → add resin and cover.

During staining, slides should not be left in safranin solutions longer than 48 hours as precipitates may form in the specimen or excess stain will be difficult to remove. Fast green is used regressively: it replaces safranin in cellulosic cell walls. If the fast green works faster than 20–30 seconds for most material, further dilute the working solution with absolute ethanol. Test slides will quickly show whether the tissues have spent too much or too little time in fast green.

Histomount (National Diagnostics) has proven to be the best mountant of the many tested, and its solvent is advertized as safe to breathe. Drying the slides on a

43 °C warming tray requires perhaps one day longer than toluene based mountants, but the specimens are much less likely to take in gas bubbles than with other mountants.

Resin-embedded thin sections (1–3 μm), stained in the metachromatic Toluidine blue-O, are becoming common in light microscopical investigations. These techniques are well covered in Feder & O'Brien (1968), Berlyn & Miksche (1976), and O'Brien & McCully (1981). They are adapted from electron microscopical procedures and will be discussed under that section below.

HAND SECTIONING AND WET MOUNT TECHNIQUES

Until recently, wet mount techniques have been neglected. There is no doubt that serial sectioning will remain the technique of choice for studies of vasculature patterns, vascular development, and other phenomena where detailed three-dimensional reconstructions are essential. But it is often more direct and informative to use microdissections, clearings, wet mounts and smears. Many floral buds have such lightly pigmented floral organs that little prior clearing is necessary. Dehydrating fluids alone may be sufficient to render the specimen nearly transparent. In other cases, opaque materials must first be removed by clearing methods. See Schmid (1977) for a procedure using Stockwell's bleach that effectively removes dark tanniniferous deposits. In general, avoid any treatment more drastic than is necessary to render tissues visible.

Hoyer's mountant, reintroduced by Anderson (1954; Kearns & Inouye 1993), has a formula that includes chloral hydrate, glycerine, gum arabic, and water. It is excellent for permanent mounts and will clear small whole specimens or thick slices not containing deposits of dark phenolics. King & Robinson (1970) have used it to discern a number of anther and other floral characters useful in classification of the Asteraceae. Glycerine jelly, long used by palynologists (see Traverse 1988), can be used unstained or carrying a stain for excellent permanent mounts of pollen or small organs. Ogden *et al.* (1974) describe a modification of glycerine jelly, Calberla's Solution, that contains basic fuchsin.

Over the past century, relatively few mountants for small whole specimens have reached widespread use (Herr 1992a). Glycerine, in wide use since the mid-

nineteenth century, is not satisfactory for use with metachromatic dyes. Alcoholic solutions (including glycerine) cause these dyes to lose their metachromatic properties and to be mostly leached from all tissues. Stains such as toluidine blue are effectively metachromatic in water mounts, which are necessarily temporary. Therefore this technique is often restricted to student use in plant anatomy teaching.

The use of concentrated calcium chloride as a mountant has been reintroduced by Herr (1992a), and he reviewed its history of use that began in the mid-nineteenth century. When used with specimens dyed in phloroglucinol or toluidine blue, lignified walls are dyed red or blue, respectively, but no coloration is retained in cellulosic walls. In fact, it was this bleaching of dyes, plus a fascination with glycerine and glycerine jelly as mountants, that led to the abandonment of calcium chloride for this purpose. However, since calcium chloride does permit permanent mounts, I explored its potential to make even better specimen preparations.

Calcium chloride is a hygroscopic salt that usually does not cause shrinkage or cell collapse of fresh, fixed, or restored materials. It is heavy (specific gravity 2.15), highly soluble (74 g per 100 ml of water), and has a refractive index of 1.52, the same as crown glass (Lide 1991).

The following techniques are offered as excellent ways to make informative 'amplitude objects' for microscopic examination and photomicrography, and not particularly as histochemical tests. In any event, with histochemical testing, several different tests should be used to provide greater confidence.

Beginning with the quinone-imine class of metachromatic dyes (Lillie 1977), I experimented with a wide range of dyes to assess their suitability in combination with calcium chloride or other clearing mountants. A number of these combinations showed much potential for permanence, brilliance, and sharp differentiation.

In use with calcium chloride, most of these dyes have not been mentioned in the literature on botanical microtechnique. The violet dyes, cresyl violet acetate (CVA) and thionin, seem to have been totally neglected by most botanists, perhaps because toluidine blue has been successfully promoted (Siegel 1967; O'Brien & McCully 1981) as a good wet mount dye. With these other dyes, much improved permanent, thick or whole mounts can be made, with all walls retaining their metachromatic responses indefinitely. As with other types of whole mounts, these must be stored flat.

Calcium chloride can be purchased from chemical supply sources for about $65 per kg or obtained from farmer's supply stores as liquid concentrate for about 95 cents per gallon. (It is sold as tractor tire ballast.) I use the liquid as it comes, which appears to be about 30–40% strength. The actual concentration is not important since you add additional drops at the edge of the cover glass several times for about 2 weeks until equilibration with ambient humidity occurs. Use of a warming tray is not useful since equilibration with atmospheric moisture works best at ambient temperature. The exposed liquid at the edge of the cover glass eventually skins over and stabilizes. Ringing is not necessary. Herr (1992a) reports that mid-nineteenth century hand sections, mounted in calcium chloride, are still extant in the Leiden museum. They are in good histological condition, and the salt remains uncrystallized.

Specimens to be mounted in calcium chloride can be thin hand slices (20–150 μm) or small whole mounts (ovules or anthers) and can be stained either fresh, fixed or restored. They can be used directly from alcohol storage or from water. The dye can be applied by the drop to the specimen on a flat slide, depression slide, or spot plate and is left for 1–5 minutes according to experience and dye concentration. Thionin or CVA can be made up as dilute aqueous or dilute ethanol solutions at about 0.1 to 0.01% strength. Ethanolic solutions (10–30%) promote better dye solubility and penetration of the specimen. Concentration is not important within these limits; the length of time in the dye will vary with dye concentration and with the specimen, but the dye does not tend to overstain badly. If overstained, destain with dilute ethanol, then transfer the specimen directly to several drops of calcium chloride on a slide and apply a cover glass. With both CVA and thionin, the general violet appearance, somewhat metachromatic, transforms quickly. Xylem elements, endodermal thickenings and cuticle turn bright blue, cellulose walls yellow-tan, and collenchyma red-tan. This is as wide a hypsochromic shift as occurs in any metachromatic substance (Baker 1958). No fading is discernible after several months. All cell walls retain coloration well enough to see layering, pitting, and other features.

Calcium chloride, with its high refractive index, has the effect of markedly reducing the microscopic depth of field (Herr 1992a). It is possible to focus on a cut surface, or some internal plane and see detail with little interference from layers above or below. Darkground, polarization, and other usual techniques may also be very useful.

CLEARING

Rendering whole plant tissues or organs transparent is an informative way to inspect a feature of interest in its structural context. Cleared squashes and smears of anthers have been used extensively, especially to ascertain stages of meiosis and pollen development (Palser 1975). Techniques suggested by Herr, discussed below, are at the core of a renewed interest in dissection and clearing of small anatomical parts. As described for ovules, summarized below, the techniques also prove useful for examination of anther slices or whole anthers at various stages of their development.

The introduction of $4\frac{1}{2}$ fluid by Herr (1971) has provided embryologists with a very useful tool for examining whole embryos. Gynoecia are fixed in 50% ethanolic fixatives (FAA or FPA) and then stored in 70% ethanol. The parts are immersed in $4\frac{1}{2}$ fluid, made up as follows: lactic acid (85% strength), chloral hydrate, phenol, clove oil, and xylene (2:2:2:2:1, by weight). Histoclear can be substituted for xylene with identical results (J.M. Herr Jr, personal communication; Rudall & Clark 1992). Tissues appear nearly transparent after 24 hours immersion at room temperature. Ovaries of some species need a pretreatment of lactic acid (85%) for 24 hours at room temperature in order to make the ovules discernible using phase contrast optics. Chloral hydrate, it should be noted, is a restricted drug in many places and an institutional drug license may be required for its purchase.

Using a dissecting microscope, Herr (1971, 1974) recommends that ovules be dissected from ovaries and placed on 'Raj slides', prepared as described below. I have found that this procedure also works well for anther slices. Two cover glasses are cemented to a slide so as to leave a trough between them about 1 cm wide. The specimens are placed with a drop of the clearing fluid between the cemented covers and a third cover is placed, supported on either end, over the specimens and fluid. If the support covers are the correct thickness, the top cover is prevented from exerting too much pressure on the specimens.

Using phase contrast or Nomarski optics, optical sectioning is readily accomplished. The refractive index of the fluid is such that little or no interference is produced by layers above or below the plane of focus. The effect of $4\frac{1}{2}$ fluid is to reduce differences in the refractive properties of organelles, and to reduce depth of field so that only one cell layer can be discerned at a time. Many cytological details can be observed and documented photographically. On occasion, fixed specimens containing dark tanniniferous deposits must be pretreated with Stockwell's bleach (Johansen 1940: 85; Schmid 1977) before using $4\frac{1}{2}$ fluid.

Later, Herr (1972, 1973) introduced some modifications to the formula. 'I_2KI-$4\frac{1}{2}$' is made by adding 100 mg of iodine and 500 mg of potassium iodide to 9 gm of the original mixture. In addition to detection of starch grains, the mixture may sharpen the structural features since I_2KI is known to have a light staining effect on cell walls. Primary meristems were not well cleared with the original recipe but are rendered much clearer with the I_2KI modification. Vascular elements and the sporogenous tissue boundaries become very clear.

'BB-$4\frac{1}{2}$' is made by adding benzyl benzoate one part by weight to the original mixture. After tissues are immersed in this mixture for about 2 weeks, definition of cytological features is diminished but cell outlines become sharply defined. This can be quite useful for determining architectural aspects of the investigated organ. The potassium permanganate modification ('PP-$4\frac{1}{2}$') is made by dissolving 3 mg potassium permanganate in a gram of the original fluid. It should be mixed immediately before use since it is not stable (a precipitate is formed under refrigerated storage which diminishes the fluid's clearing power). Fresh PP-$4\frac{1}{2}$ sharpens the image under both Nomarski and phase contrast imagery where phase birefringence of tissues above and below the plane of focus tend to diminish the clarity of the image. This can be a problem where advancing integuments obscure the image of the megasporocyte. Adding both potassium permanganate and benzyl benzoate (PPBB-$4\frac{1}{2}$) completely eliminated the image of the advancing integument from obscuring the details of embryogeny (Herr 1973).

One difficulty with Herr's fluids is their tendency to turn dark with time, due to the photooxidation of clove oil. Clove oil can be replaced in these formulas with dibutyl phthalate, a substitute with similar properties to clove oil but resistant to darkening (Herr 1992b).

For their investigation of patterns of enthothecial thickenings Manning & Goldblatt (1990) and Manning (this volume) removed anthers from herbarium specimens and placed them in 60% lactic acid at 90°C until soft, usually 2–4 hours. The cleared specimens were mounted unstained in glycerine after being mechanically macerated with needles to see separate cells. Alternatively, specimens were mounted in glycerine jelly to which basic fuchsin was added. The preparations were viewed with Nomarski interference optics since bright field optics would not show adequate detail. Locules were carefully separated and flattened to see the arrangement of endothecial cells.

Microscopists without access to phase or interference contrast equipment should try staining techniques. Stelly *et al.* (1984) suggested bulk staining of ovaries and ovules using Mayer's hemalum and dehydrating to methyl salicylate. This alternative to Herr's methods provides good contrast with bright field optics.

Numerous clearing techniques have been collected and reviewed in bibliographies by Lersten (1967) and Gardner (1975). The use of sodium hydroxide clearing followed by Chlorazol black E staining as suggested by Lersten (1986) warrants experimentation with whole mounts and thick sections of anthers. Rao (1977) published a protocol for reviving herbarium specimens for anatomical dissection using glycerine, acetic acid, EDTA, lauryl sulfate, and water.

ELECTRON MICROSCOPY

Electron microscopy has provided the opportunity to examine tissues with up to two orders of magnitude more magnification and detail than with light microscopy. Its development stimulated development of a new body of preparative techniques. Experimentation with new reagents, embedding resins, and other materials began in the 1940s and accelerated in the 1950s when these relatively expensive machines became more common.

As practised today, there are two types. Transmission electron microscopy (TEM) involves resin embedment of small specimens and their sectioning using an ultramicrotome with a diamond or glass knife. Scanning electron microscopy (SEM) allows the observation of surfaces of whole specimens with much greater depth of focus than with any kind of optical microscopy.

Because the use of EM requires the placement of specimens under high vacuum, and because the detail observed and general magnification is so much higher, specimen preparation calls for even more care than is usually brought to bear in careful light microscopical preparations.

Since 1963, preparation of most tissues, including anthers, has called for the use of glutaraldehyde fixation followed by osmium tetroxide post-fixation. Most of today's protocols are some modification of this procedure. Glutaraldehyde has the capacity to cross-link protein and may also react with lipids, carbohydrates, and nucleic acids. It may be mixed with up to 4% formaldehyde for more rapid penetration. Osmium tetroxide seems to work as a secondary fixative by reacting with lipid moieties. It oxidizes fatty acid bonds and is reduced to black metallic osmium, adding density and contrast to tissues (Hayat 1975; O'Brien & McCully 1981; Bozzola & Russell 1992).

Both glutaraldehyde and osmium tetroxide are aggressive fixatives that will readily kill and fix exposed surface tissues including those of the technician. Glutaraldehyde can cause contact dermatitis, and it is essential that procedures using it be carried out in a well-ventilated hood. Barber & Clayton (1985) provide detailed information on evaluation and avoidance of hazards in EM work.

Plant tissues are notorious for impeding the penetration of fixative due to hydrophobic cuticles and cell walls. Fixation may take as little as 2 hours or extend to 18 hours for some plant material. As a starting point, Bozzola & Russell (1992) recommend using 2–4% phosphate-buffered glutaraldehyde or a formaldehyde-glutaraldehyde mixture followed by buffered rinses. Secondary fixation is accomplished with 1–4% osmium tetroxide, followed by an ethanol dehydration series, propylene oxide transitional solvent, and a Spurr's (or other epoxy) resin infiltration series. Embedding and curing are carried out at 60°C.

These authors recommend using the same fixatives for SEM as for TEM. As a starting point, rely on established protocols where good results have been obtained by other workers on similar material. Polowick &

Sawhney (1993) studied pollen development in *Lycopersicon esculentum* anthers using TEM procedures similar to those described above. As an example of an investigation of stamen ontogeny where stamens of *L. esculentum* were prepared for SEM investigations, consult Polowick *et al.* (1990).

Personally, I have found that excellent SEM preparations can be obtained from specimens that have been fixed in FAA or FPA, CRAF, or other coagulant fixatives as used in light microscopy. These formulas often cannot be avoided when examining field-collected or stored material. It is essential, however, to use critical point drying (or modification, see below). Otherwise, cell walls of surface tissues are prone to shrink or collapse if they are not dehydrated without introducing shrinkage or distortion. Convex or papillate cells with thin cuticles are especially vulnerable to such damage.

The detailed and well-illustrated text by Bozzola & Russell (1992), covers all aspects of EM work and includes an emphasis on safe handling of all hazardous reagents and equipment. Dykstra (1992) competently covers theory, techniques, and troubleshooting as the subtitle indicates. The works by Flegler *et al.* (1993), and Parsons (1970) are shorter but useful guides for the beginner.

A major part of success in gaining competence in EM work is the capacity to recognize and avoid propagating artifacts in one's images. A most important aspect, and companion to safety, is cleanliness. Dirt and lint particles loom much larger in EM preparations due to the greater magnification. Bench surfaces should be freshly washed before working on specimens and all beem capsules, stubs, forceps, and other tools should be stored in dust-free containers. Laboratory atmospheres with high-volume ventilation can often be made much cleaner by fastening furnace filters over the air inlets.

An instructive set of contributions considering the nature of artifacts and how to avoid them in EM work was assembled by Crang & Klomparens (1988). Among these, the work by Bowers & Maser (1988) describes artifacts in fixation, explaining that fixation for specific chemical entities may produce less than the best results for morphology. They describe and illustrate by comparison, the effects of oxidizers, cross-linkers, coagulators, freezing, dehydration, and freeze-substitution, buffers, temperatures, and timing. Bullock (1984) and

Hopwood (1985) also discussed these subjects in some detail.

Artifact evaluation in dehydration and epoxy embedding for TEM work was discussed by Mollenhauer (1988). He noted that shrinkage is a major problem. The larger the vacuoles in a tissue, the greater the tendency toward shrinkage; a greater problem with plant tissues than with animal. Using small steps in dehydration series solvents is helpful, but the problem has not yet been entirely solved.

Mollenhauer reviews numerous combinations of resins, diluents, hardeners, catalysts, flexibilizers and accelerators, and their effects on photographic contrast. Good photographic documentation is critical to 'selling' one's science. Wergin & Pooley (1988), Berlyn & Miksche (1976), and O'Brien & McCully (1981) discuss basic procedures and improvements to photographic presentation.

Specimen preparation artifacts in SEM investigations were covered by Crang (1988). While it is customary in some laboratories to use the same fixation protocols for SEM and TEM, this is usually not necessary. The often-used osmium tetroxide is not necessary in many cases. Crang (1988) described critical point drying and its hazards in detail. The high pressures used in critical point drying can be avoided by using 'not-quite critical point drying' described by Boyde & Maconnachie (1983). Their procedure is simpler and safer and is claimed to produce the same acceptable specimen appearance.

ALTERNATIVE MICROSCOPICAL TECHNIQUES

I have counted at least 12 new forms of microscopy that have emerged or become more widely used during the past 15 years. Their descriptions are beyond the scope of this review. The Education Committee of the Microscopy Society of America maintains a comprehensive list of most books in print dealing with all fields of microscopy, as well as numerous journals. A few important examples of other microscopical techniques dealing with reproductive tissues are given below.

Fluorescence microscopy has been used in histochemistry for reliable localization of small amounts of compounds tagged with fluorescent dyes. Specimens are illuminated with violet or ultraviolet light and certain compounds (lignin) and dyes will fluoresce, that is, emit light energy that has shifted to longer, visible wavelengths. Kho Baer (1968) have examined squashed preparations of stigmas for pollen tubes. The fluorescence of pollen tubes in the presence of aniline blue suggests a good technique for incompatibility studies. Heslop–Harrison & Heslop–Harrison (1970) and Knox & Heslop–Harrison (1970) used enzymatically induced fluorogenic compounds in studying pollen development and viability. While studying the development of endothecium of *Chenopodium rubrum*, Fossard (1969) carried out detailed tests on the effects of a number of fixatives on the fluorescence image of the tissue. Chromic acid fixatives were found to give the best fluorescence image of lignified cells.

In a study of *Cirsium horridulum* stamens, Pesacreta *et al.* (1993) used confocal microscopy to detect autofluorescent cells in the 'viscoelastic filaments'. The autofluorescent cell wall materials were not identified. Using equipment not yet common to developmental investigations, the confocal scanning laser microscope, Fredrikson *et al.* (1988) studied embryology in the orchid genus *Dactylorhiza*. Theirs is the first study to use Herr's $4\frac{1}{2}$ fluid in conjunction with this form of microscopy and the resulting photomicrographs are exceptionally clear.

HISTOCHEMISTRY, DEVELOPMENT AND MOLECULAR TECHNIQUES

Numerous recently developed methods are now being used to elucidate the biochemical and genetic basis behind the known structural complexity in developing and mature anthers. Developmental processes leading to the release of microgametophytes are numerous and interact in ways that provide the investigator with a formidable challenge.

Gould & Lord (1988) used surface marking techniques to determine the sites of meristematic activity in expanding anthers. Using activated charcoal in a water matrix, they marked the dorsum of anthers of *Lilium longiflorum* in living flowers. They discovered that distribution of spatial growth continually changes with time, conforming to a 'waveform' pattern of expansion. Integrating this study with thin sections and SEM, they noted that the waveform phenomenon could not have been discovered by any sort of sectioning procedures used alone. Silk (1984) has suggested that surface marking experiments should be part of the empirical basis in any study of morphogenesis. Jeffrey Hill (this volume) describes the use of dental impression material to take casts of living surfaces in a non-destructive manner.

The tapetum has been subject to a number of inquiries. It has long been suspected of playing a major role in pollen development even before there was significant evidence to support the idea (see Heslop–Harrison 1968). Heslop–Harrison & Heslop–Harrison (1970) were the first to adapt enzymatically induced fluorescence to the study of pollen viability. The nuclei and the condition of the vacuoles was clearly demonstrated. The technique was discussed in detail in Knox & Heslop–Harrison (1970). Heslop–Harrison (1968) used aceto-orcein smears and phase contrast microscopy to establish the ontogenetic stages of *Lilium* pollen. Also, by using SEM, TEM, and fresh frozen material sectioned with a cryostat, he was able to establish that the tapetum is the source of various pollen wall building components including the oils and carotenoids that color the pollen wall. Only the freezing technique made it possible to handle the highly turgid tapetal cells without disruption.

Immunological characterization of tapetal proteins has been accomplished by Wang *et al.* (1992). They also established (Wang *et al.* 1993), that sporopollenin, flavonoids, and proteins are secreted into the anther locule and deposited in the 'highly specialized pollen wall'. Beerhues *et al.* (1989) have developed techniques for assaying flavanone 3-hydroxylase and dihydroflavonol oxygenase in anthers of *Tulipa*.

A problem facing botanists investigating pollen ontogeny is the selection of a series of floral buds at the right stages for investigation. Nave & Sawhney (1986) selected nine stages of anther development in *Petunia hybrida* that matched corresponding stages of corolla development. They investigated changes in the isozymes of four enzyme systems: esterases, peroxidases, alcohol dehydrogenases and malate dehydrogenases. Resin sectioned specimens were cut at 1 μm and stained in toluidine blue to monitor the presence of the desired stages. Gels were examined for the presence of the four enzyme systems for the nine stages. Histochemical localization was also accomplished (Sawhney & Nave 1986) using cryosectioning. This involved encasing specimens in 15% gelatin with 0.8% dimethylsulfoxide and freezing

at −70°C before sectioning. The histochemical localization for the four enzymes was accomplished using the same stains as used for the gels. With this technique distinctive patterns appear, such as the unusually heavy concentration of malate dehydrogenase in the stomium and substomial area in their stage three anthers, and its sporadic occurrence during later stage anthers.

Using glycol methacrylate-embedded and sectioned anthers of *Oryza sativa*, Raghavan (1989) demonstrated differential gene expression of mRNAs and a cloned histone gene. He applied the annealing reaction with [3H]poly(U) and used histone gene probes that were annealed to the sections.

Koltunow *et al.* (1990) were also able to demonstrate spatial and temporal gene expression patterns during anther development in tobacco. They divided flowers into 12 stages recognizable by bud length and morphological markers. This comprehensive investigation provides keys to the literature of numerous molecular techniques including polysomal RNA isolation, DNA isolation and labeling, DNA-excess hybridization experiments, gel blot studies, synthesis of single-stranded probes, construction and screening of a cDNA library, and histochemical staining for a GUS gene library. See also Cox & Goldberg (1988) and Jokufu & Goldberg (1988) for detailed techniques. Gallagher (1992) provides techniques for the using the GUS gene as a reporter of gene expression. Liedl *et al.* (1993) provide details of experiments with various clearing methods useful in locating GUS activity in pollen tubes in *Lycopersicon* pistils. Herr's fluids allowed complete clearing of the reproductive tissues allowing accurate location of pollen tubes expressing GUS activity. Liedl *et al.* (1993) also provide strategies for obtaining the restricted drug, chloral hydrate. For detailed protocols for isozyme techniques, see Ting *et al.* (1975), and Brewer (1970). With these approaches, a mosaic of control programs have been proposed in anthers that are partitioned with respect to cell type and developmental stage.

Tissue printing has been promoted recently as a useful advance in histochemical localization of proteins, enzyme activity, nucleic acids, and gene expression products, among other uses (Reid *et al.* 1992). Membranes such as nitrocellulose, Nytran, Genescreen, and Immobilon effectively bind proteins and nucleic acids. A freshly cut piece of plant tissue is placed firmly, or pressed against the membrane. This allows transfer of cell contents without lateral movement, and with reference to histological features. Detection of localized enzymes is accomplished using reactions that generate insoluble products. Nucleic acids can be detected through hybridization techniques. SEM imaging or X-ray microanalysis can be performed directly on the dried, imprinted membrane (Reine 1994). Reid *et al.* (1992) provide a comprehensive manual of techniques.

Vergne *et al.* (1987) developed a rapid assessment technique for determining the pollen development stage in wheat and maize inflorescences. Anthers from flowers of suspected desired stages are squashed into, or dissected in, a solution of 4'-6-diamidino 2-phenylindole (DAPI). This nuclear fluorochrome can be applied directly to fresh microspores on a slide, a cover glass added, and then examined immediately under epifluorescence optics. The technique was used by Devallee & Dumas (1988) who were able to clearly determine the condition and stage of maize nuclei. Pollen ontogeny was readily and accurately monitored while the authors studied proteins from known stage anthers using isoelectric focusing and electrophoresis.

The expanding field of culturing of immature anthers on synthetic media has been reviewed by Dunwell (1985). Anther culture is useful in studies of the diversion of immature gametophytes from normal developmental pathways toward that of a sporophyte. One can also examine the effects of differences in ploidy levels. Homozygotic (haploid) plant or callus induction from anther culture has wide application in plant breeding; it can contribute to the search for methods of single cell culture analogous to microorganism techniques.

POLLEN AND POLLINATION

Photographic documentation of pollinator–flower interactions in the field is often an essential practice in pollination biology. All aspects of the topic, including single-lens reflex cameras, films, metering, lenses, closeup and telephoto techniques are detailed in Shaw's (1987) treatment of nature photography.

Pollination biology as an integrative discipline including breeding systems, co-evolutionary adaptations, and ecology is covered by Faegri & van der Pijl (1979), Real (1983) and Jones & Little (1983). The phenomenon of secondary pollen presentation has been thoroughly reviewed by Yeo (1993). Simple experiments documenting scent production from anthers have been described by D'Arcy *et al.* (1990).

For a comprehensive compendium on all scientific techniques addressing this field, the recent work by Kearns & Inouye (1993) is a major contribution. Coverage includes approaches to the study of inflorescences, flowers and their parts, breeding systems, pollinators, sugar and other chemistry, and environmental measurements as well as anthers and pollen.

Pollen quantification is well covered by Kearns & Inouye. Anthers may be sampled to answer such questions as pollen availability versus efficiency of transfer mechanisms; pollen/ovule ratios, or pollen production versus flower or inflorescence size. For counting or analyzing content, anthers or flowers may be removed and bagged or placed in vials. Alternatively, anthers or small flowers may be collected with small forceps, placed in small polypropylene centrifuge tubes, covered with 70% ethanol, and macerated or sonicated. Collected pollen may be placed in a known quantity of ethanol, the ethanol evaporated, and the pollen counted microscopically.

Pollen collected from herbarium specimens may be rehydrated in Aerosol O-T. Dehisced or undehisced anthers may be rubbed over a 0.2 mm mesh screen, and washed into a vial with ethanol. Pollinia or anthers may be mounted in polyvinyl lactophenol. For cross-breeding, one uses an artist's brush, a toothpick, or taps anthers onto a glass slide.

Dry pollen may be removed from poricidal anthers by simulating buzz-pollination. Anthers are touched with a vibrating tine of a 512 Hz tuning fork, which is mid-range for vibrating insect wing musculature. Expelled pollen is collected on a black Bakelite ashtray and removed with an artist's brush (W. G. D'Arcy, personal communication). An electric toothbrush or a bench vibrator may expell pollen more effectively in some species.

For microscopic study of materials delivered with pollen, the pollen should be dusted onto glass slides moistened with dilute glycerine and detergent. These can be viewed directly under the microscope. Plane polarized light is useful if crystalline material is being examined. The sample can be rinsed from the slide with water into a vial or centrifuge tube.

Sampling pollen to test for viability using *in vivo* and *in vitro* tests is also discussed by Kearns & Inouye (1993). They provide formulas for germination trials using sucrose, sucrose/agar, or sucrose/gelatin. For some species, more elaborate recipes are provided that include boron and other salts necessary to induce germination of viable pollen grains.

The field of palynology has developed specific methodologies for the fields of petroleum exploration, pleistocene climatology, plant systematics, and plant evolutionary studies based on examination of pollen extending back to Mesozoic and Paleozoic strata. Traverse (1988, ch. 17) thoroughly reviews techniques in the field of paleopalynology. He describes variations in exine thickness and the effects of chemical treatments on pollen. He includes a section on the productivity of pollen per flower and per inflorescence for North American forests. The Appendix, 'Palynological laboratory techniques', is comprehensive.

Palynomorph preparation procedures used in laboratories of the U.S. Geological Survey have been described by Doher (1980). Phipps & Playford (1984) discuss techniques for extraction of palynomorphs from sediments. The text edited by Kummel & Raup (1965) on paleontological techniques has a relevant section, especially pages 598–613.

Collection of pollen on sticky slides for recognition of species has been described by Traverse (1988). D'Arcy *et al.* (this volume) used the same technique to document dispersal of crystals with pollen. Exine character recognition is usually poor with untreated pollen since the surfaces are obscured with oil drops and other 'intercalary inclusions'. Palynologists have traditionally used 'artificial fossilization' to clean the surfaces of the resistant exine before making permanent reference slides. Small flowers or anthers can be boiled in 10% KOH which hydrolyzes the cellulose and lyses the protoplasts. Resulting exines are colorless or light yellow. The grains can be stained in Safranin-O, basic fuchsin or other red stains. Optical systems (and human vision) are most acute in green light so such filtration is often used for maximum contrast when viewing red-stained pollen. Erdtman (1960) introduced the acetolysis technique which provides excellent translucent specimens of pollen that have a well-developed sporopollenin exine. Cellulose is completely hydrolyzed and the exine is darkened for good contrast in microscopy. This technique is especially apt for concentrating dispersed pollen from peat and other organic sediments.

Pollen taken from freshly dehisced anthers can be assayed for starch using the method of Baker & Baker (1979, 1982). They recommend mounting the samples in iodine dissolved in polyvinyl lactophenol.

CONCLUSION

An overview of methodology such as this demonstrates that a widening range of tools and procedures is at our disposal for studying all aspects of reproductive biology. In addition, traditional as well as newer cellular and molecular techniques are producing data that are expanding our concepts of genetics, function, structure, and systematics in integrated ways. Intellectual advances are generally due to an upward ratchet: the gradual accumulation of facts and the equally incremental improvements in ideas. A review of the empirical literature over the past century demonstrates this relationship. Jacob Bronowski (1978: 77) has argued for the primacy of the intellectual process – that finding or identifying a good problem is more difficult than solving one. But I would argue that well-honed investigative skills, and an understanding of their limits, remain central to the advance of our science.

ACKNOWLEDGMENTS

I am grateful for critical reviews of drafts of the manuscript by J.J. Bozzola, W.G. D'Arcy, J.M. Herr, Jr, W.A. Jensen, and W.J. Sundberg.

LITERATURE CITED

Anderson, L.E. 1954. Hoyer's solution as a rapid permanent mounting medium for bryologists. Bryologist 57: 242–244.

Baker, H.G. & I. Baker. 1979. Starch in angiosperm pollen grains and its significance. Amer. J. Bot. 66: 591–600.

Baker, H.G. & I. Baker. 1982. Starchy and starchless pollen in Onagraceae. Ann. Missouri Bot. Gard. 69: 748–754.

Baker, J.R. 1958. Principles of Biological Microtechnique. A Study of Fixation and Dying. Wiley, New York.

Barber, V.C. & D.L. Clayton. 1985. Electron Microscopy Safety Handbook. San Francisco Press, San Francisco.

Beerhues, L., G. Forkmann, H. Schopker, G. Stotz &

R. Wiermann. 1989. Flavanone 3-hydroxylase and dihydroflavonol oxygenase activities in anthers of *Tulipa*. J. Pl. Physiol. 133: 743–746.

Bennett, H.S., A.D. Wyrick, S.W. Lee & J.H. McNeil. 1976. Science and art in preparing tissues embedded in plastic for light microscopy, with special reference to glycol methacrylate, glass knives and simple stains. Stain Techn. 51: 71–97.

Berlyn, G.P. & J.P. Miksche. 1976. Botanical Microtechnique and Cytochemistry. Iowa State University Press, Ames.

Bhandari, N.N. 1984. The microsporangium. pp. 53–121 in Embryology of Angiosperms (edited by B.M. Johri). Springer-Verlag, Berlin.

Bierhorst, D.W. 1971. Morphology of Vascular Plants. Macmillan, New York.

Bold, H.C., C.J. Alexopoulos & T. Delevoryas. 1987. Morphology of Plants and Fungi. Harper & Row, New York.

Bowers, B. & M. Maser. 1988. Artifacts in fixation for transmission electron microscopy. pp. 13–42 in Artifacts in Biological Electron Microscopy (edited by R.F.E. Crang & K.L. Klomparens). Plenum Press, New York.

Boyde, A. & E. Maconnachie. 1983. Not quite critical point drying. pp. 71–73 in SEM (edited by J.-P. Revel, T. Barnard & G.H. Haggis). O'Hare, Illinois.

Bozzola, J. & L. Russell. 1992. Electron Microscopy. Principles and Techniques for Biologists. Jones & Bartlett, Boston.

Bracegirdle, B. 1978. A History of Microtechnique. Heinemann, London.

Bradbury, S. 1989. An Introduction to the Optical Microscope. Royal Microscopical Society; Oxford University Press, New York.

Brewer, G.J. 1970. An Introduction to Isozyme Techniques. Academic Press, New York.

Bronowski, J. 1978. Magic, Science, and Civilization. Columbia University Press, New York.

Bullock, G.R. 1984. The current status of fixation for electron microscopy: a review. J. Microsc. 133: 1–15.

Clark, G. (ed.) 1973. Staining Procedures, 3rd ed. Biological Stain Commission, Baltimore.

Conn, H.J., M.A. Darrow & V.M. Emmel. 1960. Staining Procedures Used by the Biological Stain Commission, 2nd ed. Biological Stain Commission; Williams & Wilkins, Baltimore.

Cox, K.H. & Goldberg, R.B. 1988. Analysis of plant gene expression. pp. 1–34 in Plant Molecular Biology: A Practical Approach (edited by C.H. Shaw) IRL Press. Oxford.

Crang, R.F.E. 1988. Artifacts in specimen preparation for scanning electron microscopy. pp. 107–129 in Artifacts in Biological Electron Microscopy (edited by R.F.E. Crang & K.L. Klomparens). Plenum Press, New York.

Crang, R.F.E. & K.L. Klomparens. 1988. Artifacts in Biological Electron Microscopy. Plenum Press, New York.

Culling, C.F.A. 1974. Modern Microscopy. Elementary Theory and Practice. Butterworths, Woburn, Mass.

D'Arcy, W.G., N.S. D'Arcy & R.C. Keating. 1990. Scented anthers in the Solanaceae. Rhodora 92: 50–53.

Davis, G.L. 1966. Systematic Embryology of the Angiosperms. Wiley, New York.

Delly, J. 1989. Photography Through the Microscope. Eastman Kodak, Rochester, New York.

Delvallee, I. & C. Dumas. 1988. Anther development in *Zea mays*: changes in protein, peroxidase, and esterase patterns. J. Pl. Physiol. 132: 210–217.

Determann, H. & F. Lepusch. 1974. The Microscope and its Application. Ernst Leitz, Wetzlar.

Doher, L.I. 1980. Palynomorph preparation procedures currently used in the paleontology and stratigraphy laboratories. U.S. Geol. Surv. Circ. no. 830.

Dunwell, J.M. 1985. Anther and ovary culture. pp. 1–44 in Cereal Tissue and Cell Culture (edited by S.W.J. Bright & M.G.K. Jones). W. Junk, Boston.

Dykstra, M.J. 1992. Biological Electron Microscopy. Theory, Techniques and Troubleshooting. Plenum Press, New York.

Erdtman, G. 1960. The acetolysis method. A revised description. Svensk Bot. Tidskr. 54: 561–564.

Feder, N. & T.P. O'Brien. 1968. Plant microtechnique: some principles and new methods. Amer. J. Bot. 55: 123–142.

Faegri, K. & L. van der Pijl. 1979. The Principles of Pollination Ecology. Pergamon Press, New York.

Flegler, S.L., J.W. Heckman, Jr. & K.L. Klomparens. 1993. Scanning and Transmission Electron Microscopy. An Introduction. W.H. Freeman, New York.

Forman, L. & D. Bridson. 1989. The Herbarium Handbook. Royal Botanic Gardens, Kew.

Fosberg, F.R. & M.-H. Sachet. 1965. Manual for tropical herbaria. Int. Bur. Plant Taxon & Nomenc., Regnum Veget. No. 39. Utrecht.

Fossard, R.A., de. 1969. Development and histochemistry of the endothecium in the anthers of *in vitro* grown *Chenopodium rubrum* L. Bot. Gaz. 130: 10–22.

Foster, A.S. 1949. Practical Plant Anatomy. Van Nostrand, Princeton, NJ.

Fredrikson, M., K. Carlsson & O. Franksson. 1988. Confocal scanning laser microscopy, a new technique used in an embryological study of *Dactyloriza maculata* (Orchidaceae). Nord. J. Bot. 8: 369–374.

Gahan, P.B. 1984. Plant Histochemistry and Cytochemistry. An Introduction. Academic Press, New York.

Gallagher, S.R. (ed.) 1992. GUS Protocols: Using the GUS Gene as a Reporter of Gene Expression. Academic Press, New York.

Gardner, R.O. 1975. An overview of botanical clearing technique. Stain Tech. 50: 99–105.

Gifford, E.M. & A.S. Foster. 1989. Morphology and Evolution of Vascular Plants. W.H. Freeman, New York.

Gould, K.S. & E.M. Lord. 1988. Growth of anthers in *Lilium longiflorum*. A kinematic analysis. Planta 173: 161–171.

Gray, P. 1954. The Microtomists Formulary and Guide. Blakiston, New York.

Hayat, M.A. 1975. Positive Staining for Electron Microscopy. Van Nostrand Reinhold, New York.

Herr, J.M., Jr. 1971. A new clearing-squash technique for the study of ovule development in angiosperms. Amer. J. Bot. 58: 785–790.

Herr, J.M., Jr. 1972. Applications of a new clearing technique for the investigation of vascular plant morphology. J. Elisha Mitchell Sci. Soc. 88: 137–143.

Herr, J.M., Jr. 1973. The use of Nomarski Interference microscopy for the study of structural features in cleared ovules. Acta Bot. Indica 1: 35–40.

Herr, J.M., Jr. 1974. A clearing-squash technique for the study of ovule and megagametophyte development in angiosperms. pp. 230–235 in Vascular Plant Systematics (edited by A.E. Radford, W.C. Dickison, J.R. Massey & C.R. Bell). Harper & Row, New York.

Herr, J.M., Jr. 1984. Embryology and taxonomy. pp. 647–696 in Embryology of Angiosperms (edited by B.M. Johri). Springer-Verlag, New York.

Herr, J.M., Jr. 1992a. New uses for calcium chloride solution as a mounting medium. Biotech. Histochem. 67: 9–13.

Herr, J.M., Jr. 1992b. Recent advances in clearing techniques for study of ovule and female gametophyte development. pp. 149–154 in Angiosperm Pollen and Ovules (edited by E. Ottaviano, D.L. Mulcahy, M. Sari Gorla, & G. Bergamini Mulcahy). Springer-Verlag, New York.

Heslop-Harrison, J. 1968. Tapetal origin of pollen coat substances in *Lilium*. New Phytol. 67: 779–786.

Heslop-Harrison, J. & Y. Heslop-Harrison. 1970. Evaluation of pollen viability by enzymatically induced fluorescence: intracellular hydrolysis of fluorescein diacetate. Stain Tech. 45: 115–120.

Hicks, A.J. & P.H. Hicks. 1978. A selected bibliography of plant collection and herbarium curation. Taxon 27: 63–99.

Hopwood, D. 1985. Cell and tissue fixation, 1972–1982. Histochem J. 17: 389–442.

Jensen, W.A. 1962. Botanical Histochemistry. W.H. Freeman, San Francisco.

Johansen, D.A. 1940. Plant Microtechnique. McGraw-Hill, New York.

Jokufu, K.D. & R.B. Goldberg. 1988. Analysis of plant gene structure. pp. 37–66 in Plant Molecular Biology: A Practical Approach (edited by C.H. Shaw). IRL Press, Oxford.

Jones, C.E. & R.J. Little. 1983. Handbook of Experimental Pollination Biology. Van Nostrand Reinhold, New York.

Kearns, C.A. & D.W. Inouye. 1993. Techniques for Pollination Biologists. University of Colorado Press, Niwot, CO.

Keating, R.C. 1969. A serum adhesive for plant microtechnique. Turtox News 47: 165.

Kho, Y.O. & J. Baer. 1968. Observing pollen tubes by means of fluorescence. Euphytica 17: 298–303.

King, R.M. & H. Robinson. 1970. The new synantherology. Taxon 19: 6–11.

Knox, R.B. & J. Heslop-Harrison. 1970. Direct demonstration of the low permeability of the angiosperm meiotic tetrad using a fluorogenic ester. Z. Pflanzenphysiol. 62: 451–459.

Koltunow, A.M., J. Truettner, C.H. Cox, M. Wallroth, and R.B. Goldberg. 1990. Different temporal and spatial gene expression patterns occur during anther development. Plant Cell 2: 1201–1224.

Kummel, B. & D. Raup. (eds.) 1965. Handbook of Paleontological Techniques. (esp. pp. 598–613.) W.H. Freeman, San Francisco.

Lersten, N.R. 1967. An annotated bibliography of botanical clearing methods. Iowa State J. Sci. 41: 481–486.

Lersten, N.R. 1986. Modified clearing method to show sieve tubes in minor veins of leaves. Stain Tech. 61: 231–234.

Lide, D.R. (ed.). 1991. CRC Handbook of Chemistry and Physics. CRC Press, Boca Raton.

Liedl, B.E., S. McCormick & M.A. Mutschler. 1993. A clearing technique for histochemical location of GUS activity in pollen tubes and ovules of *Lycopersicon*. Plant Molec. Biol. Rep. 11: 194–201.

Lillie, R.D. 1977. H.J. Conn's Biological Stains: A Handbook on the Nature and Uses of Dyes Employed in the Biological Laboratory, 9th ed. Williams & Wilkins, Baltimore.

Malies, H.M. 1959. Applied Microscopy and Photomicrography. Fountain Press, London.

Manning, J.C. & P. Goldblatt. 1990. Endothecium in Iridaceae and its systematic implications. Amer. J. Bot. 77: 527–532.

Mollenhauer, H.H. 1988. Artifacts caused by dehydration and epoxy embedding in transmission electron microscopy. pp. 43–64 in Artifacts in Biological Electron Microscopy (edited by R.F.E. Crang & K.L. Klomparens). Plenum Press, New York.

Mollring, F.K. 1981. Microscopy from the Very Beginning. Carl Zeiss, Oberkochen.

Nave, E.B. & V.K. Sawhney. 1986. Enzymatic changes in post-meiotic anther development in *Petunia hybrida*. I. Anther ontogeny and isozyme analysis. J. Pl. Physiol. 135: 451–465.

O'Brien, T.P. & M.E. McCully. 1981. The Study of Plant Structure. Principles and Selected Methods. Termarcarphi, Melbourne.

Ogden, E.C., Raynor, G.S., Hayes, J.V., Lewis, D.M. & J.H. Haines. 1974. Manual for Sampling Airborne Pollen. Hafner, New York.

Palser, B.F. 1975. The bases of angiosperm phylogeny: embryology. Ann. Missouri Bot. Gard. 62: 621–646.

Parsons, D.F. 1970. Biological Techniques in Electron Microscopy. Academic Press, New York.

Patzelt, W.J. 1974. Polarized Light Microscopy. Ernst Leitz, Wetzlar.

Pesacreta, T.C., V.I. Sullivan & K.H. Hasenstein. 1993. The connective base of *Cirsium horridulum*: description and comparison with the viscoelastic filament. Amer. J. Bot. 80: 411–418.

Phipps, D. & G. Playford. 1984. Laboratory techniques for extraction of palynomorphs from sediments. Pap. Dept. Geol. Univ. Queensland 11: 1–23.

Polowick, P.L. & V.K. Sawhney. 1993. An ultrastructural study of pollen development in tomato (*Lycopersicon esculentum*). I. Tetrad to early binucleate microspore stage. Can. J. Bot. 1039–1047.

Polowick, P.L., R. Bolaria & V.K. Sawhney. 1990. Stamen ontogeny in the temperature-sensitive 'stamenless-2' mutant of tomato (*Lycopersicon esculentum* L.) New Phytol. 115: 625–631.

Purvis, M.J., D.C. Collier & D. Walls. 1966. Laboratory Techniques in Botany, 2nd ed. Butterworths, London.

Radford, A.E., W.C. Dickison, J.R. Massey & C.R. Bell. 1974. Vascular Plant Systematics. Harper & Row, New York.

Raghavan, V. 1989. mRNAs and a cloned histone gene are differentially expressed during anther and pollen development in rice (*Oryza sativa* L.). J. Cell Sci. 92: 217–229.

Rao, C.K. 1977. A technique for the revival of herbarium specimens for floral dissections and anatomical studies. Curr. Sci. 46: 720

Raven, P.H., R.F. Evert & S.E. Eichhorn. 1992. Biology of Plants, 5th ed. Worth, New York.

Rawlins, D.J. 1992. Light Microscopy: Introduction to Biotechniques. Royal Microscopical Society, London.

Real, L. (ed.) 1983. Pollination Biology. Academic Press, New York.

Reid, P.D., R.F. Pont-Lezica, E. del Campillo & R. Taylor. 1992. Tissue Printing. Tools for the Study of Anatomy, Histochemistry and Gene Expression. Academic Press, New York.

Reine, B.A. 1994. Tissue printing for scanning electron microscopy and microanalysis. Microsc. Today 94: 7.

Rudall, P.J. & L. Clark. 1992. The megagametophyte in Labiatae. pp. 65–84 in Advances in Labiatae Science (edited by R.M. Harley & T. Reynolds). Royal Botanic Gardens, Kew.

Sass, J. 1958. Botanical Microtechnique, 3rd ed. Iowa State University Press. Ames.

Sawhney, V.K. & E.B. Nave. 1986. Enzymatic changes in post-meiotic anther development in *Petunia hybrida*. II. Histochemical localization of esterase, peroxidase, malate- and alcohol dehydrogenase. J. Pl. Physiol. 125: 467–473.

Schmid, R. 1977. Stockwell's bleach, an effective remover of tannins from plant tissues. Bot. Jahrb. Syst. 98: 278–287.

Shaw, J. 1987. Closeups in Nature. Amphoto, New York.

Siegel, I. 1967. Toluidine Blue O and Naphthol Yellow S; a highly polychromatic general stain. Stain Tech. 42: 29–30.

Silk, W.K. 1984. Quantitative descriptions of development. Ann. Rev. Pl. Physiol. 35: 479–518.

Smith, R.F. 1993. Microscopy and Photomicrography. A Working Manual, 2nd ed. CRC Press. Boca Raton.

Spencer, M. 1982. Fundamentals of Light Microscopy. Cambridge University Press. New York.

Stelly, D.M., S.J. Peloquin, R.G. Palmer & C.F. Crane. 1984. Mayer's hemalum-methyl salicylate: a stain-clearing technique for observations within whole ovules. Stain Tech. 59: 155–161.

Sterling, C. 1964. Comparative morphology of the carpel in the Rosaceae. I Prunoideae: *Prunus*. Amer. J. Bot. 51: 36–44.

Ting, I.P., I. Fuhr, R. Curry, W.C. Zschoche. 1975. Malate dehydrogenase isozymes in plants: preparation, properties and biological significance. pp. 369–384 in Isozymes, Vol. 2. Physiological Function (edited by C.L. Markert). Academic Press, New York.

Tomlinson. P.B. 1965. Collecting botanical material in fluid preservatives, with special reference to the tropics. pp. 117–119 in Manual for Tropical Herbaria (edited by F.R. Fosberg & M.-H. Sachet). Int. Bur. Plant Taxon. & Nom., Regn. Veget. No. 39, Utrecht.

Traverse, A. 1988. Paleopalynology. Unwin Hyman, Boston.

Venning, F. 1958. Handbook of Advanced Plant Microtechnique. W.C. Brown, Dubuque.

Vergne, P., I. Delvallee & C. Dumas. 1987. Rapid assessment of microspore and pollen development stage in wheat and maize using DAPI and membrane permeabilization. Stain Tech. 62: 299–304.

Wang, C.-S., L.L. Walling, K.J. Eckhard & E.M. Lord. 1992. Immunological characterization of a tapetal protein in developing anthers of *Lilium longiflorum*. Pl. Physiol. 99: 822–829.

Wang, C.-S., L.L. Walling, K.J. Eckhard & E.M. Lord. 1993. Characterization of an anther-specific glycoprotein in *Lilium longiflorum*. Amer. J. Bot. 80: 1155–1161.

Wergin, W.P. & C.D. Pooley. 1988. Photographic and interpretive artifacts. pp. 175–204 in 1988. Artifacts in Biological Electron Microscopy (edited by R.F.E. Crang & K.L. Klomparens). Plenum Press, New York.

Wilson, M.B. 1976. The Science and Art of Basic Microscopy. Amer. Soc. Med. Technol., Bellaire, Texas.

Yeo, P.F. 1993. Secondary Pollen Presentation. Form, Function and Evolution. Springer-Verlag, New York.

13 · A bibliography of stamen morphology and anatomy

ANNA H. LYNCH AND MARY GREGORY

This bibliography on stamens was prepared as a useful accompaniment to the original papers in this volume. We at first thought it could be compiled solely from the computerized database of plant anatomical literature maintained at the Jodrell Laboratory, but it soon became clear that this alone would not be adequate, for two main reasons.

First, the database, which was started on cards by Dr C.R. Metcalfe in the 1930s, was originally mainly devoted to vegetative anatomy, although by the 1960s the coverage had become more comprehensive. Second, the notes and keywords used at Kew were not sufficiently precise for identifying the contents of papers on floral anatomy to permit listing of each work under the categories we chose for this bibliography. Therefore we have re-checked as many of the articles as possible and have added to the list, especially ones from the first part of this century. Limitations of time have prevented us from seeing every work, and this means that some are less fully indexed than others.

Most pre-1900 articles have been omitted, except for a short list of significant publications, which are not indexed further. Taxonomic textbooks and monographs, reference works such as 'Die natürlichen Pflanzenfamilien', and floras often contain valuable information on stamens, but they are outside the scope of this bibliography, as are articles on groups other than angiosperms.

Most aspects of stamen morphology and anatomy are covered, including the tapetum, but not details of cell division, microsporogenesis or pollen. Also excluded are anther culture, most papers on pollination biology and cladistic analyses, and ones dealing with techniques. We have included papers of a general nature, for example ones discussing homologies, and those dealing with androecial anatomy of angiosperm families and genera, but we have omitted the numerous papers on the anatomy of one or two individual species unless they are of special interest. For further information on individual taxa readers should consult general works, such as Davis 1966 (234), Johri & Kapil 1953 (570), and Leinfellner 1956 (686).

The indexing is by subject (p. 323) and also by plant family (p. 325), mostly as recognized at Kew (see Brummitt, R.K. 1992. Vascular plant families and genera. Royal Botanic Gardens, Kew.), except that Liliaceae s.l. includes all the segregate families, such as Asphodelaceae, Hyacinthaceae, etc.

In the subject index, articles are not indexed under Endothecium or Tapetum if those topics are only briefly mentioned. In the list of references all authors whose names include Van have been listed under Van, although this is not the practice in some countries.

SELECTED REFERENCES TO OLDER LITERATURE (NOT INDEXED)

Bonnier, G. 1879. Les nectaires. Etude critique, anatomique et physiologique. Ann. Sci. Nat., Bot., sér. 6, 8: 5–212.

Čelakovský, L.J. 1875. Ueber den 'eingeschalteten' epipetalen Staubgefässkreis. Flora 58: 481–489, 497–504, 513–524.

Čelakovský, L.J. 1878. Teratologische Beiträge zur morphologischen Deutung des Staubgefässes. Jahrb. Wiss. Bot. 11: 124–174 + Plates V–VII.

Čelakovský, L.J. 1884. Untersuchungen über die Homologien der generativen Produkte der Fruchtblätter bei den Phanerogamen und Gefässkryptogamen. Jahrb. Wiss. Bot. 14: 291–378 + Plates XIX–XXI.

Čelakovský, L.J. 1894 [1895]. Das Reductionsgesetz der

Blüthen, das Dédoublement und die Obdiplostemonie. Sber. K. Böhm. Ges. Wiss., Math.-Nat. Cl. (3): 140 pp. + 5 plates.

Čelakovský, L.J. 1896, 1900. Über den phylogenetischen Entwickelungsgang der Blüthe und über den Ursprung der Blumenkrone. Sber. K. Böhm. Ges. Wiss., Math.-Nat. Cl. (40): 1–91 (1896); (3): 1–221 (1900).

Chatin, A. 1855. Sur les types obdiplostémone et diplostémone direct, ou de l'existence et des caractères de deux types symetriques distinctes chez les fleurs diplostémones. Bull. Soc. Bot. Fr. 2: 615–624.

Chatin, A. 1870. Causes de la déhiscence des anthères. C.R. Acad. Sci., Paris, sér. D, 70: 201–203, 410–413, 644–648.

Chester, G.D. 1897. Bau und Funktion der Spaltöffnungen auf Blumenblättern und Antheren. Ber. Dtsch. Bot. Ges. 15: 420–431.

Clos, D. 1866. La feuille florale et l'anthère. Mém. Acad. Sci. Toulouse, sér. 6, 4: 141–158.

Clos, D. 1876? La feuille florale et le filet staminal. Mém. Acad. Sci. Toulouse, sér. 7, 9: 30 pp.

Clos, D. 1889. Individualité des faisceaux fibro-vasculaires des appendices des plantes. Mém. Acad. Sci. Toulouse 11: 248–267.

Delpino, F. 1867. Sugli apparecchi della fecondazione nelle piante antocarpee (Fanerogame). Sommario di osservazione fatte negli anni 1865–1866. Atti Soc. Ital. Sci. Nat. 10(3): 39 pp.

Delpino, F. 1883. Teoria generale della fillotassi. Atti. R. Univ. Genova 4: 2 et seq.

Delpino, F. 1890–1891. Contribuzione alla teoria della pseudanzia. Malpighia 4: 302–312 + Plate X.

Dickson, A. 1866. On diplostemonous flowers; with some remarks upon the position of the carpels in the Malvaceae. Trans. Bot. Soc. Edinb. 8: 86–107 + Plate I.

Eichler, A.W. 1875–1878. Blüthendiagramme. W. Engelmann, Leipzig.

Engler, A. 1876. Beiträge zur Kenntniss der Antherenbildung der Metaspermen. Jahrb. Wiss. Bot. 10: 275–316 + Plates XX–XXIV.

Goebel, K. 1882. Beiträge zur Morphologie und Physiologie des Blattes. III. Ueber die Anordnung der Staubblätter in einigen Blüten. Bot. Ztg. 40: 352–364, 368–379, 385–394, 400–413.

Goebel, K. 1886. Beiträge zur Kenntniss gefüllter Blüthen. Jahrb. Wiss. Bot. 17: 207–296 + Plates XI–XV.

Grélot, P. 1897. Sur les faisceaux staminaux. Rev. Gén. Bot. 9: 273–281.

Grélot, P. 1898. Recherches sur le système libéroligneux

floral des Gamopétales bicarpellées. Thèse, Paris. Masson, Paris. 153 pp.

Henslow, G. 1888. On the origin of floral structures through insects and other agencies. Kegan, Paul & Trench, London.

Henslow, G. 1891. On the vascular systems of floral organs, and their importance in the interpretation of the morphology of flowers. J. Linn. Soc. (Bot.) 28: 151–197 + Plates XXIII–XXXII.

Hofmeister, W. 1868. Allgemeine Morphologie der Gewächse. Leipzig.

Knuth, P. 1898–1905. Handbuch der Blütenbiologie. 3 vols. W. Engelmann, Leipzig. (Eng. transl. by J.R. Ainsworth Davis. 1906–1909. Handbook of flower pollination. Clarendon Press, Oxford.)

Koehne, E. 1869. Über Blüthenentwicklung bei den Compositen. Inaug.-Diss., Berlin. T.C.F. Enslin, Berlin. 71 pp.

Leclerc du Sablon. 1885. Recherches sur la structure et la déhiscence des anthères. Ann. Sci. Nat., Bot., sér. 7, 1: 97–134 + 4 plates.

Masters, M.T. 1869. Vegetable Teratology. R. Hardwicks, London. 534 pp.

Mohl, H. von. 1830. Ueber die fibrosen Zellen der Antheren. Flora 13: 697–712, 715–728, 729–742.

Mohl, H. von. 1845. Beobachtungen über die Umwandlung von Antheren in Carpelle. pp. 28–44 in: Vermischte Schriften botanischen Inhalts. L.F. Fues, Tubingen.

Molliard, M. 1896. Homologie du massif pollinique et de l'ovule. Rev. Gén. Bot. 8: 273–283.

Moquin-Tandon, A. 1826. Essai sur les dédoublements ou multiplications d'organes dans les végétaux. Montpellier.

Müller, H. 1873. Die Befruchtung der Blumen durch Insekten und die gegenseitigen Anpassungen beider. W. Engelmann, Leipzig. (Eng. transl. by W. D'Arcy Thompson. 1883. The Fertilisation of Flowers. London.)

Payer, J.-B. 1857. Traité d'organogénie comparée de la fleur. Masson, Paris. 2 vols. (Reprint 1966 by J. Cramer, Lehre.)

Penzig, O. 1890–1894. Pflanzen-Teratologie systematisch geordnet. A. Ciminago, Genua. 540 pp. (2nd edn. 1921–1922. 3 vols. Borntraeger, Berlin.)

Purkinje, J.E. [Purkyne, J.E.] 1830. De cellulis antherarum fibrosis nec non de granorum pollinarium formis. Grueson, Breslau. 58 pp.

Schinz, H. 1883. Untersuchungen über den Mechanismus des Aufspringens der Sporangien und Pollensäcke. Diss., Zurich. 46 pp.

Schumann, K. 1887. Beiträge zur vergleichenden Blüthenmorphologie. Jahrb. Wiss. Bot. 18: 133–193 + Plates IV–V.

Schumann, K. 1890. Neue Untersuchungen über den Blüthenanschluss. W. Engelmann, Leipzig. 519 pp.

Schwendener, S. 1899. Über den Öffnungsmechanismus der Antheren. Sber. Preuss. Akad. Wiss., Berlin, Phys.-Math. Kl. (6): 101–107.

Sprengel, C.K. 1793. Das entdeckte Geheimnis der Natur im Bau und in der Befruchtung der Blumen. Berlin.

Steinbrinck, C. 1895. Grundzüge der Oeffnungsmechanik von Blütenstaub- und einigen Sporenbehältern. Bot. Jaarboek, Gent 7; 222–358.

Steinbrinck, C. 1899. Ueber den hygroskopischen Mechanismus von Staubbeuteln und Pflanzenhaare. pp. 165–183 in: Botanische Untersuchungen. S. Schwendener zum 10 Februar 1899 dargebracht. Borntraeger, Berlin.

Van Teighem, P. 1871. Recherches sur la structure du pistil et sur l'anatomie comparée de la fleur. Mém. Acad. Sci. Fr. 21: 1–261 + 12 plates.

Warming, E. 1873. Untersuchungen über Pollen bildende Phyllome und Kaulome. Hanstein's Bot. Abh. 2(2): 1–90.

Warming, E. 1878. De l'ovule. Ann. Sci. Nat., Bot., sér. 6, 5: 177–266.

REFERENCES (INDEXED: Subject p. 323; Systematic p. 325)

1. Abadie, M., M.-T. Cerceau-Larrival & F. Roland-Heydacker. 1980–1 [1982]. Quelques aspects de la production des tapis dans les anthères de deux ombellifères: *Turgenia latifolia* (L.) Hoffm. et *Hydrocotyle mexicana* Cham. et Schlecht. Ann. Sci. Nat., Bot., sér. 13, 2–3: 149–162.

2. Abadie, M. & M. Hideux. 1979, 1980. L'anthère de *Saxifraga cymbalaria* L. ssp. *huetiana* (Boiss.) Engl. et Irmsch. en microscopie électronique (M.E.B. et M.E.T.). 1. Généralités. Ontogenèse des orbicules. Ann. Sci. Nat., Bot., sér. 13, 1: 199–233 (1979). 2. Ontogenèse du sporoderme. Ibid. sér. 13, 1: 237–281 (1980).

3. Abadie, M. & M. Hideux. 1983. Les anthères de *Saxifraga clusii* Gouan et de *Saxifraga sempervivum* C. Koch. Quelques phases de l'ontogenèse des cellules tapétales et sporales. Ann. Sci. Nat., Bot., sér. 13, 5: 71–95.

4. Abbe, E.C. 1935, 1938. Studies in the phylogeny of the Betulaceae. I. Floral and inflorescence anatomy and morphology. Bot. Gaz. 97: 1–67 (1935). II. Extremes in the range of variation of floral and inflorescence morphology. Bot. Gaz. 99: 431–469 (1938).

5. Abramian, L.Kh. 1979. Ultrastructure of anthers of *Cerasus vulgaris* Mill. Biol. Zh. Arm. 32: 1008–1015. [Russ.] [Not seen.]

6. Abreu, I., A. Santos & R. Salema. 1981. Fine structure of protein crystals and bacilliform-type virus in anthers of *Lycopersicon esculentum* Mill. Bol. Soc. Brot., sér. 2, 55: 9–17.

7. Addicott, F.T. 1982. Abscission. University of California Press, Berkeley. 369 pp.

8. Agarwal, A. & P.K.K. Nair. 1976 [1977]. A study of anther morphology and dehiscence in some grasses. New Bot. 3: 154–155.

9. Ahurmetiv, A.A. 1982. Structure and development of the anther, microsporogenesis and the formation of pollen in *Meristotropis bucharica* (Rgl.) Krug. Uzb. Biol. Zh. (3): 35–38. [Russ.] [Not seen.]

10. Albertini, L., H. Grenet-Auberger & A. Souvré. 1981. Polysaccharides and lipids in microsporocytes and tapetum of *Rhoeo discolor* Hance. Cytochemical study. Acta Soc. Bot. Pol. 50: 21–28.

11. Albertini, L., A. Souvré & J.C. Audran. 1987. Le tapis de l'anthère et ses relations avec les microsporocytes et les grains de pollen. Rev. Cytol. Biol. Vég., Bot., 10: 211–242.

12. Aleksandrov, V.G. & A.V. Dobrotvorskaya. 1957. The formation of stamens and of a fibrous layer in anthers. Bot. Zh. SSSR 43: 1473–1490. [Russ.]

13. Aleksandrov, V.G. & A.V. Dobrotvorskaya. 1958. On the floral parts metamorphosis and morphological nature of the stamen. Probl. Bot. 3: 248–270. [Russ.; Eng. summ.]

14. Aleksandrov, V.G. & A.V. Dobrotvorskaya. 1960. The formation of stamens and of the fibrous layer in the anthers of some plants. Bot. Zh. SSSR 45: 823–831. [Russ.; Eng. summ.]

15. Allorge, L. 1975. Rattachement de la tribu des Allamadées aux Echitoidées (Apocynacées). Adansonia, sér. 2, 15: 273–276.

16. Al Nowaihi, A.S. & S.F. Khalifa. 1971 [1972]. Studies on some taxa of the Geraniales. I. Floral morphology of *Averrhoa carambola* L., *Oxalis cernua* Thunb. and *O. corniculata* L. with reference to the nature of the staminodes. Proc. Indian Natn. Sci. Acad., B, 37: 189–198.

17. Amaral, M.C.E. 1991. Phylogenetische Systematik der Ochnaceae. Bot. Jahrb. 113: 105–196.

18. Anderberg, A. 1982. The genus *Anvillea* (Compositae). Nord. J. Bot. 2: 297–305.

19. Anderberg, A.A., P.-O. Karis & G. El-Ghazaly. 1992. *Cratystylis* an isolated genus of the Asteraceae-Cichorioideae. Aust. Syst. Bot. 5: 81–94.

20. Anderson, G.J. & D.E. Symon. 1989. Functional dioecy and andromonoecy in *Solanum*. Evolution 43: 204–219.

21. Anderson, L.C. 1970. Floral anatomy of *Chrysothamnus* (Astereae, Compositae). Sida 3: 466–503.

22. Andronova, N.N. 1984. The structure of the anther and pollen development in the Rubiaceae. Bot. Zh. SSSR 69: 43–54. [Russ.; Eng. summ.]

23. Arber, A. 1913. On the structure of the androecium in *Parnassia* and its bearing on the affinities of the genus. Ann. Bot. 27: 491–510 + Plate XXXVI.

24. Arber, A. 1915. The anatomy of the stamens in certain Indian species of *Parnassia*. Ann. Bot. 29: 159–160.

25. Arber, A. 1926, 1927, 1929. Studies in the Gramineae. I. The flowers of certain Bambuseae. Ann. Bot. 40: 447–609 (1926). II. Abnormalities in *Cephalostachyum virgatum* Kurz, and their bearing on the interpretation of the bamboo flower. Ibid. 41: 47–74 (1927). VI. 1. *Streptochaeta*. 2. *Anomochloa*. 3. *Ichnanthus*. Ibid. 43: 35–53. VIII. On the organization of the flower in the bamboo. Ibid.: 765–781 (1929).

26. Arber, A. 1931, 1932. Studies in floral morphology. I. On some structural features of the cruciferous flowers. New Phytol. 30: 11–41. II. On some normal and abnormal crucifers: with a discussion on teratology and atavism. Ibid.: 172–203. III. On the Fumarioideae, with special reference to the androecium. Ibid.: 317–354 (1931). IV. On the Hypecoideae, with special reference to the androecium. Ibid. 31: 145–173 (1932).

27. Arber, A. 1933. Floral anatomy and its morphological interpretation. New Phytol. 32: 231–242.

28. Arber, A. 1937. The interpretation of the flower: a study of some aspects of morphological thought. Biol. Rev. 12: 157–184.

29. Arber, A. 1937, 1939, 1940. Studies in flower structure. III. On the 'corona' and androecium in certain Amaryllidaceae. Ann. Bot. (NS.) 1: 293–304 (1937). V. On the interpretation of the petal and 'corona' in *Lychnis*. Ibid. 3: 337–346 (1939). VI. On the residual vascular tissue in the apices of reproductive shoots, with special reference to *Lilaea* and *Amherstia*. Ibid. 4: 617–627 (1940).

30. Arber, A. 1950. The natural philosophy of plant form. pp. 33–58 in: Plant Morphology (by A. Arber). Cambridge University Press, Cambridge.

31. Armstrong, J.E. 1986. Comparative floral anatomy of Solanaceae: a preliminary survey. pp. 101–113 in: Solanaceae. Biology and systematics (edited by W.G. D'Arcy). Columbia University Press, New York.

32. Armstrong, J.E. 1992. Lever action anthers and the forcible shedding of pollen in *Torenia* (Scrophulariaceae). Amer. J. Bot. 79: 34–40.

33. Armstrong, J.E. & A.K. Irvine. 1990. Function of staminodia in the beetle-pollinated flowers of *Eupomatia laurina*. Biotropica 22: 429–431.

34. Armstrong, J.E. & S.C. Tucker. 1986. Floral development in *Myristica* (Myristicaceae). Amer. J. Bot. 73: 1131–1143.

35. Armstrong, J.E. & T.K. Wilson. 1978. Floral morphology of *Horsfieldia* (Myristicaceae). Amer. J. Bot. 65: 441–449.

36. Arnal, C. 1945. Recherches Morphologiques et Physiologiques sur la Fleur des Violacées. Diss. Berthier, Dijon.

37. Aronne, G., C.C. Wilcock & P. Pizzolongo. 1993. Pollination biology and sexual differentiation of *Osyris alba* (Santalaceae) in the Mediterranean region. Pl. Syst. Evol. 188: 1–16.

38. Arora, K. & B. Tiagi. 1977. The role of the endothecium in the identification of umbellifers. Curr. Sci. 46: 531.

39. Artopoeus, A. 1903. Über den Bau und die Öffnungsweise der Antheren und die Entwickelung der Samen der Erikaceen. Flora 92: 309–345.

40. Asplund, E. 1920. Studien über die Entwicklungsgeschichte der Blüten einiger Valerianaceen. K. Sven. Vetenkapsakad. Handl. 61 (3): 1–66.

41. Audran, J.C. & M. Batcho. 1981. Cytochemical and infrastructural aspects of pollen and tapetum ontogeny in *Silene dioica* (Caryophyllaceae). Grana 20: 65–80.

42. Audran, J.C. & L. Dan Dicko-Zafimahova. 1992. Aspects ultrastructuraux et cytochimiques du tapis staminal chez *Calotropis procera* (Asclepiadaceae). Grana 31: 253–272. [Eng. summ.]

43. Ayers, T.J. 1990. Systematics of *Heterotoma* (Campanulaceae) and the evolution of nectar spurs in the New World Lobelioideae. Syst. Bot. 15: 296–327.

44. Aziz, P. 1978. Initiation of procambial strands in the primordia of stamens and carpel of *Triticum aestivum* L. Pak. J. Sci. Indust. Res. 21 (1): 12–16. [Not seen.]

45. Aziz, P. 1981. Initiation of primary vascular elements in the stamens and carpel of *Triticum aestivum* L. Bot. J. Linn. Soc. 82: 69–79.

46. Bacon, J.D. 1989. Systematics of *Nama*

(Hydrophyllaceae): reevaluation of the taxonomic status of *Lemmonia californica*. Aliso 12: 327–333.

47. Baehni, C. & C.E.B. Bonner. 1948. La vascularisation des fleurs chez les Lopezieae (Onagracées). Candollea 11: 305–322.

48. Bahadur, B., A. Chaturvedi & N.R. Swamy. 1990. SEM studies of pollen in relation to enantiostyly and heteranthery in *Cassia* (Caesalpinaceae). J. Palynol. 26: 7–22. [Also in: Silver Jubilee Commem. Vol.: Current perspectives in palynological research (edited by S. Chanda).]

49. Bailey, I.W., & C.G. Nast. 1943, 1945. The comparative morphology of the Winteraceae. I. Pollen and stamens. J. Arnold Arbor. 24: 340–346 (1943). V. Summary and conclusions. Ibid. 26: 37–47 (1945).

50. Bailey, I.W., C.G. Nast & A.C. Smith. 1943. The family Himantandraceae. J. Arnold Arbor. 24: 190–206.

51. Bailey, I.W. & A.C. Smith. 1942. Degeneriaceae, a new family of flowering plants from Fiji. J. Arnold Arbor. 23: 356–365.

52. Bailey, I.W. & B.G.L. Swamy. 1949. Morphology and relationships of *Austrobaileya*. J. Arnold Arbor. 30: 211–226.

53. Banerjee, U.C. 1967. Ultrastructure of the tapetal membrane of grasses. Grana Palynol. 7: 365–377.

54. Banerji, M.L. 1977. The laminar stamen and what after that? Acta Bot. Indica 5: 1–7.

55. Barboza, G. & A.T. Hunziker. 1991. Estudios sobre Solanaceae XXXI. Peculiaridades del androceo de interes taxonomico en *Solanum*. Kurtziana 21: 185–194. [Eng. summ.]

56. Barlow, B.A. 1987. *Regelia punicea* (N. Byrnes) Barlow, comb. nov. (Myrtaceae) from the Northern Territory: phytogeographic implications. Brunonia 9: 89–97.

57. Barlow, P.W. & J.A. Sargent. 1975. The ultrastructure of the hair cells on the anther of *Bryonia dioica*. Protoplasma 83: 351–364.

58. Barnard, C. 1957, 1958, 1960. Floral histogenesis in the monocotyledons. I. The Gramineae. Aust. J. Bot. 5: 1–20. II. The Cyperaceae. Ibid.: 115–128 (1957). III. The Juncaceae. Ibid. 6: 285–298 (1958). IV. The Liliaceae. Ibid. 8: 213–225 (1960).

59. Barnard, C. 1961. The interpretation of the angiosperm flower. [Presidential address.] Aust. J. Sci. 24: 64–72.

60. Batenburg, L.H. & B.M. Moeliono. 1982. Oligomery and vasculature in the androecium of *Mollugo nud-*

icaulis Lam. (Molluginaceae). Acta Bot. Neerl. 31: 215–220.

61. Batygina, T.B. & I.I. Shamrov. 1981. The embryology of Nymphaeales and Nelumbonales, I. The development of the anther. Bot. Zh. SSSR 66: 1696–1709. [Russ.; Eng. summ.]

62. Batygina, T.B. & I.I. Shamrov. 1983. Embryology of the Nelumbonaceae and Nymphaeaceae. Pollen grain structure (some peculiar features of correlated development of the pollen grain and anther wall). Bot. Zh. SSSR 68: 1177–1183 + 2 plates. [Russ.; Eng. summ.]

63. Batygina, T.B., E.S. Teriokhin [E.S. Terekhin], G.K. Alimova & M.S. Yakovlev. 1963. Genesis of male sporangia in the families Gramineae and Ericaceae. Bot. Zh. SSSR 48: 1108–1120. [Russ.; Eng. summ.]

64. Bauer, R. 1922. Entwicklungsgeschichtliche Untersuchungen an Polygonaceenblüten. Flora 115: 272–292 + Plates I-II.

65. Baum, H. 1948. Postgenitale Verwachsung in und zwischen Karpell- und Staubblattkreisen. Sber. Akad. Wiss. Wien, Math.-Nat. Kl. Abt. I. 157: 17–38.

66. Baum, H. 1949. Beiträge zur Kenntnis der Schildform bei den Staubblättern. Öst. Bot. Z. 96: 453–466.

67. Baum, H. 1950. Lassen sich in der Ontogenese der Karpelle und Staubblätter noch Anklänge an ihre phylogenetische Entwicklung aus Telomen feststellen? Öst. Bot. Z. 97: 333–341.

68. Baum, H. 1952. Die Bedeutung der diplophyllen Übergangsblätter für den Bau der Staubblätter. Öst. Bot. Z. 99: 228–243.

69. Baum, H. 1953. Die Unabhängigkeit der diplophyllen Gestalt der Staubblattspreite von ihrer Funktion als Träger der Pollensäcke. Öst. Bot. Z. 100: 265–269.

70. Baum, H. & W. Leinfellner. 1953. Die ontogenetischen Abänderungen des diplophyllen Grundbaues der Staubblätter. Öst. Bot. Z. 100: 91–135.

71. Beardsell, D.V., E.G. Williams & R.B. Knox. 1989. The structure and histochemistry of the nectary and anther secretory tissue of the flowers of *Thryptomene calycina* (Lindl.) Stapf (Myrtaceae). Aust. J. Bot. 37: 65–80.

72. Bechtel, A.R. 1921. The floral anatomy of the Urticales. Amer. J. Bot. 8: 386–410.

73. Beck, N.G. & E.M. Lord. 1988. Breeding system in *Ficus carica*, the common fig. I. Floral diversity. Amer. J. Bot. 75: 1904–1912.

74. Becquerel, P. 1932. La déhiscence de l'anthère du lis blanc. C.R. Acad. Sci., Paris, D, 195: 165–167.

75. Beer, R. 1905. On the development of the pollen

grain and anther of some Onagraceae. Beih. Bot. Zbl. 19 (I): 286–313.

76. Beille, M.L. 1901. Recherches sur le développement floral des Disciflores. Actes Soc. Linn. Bordeaux 56: 235–410.

77. Beljaeva, L.E., E.A. Chajka & N.S. Fursa. 1978. Development of anther, ovule and gametogenesis of *Diplotaxis tenuifolia* DC. Ukrain. Bot. Zh. 35: 175–179. [Ukrain.; Eng. summ.]

78. Benl, G. 1937. Eigenartige Verbreitungseinrichtungen bei der Cyperaceengattung *Gahnia* Forst. (Die Befestigung der Früchte an den persistierenden Filamenten.) Flora Jena 131: 369–386.

79. Bennek, C. 1958. Die morphologische Beurteilung der Staub- und Blumenblätter der Rhamnaceen. Bot. Jahrb. 77: 423–457 + Plates 19–27.

80. Bennett, M.D., M.K. Rao, J.B. Smith & M.W. Bayliss. 1973. Cell development in the anther, the ovule, and the young seed of *Triticum aestivum* L. var. Chinese spring. Phil. Trans. R. Soc., Lond., B, 266: 39–81. [Not seen.]

81. Bennett, M.D. & J.R. Smith. 1979. Colchicine induced paracrystals in the tapetum of wheat anthers. J. Cell Sci. 38: 23–32.

82. Bensel, C.R. & B.F. Palser. 1975. Floral anatomy in the Saxifragaceae sensu lato. I. Introduction, Parnassioideae and Brexioideae. Amer. J. Bot. 62: 176–185. II. Saxifragoideae and Iteoideae. Ibid.: 661–675. III. Kirengeshomoideae, Hydrangeoideae and Escallonioideae. Ibid.: 676–687. IV. Baueroideae and conclusions. Ibid.: 688–694.

83. Berg. C.C. 1977. Abscission of anthers in *Cecropia* Loefl. Acta Bot. Neerl. 26: 417–419.

84. Bergann, F. & L. Bergann. 1984. Zur Entwicklungsgeschichte des Angiospermenblattes. V. Über die Anlegung von Blättern und Blütenorganen im Lichte klassischer und moderner Histogeneseforschung (Abschliessender Bericht). Biol. Zbl. 103: 655–675.

85. Berger, K. 1938. Zucker, Stärke und Verholzung in den Antheren einiger Angiospermen. Öst. Bot. Z. 87: 161–220.

86. Bersillon, G.G. 1955. Recherches sur les Papavéracées. Contribution à l'étude du développement des dicotylédones herbacées. Ann. Sci. Nat., Bot., sér. 11, 16: 225–447.

87. Bertin, R.I. & C.M. Newman. 1993. Dichogamy in angiosperms. Bot. Rev. 59: 112–152.

88. Bhandari, N.N. 1984. The microsporangium. pp. 53–121 in: Embryology of Angiosperms (edited by B.M. Johri). Springer-Verlag, Berlin.

89. Bhandari, N.N. & R. Khosla. 1982 [1983]. Development and histochemistry of anther in *Triticale* cv. Tri-1. I. Some new aspects in early ontogeny. Phytomorphology 32: 18–27.

90. Bhandari, N.N. & R. Kishori. 1973. Development of tapetal membrane and Ubisch-granules in *Nigella damascena* – a histochemical approach. Beitr. Biol. Pfl. 49: 59–72.

91. Bhandari, N.N., R. Kishori & S. Natesh. 1976. Ontogeny, cytology, and histochemistry of anther tapetum in relation to pollen development in *Nigella damascena*. Phytomorphology 26: 46–59.

92. Bhandari, N.N. & M. Sharma. 1983. Ontogenetic and histochemical studies on the anther of *Carthamus tinctorius* L. J. Palynol. 19: 153–180.

93. Bhandari, N.N. & M. Sharma. 1987. Histochemical and ultrastructural studies during anther development in *Solanum nigrum* Linn. I. Early ontogeny. Phytomorphology 37: 249–260.

94. Bhatia, D.S. & A. Chopra. 1978. Histochemical localization of polysaccharides, phosphorylase α-amylase in the developing anther of *Datura alba* Nees. pp. 15–19 in: Physiology of Sexual Reproduction in Flowering Plants (edited by C.P. Malik *et al.*). Kalyani Publ., New Delhi.

95. Bhatnagar, A.K. & R.N. Kapil. 1979 [1980]. Ontogeny and taxonomic significance of anther in *Bischofia javanica*. Phytomorphology 29: 298–306.

96. Bhatnagar, S.P. & V. Gandhi. 1980. The anther wall of *Phthirusa pyrifolia* (Loranthaceae). Bot. Jahrb. 101: 385–388.

97. Bhatnagar, S.P. & R. Garg. 1984. Endothecial thickenings in the anther wall of *Phthirusa adunca* (Loranthaceae). Pl. Syst. Evol. 146: 265–267.

98. Bhattacharjya, S.S. 1954. Ein Beitrag zur Morphologie des Androeceums von *Benincasa hispida* (Thunb.) Cogn. Ber. Dtsch. Bot. Ges. 67: 22–25.

99. Bhojwani, S.S. & S.P. Bhatnagar. 1978. The Embryology of Angiosperms, 3rd ed. Vikas Publ. House, New Dehli. [Not seen.]

100. Bhuskute, S.M. 1985. Staminal organization in Cucurbitaceae. II. *Bryonopsis laciniosa* (L.) Naud. J. Pl. Anat. Morph. 2 (2): 53–57.

101. Bhuskute, S.M., K.H. Makde & P.K. Deshpande. 1986. Staminal organization in *Coccinia grandis* (L.) Voigt. Ann. Bot. 57: 415–418.

102. Biddle, J.A. 1979. Anther and pollen development in garden pea and cultivated lentil. Can. J. Bot. 57: 1883–1900.

103. Bino, R.G. 1985. Histological aspects of microsporo-

genesis in fertile, cytoplasmic male sterile and restored fertile *Petunia hybrida*. Theor. Appl. Genet. 69: 423–428.

104. Bino, R.J. 1985. Ultrastructural aspects of cytoplasmic male sterility in *Petunia hybrida*. Protoplasma 127: 230–240.

105. Blaser, H.W. 1941. Studies in the morphology of the Cyperaceae. I. Morphology of flowers. A. Scirpoid genera. Amer. J. Bot. 28: 542–551. B. Rhynchosporoid genera. Ibid.: 832–838.

106. Bloembergen, S. 1939. A revision of the genus *Alangium*. Bull. Jard. Bot. Buitenzorg III, 16: 139–235.

107. Bode, H.R. 1938. Das Stern-Endothecium als ein wichtiges Erkennungsmerkmal im mikroskopischen Drogenbild von Flores Verbasci. Pharm. Zentralhalle 79: 681–683. [Not seen.]

108. Bogle, A.L. 1970. Floral morphology and vascular anatomy of the Hamamelidaceae: the apetalous genera of Hamamelidoideae. J. Arnold Arbor. 51: 310–366.

109. Bogle, A.L. 1986. The floral morphology and vascular anatomy of the Hamamelidaceae: subfamily Liquidambaroideae. Ann. Missouri Bot. Gard. 73: 325–347.

110. Bogle, A.L. 1989. The floral morphology, vascular anatomy, and ontogeny of the Rhodoleioideae (Hamamelidaceae) and their significance in relation to the 'lower' hamamelids. pp. 201–220 in: Evolution, Systematics, and Fossil History of the Hamamelidae, Vol. 1. Introduction and 'Lower' Hamamelidae (edited by P.R. Crane & S. Blackmore). Clarendon Press, Oxford.

111. Boke, N.H. 1949. Development of the stamens and carpels in *Vinca rosea* L. Amer J. Bot. 36: 535–547.

112. Bond, C.J. 1915. On the primary and secondary sex characters of some abnormal *Begonia* flowers and on the evolution of the monoecious condition in plants. J. Genet. 4: 341–352 + Plates XVI–XVII.

113. Bonner, C.E.B. 1948. The floral vascular supply in *Epilobium* and related genera. Candollea 11: 277–303.

114. Bonner, L.J. & H.G. Dickinson. 1989. Anther dehiscence in *Lycopersicon esculentum* Mill. I. Structural aspects. New Phytol. 113: 97–115.

115. Bonnet, J. 1912. Recherches sur l'évolution des cellules nourricières du pollen chez les angiospermes. Arch. f. Zellf. 7: 605–722.

116. Bosch, E. 1947. Blütenmorphologische und zytologische Untersuchungen an Palmen. Ber. Schweiz. Bot. Ges. 57: 37–100.

117. Bowman, J.L., D.R. Smyth & E.M. Meyerowitz. 1989. Genes directing flower development in *Arabidopsis*. Plant Cell 1: 37–52.

118. Brenner, W. 1922. Zur Kenntnis der Blutenentwicklung einiger Juncaceen. Acta Soc. Sci. Fenn. 50 (4): 1–37.

119. Brett, J.F. & U. Posluszny. 1982. Floral development in *Caulophyllum thalictroides* (Berberidaceae). Can. J. Bot. 60: 2133–2141.

120. Brizicky, G.K. 1959. Variability in the floral parts of *Gomortega* (Gomortegaceae). Willdenowia 2: 200–207.

121. Brouland, M. 1935. Recherches sur l'anatomie florale des Renonculacées. Le Botaniste 27: 1–278.

122. Brown, D.K. & R.B. Kaul. 1981. Floral structure and mechanism in Loasaceae. Amer. J. Bot. 68: 361–372.

123. Brown, E.G.S. 1935. The floral mechanism of *Saurauja subspinosa* Anth. Trans. Proc. Bot. Soc. Edinb. 31: 485–497.

124. Brown, W. 1938. The bearing of nectaries on the phylogeny of flowering plants. Proc. Amer. Phil. Soc. 79: 549–595.

125. Brunkener, L. 1975. Beiträge zur Kenntnis der frühen Mikrosporangienentwicklung der Angiospermen. Svensk Bot. Tidskr. 69: 1–27.

126. Buchmann, S.L. 1983. Buzz pollination in angiosperms. pp. 73–113 in: Handbook of Experimental Pollination Biology (edited by C.E. Jones & R.J. Little). Van Nostrand Reinhold, New York.

127. Buchmann, S.L. & M.D. Buchmann. 1981. Anthecology of *Mouriri myrtilloides* (Melastomataceae: Memecyleae), an oil flower in Panama. Reprod. Bot. Suppl. to Biotropica 13: 7–24.

128. Buchmann, S.L., C.E. Jones & L.J. Colin. 1977. Vibratile pollination of *Solanum douglasii* and *S. xanti* (Solanaceae) in southern California. Wasmann J. Biol. 35: 1–25. [Not seen.]

129. Buchner, R. & C. Puff. 1993. The genus complex *Danais-Schismatoclada-Payera* (Rubiaceae). Character states, generic delimitation and taxonomic position. Bull. Mus. Natn. Hist. Nat. Paris, sect. B, Adansonia 15: 23–74.

130. Budell, B. 1964. Untersuchungen der Antherenentwicklung einiger Blütenpflanzen. Z. Bot. 52: 1–28.

131. Bunniger, L. 1972. Untersuchungen über die morphologische Natur des Hypanthiums bei Myrtales- und Thymelaeales-Familien. II. Myrtaceae. III. Vergleich mit den Thymelaeaceae. Beitr. Biol. Pfl. 48: 79–156.

132. Bunniger, L. & F. Weberling. 1968. Untersuchungen über die morphologische Natur des Hypanthiums bei

Myrtales-Familien. I. Onagraceae. Beitr. Biol. Pfl. 44: 447–477.

133. Bünning, E. 1930. Die Reizbewegungen der Staubblätter von *Sparmannia africana*. Protoplasma 11: 49–84. Paper with similar title in: Proc. K. Ned. Akad. Wet. Amsterdam 33: 284–294.

134. Buntman, D.J. & H.T. Horner. 1983. Microsporogenesis of normal and male sterile (MS3) mutant soybean (*Glycine max*). Scanning Electron Microsc. 1983/2: 913–922.

135. Burck, W. 1906 [1907]. On the influence of the nectaries and other sugarcontaining tissues in the flower on the opening of the anthers. Rec. Trav. Bot. Neerl. 3: 163–172.

136. Burger, W.C. 1977 [1978]. The Piperales and the monocots. Alternate hypotheses for the origin of monocotyledonous flowers. Bot. Rev. 43: 345–393.

137. Burns-Balogh, P. & P. Bernhardt. 1985. Evolutionary trends in the androecium of the Orchidaceae. Pl. Syst. Evol. 149: 119–134.

138. Burns-Balogh, P. & P. Bernhardt. 1988. Floral evolution and phylogeny in the tribe Thelymitreae (Orchidaceae: Neottioideae). Pl. Syst. Evol. 159: 19–47.

139. Buss, P.A., D.F. Galen & N.R. Lersten. 1969. Pollen and tapetum development in *Desmodium glutinosum* and *D. illinoense* (Papilionoideae; Leguminosae). Amer. J. Bot. 56: 1203–1208.

140. Buss, P.A. & N.R. Lersten. 1972. Crystals in tapetal cells of the Leguminosae. Bot. J. Linn. Soc. 65: 81–85.

141. Buss, P.A. & N.R. Lersten. 1975. Survey of tapetal nuclear number as a taxonomic character in Leguminosae. Bot. Gaz. 136: 388–395.

142. Bütow, R. 1955. Die Entwicklung der *Pulsatilla*-Anthere. Z. Bot. 43: 423–449.

143. Buxbaum, F. 1961. Vorläufige Untersuchungen über Umfang, systematische Stellung und Gliederung der Caryophyllales (Centrospermae). Beitr. Biol. Pfl. 36: 1–56.

144. Canright, J.E. 1952. The comparative morphology and relationships of the Magnoliaceae. I. Trends of specialization in the stamens. Amer. J. Bot. 39: 484–497.

145. Carlquist, S. 1969 [1970]. Toward acceptable evolutionary interpretations of floral anatomy. Phytomorphology 19: 332–362.

146. Carniel, K. 1962. Beiträge zur Entwicklungsgeschichte des sporogenen Gewebes der Gramineen und Cyperaceen II. Cyperaceae. Öst. Bot. Z. 109: 81–95.

147. Carniel, K. 1963. Das Antherentapetum. Ein kritischer Überblick. Öst. Bot. Z. 110: 145–176.

148. Carniel, K. 1963. Beiträge zur Entwicklungsgeschichte des sporogenen Gewebes und des Tapetums bei einigen Dipsacaceen. Öst. Bot. Z. 110: 547–555.

149. Carniel, K. 1969. Beiträge zur Entwicklungsgeschichte des Antherentapetums in der Gattung *Oxalis*. I. *Oxalis rosea* und *O. pubescens*. Öst. Bot. Z. 116: 423–429. II. *Oxalis acetosella*. Ibid. 117: 201–204.

150. Carniel, K. 1971. Zur Kenntnis der Ubischkörper von *Heleocharis palustris*. Öst. Bot. Z. 119: 496–502.

150a. Carolin, R.C. 1959. Floral structure and anatomy in the family Goodeniaceae Dumort. Proc. Linn. Soc. NSW 84: 242–255.

151. Carolin, R.C. 1960. Floral structure and anatomy in the family Stylidiaceae Swartz. Proc. Linn. Soc. NSW 85: 189–196.

152. Carolin, R.C. 1960. The structures involved in the presentation of pollen to visiting insects in the order Campanuales. Proc. Linn. Soc. NSW 85: 197–207.

153. Carr, D.J. & S.G.M. Carr. 1987. *Eucalyptus* II. The Rubber Cuticle, and other Studies of the Corymbosae. Phytoglyph Press, Canberra. 372 pp.

154. Carr, S.G.M. & D.J. Carr. 1990. Cuticular features of the Central Australian bloodwoods *Eucalyptus*, section Corymbosae (Myrtaceae). Bot. J. Linn. Soc. 102: 123–156.

155. Carraro, L. & G. Lombardo. 1976 [1977]. Tapetal ultrastructural changes during pollen development. II. Studies on *Pelargonium zonale* and *Kalanchoe obtusa*. Caryologia 29: 339–344. III. Studies on *Gentiana acaulis*. Ibid.: 345–349.

156. Castro, M.A. 1981 Ontogenia de la dehiscencia de anteras y microsporogenesis en *Ocotea acutifolia* (Lauraceae). Bol. Soc. Argent. Bot. 20: 31–42.

157. Catling, P.M. 1990. Auto-pollination in the Orchidaceae. pp. 121–58 in: Orchid Biology, Reviews and Perspectives, V (edited by J. Arditti). Timber Press, Portland, Oregon.

158. Catling, P.M. & V.R. Catling. 1991. Anther-cap retention in *Tipularia discolor*. Lindleyana 6: 113–116.

159. Cejp, K. 1925. Beitrag zur vergleichenden Morphologie der dimerischen Blüten. Beih. Bot. Zbl. 41(1): 128–164.

160. Čelakovský, L.J. 1898. Ueber petaloid umgebildete Staubgefässe von *Philadelphus coronarius* und von *Deutzia crenata*. Öst. Bot. Z. 48: 371–380, 416–419 + Plate X.

161. Cerceau-Larrival, M.T. & F. Roland–Heydacker. 1976. Ontogénie et ultrastructure de pollen d'Ombel-

liferes. Tapis et corps d'Ubisch. C.R. Acad. Sci., Paris, sér. D, 283: 29–32.

162. Chadefaud, M. 1954. Sur l'obdiplostémonie. C.R. Acad. Sci., Paris 239: 1673–1675.

163. Chadefaud, M. 1956. La fleur et des pieces florales des Crucifères d'après quelques structures teratologiques. Bull. Soc. Bot. Fr. 103: 454–460.

164. Chakravarty, H.L. 1949. Morphology of the staminate flowers of bottle gourd, *Lagenaria leucantha* (Duch.) Rusby. New Phytol. 48: 448–452.

165. Chakravarty, H.L. 1958 [1959]. Morphology of the staminate flowers in the Cucurbitaceae with special reference to the evolution of the stamen. Lloydia 21: 49–87.

166. Chakravarty, H.L. & K.S. Gupta. 1951. Morphology of the male flower of *Coccinia cordifolia* (Linn.) Cogn. and its possible systematic position in the family. Bull. Bot. Soc. Bengal 5: 31–50.

167. Chandra, S. 1987 [1988]. Development of male gametophyte in some Scrophulariaceae. Geophytology 17: 199–203.

168. Chapman, G.P. 1987. The tapetum. pp. 111–125 in: Pollen: Cytology and Development (edited by K.L. Giles & J. Prakash). Intn. Rev. Cytol. 107. Academic Press, Orlando.

169. Chaudhury, M.R. & M.L. Banerji. 1987. The bidirectional phylogeny of the laminar stamen – further observation. Bull. Bot. Soc. Bengal 41: 13–18.

170. Chaudry, B. & M.R. Vijayaraghavan. 1992. Structure and function of the anther gland in *Prosopis juliflora* (Leguminosae, Mimosoideae): a histochemical analysis. Phyton (Horn, Austria) 321: 1–7.

171. Chauhan, E. 1979. Role of insoluble polysaccharides in the developing stamen and staminode of *Coronopus didymus* (L.) Sm. New Botanist 6: 111–120.

172. Chauhan, E. 1981 [1982]. Cytochemical studies on the developing androecium of *Coronopus didymus* (L.) Sm. – staminode. Beitr. Biol. Pfl. 56: 19–24.

173. Chauhan, E. 1982. Cytochemical studies on the developing androecium of *Coronopus didymus* (L.) Sm. – Anther. Beitr. Biol. Pfl. 57: 137–155.

174. Chauhan, S.V.S. 1977. Dual role of the tapetum. Curr. Sci. India 16(19): 674. [Not seen.]

175. Chauhan, S.V.S. 1979 [1980]. Development of endothecium in relation to tapetal behaviour in some male sterile plants. Phytomorphology 29: 245–251.

176. Chauhan, S.V.S. 1980. Effect of maleic hydrazide, FW-450 and Dalapon on anther development in *Capsicum annuum* L. J. Indian Bot. Soc. 59: 133–136.

177. Chauhan, S.V.S. 1986. Studies on pollen abortion in some Solanaceae. pp. 505–532 in: Solanaceae. Biology and systematics (edited by W.G. D'Arcy). Columbia University Press, New York.

178. Chauhan, S.V.S., R.K. Dhingra & T. Kinoshita. 1982. Effect of phytogametocidal compounds on the development of endothecium in some plants. Jpn. J. Breeding 32: 139–145. [Eng.] [Not seen.]

179. Chauhan, S.V.S., J.N. Srivastava & T. Kinoshita. 1981. Histological and histochemical changes in the anthers of some diseased vegetable crops. J. Fac. Agric. Hokkaido Univ. 60: 152–158. [Eng.]

180. Chen, B. 1985. Studies on *Magnolia coco* (Lour.) DC. Acta Scient. Nat. Univ. Sunyatseni (3): 82–88. [Chin.; Eng. summ.]

181. Chen, Z.-K., F.-H. Wang & F. Zhou. 1988. The ultrastructural aspect of tapetum and Ubisch bodies in *Anemarrhena asphodeloides*. Acta Bot. Sin. 30: 1–5 + 3 plates. [Chin.; Eng. summ.]

182. Chen, Z.-K., F.-H. Wang & F. Zhou. 1988. On the origin, development and ultrastructure of the orbicules and pollenkitt in the tapetum of *Anemarrhena asphodeloides* (Liliaceae). Grana 27: 273–282.

183. Cheng, P.-C., Y.-K. Cheng & C.-S. Huang. 1981. The structure of anther cuticle in cultivated rice (*Oryza sativa* cultivar Taichung 65). Natn. Sci. Counc. Mon. 9: 983–994. [Chin.; Eng. summ.] [Not seen.]

184. Cheng, P.-C., R.I. Greyson & D.B. Walden. 1979. Comparison of anther development in a genic male-sterile (ms-10) and male-fertile corn (*Zea mays*) from light microscopy and scanning electron microscopy. Can. J. Bot. 57: 578–596. Also in: Natn. Sci. Counc. Mon. 7: 463–484. [Chin.; Eng. summ.]

185. Cheng, P.-C., R.I. Greyson & D.B. Walden. 1983. Organ initiation and the development of unisexual flowers in the tassel and ear of *Zea mays*. Amer. J. Bot. 70: 450–462.

186. Cheng, P.-C., R.I. Greyson & D.B. Walden. 1986. The anther cuticle of *Zea mays*. Can. J. Bot. 64: 2088–2097.

187. Cheng, P.-C. & M.-I. Lin. 1980. Cytological studies on the stamen of *Caltha palustris*: I. Anther and filament morphology, ultrastructural studies of their cuticle and orbicule. Natn. Sci. Counc. Mon. 8: 1113–1140. [Chin.; Eng. summ.] [Not seen.]

188. Cheng, Y.-K. & C.-S. Huang. 1980. Studies on cytoplasmic-genetic male sterility of cultivated rice (*Oryza sativa*): 2. Morphological-histological investigation on functional male sterility. J. Agric. Res. China 29(2): 67–80. [Eng.] [Not seen.]

189. Cheng, Y.-K., C.-S. Huang & K.-L. Lai. 1982. Studies on cytoplasmic-genetic male sterility in cultivated rice (*Oryza sativa*): 5. Comparison of anther development in male sterile and male fertile by transmission and scanning electron microscopy. Natn. Sci. Counc. Mon. 10: 512–534. [Chin.; Eng. summ.]. [Not seen.]

190. Chennaveeraiah, M.S. & P.M. Shivakumar. 1983. Pollen bud formation and its role in *Ophiorrhiza* spp. Ann. Bot. 51: 449–452.

191. Chernik, V.V. 1975. Arrangement and reduction of perianth and androecium parts in representatives of Ulmaceae Mirbel and Celtidaceae Link. Bot. Zh. SSSR 60: 1561–1573. [Russ.; Eng. summ.]

192. Chodat, R. & G. Balicka. 1893. Remarques sur la structure des Tremandracées. Bull. Herb. Boissier 1: 344–353.

193. Cholakhyan, D.P. & L.K. Abramyan. 1982. Cell ultrastructure of various *Cerasus avium* (Rosaceae) anther layers. Biol. Zh. Arm. 35: 344–352. [Russ.; Eng. summ.] [Not seen.]

194. Chou, Y.-L. 1952 [1952–53]. Floral morphology of three species of *Gaultheria*. Bot. Gaz. 114: 198–221.

195. Christensen, J.E., H.T. Horner Jr & N.R. Lersten. 1972. Pollen wall and tapetal orbicular wall development in *Sorghum bicolor* (Gramineae). Amer. J. Bot. 59: 43–48.

196. Chupov, V.S. 1986. Some features of the evolution of stamen and perianth parts in angiosperms. Bot. Zh. SSSR 71: 323–333. [Russ.; brief Eng. summ.]

197. Chupov, V.S. 1990. Some taxonomically and phylogenetically important characters of the stamen structure. Bot. Zh. SSSR 75: 965–973. [Russ. only]

198. Ciampolini, F., M. Nepi & E. Pacini. 1993. Tapetum development in *Cucurbita pepo*. Pl. Syst. Evol., Suppl. 7: 13–22.

199. Clark, C. 1979. Ultraviolet absorption by flowers of the Eschscholzioideae (Papaveraceae). Madroño 26: 22–25.

200. Clausen, P. 1927. Ueber das Verhalten des Antheren-Tapetums bei einigen Monokotylen und Ranales. Bot. Arch. 18: 1–27.

201. Clément, C. & J.-C. Audran. 1992. Apports de la cytochimie à la connaissance des orbicules dans l'anthère de *Lilium* (Liliacées). 1. Le coeur orbiculaire. Bull. Soc. Bot. Fr., Lett. bot. 139: 369–376.

202. Clément, C. & J.-C. Audran. 1993. Cytochemical and ultrastructural evolution of orbicules in *Lilium*. Pl. Syst. Evol., Suppl. 7: 63–74.

203. Cocucci, A.E. 1975. Estudios en el genero *Prosopan-che* (Hydnoraceae). II. Organizacion de la flor. Kurtziana 8: 7–15. [Eng. summ.]

204. Coen, E.S. 1991. The role of homeotic genes in flower development and evolution. Ann. Rev. Pl. Physiol. Mol. Biol. 42: 241–279.

205. Coen, E.S. & E.M. Meyerowitz. 1991. The war of the whorls: genetic interactions controlling flower development. Nature 353 (Sept. 5): 31–37.

206. Coetzee, H. & P.J. Robbertse. 1985. Pollen and tapetal development in *Securidaca longepedunculata*. S. Afr. J. Bot. 51: 111–124.

207. Cohen, L.I. 1968. Development of the staminate flower in the dwarf mistletoe, *Arceuthobium*. Amer. J. Bot. 55: 187–193.

208. Colhoun, C.W. & M.W. Steer. 1981. Microsporogenesis and the mechanism of cytoplasmic male sterility in maize. Ann. Bot. 48: 417–424.

209. Copeland, H.F. 1941. Further studies in Monotropoideae. Madroño 6: 97–119.

210. Copeland, H.F. 1943. A study, anatomical and taxonomic, of the genera of Rhododendroideae. Amer. Midl. Naturalist 30: 533–625.

211. Copeland, H.F. 1947. Observations on the structure and classification of the Pyroleae. Madroño 9: 65–102.

212. Copeland, H.F. 1953. Observations on the Cyrillaceae particularly on the reproductive structures of North American species. Phytomorphology 3: 405–411.

213. Corner, E.J.H. 1946. Centrifugal stamens. J. Arnold Arbor. 27: 423–437.

214. Cornet, B. 1989. The reproductive morphology and biology of *Sanmiguelia lewisii*, and its bearing on angiosperm evolution in the Late Triassic. Evol. Trends Plants 3: 25–51.

215. Costerus, J.C. 1915. Das Labellum und das Diagram der Zingiberaceen. Ann. Jard. Bot Buitenzorg, sér. 2, 14: 95–108.

216. Cousin, M.-T. 1979. Tapetum and pollen grains of *Vinca rosea* (Apocynaceae). Ultrastructure and investigations with scanning electron microscope. Grana 18: 115–128.

217. Cousin, M.-T. 1980. Changes induced by mycoplasma-like organisms (M.L.O.), etiologic agents of the Stolbur disease in the different tissues of the anther of *Vinca rosea* L. (Apocynaceae). Grana 19: 199–125.

218. Crane, P.R., E.M. Friis & K.R. Pedersen. 1989. Reproductive structure and function in Cretaceous Chloranthaceae. Pl. Syst. Evol. 165: 211–226.

219. Crepet, W.L. & D.L. Dilcher. 1977. Investigations

of angiosperms from the Eocene of North America: a mimosoid inflorescence. Amer. J. Bot. 64: 714–725.

220. Crepet, W.L., D.L. Dilcher & F.W. Potter. 1974. Eocene angiosperm flowers. Science 185: 781–782.

221. Crone, W. & E.M. Lord. 1993. Flower development in the organ number mutant *Clavata 1–1* of *Arabidopsis thaliana* (Brassicaceae). Amer. J. Bot. 80: 1419–1426.

222. Cuellar, H.S. 1967. Description of a pollen release mechanism in the flower of the Mexican hackberry tree, *Celtis laevigata*. Southwest. Nat. 12: 471–474.

223. Cugnac, A. de & F. Obaton. 1935. Sur l'allongement des filets staminaux chez les Graminées. Rev. Gén. Bot. 47: 657–680 + Plates VII–IX.

224. D'Amato, F. 1984. Role of polyploidy in reproductive organs and tissues. pp. 519–566 in: Embryology of Angiosperms (edited by B.M. Johri). Springer-Verlag, Berlin.

225. Däniker, A.E. 1959. Was ist eine Blüte? Betrachtungen über das Wesen und die Organisation der Blüten. Ein Beitrag zur Phylogenie der Angiospermen. Mitt. Bot. Mus. Univ. Zürich 214: 379–389.

226. D'Arcy, W.G. 1992. Solanaceae of Madagascar: form and geography. Ann. Missouri Bot. Gard. 79: 29–45.

227. D'Arcy, W.G., N.S. D'Arcy & R.C. Keating. 1990. Scented anthers in the Solanaceae. Rhodora 92: 50–53.

228. Darwin, C. 1862. On the various contrivances by which British and foreign orchids are fertilized by insects, and on the good effects of intercrossing. John Murray, London. 365 pp.

229. Daumann, E. 1930. Das Blütennektarium von *Magnolia* und die Futterkörper in der Blüte von *Calycanthus*. Planta 11: 108–116.

230. Daumann, E. 1931. Zur morphologischen Wertigkeit der Blütennektarien von *Laurus*. Beih. Bot. Zbl. 48 (1): 209–213.

231. Daumann, E. 1931. Zur Phylogenie der Diskusbildungen. Beiträge zur Kenntnis der Nektarien II. (*Hydrocharis, Sagittaria, Sagina.*) Beih. Bot. Zbl. 48 (1): 183–208.

232. Davidson, C. 1973. An anatomical and morphological study of Datiscaceae. Aliso 8: 49–110.

233. Davis, G.L. 1961. The life history of *Podolepis jaceoides* (Sims) Voss. I. Microsporogenesis and male gametogenesis. Phytomorphology 11: 86–97.

234. Davis, G.L. 1966. Systematic Embryology of the Angiosperms. John Wiley, New York. 528 pp.

235. Davis, G.L. 1968. Apomixis and abnormal anther

development in *Calotis lappulacea* Benth. (Compositae). Aust. J. Bot. 16: 1–17.

236. Davis, T.A. 1966. Floral structure and stamens in *Bombax ceiba* L. J. Genet. 59: 294–328.

237. Davis, T.A. & A. Kundu. 1965. Floral structure and stamens in *Ceiba pentandra* (Linn.) Gaertn. J. Bombay Nat. Hist. Soc. 62: 394–411.

238. Davis, T.A. & K.O. Mariamma. 1965. The three kinds of stamens in *Bombax ceiba* L. (Bombacaceae). Bull. Jard. Bot. Etat Bruxelles 35: 185–211.

239. Dayanandan, P. & P.B. Kaufman. 1976. Trichomes of *Cannabis sativa* L. (Cannabaceae). Amer. J. Bot. 63: 578–591.

240. De Block, M. & D. Debrouwer. 1993. Engineered fertility control in transgenic *Brassica napus* L.: histochemical analysis of anther development. Planta 189: 218–225.

241. De Candolle, A.-P. 1817. Considérations générales sur les fleurs doubles, et en particulier sur celles de la famille des Renonculacées. Mem. Phys. Chim. Soc. Arcueil 3: 385–404.

242. Dell, B. 1981. Male sterility and anther wall structure in copper-deficient plants. Ann. Bot. 48: 599–608.

243. DeMaggio, A.E. & C.L. Wilson. 1986. Floral structure and organogenesis in *Podophyllum peltatum* (Berberidaceae). Amer. J. Bot. 73: 21–32.

244. Demeter, K. 1922. Vergleichende Asclepiadaceenstudien. Flora 115: 130–176.

245. Deroin, T. 1991. La vascularisation florale des Magnoliales: première approche experimental de son rôle au cours de la pollinisation. C.R. Acad. Sci., Paris, sér. III, 312: 355–360.

246. Deroin, T. 1991. Anatomie comparée de l'anthère des Annonacées. Confirmation de la nature diplophylle de l'étamine. C.R. Acad. Sci., Paris, sér. III, 313: 627–632. [Eng. summ.]

247. Derstine, K.S. & S.C. Tucker. 1991. Organ initiation and development of inflorescences and flowers of *Acacia baileyana*. Amer. J. Bot. 78: 816–832.

248. Deshpande, P.K., S.M. Bhuskute & K.H. Makde. 1986. Microsporogenesis and male gametophyte in some Cucurbitaceae. Phytomorphology 36: 145–150.

249. De Vaal, J., M.G. Gilliland & A.R.A. Noel. 1978. Development of the endothecium in *Bulbine*. Proc. S. Afr. Electron Microsc. Soc. 8: 87–88. [Not seen.]

250. Dhaliwal, H.S. & B.L. Johnson. 1976. Anther morphology and origin of the tetraploid wheats. Amer. J. Bot. 63: 363–368.

251. Dharamadhaj, P. & N. Prakash. 1978. Development

of the anther and ovule in *Capsicum* L. Aust. J. Bot. 26: 433–439.

252. Dhillon, M. 1981. Floral anatomy of two species of *Swartzia* and its bearing on its systematic position. Feddes Repert. 92: 689–693.

253. Diao, X.-M. & H. Zhi. 1991. A cytomorphological study on the anther opening of foxtail millet and wheat. Acta Bot. Sin. 33: 240–242 + 1 plate. [Chin.; Eng. summ.]

254. Dickinson, H.G. 1973. The role of plastids in the formation of pollen grain coatings. Cytobios 8: 25–40.

255. Dickison, W.C. 1970. Comparative morphological studies in Dilleniaceae, VI. Stamens and young stem. J. Arnold Arbor. 51: 403–422.

256. Dickison, W.C. 1972. Observations on the floral morphology of some species of *Saurauia*, *Actinidia* and *Clematoclethra*. J. Elisha Mitchell Sci. Soc. 88: 43–54.

257. Dickison, W.C. 1975. Studies on the floral anatomy of the Cunoniaceae. Amer. J. Bot. 62: 433–447.

258. Dickison, W.C. 1975. Floral morphology and anatomy of *Bauera*. Phytomorphology 25: 69–76.

259. Dickison, W.C. 1978. Comparative anatomy of Eucryphiaceae. Amer. J. Bot. 65: 722–735.

260. Dickison, W.C. 1989. Comparisons of primitive Rosidae and Hamamelidae. pp. 47–73 in: Evolution, Systematics, and Fossil History of the Hamamelidae. Vol. 1: Introduction and 'Lower' Hamamelidae (edited by P.R. Crane & S. Blackmore). Clarendon Press, Oxford.

261. Dickison, W.C. 1990. The morphology and relationships of *Medusagyne* (Medusagynaceae). Pl. Syst. Evol. 171: 27–55.

262. Dickison, W.C. 1990. A study of the floral morphology and anatomy of the Caryocaraceae. Bull. Torrey Bot. Cl. 117: 123–137.

263. Dickison, W.C. 1993. Floral anatomy of the Styracaceae, including observations on intra-ovarian trichomes. Bot. J. Linn. Soc. 112: 223–255.

264. Dickison, W.C. & R.B. Miller. 1993. Morphology and anatomy of the malagasy genus *Physena* (Physenaceae), with a discussion of the relationships of the genus. Bull. Mus. Natn. Hist. Nat. Paris, sect. B, Adansonia 15: 85–106.

265. Diels, L. 1916. Käferblumen bei den Ranales und ihre Bedeutung für die Phylogenie der Angiospermen. Ber. Dtsch. Bot. Ges. 34: 758–774.

266. Dilcher, D.L., P.S. Herendeen & F. Hueber 1992. Fossil *Acacia* flowers with attached anther glands from Dominican Republic amber. pp. 33–42 in: Advances in Legume Systematics. 4. The Fossil Record (edited by P.S. Herendeen & D.L. Dilcher). Royal Botanic Gardens, Kew. 326 pp.

267. Ding Hou. 1981. Florae malesianae praecursores LXII. On the genus *Thottea* (Aristolochiaceae). Blumea 27: 301–332.

268. Donato, A.M. 1991. Anatomia floral de *Chorisia speciosa* St Hil. (Bombacaceae). Bradea 5 (49): 455–477. [Eng. summ.]

269. Dormer, K.J. 1962. The fibrous layer in the anthers of Compositae. New Phytol. 61: 150–153.

270. Dowding, P. 1987 Wind pollination mechanisms and aerobiology. pp. 421–437 in: Pollen: Cytology and Development. (edited by K.L. Giles & J. Prakash) Intn. Rev. Cytol. 107. Academic Press, Orlando.

271. Dressler, R.L. 1981. The Orchids: Natural History and Classification. Harvard University Press, Cambridge, Mass. 332 pp.

272. Drinnan, A.N., P.R. Crane, E.M. Friis & K.R. Pedersen. 1990. Lauraceous flowers from the Potomac Group (Mid-Cretaceous) of eastern North America. Bot. Gaz. 151: 370–384.

273. Drinnan, A.N. & P.Y. Ladiges. 1989. Corolla and androecium development in some *Eudesmia* eucalypts (Myrtaceae). Pl. Syst. Evol. 165: 239–254.

274. Dundas, I.S., K.B. Saxena & D.E. Byth. 1981. Microsporogenesis and anther wall development in male-sterile and fertile lines of pigeon pea (*Cajanus cajan* (L.) Millsp.). Euphytica 30: 431–435.

275. Dupuy, P. 1962. Diplophyllie et scyphogénie mériphylle chez les étamines virescentes de *Tropaeolum majus*. C.R. Soc. Biol. 156: 362–365.

276. Dupuy, P. & M. Guédès. 1980. Documents tératologiques pour servir à l'étude morphologique des angiospermes. Bull. Mus. Natn. Hist. Nat., Paris, sér. 4, 2B: 83–144.

277. Dwivedi, N.K. & B.M. Joshi. 1986. Dual origin and dimorphism of anther tapetum in *Melissa parviflora* Benth. Acta Bot. Indica 14: 57–60.

278. Dyki, B. & J. Hoser–Krause. 1990. The influence of temperature on the morphology and anatomy of anthers of a male sterile hybrid *Brassica oleracea* var. *italica* Plenck × *B. oleracea* var. *botrytis* L. Genet. Pol. 31: 115–121.

279. Echlin, P. 1971. The role of the tapetum during microsporogenesis of angiosperms. pp. 41–61 in: Pollen: Development and Physiology (edited by J. Heslop-Harrison). Butterworths, London. [Not seen.]

280. Echlin, P. 1971. Production of sporopollenin by the

tapetum. pp. 220–247 in: Sporopollenin (edited by J. Brooks, P.R. Grant, P.M. Muir, P. Van Gijzel & G. Shaw). Academic Press, London.

281. Echlin, P. & H. Godwin. 1968. The ultrastructure and ontogeny of pollen in *Helleborus foetidus* L. I. The development of the tapetum and Ubisch bodies. J. Cell Sci. 3: 161–174 + 7 plates.

282. Eckardt, T. 1959 [1960]. Das Blütendiagramm von *Batis* P.Br. Ber. Dtsch. Bot Ges. 72: 411–418.

283. Eckardt, T. 1963. Some observations on the morphology and embryology of *Eucommia ulmoides* Oliv. J. Indian Bot. Soc. (Maheshwari Comm. Vol.) 42A: 27–34.

284. Eckardt, T. 1971. Anlegung und Entwicklung der Blüten von *Gyrostemon ramulosus* Desf. Bot. Jahrb. 90: 434–446.

285. Eckert, G. 1966. Entwicklungsgeschichtliche und blütenanatomische Untersuchungen zum Problem der Obdiplostemonie. Bot. Jahrb. 85: 523–604. [Eng. summ. 599–601.]

286. Edlin, H.L. 1935. A critical revision of certain taxonomic groups of the Malvales. I, II. New Phytol. 34: 1–20, 122–143.

287. Edmonds, J.M. 1982. Epidermal hair morphology in *Solanum* L. section Solanum. Bot. J. Linn. Soc. 85: 153–167.

288. Edwards, J. & J.R. Jordan. 1992. Reversible anther opening in *Lilium philadelphicum* (Liliaceae): a possible means of enhancing male fitness. Amer. J. Bot. 79: 144–148.

289. Eggers, O. 1935. Über die morphologische Bedeutung des Leitbündelverlaufes in den Blüten der Rhoeadalen und über das Diagramm der Cruciferen und Capparidaceen. Planta 24: 14–58.

290. Ehrenberg, L. 1945. Zur Kenntnis der Homologienverhältnisse in der angiospermen Blüte. Bot. Notiser 438–444.

291. El Ghazaly, G. 1980. Palynology of Hypochoeridinae and Scolyminae (Compositae). Opera Bot. 58: 1–48.

292. El Ghazaly, G. 1985. On pollen morphology and floral micromorphology of South American *Hypochoeris* L. Arab. Gulf J. Sci. Res. 3: 425–436.

293. El Ghazaly, G. & S. Nilsson. 1991. Development of tapetum and orbicules of *Catharanthus roseus* (Apocynaceae). pp. 317–329 in: Pollen and Spores (edited by S. Blackmore & S.H. Barnes). Syst. Assoc. spec. vol. 44. Clarendon Press, Oxford.

294. Eliasson, U.H. 1988. Floral morphology and taxonomic relations among the genera of Amaranthaceae

in the New World and the Hawaiian Islands. Bot. J. Linn. Soc. 96: 235–283.

295. Elvers, I. 1984. Fluorescent *Tussilago* anthers. Nord J. Bot. 4: 61–63.

296. Emberger, L. 1939. Recherches sur la fleur des Polygonacées. Rev. Gén. Bot. 51: 581–599.

297. Emberger, L. 1944. Observations sur la méiomérie de *Colchicum autumnale* L. et sur l'origine de la corolle et des fleurs à androcée épitépale. Rec. Trav. Inst. Bot. Montpellier 1: 5–15.

298. Endress, P.K. 1967. Systematische Studie über die verwandtschaftlichen Beziehungen zwischen den Hamamelidaceen und Betulaceen. Bot. Jahrb. 87: 431–525.

299. Endress, P.K. 1969. Gesichtspunkte zur systematischen Stellung der Eupteleaceen (Magnoliales). Untersuchungen über Bau und Entwicklung der generativen Region bei *Euptelea polyandra* Sieb. et Zucc. Ber. Schweiz. Bot. Ges. 79: 229–278. [Eng. summ. 262.]

300. Endress, P.K. 1976. Die Androeciumanlage bei polyandrischen Hamamelidaceen und ihre systematische Bedeutung. Bot. Jahrb. 97: 436–457.

301. Endress, P.K. 1977. Über Blütenbau und Verwandtschaft der Eupomatiaceae und Himantandraceae (Magnoliales). Ber. Dtesch. Bot. Ges. 90: 83–103.

302. Endress, P.K. 1978. Blütenontogenese, Blütenabgrenzung und systematische Stellung der perianthlosen Hamamelidoideae. Bot. Jahrb. 100: 249–317.

303. Endress, P.K. 1980. The reproductive structures and systematic position of the Austrobaileyaceae. Bot. Jahrb. 101: 393–433.

304. Endress, P.K. 1980. Floral structure and relationships of *Hortonia* (Monimiaceae). Pl. Syst. Evol. 133: 199–221.

305. Endress, P.K. 1980. Ontogeny, function and evolution of extreme floral construction in Monimiaceae. Pl. Syst. Evol. 134: 79–120.

306. Endress, P.K. 1983. The early floral development of *Austrobaileya*. Bot. Jahrb. 103: 481–497.

307. Endress, P.K. 1984. The role of inner staminodes in the floral display of some relic Magnoliales. Pl. Syst. Evol. 146: 269–282.

308. Endress, P.K. 1984. The flowering process in the Eupomatiaceae (Magnoliales). Bot. Jahrb. 104: 297–319.

309. Endress, P.K. 1985 Stamenabszission und Pollenpräsentation bei Annonaceae. Flora 176: 95–98.

310. Endress, P.K. 1986. Floral structure, systematics, and

phylogeny in Trochodendrales. Ann. Missouri Bot. Gard. 73: 297–324.

311. Endress, P.K. 1986. Reproductive structures and phylogenetic significance of extant primitive angiosperms. Pl. Syst. Evol. 152: 1–28.

312. Endress, P.K. 1987. The early evolution of the angiosperm flower. Trends Ecol. Evol. 2: 300–304.

313. Endress, P.K. 1987. Floral phyllotaxis and floral evolution. Bot. Jahrb. 108: 417–438.

314. Endress, P.K. 1987. The Chloranthaceae: reproductive structures and phylogenetic position. Bot. Jahrb. 109: 153–226.

315. Endress, P.K. 1989. Chaotic floral phyllotaxis and reduced perianth in *Achlys* (Berberidaceae). Bot. Acta 102: 159–163.

316. Endress, P.K. 1989. Aspects of evolutionary differentiation of the Hamamelidaceae and the lower Hamamelididae. Pl. Syst. Evol. 162: 193–211.

317. Endress, P.K. 1989. Phylogenetic relationships in the Hamamelidoideae. pp. 227–248 in: Evolution, Systematics and Fossil History of the Hamamelidae. Vol. 1: Introduction and 'Lower' Hamamelidae (edited by P.R. Crane & S. Blackmore). Clarendon Press, Oxford.

318. Endress, P.K. 1990. Patterns of floral construction in ontogeny and phylogeny. Biol. J. Linn. Soc. 39: 153–175.

319. Endress, P.K. 1990. Evolution of reproductive structures and functions in primitive angiosperms (Magnoliidae). Mem. N.Y. Bot. Gard. 55: 5–34.

320. Endress, P.K. 1992. Primitive Blüten: Sind Magnolien noch zeitgemäss? Stapfia 28: 1–10.

321. Endress, P.K. 1992. Evolution and floral diversity: the phylogenetic surroundings of *Arabidopsis* and *Antirrhinum*. Int. J. Plant Sci. 153(3): S106–S122.

322. Endress, P.K. 1992. Protogynous flowers in Monimiaceae. Pl. Syst. Evol. 181: 227–232.

323. Endress, P.K. 1994. Shapes, sizes and evolutionary trends in stamens of Magnoliidae. Bot. Jahrb. 115: 429–460.

324. Endress, P.K. & E.M. Friis. 1991. *Archamamelis*, hamamelidalean flowers from the Upper Cretaceous of Sweden. Pl. Syst. Evol. 175: 101–114.

325. Endress, P.K. & L.D. Hufford. 1989. The diversity of stamen structures and dehiscence patterns among Magnoliidae. Bot. J. Linn. Soc. 100: 45–85.

326. Endress, P.K. & D.H. Lorence. 1983. Diversity and evolutionary trends in the floral structure of *Tambourissa* (Monimiaceae). Pl. Syst. Evol. 143: 53–81.

327. Endress, P.K. & F.B. Sampson. 1983. Floral structure and relationships of the Trimeniaceae (Laurales). J. Arnold Arbor. 64: 447–473.

328. Endress, P.K. & S. Stumpf. 1990. Nontetrasporangiate stamens in the angiosperms: structure, systematic distribution and evolutionary aspects. Bot. Jahrb. 112: 193–240.

329. Endress, P.K. & S. Stumpf. 1991. The diversity of stamen structure in 'Lower' Rosidae (Rosales, Fabales, Proteales, Sapindales). Bot. J. Linn. Soc. 107: 217–293.

330. Endress, P.K. & P. Voser. 1975. Zur Androeciumanlage und Antherenentwicklung bei *Caloncoba echinata* (Flacourtiaceae). Pl. Syst. Evol. 123: 241–253.

331. Erbar, C. 1988. Early developmental patterns in flowers and their value for systematics. pp. 7–23 in: Aspects of Floral Development (edited by P. Leins, S.C. Tucker & P.K. Endress). J. Cramer, Berlin.

332. Erbar, C. 1993. Studies on the floral development and pollen presentation in *Acicarpha tribuloides* with a discussion of the systematic position of the family Calyceraceae. Bot. Jahrb. Syst. 115: 325–350.

333. Erbar, C. & P. Leins. 1982. Zur Spirale in Magnolien-Blüten. Beitr. Biol. Pfl. 56: 225–241.

334. Erbar, C. & P. Leins. 1983. Zur Sequenz von Blütenorganen bei einigen Magnoliiden. Bot. Jahrb. 103: 433–449.

335. Erbar, C. & P. Leins. 1985. Studien zur Organsequenz in Apiaceen-Blüten. Bot. Jahrb. 105: 379–400. [Eng. summ.]

336. Erickson, E.H., M.B. Garment & C.E. Peterson. 1982. Structure of cytoplasmic male-sterile and fertile carrot flowers. J. Amer. Soc. Hort. Sci. 107: 698–706.

337. Erickson, R.O. 1948. Cytological and growth correlations in the flower bud and anther of *Lilium longiflorum*. Amer. J. Bot. 35: 729–739.

338. Ernst, A. & E. Schmid. 1913. Über Blüte und Frucht von *Rafflesia*. Ann. Jard. Bot. Buitenzorg, sér. 2, 12: 1–58 + Plates I–VIII.

339. Ernst, W.R. 1967. Floral morphology and systematics of *Platystemon* and its allies *Hesperomecon* and *Meconella* (Papaveraceae: Platystemonoideae). Univ. Kansas Sci. Bull. 47: 25–70.

340. Ernst-Schwarzenbach, M. 1945. Zur Blütenbiologie einiger Hydrocharitaceen. Ber. Schweiz. Bot. Ges. 55: 33–69.

341. Ernst-Schwarzenbach, M. 1956. Kleistogamie und Antherenbau in der Hydrocharitaceen-Gattung *Ottelia*. Phytomorphology 6: 296–311.

342. Evans, L.S. & J. Van't Hof. 1975. Is polyploidy necessary for tissue differentiation in higher plants? Amer. J. Bot. 62: 1060–1064.

343. Evans, T.M. & G.K. Brown. 1989. Plicate staminal filaments in *Tillandsia* subgenus *Anoplophytum* (Bromeliaceae). Amer. J. Bot. 76: 1478–1485.

344. Evans, T.M. & G.K. Brown. 1990. Plicate staminal filaments in *Tillandsia* subgenus *Anoplophytum*. J. Bromel. Soc. 40(1): 11–15.

345. Eyde, R.H. 1968. Flowers, fruits, and phylogeny of Alangiaceae. J. Arnold Arbor. 49: 167–192.

346. Eyde, R.H. 1975 [1976]. The bases of angiosperm phylogeny: floral anatomy. Ann. Missouri Bot. Gard. 62: 521–537.

347. Eyde, R.H. 1977 [1978]. Reproductive structures and evolutions in *Ludwigia* (Onagraceae). I. Androecium, placentation, merism. Ann. Missouri Bot. Gard. 64: 644–655.

348. Eyde, R.H., D.H. Nicolson & P. Sherwin. 1967. A survey of floral anatomy in Araceae. Amer. J. Bot. 54: 478–497.

349. Faden, R.B. 1992. Floral attraction and floral hairs in the Commelinaceae. Ann. Missouri Bot. Gard. 79: 46–52.

350. Fagerlind, F. 1945. Bau der floralen Organe bei der Gattung *Langsdorffia*. Svensk Bot. Tidskr. 39: 197–210.

351. Fagerlind, F. 1945. Blüte und Blütenstand der Gattung *Balanophora*. Bot. Notiser 330–350.

352. Fagerlind, F. 1959. Development and structure of the flower and gametophytes in the genus *Exocarpos*. Svensk Bot. Tidskr. 53: 257–282.

353. Falck, K. 1910. Über die Syngenesie der *Viola*-Anthere. Svensk Bot. Tidskr. 4: 85–90.

354. Fallen, M.E. 1986. Floral structure in the Apocynaceae: morphological, functional, and evolutionary aspects. Bot. Jahrb. 106: 245–286.

355. Fang, J. & S.-Z. Zhang. 1988. Tapetum dimorphism and its histochemical study in sweet pepper (*Capsicum annuum* L. var. *grossum*). Acta bot. Sin. 30: 352–356 + 2 plates. [Chin.; Eng. summ.]

356. Febulaus, G.N.V. & T. Pullaiah. 1992 [1993]. Embryological studies in *Panicum repens* Linn. Phytomorphology 42: 125–131.

357. Fedortschuk, W. 1932. Entwicklung und Bau des männlichen Gametophyten bei den Arten der Convolvulaceen-Gattung *Quamoclit*. (Beispiel einer frühen, nicht hybriden Sterilität.) Planta 16: 554–574.

358. Feehan, J. 1985. Explosive flower opening in ornithophily: a study of pollination mechanisms in some Central African Loranthaceae. Bot. J. Linn. Soc. 90: 129–144.

359. Feldhofen, E. 1933. Beiträge zur physiologischen Anatomie der nuptialen Nektarien aus den Reihen der Dikotylen. Beih. Bot. Zbl. 50(1): 459–634 + Plates II–XXXI.

360. Findlay, N. & G.P. Findlay. 1984. Movement of potassium ions in the motor tissue of *Stylidium*. Aust. J. Pl. Physiol. 11: 451–457.

361. Findlay, N. & G.P. Findlay. 1989. The structure of the column in *Stylidium*. Aust. J. Bot. 37: 81–101.

362. Fisel, K.J. & F. Weberling. 1990. Untersuchungen zur Morphologie und Ontogenie der Blüten von *Tovaria pendula* Ruiz & Pavón und *Tovaria diffusa* (Macfad.) Fawcett & Rendle (Tovariaceae). Bot. Jahrb. 111: 365–387.

363. Fisher, J.E. 1972. The transformation of stamens to ovaries and of ovaries to inflorescences in *Triticum aestivum* L. under short-day treatment. Bot. Gaz. 133: 78–85.

364. Fisher, M.J. 1928. The morphology and anatomy of the flower of Salicaceae. I, II. Amer. J. Bot. 15: 307–326, 372–394.

365. Fitzgerald, M.A., D.M. Calder & R.B. Knox. 1993. Secretory events in the freeze-substituted tapetum of the orchid *Pterostylis concinna*. Pl. Syst. Evol., Suppl. 7: 53–62.

366. Fleurat-Lessard, P. & B. Millet. 1984. Ultrastructural features of cortical parenchyma cells ('motor cells') in stamen filaments of *Berberis canadensis* Mill. and tertiary pulvini of *Mimosa pudica* L. J. Exp. Bot. 35: 1332–1341.

367. Flores, E.M. & M.F. Moseley. 1990. Anatomy and aspects of development of the staminate inflorescences and florets of seven species of *Allocasuarina* (Casuarinaceae). Amer. J. Bot. 77: 795–808.

368. Foster, A.S. 1963. The morphology and relationships of *Circaeaster*. J. Arnold Arbor. 44: 299–321.

369. Fourcroy, M. 1952. Au sujet de l'appareil conducteur de l'anthère du *Lilium candidum* L. Bull. Soc. Bot. Fr. 99: 69–72.

370. Fourcroy, M. & S. Rivière. 1951. Inversion d'un faisceau cribro-vasculaire dans l'anthère introrse du lis. C.R. Acad. Sci., Paris 233: 192–194.

371. Franz, E. 1908. Beiträge zur Kenntnis der Portulacaceen und Basellaceen. Diss. Halle. 50pp. Also in Bot. Jahrb. 42, Beibl. 97: 1–46.

372. French, J.C. 1985. Patterns of endothecial wall thickenings in Araceae: subfamilies Pothoideae and Monsteroideae. Amer. J. Bot. 72: 472–486.

373. French, J.C. 1985. Patterns of endothecial wall thickenings in Araceae: subfamilies Calloideae, Lasioideae, and Philodendroideae. Bot. Gaz. 146: 521–533.

374. French, J.C. 1986. Patterns of stamen vasculature in the Araceae. Amer. J. Bot. 73: 434–449.

375. French, J.C. 1986. Patterns of endothecial wall thickenings in Araceae: subfamilies Colocasioideae, Aroideae, and Pistioideae. Bot. Gaz. 147: 166–179.

376. Freudenstein, J.V. 1991. A systematic study of endothecial thickenings in the Orchidaceae. Amer. J. Bot. 78: 766–781.

377. Friedrich, H.-C. 1955–56. Studien über die natürliche Verwandtschaft der Plumbaginales und Centrospermae. Phyton (Austria) 6: 220–263.

378. Friis, E.M. 1984. Preliminary report of Upper Cretaceous angiosperm reproductive organs from Sweden and their level of organization. Ann. Missouri Bot. Gard. 71: 403–418.

379. Friis, E.M. 1985. Structure and function in Late Cretaceous angiosperm flowers. Biol. Skr. K. Dan. Vidensk Selsk. 25: 1–37.

380. Friis, E.M. & P.R. Crane. 1989. Reproductive structures of Cretaceous Hamamelidae. pp. 155–174 in: Evolution, Systematics and Fossil History of the Hamamelidae. Vol. 1: Introduction and 'Lower' Hamamelidae (edited by P.R. Crane & S. Blackmore). Clarendon Press, Oxford.

381. Friis, E.M, P.R. Crane & K.R. Pedersen. 1986. Floral evidence for Cretaceous chloranthoid angiosperms. Nature 320: 163.

382. Friis, E.M., P.R. Crane & K.R. Pedersen. 1991. Stamen diversity and *in situ* pollen of Cretaceous angiosperms. pp. 197–224 in: Pollen and Spores (edited by S. Blackmore & S.H. Barnes). Clarendon Press, Oxford.

383. Friis, E.M. & P.K. Endress. 1990. Origin and evolution of angiosperm flowers. Adv. Bot. Res. 17: 99–162.

384. Fuchs, A. 1938. Beiträge zur Embryologie der Thymelaeaceae. Öst. Bot. Z. 87: 1–41.

385. Fuchs, E., D. Atsmon & A.H. Halevy. 1977. Adventitious staminate flower formation in gibberellin treated gynoecious cucumber plants. Plant Cell Physiol. 18: 1193–1201.

386. Gabara, B. 1975. Characterization of fibrous compound of Golgi vesicles in tapetal cells of *Delphinium*. Protoplasma 86: 159–168.

387. Gale, R.M.O. & S.J. Owens. 1983. Cell distribution and surface morphology in petals, androecia and styles of Commelinaceae. Bot. J. Linn. Soc. 87: 247–262.

388. Galetto, L. 1993. El nectario y la composicion quimica del nectar en *Plumbago auriculata* y *P. caerulea* (Plumbaginaceae). Bol. Soc. Argent. Bot. 29: 67–72.

389. Galil, J. & L. Meiri. 1981. Number and structure of anthers in fig syconia in relation to behaviour of the pollen vectors. New Phytol. 88: 83–87.

390. Galle, P. 1977. Untersuchungen zur Blütenentwicklung der Polygonaceen. Bot. Jahrb. 98: 449–489.

391. Gandhi, K.N. & R.D. Thomas. 1983. A note on the androecium of the genus *Croton* and flowers in general of the family Euphorbiaceae. Phytologia 54: 6–8.

392. Gassner, G.G. 1941. Über Bau der männlichen Blüten und Pollenentwicklung einiger Palmen der Unterfamilie der Ceroxylinae. Beih. Bot. Zbl. 61A: 237–276 + Plates XIII–XV.

393. Gauthier, R. & J. Arros. 1963. L'anatomie de la fleur staminée del'*Hillebrandia sandwicensis* Oliver et la vascularisation de l'étamine. Phytomorphology 13: 115–127.

394. Gavaudan, P. & Dupuy, P. 1965. Sur l'expression de l'intersexualité dans les étamines tératologiques et normales des angiospermes. C.R. Acad. Sci., Paris 260: 4568–4571.

395. Gelius, L. 1967. Studien zur Entwicklungsgeschichte an Blüten der Saxifragales sensu lato mit besonderer Berücksichtigung des Androeceums. Bot. Jahrb. 87: 253–303. [Eng. summ. 291–292.]

396. Gemmeke, V. 1982. Entwicklungsgeschichtliche Untersuchungen an Mimosaceen-Blüten. Bot. Jahrb. 103: 185–210.

397. George, A.S. 1991. New taxa, combinations and typifications in *Verticordia* (Myrtaceae: Chamelaucieae). Nuytsia 7: 231–394.

398. Gerenday, S.A. & J.C. French. 1988. Endothecial thickenings in anthers of porate monocotyledons. Amer. J. Bot. 75: 22–25.

399. Gerrath, J.M. & U. Posluszny. 1988. Comparative floral development in some members of the Vitaceae. pp. 121–131 in: Aspects of Floral Development (edited by P. Leins, S.C. Tucker & P.K. Endress). J. Cramer, Berlin.

400. Ghazanfar, S.A. 1990. Staminal dimorphism in *Cassia* s.l. (Caesalpiniaceae) in West Africa. Mitt. Inst. Allg. Bot. Hamburg 23b: 731–738.

401. Gibbs, P.E., J. Semir & N.D. da Cruz. 1988. A proposal to unite the genera *Chorisia* Kunth and *Ceiba* Miller (Bombacaceae). Notes R. Bot. Gard. Edinb. 45: 125–136.

402. Glück, H. 1919. Blatt- und blütenmorphologische Studien. G. Fischer, Jena. 696 pp.

403. Godley, E.J. 1980. Unilocular anthers in the Carmichaelieae (Papilionaceae). N.Z. J. Bot. 18: 449–450.

404. Goebel, K. 1924. Die Entfaltungsbewegungen der Pflanzen und deren teleologische Deutung, 2nd edn. G. Fischer, Jena. 565 pp.

405. Goebel, K. 1928. Vegetative Verzweigung und Blutenbildung. In: Organographie der Pflanzen. 3rd edn. G. Fischer, Jena.

406. Goebel, K. 1928–1931. Organographie der Pflanzen. 3rd edn. G. Fischer, Jena. (Also 1900 (Pt. I) & 1905 (Pt. 2) Organography of Plants. Clarendon Press, Oxford.)

407. Goldberg, R.B., T.P. Beals & P.M. Sanders. 1993. Anther development: basic principles and practical applications. Pl. Cell 5: 1217–1229.

408. Golinski, J. 1893. Ein Beitrag zur Entwicklungsgeschichte des Androeceums und des Gynaeceums der Gräser. Bot. Zbl. 55: 1–17, 65–72, 129–135 + Plates I–III.

409. Golubeva, E.A. 1989. Ultrastructure and the function of middle layers of sugar-beet anther wall. Sborn. Nauch. Trud. Prikl. Bot. Genet. Selek. 124: 78–82. [Russ.; Eng. summ.]

410. Gori, P. 1982. Accumulation of polysaccharides in the anther cavity of *Allium sativum*, clone Piemonte. J. Ultrastruct. Res. 81: 158–162.

411. Gori, P. 1982 [1983]. An ultrastructural investigation of microspores, pollen grains and tapetum in *Asparagus officinalis*. Phytomorphology 32: 277–284.

412. Gori, P. & M. Lorito. 1988 [1989]. An ultrastructural investigation of the anther wall and tapetum in *Capparis spinosa* L. var. *inermis*. Caryologia 41: 329–340.

413. Goto, N., N. Katoh & R. Kranz. 1991. Morphogenesis of floral organs in *Arabidopsis*: predominant carpel formation of the pin-formed mutant. Jpn. J. Genet. 66: 551–568. [Eng.] [Not seen.]

414. Gottsberger, G. 1970. Beiträge zur Biologie von Annonaceen-Blüten. Öst. Bot. Z. 118: 237–279.

415. Gottsberger, G. 1974. The structure and function of the primitive angiosperm flower – a discussion. Acta Bot. Neerl. 23: 461–471.

416. Gould, K.S. & E.M. Lord. 1988. Growth of anthers in *Lilium longiflorum*. A kinematic analysis. Planta 173: 161–171.

417. Govindarajalu, E. 1984. Overlooked exomorphological evidences towards the correct nomenclature of the so-called *Nechamandra alternifolia* (Roxb.) Thw. Proc. Indian Acad. Sci., Pl. Sci. 93: 7–17.

418. Grant, I., W.D. Beversdorf & R.L. Peterson. 1986. A comparative light and electron microscopic study of microspore and tapetal development in male fertile and cytoplasmic male sterile oilseed rape (*Brassica napus*). Can. J. Bot. 64: 1055–1068.

419. Grant, V. 1950. The pollination of *Calycanthus occidentalis*. Amer. J. Bot. 37: 294–297.

420. Grau, J. & A. Schwab. 1982. Mikromerkmale der Blüte zur Gliederung der Gattung *Myosotis*. Mitt. Bot. Staatssamml. München 18: 9–58 [Eng. summ.]

421. Graybosch, R.A. & R.G. Palmer. 1985. Male sterility in soybean (*Glycine max*). I. Phenotypic expression of the ms2 mutant. II. Phenotypic expression of the ms4 mutant. Amer. J. Bot. 72: 1738–1750, 1751–1764.

422. Grayum, M.H. 1991. Systematic embryology of the Araceae. Bot. Rev. 57: 167–203.

423. Green, J.W. 1980. A revised terminology for the spore-containing parts of anthers. New Phytol. 84: 401–406.

424. Gregoire, V. 1935. Sporophylles et organes floraux, tige et axe floral. Rec. Trav. Bot. Neerl. 32: 453–466.

425. Gregory, P.J. 1936. The floral morphology and cytology of *Elettaria cardamomum*. J. Linn. Soc., Bot. 50: 363–391.

426. Grevtsova, N.A. & Kh. Kh. Dzhalilova. 1986. Embryology of some *Peucedanum* species. Nauchn. Dokl. Vyssh. Shk., Biol. Nauki (1): 66–71. [Russ.; Eng. summ.] [Not seen.]

427. Grevtsova, N.A. & L.V. Korobova Semenchenko. 1981. Embryology of Ranunculaceae. 2. Development of male generative sphere in *Actaea acuminata*. Vestn. Mosk. Univ. Biol., ser. 16, (2): 17–21. [Russ.; Eng. summ.] [Not seen.]

428. Greyson, R.I., D.B. Walden & P.C. Cheng. 1980. LM, TEM and SEM observations of anther development in the genic male sterile (ms9) mutant of corn *Zea mays*. Can. J. Genet. Cytol. 22: 153–166.

429. Gross, H. 1913. Beiträge zur Kenntnis der Polygonaceen. Bot. Jahrb. 49: 234–339.

430. Grove, A.R. 1941. Morphological study of *Agave lechuguilla*. Bot. Gaz. 103: 354–365.

431. Grushvitzky, I.V. & N.T. Skvortsova. 1973. *Scheffleropsis* Ridl. – the new genus for the North Vietnam flora and representative of the oldest tribe of Araliaceae. Bot. Zh. SSSR 58: 1492–1503. [Russ.]

432. Guédès, M. 1965. Homologies des pièces florales chez *Tulipa gesneriana* L. Rev. Gén. Bot. 72: 289–322.

433. Guédès, M. 1965. Sur la morphologie des staminocarpelles. C.R. Acad. Sci., Paris 260: 2064–2067.

434. Guédès, M. 1966. Stamen, carpel and ovule. The

teratological approach to their interpretation. Adv. Front. Pl. Sci. 14: 43–95.

435. Guédès, M. 1966. Stamen, tepal and corona in *Narcissus*. Adv. Front. Pl. Sci. 16: 113–136.

436. Guédès, M. 1966. Le carpelle du *Prunus paniculata* Thunb. (*P. serrulata* Lindl.). Ses modifications morphologiques dans les fleurs doubles et sa signification. Flora (Jena) B, 156: 464–499.

437. Guédès, M. 1966. Homologies du carpelle et de l'étamine chez *Tulipa gesneriana* L. Öst. Bot. Z. 113: 47–83.

438. Guédès, M. 1967. The cruciferous flower. Adv. Front. Pl. Sci. 18: 169–218.

439. Guédès, M. 1968. Homologies de l'étamine et du carpelle chez *Veronica subsessilis* (Miq.) Carr. (*Veronica longifolia* L. var. *subsessilis* Miq.). Bot. Jahrb. 88: 382–409.

440. Guédès, M. 1969. Homologies de l'étamine et du carpelle chez *Papaver orientale* L. C.R. Acad. Sci., Paris, D, 268: 926–929.

441. Guédès, M. 1969. Contribution à la morphologie du phyllome. Cellule 67: 343–365.

442. Guédès, M. 1972. Stamen-carpel homologies. Flora 161: 184–208.

443. Guédès, M. 1972. Contribution à la morphologie du phyllome. Mem. Mus. Natn. Hist. Nat., Paris (N.S.) sér. B, 21: 1–179.

444. Guédès, M. & P. Dupuy. 1963. Dégradation des caractères diplophylle et pelté chez les pièces hermaphrodites de *Sempervivum tectorum* L., subsp. *eutectorum* Wettst. Bull. Soc. Bot. Fr. 110: 282–296.

445. Guédès, M. & P. Dupuy. 1979. Teratological modifications and the meaning of flower parts. Vistas Pl. Sci. (edited by T.M. Varghese) 5: 75–131.

446. Guérin, P. 1917. Sur l'étamine et le développement du pollen des Sauges. C.R. Acad. Sci., Paris 165: 1009–1012.

447. Guérin, P. 1919. Développement de l'anthère et du pollen des Labiées. C.R. Acad. Sci., Paris 168: 182–185.

448. Guérin, P. 1926. Le développement de l'anthère chez les Gentianacées. Bull. Soc. Bot. Fr. 73: 5–18.

449. Guérin, P. 1927. Le développement de l'anthère et du pollen chez les Liliacées (*Sansevieria, Liriope, Ophiopogon, Peliosanthes*). Bull. Soc. Bot. Fr. 74: 102–107.

450. Gunn, C.R. & J. Kluve. 1976. Androecium and pistil characters for Tribe Vicieae (Fabaceae). Taxon 25: 563–575.

451. Gunthart, A. 1917. Über die Entwicklung und

Entwicklungsmechanik der Cruciferenblüte und ihre Funktion unter natürlichen und künstlichen Bedingungen. Beih. Bot. Zbl. 35 (1): 60–170.

452. Gupta, S.C. & R. Ahluwalia. 1979. The anther and ovule of *Nelumbo nucifera* – a reinvestigation. J. Indian Bot. Soc. 58: 177–182.

453. Gupta, S.C. & K. Nanda. 1973. Fibrous endothecium, tapetum, and pollen development in *Belamcanda chinensis* DC. Bot. Gaz. 134: 125–129.

454. Gupta, S.C. & K. Nanda. 1978. Ontogeny and histochemistry of dimorphic tapetum in *Tecoma stans* anthers. Bull. Soc. Bot. Fr. 125, Act. Bot. 1/2: 129–134.

455. Gupta, S.C. & K. Nanda. 1978. Studies in the Bignoniaceae. I. Ontogeny of dimorphic anther tapetum in *Pyrostegia*. Amer. J. Bot. 65: 395–399.

456. Gupta, S.C. & K. Nanda. 1983. Histochemical studies in the Bignoniaceae. I. Total carbohydrates of insoluble polysaccharides in *Pyrostegia* anthers. Beitr. Biol. Pfl. 58: 237–242.

457. Gurudeva, M.R. & B. Gowda. 1988. Development of male gametophyte and origin of tapetum in *Scutellaria discolor* Wall ex Bentham. Curr. Sci. 57: 148–150.

458. Guttridge, C.G., J.F. Reeves & W.W. George. 1980. Scanning electron microscopy of healthy and aborted strawberry anthers. Hort. Res. 20: 79–82.

459. Guyot, M. & P. Dupuy. 1963. Les étamines virescentes de Crucifères: un cas particulier de diplophyllie. Bull. Soc. Bot. Fr. 110: 210–216.

460. Haas, P. 1976. Morphologische, anatomische und entwicklungsgeschichtliche Untersuchungen an Blüten und Früchten hochsukkulenter Mesembryanthemaceen–Gattungen – ein Beitrag zu ihrer Systematik. Diss. Bot. 33: 1–256. [Not seen.]

461. Haber, E. & J.E. Cruise. 1974. Generic limits in the Pyroloideae (Ericaceae). Can. J. Bot. 52: 877–883.

462. Haber, J.M. 1959 [1960]. The comparative anatomy and morphology of the flowers and inflorescences of the Proteaceae. I. Some Australian taxa. Phytomorphology 9: 325–358.

463. Haberlandt, G. 1901; 1906. Sinnesorgane in Pflanzenreich, zur Perception mechanischer Reize. 1st & 2nd edns. W. Engelmann, Leipzig.

464. Hagerup, O. 1950. Rain pollination. Kgl. Vidensk. Selsk. Biol. Medd. 18 (5): 1–19.

465. Håkansson, A. 1921. Beiträge zur Entwicklungsgeschichte der Taccaceen. Bot. Notiser 189–220, 257–268. [Eng. summ.]

466. Håkansson, A. 1923. Studien über Entwicklung der

Umbelliferen. Acta Univ. Lund 18: 1–120. [Not seen.]

467. Hakki, M.I. 1972. Blütenmorphologische und embryologische Untersuchungen an *Chenopodium capitatum* und *Chenopodium foliosum* sowie weiteren Chenopodiaceae. Bot. Jahrb. 92: 178–330.

468. Halldén, C., G. Karlsson, C. Lind, I.M. Moller & W.K. Heneen. 1991. Microsporogenesis and tapetal development in fertile and cytoplasmic male-sterile sugar beet (*Beta vulgaris* L.). Sex. Pl. Reprod. 4: 215–225.

469. Hamann, U. 1962. Beitrag zur Embryologie der Centrolepidaceae mit Bemerkungen über den Bau der Blüten und Blütenstande und die systematische Stellung der Familie. Ber. Dtsch. Bot. Ges. 75: 153–171.

470. Hamel, M.-C. 1989. Contribution bibliographique à la connaissance botanique des Apocynacées. IV. Relations avec les familles affines dans l'ordre des Gentianales. Rev. Cytol. Biol. Vég., Bot. 12: 139–158.

471. Ha Ngoc, K.A. 1980. Action de l'acide 2, 4–dichlorophénoxyacétique et de l'acide 2,4,5-trichlorophénoxyacétique sur la croissance des filets staminaux du *Zea mays*. Étude de l'activité saccharasique et des modifications ultrastructurales des parois cellulaires au cours de leur allongement. Can. J. Bot. 58: 1–13.

472. Hannig, E. 1909. Über den Öffnungsmechanismus der Antheren. Jahrb. Wiss. Bot. 47: 186–218.

473. Harling, C. 1946. Studien über den Blütenbau und die Embryologie der Familie Cyclanthaceae. Svensk Bot. Tidskr. 40: 257–272.

474. Harney, P.M. & H.C.C. Kung. 1967. Development of non-dehiscent anthers in partially male sterile plants of *Pelargonium* × *hortorum* Bailey cv. 'Jacqueline'. Can. J. Genet. Cytol. 9: 359–366.

475. Harris, J.A. 1905. The dehiscence of anthers by apical pores. Ann. Rep. Missouri Bot. Gard. 16: 167–257.

476. Harris, J.A. 1906. The anomalous anther-structure of *Dicorynia*, *Duparquetia*, and *Strumpfia*. Bull. Torrey Bot. Cl. 33: 223–228.

477. Harrison, H.K. 1968 [1969]. Contributions to the study of the genus *Eriastrum*. I. The corolla and androecium. Phytomorphology 18: 393–402.

478. Haskell, G. 1949. Variation in the number of stamens in the common chickweed. J. Genet. 49: 291–301.

479. Hassan, A., B.M. Mansour, N. Toema & M.K. Nada. 1982. Histological and cytological studies on ten hybrid tea rose varieties. Egypt. J. Hort. 9 (1): 23–30.

480. Haulik, T.K. & L.C. Holtzhausen. 1988. Anatomy of staminate flower ontogeny of the pecan as determined

by scanning electron microscopy. S. Afr. J. Pl. Soil 5: 205–208. [Not seen.]

481. Hausbrandt, L. & W. Golinowski. 1975 [1976]. Rozwój warstw ściennych i otwieranie się pylnika u kilku dzikich gatunków ziemniaka (*Solanum chacoense* Bitt., *S. phureja* Juz. et Buk., *S. giberulosum* Juz. et Buk., *S. commersonii* Dun.). Acta Agrobot., Warsaw 28: 89–93.

482. Hayashi, Y. 1960. On the microsporogenesis and pollen morphology in the family Magnoliaceae. Sci. Rep. Tôhoku Univ., ser 4, Biol. 26: 45–52.

483. He, G., Z. Zhang & Y. Lin. 1983. A study on the fine structures of staminate development in rice (*Oryza sativa* L.). III. Origin and development of the tapetal sporopollenin bodies. Acta Scient. Nat. Univ. Sunyatseni (3): 75–82. [Chin.; Eng. summ.]

484. Hegde, R.R. 1985. Histochemical studies on *Datura* anthers. J. Palynol. 21: 15–20.

485. Hegde, R.R. & L. Andrade. 1982. Anther development in *Datura*: distribution of proteins, esterase and adenosine triphosphatase. Pl. Sci. Lett. 28: 95–101.

486. Heimlich, L.F. 1927. The development and the anatomy of the staminate flower of the cucumber. Amer. J. Bot. 14: 227–237.

487. Heinig, K.H. 1951. Studies in the floral morphology of the Thymelaeaceae. Amer. J. Bot. 38: 113–132.

488. Heinsbroek, P.G. & W.A. Van Heel. 1969. Note on the bearing of the pattern of vascular bundles on the morphology of the stamens of *Victoria amazonica* (Poepp.) Sowerby. Proc. K. Ned. Akad. Wet., Amsterdam, C. 72: 431–444.

489. Heller, J. 1933. Über Verholzungen in der Blütenregion windblütiger Gewächse. Beih. Bot. Zbl. 51 (1): 517–523.

490. Heller, J. 1935. Über die Staubblätter einiger windblütiger Holzgewächse. Beih. Bot. Zbl. 53 (1): 82–94.

491. Hendrychová-Tomková, J. & H.B. Nguyen Thi. 1982. The persistence and lignin-like appearance of the primary cell wall of microsporocytes in malesterile (cms) sweet pepper, *Capsicum annuum* L. Biol. Plantarum 24: 440–445.

492. Hennig, L. 1929. Beiträge zur Kenntnis der Resedaceenblüte und frucht. Planta 9: 507–563.

493. Herich, R. 1983. Study of specific programs of cytodifferentiation and degradation of tapetal cells. Acta Fac. Rerum Nat. Univ. Comen. Physiol. Pl. 19: 1–8. [Eng.] [Not seen.]

494. Heslop-Harrison, J. 1963. Ultrastructural aspects of differentiation in sporogenous tissue. pp. 315–340 in:

Symp. Soc. Exp. Biol. No. 17: Cell Differentiation. Cambridge University Press, London.

495. Heslop-Harrison, J. 1968. Tapetal origin of pollen coat substances in *Lilium*. New Phytol. 67: 779–786.

496. Heslop-Harrison, J. 1969. An acetolysis-resistant membrane investing tapetum and sporogenous tissue in the anthers of certain Compositae. Can. J. Bot. 47: 541–542.

497. Heslop-Harrison, J. 1972. Sexuality of angiosperms. pp. 133–289 in: Plant Physiology, Vol. VIC (edited by F.C. Steward). Academic Press, New York & London.

498. Heslop-Harrison, J.S., Y. Heslop-Harrison & B.J. Reger. 1987. Anther-filament extension in *Lilium*: potassium ion movement and some anatomical features. Ann. Bot. 59: 505–515.

499. Hess, D. 1983. Die Blüte. Verlag Eugen Ulmer, Stuttgart. 458 pp.

500. Hesse, M. 1978. Entwicklungsgeschichte und Ultrastruktur des Pollenkitts bei *Tilia* (Tiliaceae). Pl. Syst. Evol. 129: 13–30.

501. Hesse, M. 1986. Orbicules and the ektexine are homologous sporopollenin concretions in Spermatophyta. Pl. Syst. Evol. 153: 37–48.

502. Hesse, M. 1993. Pollenkitt development and composition in *Tilia platyphyllos* (Tiliaceae) analysed by conventional and energy filtering TEM. Pl. Syst. Evol., Suppl. 7: 39–52.

503. Hesse, M. & J. Greilhuber. 1975. Ultrastructural and onotogenetic aspects of the 'inner anther tapetum' in *Rhinanthus* (Scrophulariaceae). Linzer Biol. Beitr. 7: 257–276.

504. Hesse, M. & M.W. Hess. 1993. Recent trends in tapetum research. A cytological and methodological review. Pl. Syst. Evol., Suppl. 7: 127–145.

505. Hideux, M. & M. Abadie. 1980–1 [1981]. The anther of *Saxifraga cymbalaria* L. ssp. *huetiana* (Boiss.) Engl. and Irmsch.; a study by electron microscopy (S.E.M. and T.E.M.). 3. Dynamics of the relationships between tapetal and sporal cells. Ann. Sci. Nat., Bot., sér. 13, 2-3: 27–37.

506. Hideux M. & M. Abadie. 1985. Cytologie ultrastructurale de l'anthère de *Saxifraga*. I. Période d'initiation des précurseurs des sporopollénines au niveau des principaux types exiniques. Can. J. Bot. 63: 97–112.

507. Hiepko, P. 1964. Das zentrifugale Androeceum der Paeoniaceae. Ber. Dtsch. Bot. Ges. 77: 427–435.

508. Hiepko, P. 1965. Vergleichend-morphologische und entwicklungs-geschichtliche Untersuchungen über das Perianth bei den Polycarpicae I. Bot. Jahrb. 84: 359–426. II. Ibid.: 427–508. [Eng. summ.]

509. Hilderbrand, F. 1870. Ueber die Bestäubungsvorrichtungen bei den Fumariaceen. Jahrb. Wiss. Bot. 7: 423–471 + Plates XXIX–XXXI.

510. Hill, B.S. & G.P. Findlay. 1981. The power of movement in plants: the role of osmotic machines. Q. Rev. Biophys. 14: 173–222. [Not seen.]

511. Hillson, C.J. 1959. Comparative studies of floral morphology of the Labiatae. Amer. J. Bot. 46: 451–459.

512. Hirmer, M. 1918. Beiträge zur Morphologie polyandrischen Blüten. Flora 110: 140–192 + Plates I–XI.

513. Hiroi, K., A. Al Mamun, T. Wada & Y. Takeoka. 1991. A study on the interspecific variation of spikelet structure in the genus *Oryza*. Jpn. J. Crop Sci. 60: 153–160. [Eng.]

514. Hjelmqvist, H. 1948. Studies on the floral morphology and phylogeny of the Amentiferae. Bot. Notiser Suppl. vol. 2(1): 1–171.

515. Hoefert, L.L. 1971. Ultrastructure of tapetal cell ontogeny in *Beta*. Protoplasma 73: 397–406.

516. Hoefert, L.L. & K. Esau. 1975. Plastid inclusions in epidermal cells of *Beta*. Amer. J. Bot. 62: 36–40.

517. Hofmann, U. 1973. Morphologische Untersuchungen zur Umgrenzung und Gliederung der Aizoaceen. Bot. Jahrb. 93: 247–324.

518. Homedes, J. 1928. Datos para una interpretación endocrina de las células de tapiz de los sacos polinicos. Bol. R. Soc. Esp. Hist. Nat. 28: 315–320.

519. Horne, A.S. 1909. The structure and affinities of *Davidia involucrata* Baill. Trans. Linn. Soc., Lond., ser. 2, 7: 303–326 + Plates 31–33.

520. Horne, A.S. 1914. A contribution to the study of the evolution of the flower, with special reference to the Hamamelidaceae, Caprifoliaceae and Cornaceae. Trans. Linn. Soc., Lond., ser. 2, 8: 239–309 + Plates 28–30.

521. Horner, H.T. 1977. A comparative light- and electron-microscopic study of microsporogenesis in male-fertile and cytoplasmic male-sterile sunflower (*Helianthus annuus*). Amer. J. Bot. 64: 745–759.

522. Horner, H.T., A.P. Kausch & B.L. Wagner. 1981. Growth and change in shape of raphide and druse calcium oxalate crystals as a function of intracellular development in *Typha angustifolia* L. (Typhaceae) and *Capsicum annuum* L. (Solanaceae). Scanning Electron Microsc. 1981, 3: 251–262.

523. Horner, H.T. & B.L. Wagner. 1980. The association of druse crystals with the developing stomium of

Capsicum annuum (Solanaceae) anthers. Amer. J. Bot. 67: 1347–1360.

524. Horner, H.T. & B.L. Wagner. 1992. Association of four different calcium crystals in the anther connective tissue and hypodermal stomium of *Capsicum annuum* (Solanaceae) during microsporogenesis. Amer. J. Bot. 79: 531–541.

525. Howard, R.A. 1948. The morphology and systematics of the West Indian Magnoliaceae. Bull. Torrey Bot. Cl. 75: 335–357.

526. Hu, S.-Y., L.-Y. Xu & C.-Y. Ma. 1992. Anther wall of *Cymbidium goeringii* with special reference to the tapetum. Acta Bot. Sin. 34(8): 581–587 + 3 plates. [Chin.; Eng. summ.]

527. Huber, E, 1953. Beitrag zur anatomischen Untersuchung der Antheren von *Saintpaulia*. Sber. Akad. Wiss. Wien, Math.-Nat. Kl. I, 162: 227–234. [Not seen.]

528. Huber, H. 1963. Die Verwandtschaftsverhältnisse der Rosifloren. Mitt. Bot. Staatssamml. München 5: 1–48.

529. Huber, K.A. 1977. Morphologische und entwicklungsgeschichtliche Untersuchungen an Blüten und Blütenständen von Solanaceen und von *Nolana paradoxa* Lindl. (Nolanaceae). Diss. Bot. 55. J. Cramer, Vaduz. 486 pp.

530. Hufford, L. 1990. Androecial development and the problem of monophyly of Loasaceae. Can. J. Bot. 68: 402–419.

531. Hufford, L. 1992. Floral structure of *Besseya* and *Synthyris* (Scrophulariaceae). Int. J. Pl. Sci. 153: 217–229.

532. Hufford, L.D. 1980. Staminal vascular architecture in five dicotyledonous angiosperms. Proc. Iowa Acad. Sci. 87: 96–102.

533. Hufford, L.D. & P.K. Endress. 1989. The diversity of anther structures and dehiscence patterns among Hamamelididae. Bot. J. Linn. Soc. 99: 301–346.

534. Huynh, K.-L. 1982. La fleur mâle de quelques espèces de *Pandanus* subg. *Lophostigma* (Pandanaceae) et sa signification taxonomique, phylogénique et évolutive. Beitr. Biol. Pfl. 57: 15–83.

535. Huynh, K.-L. 1983. Carpellodes or staminodes? Problems in the genus *Pandanus* (Pandanaceae), and their taxonomic significance. Bot. J. Linn. Soc. 87: 177–192.

536. Huynh, K.-L. 1983. The taxonomic significance of the anther structure in the genus *Pandanus* (Pandanaceae) with reference to *Pandanus* sect. Martellidendron. Webbia 37: 141–148.

537. Huynh, K.-L. 1991. New data on the taxonomic position of *Pandanus eydouxia* (Pandanaceae), a species of the Mascarene Islands. Bot. Helv. 101: 29–37.

538. Huynh, K.-L. 1992. The flower structure in the genus *Freycinetia* Pandanaceae (2). Early differentiation of the sex organs, especially the staminodes, and further notes on the anthers. Bot. Jahrb. 114: 417–441.

539. Igersheim, A. 1993. Floral development and secondary pollen presentation in *Petromarula* Vent. ex Hedwig f. (Campanulaceae). Bot. Jahrb. Syst. 115: 301–313.

540. Ihlenfeldt, H.-D. 1960. Entwicklungsgeschichtliche, morphologische und systematische Untersuchungen an Mesembryanthemen. Feddes Repert. 63: 1–104.

541. Imaichi, R. & K. Okamoto. 1992. Comparative androecium morphogenesis of *Sicyos angulatus* and *Sechium edule* (Cucurbitaceae). Bot. Mag. Tokyo 105: 539–548.

542. Imaichi, R. & N. Uchibori. 1992. Androecium morphogenesis and staminal vasculature of *Luffa cylindrica* (Cucurbitaceae). Bull. Fac. Agric. Tamagawa Univ. (32): 95–107. [Jap.][Not seen.]

543. Inamdar, J.A., M. Gangadhara & R.B. Bhat. 1977. Epidermal structure and ontogeny of stomata in vegetative and floral organs of *Yucca filamentosa* Linn. var. *concava* Baker. Ceylon J. Sci., Biol. Sci. 12(2): 119–124. [Not seen.]

544. Inamdar, J.A., R.C. Patel & J.S.S. Mohan. 1991. Structure and ontogeny of stomata in vegetative and floral organs of some Apocynaceae. Feddes Repert. 102: 409–423.

545. Innes, R.L., W.R. Remphrey & L.M. Lenz. 1989. An analysis of the development of single and double flowers in *Potentilla fruticosa*. Can. J. Bot. 67: 1071–1079.

546. Irish, E.E. & T. Nelson. 1989. Sex determination in monoecious and dioecious plants. Plant Cell 1: 737–744.

547. Ito, M. 1983, 1984, 1986. Studies in the floral morphology and anatomy of Nymphaeales. I. The morphology of vascular bundles in the flower of *Nymphaea tetragona* George. Acta Phytotax. Geobot. 34: 18–26 (1983). II. Floral anatomy of *Nymphaea tetragona* George. Ibid. 35: 94–102 (1984). III. Floral anatomy of *Brasenia schreberi* Gmel. and *Cabomba caroliniana* A.Gray. Bot. Mag., Tokyo 99: 169–184. IV. Floral anatomy of *Nelumbo nucifera*. Acta Phytotax. Geobot. 37: 82–96 (1986). [Eng.]

548. Ivancich, A. 1906. Der Bau der Filamente der Amen-

taceen. Öst. Bot. Z. 56: 305–309, 385–394 + Plates VII–VIII.

549. Jackson, G. 1934. The morphology of the flowers of *Rosa* and certain closely related genera. Amer. J. Bot. 21: 453–466.

550. Jacobson Paley, R. 1920. Le periplasmodium dans les anthères de l'*Arum maculatum* L. Bull. Soc. Bot. Genève, sér. 2, 12: 306–318.

551. Jacques Félix, H. 1981. Observations sur les caractères staminaux et la classification des Osbeckieae (Melastomataceae) capsulaires africaines. Adansonia, sér. 2, 20: 405–429.

552. Jaffe, M.J., C. Gibson & R. Biro. 1977. Physiological studies of mechanically stimulated motor responses of flower parts. I. Characterization of the thigmotropic stamens of *Portulaca grandiflora* Hook. Bot. Gaz. 138: 438–447.

553. Jäger, I. 1961. Vergleichend-morphologische Untersuchungen des Gefässbündelsystems peltater Nektar- und Kronblätter sowie verbildeter Staubblätter. Öst. Bot. Z. 108: 433–504.

554. Jain, B.K. 1986 [1987]. Development of the staminate flower of *Najas marina* L. Beitr. Biol. Pfl. 61: 401–410.

555. Jalan, S. 1986. The morphology and taxonomic significance of amoeboid tapetum in *Actaea spicata* Linne. Indian J. Bot. 9(1): 60–62.

556. Jalan, S. 1987. A new type of staminal vasculature in *Schisandra neglecta*. J. Indian Bot. Soc. 66: 150.

557. Janchen, E. 1950. Die sogenannte Schildform der jungen Staubgefässe. Phyton 2: 267–270.

558. Janchen, E. 1950. Die Herkunft der Angiospermen–Blüte und die systematische Stellung der Apetalen. Öst. Bot. Z. 97: 129–167.

559. Janchen, E. 1952. Die sogennante Schildblatt–Natur der Staubgefässe. Phyton 4: 224–229.

560. Janczewski, E. 1908. Sur les anthères stériles des groseilliers. Bull. Acad. Sci. Cracov. 587–596 + Plate XXIV.

561. Jaretzky, R. 1928. Bildungsabweichungen in Cruciferenblüten. Planta 5: 444–463.

562. Jaretzky, R. 1928. Histologische und karyologische Studien an Polygonaceen. Jahrb. Wiss. Bot. 69: 357–490.

563. Jassem, B., J. Macewicz & A. Majewska Sawka. 1989. Scanning electron microscope studies on the anthers of male-fertile and male-sterile sugar beet (*Beta vulgaris* L.) plants. Genet. Pol. 30: 139–144.

564. Jauch, B. 1918. Quelques points de l'anatomie et de la biologie des Polygalacées. Thèse, Univ. Genève. 42

pp. Also: Bull. Soc. Bot. Genève, sèr. 2, 10: 47–84 (1918).

565. Jazdzewska, E. 1989. Some biological traits of male-sterile sugar beet lines. II. Anatomical examinations of anthers. Hodowla Rosl. Aklim. Nasienn. 33 (1–2): 63–80. [Pol.] [Not seen.]

566. Jeffrey, C. & Y.-L. Chen. 1984. Taxonomic studies on the tribe Senecioneae (Compositae) of Eastern Asia. Kew Bull. 39: 205–446.

567. Jimenez, M.S. & D. Morales. 1978 [1979]. Histología floral y desarrollo del fruto de *Euphorbia atropurpurea* Brouss. Vieraea 8: 131–143.

568. Johnson, D.S. 1935. The development of the shoot, male flower and seedling of *Batis maritima*. Bull. Torrey Bot. Cl. 62: 19–31 + Plates 1–3.

569. Johnson, L.A.S. & B.G. Briggs. 1983. Myrtaceae – comments on comments. Taxon 32: 103–105.

570. Johri, B.M., K.B. Ambegaokar & P.S. Srivastava (eds.). 1992. Comparative Embryology of Angiosperms. 2 vols. Springer-Verlag, Berlin. 1221 pp.

571. Johri, B.M. & R.N. Kapil. 1953. Contribution to the morphology and life history of *Acalypha indica* L. Phytomorphology 3: 137–151.

572. Joppa, L.R., F.H. McNeal & J.R. Welsh. 1966. Pollen and anther development in cytoplasmic male sterile wheat (*Triticum aestivum* L.). Crop Sci. 6: 296–297. [Not seen.]

573. Joshi, A.C. 1932. Dédoublement of stamens in *Achyranthes aspera* Linn. J. Indian Bot. Soc. 11: 335–339.

574. Joshi, A.C. 1939. Vascular supply of the stamens and ovules of *Gloriosa superba*. Nature 143: 437.

575. Joshi, A.C. & V.S.R. Rao. 1934. Vascular anatomy of the flowers of four Nyctaginaceae. J. Indian Bot. Soc. 13: 169–186.

576. Juel, H.O. 1915. Untersuchungen über die Auflösung der Tapetenzellen in den Pollensäcken der Angiospermen. Jahrb. Wiss. Bot. 56: 337–364 + Plates IV–V.

577. Juliano, J.B. & E. Quisumbing. 1931. Morphology of the male flower of *Cocos nucifera* Linnaeus. Philipp. J. Sci. 45: 449–458.

578. Juncosa, A.M. 1988. Floral development and character evolution in Rhizophoraceae. pp. 83–101 in: Aspects of Floral Development (edited by P. Leins, S.C. Tucker & P.K. Endress). J. Cramer, Berlin.

579. Juncosa, A.M. & P.B. Tomlinson. 1987. Floral development in mangrove Rhizophoraceae. Amer. J. Bot. 74: 1263–1279.

580. Junell, S. 1931. Die Entwicklungsgeschichte von *Circaeaster agrestis*. Svensk Bot. Tidskr. 25: 238–270.

581. Kamelina, O.P. 1984. The anther and pollen grain

development in *Daphniphyllum macropodum* (Daphniphyllaceae). Bot. Zh. SSSR 69: 376–382 + 2 plates. [Russ.]

582. Kamelina, O.P. & K.S. Maldybekova. 1980. The development of the anther and pollen grains of *Valeriana phu* (Valerianaceae). Bot. Zh. SSSR 65: 640–646. [Russ.; Eng. summ.]

583. Kamelina, O.P. & O.B. Proskurina. 1987. Anther and pollen grain development in Elaeagnaceae family. Bot. Zh. SSSR 72: 909–917. [Russ. only.]

584. Kamelina, O.P. & N.G. Toutchina. [N.G. Tuchina]. 1982. A contribution to the embryology of the non-investigated taxa. II. The development of the anther and pollen grains in *Nandina domestica* (Berberidaceae). Bot. Zh. SSSR 67: 1459–1468. [Russ.; Eng. summ.]

585. Kamelina, O.P. & M.S. Yakovlev. 1976. The development of anther and microgametogenesis in representatives of the families Dipsacaceae and Morinaceae. Bot. Zh. SSSR 61: 932–945. [Russ.]

586. Kampny, C.M. & J.M. Canne–Hilliker. 1988. Aspects of floral development in Scrophulariaceae. Striking early differences in three tribes. pp. 147–157 in: Aspects of Floral Development (edited by P. Leins, S.C. Tucker & P.K. Endress). J. Cramer, Berlin.

587. Kang, W.J. & B. Sun. 1987. The study on development characteristics of staminate flower and pistillate flower of *Luffa cylindrica*. Acta Bot. Bor.-Occ. Sin. 7: 45–51 + 1 plate. [Chin.; Eng. summ.]

588. Kania, W. 1973. Entwicklungsgeschichtliche Untersuchungen an Rosaceenblüten. Bot. Jahrb. 93: 175–246.

589. Kapil, R.N. & S. Sethi. 1962. Gametogenesis and seed development in *Ainsliaea aptera* DC. Phytomorphology 12: 222–234.

590. Kaplan, D.R. 1967. Floral morphology, organogenesis and interpretation of the inferior ovary in *Downingia bacigalupii*. Amer. J. Bot. 54: 1274–1290.

591. Kaplan, D.R. 1968. Histogenesis of the androecium and gynoecium in *Downingia bacigalupii*. Amer. J. Bot. 55: 933–950.

592. Kapoor, L.D. & B.M. Sharma. 1964. *Argemone mexicana* Linn. II. Morphological and structural studies in some floral abnormalities. Proc. Indian. Acad. Sci. B 59: 88–95. [Not seen.]

593. Kapoor, T., M.R. Vijayaraghavan & N.K. Parulekar. 1978. Ontogeny, structure and differentiation of anther tapetum in *Celsia coromandeliana*. Phyton (Austria) 18: 209–216.

594. Karanth, K.A., P.K. Bhat & G.D. Arekal. 1979. Tap-

etum-like anther epidermis in *Zeuxine longilabris* (Lindl.) Benth. ex HK., Orchidaceae. Curr. Sci. 48: 542–543. [Not seen.]

595. Karis, P.O. 1992. *Hoplophyllum* DC., the sister group to *Eremothamnus* O. Hoffm. (Asteraceae)? Taxon 41: 193–198.

596. Karis, P.O., M. Källersjö & K. Bremer. 1992. Phylogenetic analysis of the Cichorioideae (Asteraceae), with emphasis on the Mutisieae. Ann. Missouri Bot. Gard. 79: 416–427.

597. Karlova, A.A. 1973. Development of the anther wall, microsporogenesis and formation of the male gametophyte in four species of the genus *Digitalis* L. Nauch. Dokl. Vyssh. Shk., Biol. Nauki 16(8): 55–57. [Russ.] [Not seen.]

598. Kasapligil, B. 1951. Morphological and ontogenetic studies of *Umbellularia californica* Nutt. and *Laurus nobilis* L. Univ. Calif. Publ. Bot. 25: 115–240.

599. Kaul, R.B. 1967. Development and vasculature of the flowers of *Lophotocarpus calycinus* and *Sagittaria latifolia* (Alismataceae). Amer. J. Bot. 54: 914–920.

600. Kaul, R.B. 1967. Ontogeny and anatomy of the flower of *Limnocharis flava* (Butomaceae). Amer. J. Bot. 54: 1223–1230.

601. Kaul, R.B. 1968. Floral development and vasculature in *Hydrocleis nymphoides* (Butomaceae). Amer. J. Bot. 55: 236–242.

601a. Kaul, R.B. 1968. Floral morphology and phylogeny in the Hydrocharitaceae. Phytomorphology 18: 13–35.

602. Kaul, R.B. 1969. Morphology and development of the flowers of *Bootia cordata*, *Ottelia alismoides* and their synthetic hybrid (Hydrocharitaceae). Amer. J. Bot. 56: 951–959.

603. Kaul, R.B. 1993. Meristic and organogenetic variation in *Ruppia occidentalis* and *R. maritima*. Int. J. Pl. Sci. 154: 416–424.

604. Kausik, S.B. 1941. Structure and development of the staminate flower and male gametophyte of *Enalus acoroides*. Proc. Indian Acad. Sci., B, 14: 1–16.

605. Kausik, S.B. 1941. Studies in the Proteaceae. V. Vascular anatomy of the flower of *Grevillea robusta*. Proc. Natn. Inst. Sci., India 7: 257–266.

606. Kaussmann, B. 1941. Vergleichende Untersuchungen über die Blattnatur der Kelch-, Blumen- und Staubblätter. Bot. Archiv 42: 503–572.

607. Kavaljian, L.G. 1952. The floral morphology of *Clethra alnifolia* with some notes on *C. acuminata* and *C. arborea*. Bot. Gaz. 113: 392–413.

608. Kawano, S. 1965. Anatomical studies on the

androecia of some members of the Guttiferae-Moronoboideae. Bot. Mag., Tokyo 78: 97–108.

609. Kazimierski, T. & E.M. Kazimierska. 1974 [1975]. Structure of anther heads in some *Trifolium* L. species. Acta Soc. Bot. Pol. 43: 321–329.

610. Keating, R.C. 1972 [1973]. The comparative morphology of the Cochlospermaceae. III. The flower and pollen. Ann. Missouri Bot. Gard. 59: 282–296.

611. Keijzer, C.J. 1987. The processes of anther dehiscence and pollen dispersal. I. The opening mechanism of longitudinally dehiscing anthers. II. The formation and the transfer mechanism of pollenkitt, cell-wall development of the loculus tissues and a function of orbicules in pollen dispersal. New Phytol. 105: 487–498, 499–507.

612. Keijzer, C.J. & M. Cresti. 1987. A comparison of anther tissue development in male sterile *Aloe vera* and male fertile *Aloe ciliaris*. Ann. Bot. 59: 533–542.

613. Keijzer, C.J., I.H.S. Hoek & M.T.M. Willemse. 1987. The development of the staminal filament of *Gasteria verrucosa*. Acta Bot. Neerl. 36: 271–282.

614. Keijzer, C.J. & M.T.M. Willemse. 1988. Tissue interactions in the developing locule of *Gasteria verrucosa* during microgametogenesis. Acta Bot. Neerl. 37: 475–492.

615. Keijzer, C.J. & M.T.M. Willemse. 1988. Tissue interactions in the developing locule of *Gasteria verrucosa* during microsporogenesis. Acta Bot. Neerl. 37: 493–508.

616. Kenda, G. 1950. Anthozyanstaub in der Antherenepidermis von *Papaver rhoeas*. Phyton (Austria) 2: 288–290.

617. Kenda, G. 1952–3. Stomata an Antheren. I. Anatomischer Teil. Phyton (Austria) 4: 83–96.

618. Keng, H. 1959. Androdioecism in the flowers of *Trochodendron aralioides*. J. Arnold Arbor. 40: 158–160.

619. Keng, H. 1962. Comparative morphological studies in the Theaceae. Univ. Calif. Publ. Bot. 33: 269–384.

620. Kenrick, J. & R.B. Knox. 1979. Pollen development and cytochemistry: some Australian species of *Acacia*. Aust. J. Bot. 27: 413–427.

621. Khanam, K. & M.S. Zahur. 1962. Development, growth and morphological nature of the staminal truss in *Ricinus communis*. Biologia (Lahore) 8: 163–178.

622. Khattra, S.S. & G. Singh. 1989. Histochemical studies of anther development in male sterile *Pennisetum typhoides*. Acta Bot. Indica 17: 159–162.

623. Khattra, S.S. & G. Singh. 1991. Callose behaviour in male sterile/fertile anthers of *Pennisetum typhoides*. Acta Bot. indica 19 (1): 110–111.

624. King, R.M. & H. Robinson. 1970. The new synantherology. Taxon 19: 6–11.

625. King, R.M. & H. Robinson. 1987. The genera of the Eupatorieae (Asteraceae). Monogr. Syst. Bot. Missouri Bot. Gard. 22: 1–581.

626. Kirchoff, B.K. 1983. Floral organogenesis in five genera of the Marantaceae and in *Canna* (Cannaceae). Amer. J. Bot. 70: 508–523.

627. Kirchoff, B.K. 1988. Inflorescence and flower development in *Costus scaber* (Costaceae). Can. J. Bot. 66: 339–345.

628. Kirchoff, B.K. 1988. Floral ontogeny and evolution in the ginger group of the Zingiberales. pp. 45–56 in: Aspects of Floral Development (edited by P. Leins, S.C. Tucker & P.K. Endress). J. Cramer, Berlin.

629. Kirchoff, B.K. 1991. Homeosis in the flowers of the Zingiberales. Amer. J. Bot. 78: 833–837.

630. Kirkwood, J.E. 1907. Some features of pollen-formation in the Cucurbitaceae. Bull. Torrey Bot. Cl. 34: 221–242.

631. Klackenberg, J. 1983. A reevaluation of the genus *Exacum* (Gentianaceae) in Ceylon. Nord. J. Bot. 3: 355–370.

632. Klackenberg, J. 1985. The genus *Exacum* (Gentianaceae). In Opera Bot. 84: 1–144.

633. Klopfer, K. 1968, 1969, 1970, 1971, 1972. Beiträge zur floralen Morphogenese und Histogenese der Saxifragaceae. 2. Die Blütenentwicklung von *Tellima grandiflora*. Flora (Jena) 158 B: 1–21 (1968). 3. Die Blütenentwicklung einiger *Ribes*-Arten. Wiss. Z. Päd. Hochsch. Potsdam 13: 187–205 (1969). 4. Die Blütenentwicklung einiger *Saxifraga*-Arten. Flora 159: 347–365. 5. Die Blütenentwicklung der Gattungen *Astilbe, Rodgersia, Astilboides* und *Bergenia*. Wiss. Z. Päd. Hochsch. Potsdam 14: 327–347 (1970). 6. Die Hydrangeoideen. Ibid. 15: 77–95 (1971). 7. *Parnassia palustris* und *Francoa sonchifolia*. Flora 161: 320–332 (1972).

634. Knoll, F. 1914. Zur Ökologie und Reizphysiologie des Andröceums von *Cistus salvifolius* L. Jahrb. Wiss. Bot. 54: 498–527.

635. Koch, M.F. 1930. Studies in the anatomy and morphology of the Compositae flower. I. The corolla. Amer. J. Bot. 17: 938–952. II. The corollas of the Heliantheae and Mutisieae. Ibid.: 995–1010.

636. Koevenig, J.L. 1973. Floral development and stamen filament elongation in *Cleome hassleriana*. Amer. J. Bot. 60: 122–129.

637. Komaki, M.K., K. Okada, E. Nishino & Y. Shimura. 1988. Isolation and characterization of novel mutants of *Arabidopsis thaliana* defective in flower development. Development 104: 195–203.

638. Kondorskaya, V.R. 1967. Morphologie de la fleur staminée et développement du gametophyte mâle chez *Hippophae rhamnoides* L. Nauch. Dokl. Vyssh. Shk., Biol. Nauki 10(4): 69–75. [Russ.] [Not seen.]

639. Kordyum, E.L. 1961. Microsporogenesis and specific characteristics of the development of the tapetum in some species of the genus *Vincetoxicum* Moench. Ukrain. Bot. Zh. 18 (5): 6–14. [Ukrain.; Eng. summ.]

640. Kosmath, L. 1927. Studien über das Antherentapetum. Öst. Bot. Z. 76: 235–241.

641. Kostrikova, L.N. 1976. Development of the anther and microsporogenesis in *Amorpha glabra* Desf. Vest. Mosk. Univ. ser. VI, Biol. Pochvov. 31 (3): 56–59. [Russ.; Eng. summ.] [Not seen.]

642. Kraft, E. 1917. Experimentelle und entwicklungsgeschichtliche Untersuchungen an Caryophyllaceen-Blüten. Flora 109: 283–362.

643. Krishnamurthy, K.V. & N.R. Bhat. 1959. Staminal teratology in the genus *Nicotiana*. J. Indian Bot. Soc. 38: 84–92.

644. Krjatchenko, D.D. 1925. De l'activité des chondriosomes pendant le développement des grains de pollen et des cellules nourricières du pollen dans *Lilium croceum* Chaix. Rev. Gén. Bot. 37: 193–211 + Plates VIII–XI.

645. Kronestedt, E. & P.A. Bystedt. 1981. Thread-like formations in the anthers of *Strelitzia reginae*. Nord. J. Bot. 1: 523–529.

646. Kronestedt–Robards, E.C. & J.R. Rowley. 1989. Pollen grain development and tapetal changes in *Strelitzia reginae* (Strelitziaceae). Amer. J. Bot. 76: 856–870.

647. Kubitzki, K. 1969. Monographie der Hernandiaceen. Bot. Jahrb. 89: 78–209.

648. Kubitzki, K., J.G. Rohwer & V. Bittrich (eds.). 1993. The Families and Genera of Vascular Plants. Vol. II. Flowering Plants. Dicotyledons. Magnoliid, Hamamelid and Caryophyllid families. Springer-Verlag, Berlin 653 pp.

649. Kuhn, E. 1908. Über den Wechsel der Zelltypen im Endothecium der Angiospermen. Diss., Zürich. Leemann, Zürich. [Not seen.]

650. Kuijt, J., D. Wiens & D. Coxson. 1979. A new androecial type in African *Viscum*. Acta Bot. Neerl. 28: 349–355.

651. Kumar, M. & K.S. Manilal. 1992 [1993]. Floral morphology and anatomy of *Paphiopedilum insigne* and the taxonomic status of Cypripedioids (Orchidaceae). Phytomorphology 42: 293–297.

652. Kumar, R. & G. Singh. 1987. Investigations into the causes of sterility II. *Tabebuia pentaphylla* L. Acta Bot. Hung. 33: 401–405.

653. Kumar, R. & G. Singh. 1988. Investigations into the cause of sterility. *Tecoma stans* L. Bull. Soc. Bot. Fr. 135, Lett. Bot. (2): 131–135.

654. Kumar, S. 1979. Embryological studies on Indian species of *Morus*. II. Anther dehiscence. Acta Bot. Indica 7: 176–177.

655. Kumazawa, M. 1940. Further studies on the vascular course in the male inflorescence of *Zea mays*. Vascular anatomy in maize II. Bot. Mag., Tokyo 54: 307–313. [Jpn.; French summ.]

656. Kunst, L., J.E. Klenz, J. Martinez Zapater & G.W. Haughn. 1989. AP2 gene determines the identity of perianth organs in flowers of *Arabidopsis thaliana*. Plant Cell 1: 1195–1208.

657. Kunze, H. 1979. Typologie und Morphogenese des Angiospermen-Staubblättes. Beitr. Biol. Pfl. 54: 239–304.

658. Kunze, H. 1981 [1982]. Morphogenese und Synorganisation des Bestäubungsapparates einiger Asclepiadaceen. Beitr. Biol. Pfl. 56: 133–170.

660. Kunze, H. 1984. Vergleichende Studien an Cannaceen- und Marantaceenblüten. Flora 175: 301–318.

661. Kurzweil, H. 1987, 1988. Developmental studies in orchid flowers I: Epidendroid and vandoid species. Nord. J. Bot. 7: 427–442. II: Orchidoid species. Ibid.: 443–451 (1987). III: Neottioid species. Ibid. 8: 271–282 (1988).

662. Kurzweil, H. 1990. Floral morphology and ontogeny in Orchidaceae subtribe Disinae. Bot. J. Linn. Soc. 102: 61–83.

663. Kurzweil, H. 1991. The unusual structure of the gynostemium in the Orchidaceae-Coryciinae. Bot. Jahrb. 112: 273–293.

664. Kurzweil, H. 1993. The remarkable anther structure of *Pachites bodkinii* (Orchidaceae). Bot. Jahrb. 114(4): 561–569.

665. Kurzweil, H. & A. Weber. 1991, 1992. Floral morphology of southern African Orchideae. I. Orchidinae. Nord. J. Bot. 11: 155–178 (1991). II. Habenariinae. Ibid. 12: 39–61 (1992).

666. Lacroix, C. & R. Sattler. 1988. Phyllotaxis theories and tepal-stamen superposition in *Basella rubra*. Amer. J. Bot. 75: 906–917.

667. Ladd, P.G. & Donaldson, J.S. 1993. Pollen presenters

in the South African flora. S. Afr. J. Bot. 59: 465–477.

668. Lakshmanan, K.K. & E.J. Dulcy. 1982 [1983]. Tapetum in Chenopodiaceae. Beitr. Biol. Pfl. 57: 469–479.

669. Lakshminarayana, K. 1989. Embryological studies in a few taxa of Asclepiadaceae. J. Indian Bot. Soc. 68: 29–34.

670. Lam, H.J. 1950. Stachyospory and phyllospory as factors in the natural system of the Cormophyta. Svensk Bot. Tidskr. 44: 517–534.

671. Lamond, M. & J. Veith. 1972. L'androcée synanthéré du *Kechsteineria cardinalis* (Gesneriacées). Une contribution au problème des fusions. Can. J. Bot. 50: 1633–1637.

672. Larkin, R.A. & H.O. Graumann. 1954. Anatomical structure of the alfalfa flower and an explanation of the tripping mechanism. Bot. Gaz. 116: 40–52.

673. Laser, K.D. 1974. Plastids of sieve tube members in the stamen vascular bundle of *Sorghum bicolor* (Gramineae). Protoplasma 80: 279–283.

674. Laser, K.D. & N.R. Lersten. 1972. Anatomy and cytology of microsporogenesis in cytoplasmic male-sterile angiosperms. Bot. Rev. 38: 425–450.

675. Lasseigne, A. 1979. Studies in *Cassia* (Leguminosae-Caesalpinioideae) III. Anther morphology. Iselya 1: 141–160.

676. Laubengayer, R.A. 1937. Studies in the anatomy and morphology of the polygonaceous flower. Amer. J. Bot. 24: 329–343.

677. Lavialle, P. 1926. Le développement de l'anthère et du pollen chez *Knautia arvensis* Coult. C.R. Acad. Sci., Paris 182: 77–79.

678. Lawalrée, A. 1948. Histogénèse florale et végétative chez quelques Composées. Cellule 52: 213–294 + Plates I–VI.

679. Lazarte, J.E. & B.F. Palser. 1979. Morphology, vascular anatomy and embryology of pistillate and staminate flowers of *Asparagus officinalis*. Amer. J. Bot. 66: 753–764.

680. Léandri, J. 1930. Recherches anatomiques sur les Thyméléacées. Ann. Sci. Nat., Bot., sér. 10, 12: 125–237.

681. Lehmann, N.L. & R. Sattler. 1992. Irregular floral development in *Calla palustris* (Araceae) and the concept of homeosis. Amer. J. Bot. 79: 1145–1157.

682. Lehmann, N.L. & R. Sattler. 1993. Homeosis in floral development of *Sanguinaria canadensis* and *S. canadensis* 'Multiplex' (Papaveraceae). Amer. J. Bot. 80: 1323–1335.

683. Leinfellner, W. 1954. Die petaloiden Staubblätter und ihre Beziehungen zu den Kronblättern. Öst. Bot. Z. 101: 373–406.

684. Leinfellner, W. 1955. Beiträge zur Kronblattmorphologie. V. Über den homologen Bau der Kronblattspreite und der Staubblattanthere bei *Koelreuteria paniculata*. Öst. Bot. Z. 102: 89–98.

685. Leinfellner, W. 1956. Medianstipulierte Staubblätter. Öst. Bot. Z. 103: 24–43.

686. Leinfellner, W. 1956. Die blattartig flachen Staubblätter und ihre gestaltlichen Beziehungen zum Bautypus des Angiospermen-Staubblattes. Öst. Bot. Z. 103: 247–290.

687. Leinfellner, W. 1956. Die Gefässbündelversorgung des *Lilium*-Staubblattes. Öst. Bot. Z. 103: 346–352.

688. Leinfellner, W. 1956. Inwieweit kommt der peltat-diplophylle Bau des Angiospermen Staubblattes in dessen Leitbündelanordnung zum Ausdruck? Öst. Bot. Z. 103: 381–399.

689. Leinfellner, W. 1957. Die augenfällige Diplophyllie der Violaceen-Anthere. Öst. Bot. Z. 104: 209–227.

690. Leinfellner, W. 1957. Der Bündelverlauf in der dorsifixen Anthere von *Trapa natans*. Beitr Biol. Pfl. 34: 83–87.

691. Leinfellner, W. 1958. Zur Morphologie des Melastomataceen-Staubblattes. Öst. Bot. Z. 105: 44–70.

692. Leinfellner, W. 1960. Petaloid verbildete Staubblätter von *Narcissus* als ein weiteres Beispiel für die Umbildung diplophyller in sekundär schlauch- oder schildförmige Spreiten. Öst. Bot. Z. 107: 39–44.

693. Leinfellner, W. 1961. Staubblattverwachsungen bei *Yucca filamentosa*. Öst. Bot. Z. 108: 368–378.

694. Leinfellner, W. 1962. Über die Variabilität der Blüten von *Tofieldia calyculata*. I. Zu Karpellen verbildete Staubblätter. II. Der Ersatz von Perigonblättern durch Staubblätter. III. Zusammenfassende Übersicht der vorgefunden Abweichungen. Öst. Bot. Z. 109: 1–17, 113–124, 395–430.

695. Leinfellner, W. 1963. Über die Wiederherstellung der normalen Lage der überkippten *Rhododendron*-Anthere bei petaloider Verbildung. Öst. Bot. Z. 110: 374–379.

696. Leinfellner, W. 1963. Das Perigon der Liliaceen ist staminaler Herkunft. Öst. Bot. Z. 110: 448–467.

697. Leinfellner–Baum, H. 1953. Die Peltation der Staubblätter und die Phylogenie der Angiospermen. Phyton (Horn, Austria) 5: 16–21.

698. Leins, P. 1964. Die frühe Blütenentwicklung von *Hypericum hookerianum* Wight et Arn. und *H. aegypticum* L. Ber. Dtsch. Bot. Ges. 77: 112–123.

699. Leins, P. 1964. Entwicklungsgeschichtliche Studien an Ericales-Blüten. Bot. Jahrb. 83: 57–88. [Eng. summ.]

700. Leins, P. 1965. Die Inflorescenz und frühe Blütenentwicklung von *Melaleuca nesophila* F. Muell. (Myrtaceae). Planta 65: 195–204.

701. Leins, P. 1967. Die frühe Blütenentwicklung von *Aegle marmelos* (Rutaceae). Ber. Dtsch. Bot. Ges. 80: 320–325.

702. Leins, P. 1972 [1973]. Das zentrifugale Androeceum von *Couroupita guianensis* (Lecythidaceae). Beitr. Biol. Pfl. 48: 313–319.

703. Leins, P. 1975. Die Beziehungen zwischen multistaminaten und einfachen Androeceen. Bot. Jahrb. 96: 231–237.

704. Leins, P. 1979 [1980]. Der Übergang vom zentrifugalen komplexen zum einfachen Androeceum. Ber. Dtsch. Bot. Ges. 92: 717–719.

705. Leins, P. 1988. Das zentripetale Androeceum von *Punica*. Bot. Jb. 109: 555–561.

706. Leins, P. & K. Boecker. 1981 [1982]. Entwickeln sich Staubgefässe wie Schildblätter? Beitr. Biol. Pfl. 56: 317–327.

707. Leins, P. & R. Bonnery-Brachtendorf. 1977. Entwicklungsgeschichtliche Untersuchungen an Blüten von *Datisca cannabina* (Datiscaceae). Beitr. Biol. Pfl. 53: 143–155.

708. Leins, P. & C. Erbar. 1980. Zur Entwicklung der Blüten von *Mondora crispata* (Annonaceae). Beitr. Biol. Pfl. 55: 11–22.

709. Leins, P. & C. Erbar. 1985. Ein Beitrag zur Blütenentwicklung der Aristolochiaceen, einer Vermittlergruppe zu den Monokotylen. Bot. Jahrb. 107: 343–368. [Eng. summ.]

710. Leins, P. & C. Erbar. 1991. Fascicled androecia in Dilleniidae and some remarks on the *Garcinia* androecium. Bot. Acta 104: 336–344.

711. Leins, P. & P. Galle. 1971 [1972]. Entwicklungsgeschichtliche Untersuchungen an Cucurbitaceen-Blüten. Öst. Bot. Z. 119: 531–548.

712. Leins, P. & G. Metzenauer. 1979. Entwicklungsgeschichtliche Untersuchungen an *Capparis*-blüten Bot. Jahrb. 100: 542–554.

713. Leins, P. & S. Schwitalla. 1985 [1986]. Studien an Cactaceen-Blüten I. Einige Bemerkungen zur Blütenentwicklung von *Pereskia*. Beitr. Biol. Pfl. 60: 313–323.

714. Leins, P. & P. Stadler. 1973. Entwicklungsgeschichtliche Untersuchungen am Androeceum der Alismatales. Öst. Bot. Z. 121: 51–63.

715. Leins, P. & W. Winhard. 1973. Entwicklungsgeschichtliche Studien an Loasaceen-Blüten. Öst. Bot. Z. 122: 145–165.

716. Leroy, J.-F. 1983. The origin of angiosperms: an unrecognized ancestral dicotyledon, *Hedyosmum* (Chloranthales), with a strobiloid flower is living today. Taxon 32: 169–175.

717. Lersten, N.R. 1971. A review of septate microsporangia in vascular plants. Iowa St. J. Sci. 45: 487–497.

718. Lersten, N.R. & J.D. Curtis. 1989. Polyacetylene reservoir (duct) development in *Ambrosia trifida* (Asteraceae) staminate flowers. Amer. J. Bot. 76: 1000–1005.

719. Lersten, N.R. & L.J. Eilers. 1974. Binucleate tapetum in two species of *Lysimachia* (Primulaceae). Proc. Iowa Acad. Sci. 81: 197–198.

720. Levacher, P. 1968. Sur deux particularités ontogéniques des étamines par rapport aux feuilles, sépales et pétales chez le *Paris quadrifolia* L. C.R. Acad. Sci., Paris D, 267: 418–420. [Not seen.]

721. Li, P., B.-C. Gao, F. Chen & H.-X. Luo. 1992. Studies on morphology and embryology of *Acanthochlamys bracteata*. II. The anther and ovule development. Bull. Bot. Res. N.E. For. Univ. 12: 389–398. [Chin.; Eng. summ.]

722. Liang, H.-X. & S.C. Tucker. 1990. Comparative study of the floral vasculature in Saururaceae. Amer. J. Bot. 77: 607–623.

723. Lindsey, A.A. 1940. Floral anatomy in the Gentianaceae. Amer. J. Bot. 27: 640–652.

724. Liu, N., F.-X. Wang & Z.-K. Chen. 1992. Fine structure of tapetal cells and Ubisch bodies in the anther of *Ophiopogon japonicus*. Acta Bot. Sin. 34 (1): 15–19 + 2 plates. [Chin.; Eng. summ.]

725. Lloyd, D.G. & C.J. Webb. 1986. The avoidance of interference between the presentation of pollen and stigmas in angiosperms. I. Dichogamy. N.Z. J. Bot. 24: 135–162.

726. Lodkina, M.M. 1957. Special features of development of stamens of wheat and lily in connection with the general physiology of flowers. Trudy Bot. Inst. V.L. Komarova, Akad. Nauki SSSR, ser. VII. 4: 323–377. [Russ. only.]

727. Lodkina, M.M. 1982. Structure and development of androecium in two species of *Chrozophora* (Euphorbiaceae). Bot. Zh. SSSR 67: 1271–1276. [Russ.]

728. Lombardo, G. & L. Carraro. 1976, 1976 [1977]. Tapetal ultrastructural changes during pollen develop-

ment. I. Studies on *Antirrhinum maius*. Caryologia 29: 113–125. III. Studies on *Gentiana acaulis*. Ibid.: 345–349 (1976 [1977]).

729. Lord, E.M., K.J. Eckard & W. Crone. 1989. Development of the dimorphic anthers in *Collomia grandiflora*; evidence for heterochrony in the evolution of the cleistogamous anther. J. Evol. Biol. 2(2): 81–93. [Not seen.]

730. Lou, D.-G., C.-Z. Chen & P.-C. Chou. 1985. An anatomical study on the pistil and stamen of sugar cane. Guihaia 5: 377–379 + 2 plates. [Chin.; Eng. summ.]

731. Luders, H. 1907. Systematische Untersuchungen über die Caryophyllaceen mit einfachem Diagramm. Bot. Jahrb. 40: Beibl. 91: 1–37.

732. Luo, Y. & Q. Song. 1993. A cytological observation in microsporogenesis of male sterile lines and fertile lines of millet. Acta Bot. Bor.-Occ. Sin. 13: 192–197 + 2 plates. [Chin.; Eng. summ.]

733. Luza, J.G. & V.S. Polito. 1988. Microsporogenesis and anther differentiation in *Juglans regia* L.: a developmental basis for heterodichogamy in walnut. Bot. Gaz. 149: 30–36.

734. Lyndon, R.F. 1978. Flower development in *Silene*: morphology and sequence of initiation of primordia. Ann. Bot. 42: 1343–1348.

735. Lyndon, R.F. 1979. Rates of growth and primordial initiation during flower development in *Silene* at different temperatures. Ann. Bot. 43: 539–551.

736. Ma, H., T. Li & L. Tu. 1993. The studies on embryology of *Cistanche deserticola* Ma. Acta Sci. Nat. Univ. Nei Menggu 24(1): 81–86 + 2 plates. [Chin.; Eng. summ.]

736a. Macdonald, A.D. & R. Sattler. 1973. Floral development of *Myrica gale* and the controversy over floral concepts. Can. J. Bot. 51: 1965–1975.

737. Madjd, A. & F. Roland–Heydacker. 1978. Sécrétions et dégénérescence des cellules du tapis dans l'anthère du *Soja hispida* Moench, Papilionaceae. Grana 17: 167–174.

738. Maekawa, F. 1970, 1971. Notes on the stamens of *Chloranthus japonicus*. J. Jpn. Bot. 45: 289–294. [Jpn.; Eng. summ.] Further notes on ... Ibid. 46: 198 (1971). [Jpn.]

739. Magin, N., R. Classen & C. Gack. 1989. The morphology of false anthers in *Craterostigma plantagineum* and *Torenia polygonoides* (Scrophulariaceae). Can. J. Bot. 67: 1931–1937.

740. Mahalingappa, M.S. 1975. Anther and male gameto-

phyte development in *Turnera ulmifolia* Linn. (var. *angustifolia* Willd.). Curr. Sci. 44: 640–641. [Not seen.]

741. Maheshwari, P. 1950. An Introduction to the Embryology of Angiosperms. McGraw-Hill, New York.

742. Maheswari Devi, H. & K. Lakshminarayana. 1978. Life history of *Caralluma adscendens*. Acta Bot. Indica 6: 177–184.

743. Mair, O. 1977. Zur Entwicklungsgeschichte monosymmetrischer Dicotylen-Blüten. Diss. Bot. 38. J. Cramer, Vaduz. pp. 1–105.

744. Majewska-Sawka, A., B. Jassem, J. Macewicz & M.I. Rodriguez-Garcia. 1990. An electron microscopic study of anther structure in male-fertile and male-sterile sugar beets: tapetum development. pp. 57–63 in: Polen, Esporas y sus Aplicaciones (edited by G. Blancoa *et al.*). University of Granada, Granada. [Not seen.]

745. Makde, K.H. 1982. Pollen development in the Cyperaceae. J. Indian Bot. Soc. 61: 242–249.

746. Malick, K.C. & B. Safui. 1980 [1982]. A review of the androecium in Sterculiaceae with a key to the genera. Bull. Bot. Surv. India 22: 213–216.

747. Manning, J.C. & P. Goldblatt. 1990. Endothecium in Iridaceae and its systematic implications. Amer. J. Bot. 77: 527–532.

748. Manning, J.C. & H.P. Linder. 1990. Cladistic analysis of patterns of endothecial thickenings in the Poales/Restionales. Amer. J. Bot. 77: 196–210.

749. Manning, W.E. 1948. The morphology of the flowers of the Juglandaceae. III. The staminate flowers. Amer. J. Bot. 35: 606–621.

750. Mariani, C., V. Gossele, M. de Beuckeleer, M. de Block, R.B. Goldberg, W. de Greef & J. Leemans. 1992. A chimaeric ribonuclease-inhibitor gene restores fertility to male sterile plants. Nature 357 (June 4): 384–387.

751. Markgraf, F. 1936. Blütenbau und Verwandtschaft bei den Einfachsten Helobiae. Ber. Dtsch Bot. Ges. 54: 191–229.

752. Marquardt, H., O.M. Barth & U. von Rahden. 1968. Zytophotometrische und elektronenmikroskopische Beobachtungen über die Tapetumzellen in den Antheren von *Paeonia tenuifolia*. Protoplasma 65: 407–421.

753. Marquez-Guzman, J., M. Engleman, A. Martínez-Mena, E. Martínez & C. Ramos. 1989. Anatomia reproductiva de *Lacandonia schismatica* (Lacandoniaceae). Ann. Missouri Bot. Gard. 76: 124–127.

754. Martel, E. 1908. Contribuzione all'anatomia del fiore dell'*Hedera helix*, dell'*Aralia sieboldii* e del *Cornus sanguinea*. Memorie Accad. Sci. Torino II, 58: 561–579. [Not seen.]

755. Mascarenhas, J.P. 1990. Gene activity during pollen development. Ann. Rev. Pl. Physiol. Pl. Mol. Biol. 41: 317–338.

756. Mascré, M. 1919. Sur le rôle de l'assise nourricière du pollen. C.R. Acad. Sci., Paris 168: 1120–1122. Nouvelles remarques . . . Ibid: 1214–1216.

757. Mascré, M. 1921. Recherches sur le développement de l'anthère chez les Solanacées. Contribution à l'étude de l'assise nourricière du pollen. Thèse, Paris. [Not seen.]

758. Mascré, M. 1922. Sur l'étamine des Borraginées. C.R. Acad. Sci., Paris 175: 987–989.

759. Mascré, M. 1925. Sur le periplasmodium staminal des Commélinacées. C.R. Acad. Sci., Paris 181: 1165–1166.

760. Mascré, M. 1925. Sur l'évolution de l'étamine des Commélinacées. Bull. Soc. Bot. Fr. 72: 1060–1066 + Plate XLV.

761. Mascré, M. 1928. Sur le tapis staminal et le grain de pollen de l'*Arum maculatum* L. C.R. Acad. Sci., Paris 186: 1642–1644.

762. Mascré, M. & R. Thomas. 1930. Le tapis staminal (assise nourricière du pollen) chez les angiospermes. Bull. Soc. Bot. Fr. 77: 654–664.

763. Masters, M.T. 1873. On the development of the androecium in *Cochliostema* Lem. J. Linn. Soc. 13: 204–209 + Plate IV.

764. Mattfeld, J. 1938. Das morphologische Wesen und die phylogenetische Bedeutung der Blumenblätter. Ber. Dtsch. Bot. Ges. 56: 86–116.

765. Mattfeld, J. 1938. Über eine angebliche *Drymaria* Australiens, nebst Bemerkungen über die Staminaldrüsen und die Petalen der Caryophyllaceen. Feddes Repert., Beih. 100: 147–164 + Plates VII–IX. [Bornmüller-Festschrift.]

766. Mattfeld, J. 1939. Biologische und morphologische Blütenformen bei den Caryophyllaceen. Notizbl. Bot. Gart. Berlin 14: 470–482.

767. Matthews, J.R. & E.M. Knox. 1926. The comparative morphology of the stamen in the Ericaceae. Trans. Proc. Bot. Soc. Edinb. 29: 243–281.

768. Matthews, J.R. & C.M. Maclachlan. 1929. The structure of certain poricidal anthers. Trans. Proc. Bot. Soc. Edinb. 30: 104–122.

769. Matthews, J.R. & G. Taylor. 1926. The structure and development of the stamen in *Erica hirtiflora*. Trans. Proc. Bot. Soc. Edinb. 29: 235–242.

770. Mattsson, O. 1982. The morphogenesis of dimorphic pollen and anthers in *Tripogandra amplexicaulis*. Light microscopy and growth analysis. Opera Bot. 66: 1–46.

771. Matzke, E.B. 1930. Der Einfluss einiger Bedingungen, besonders der Buntblättrigkeit, auf die Zahl der Staubblätter bei *Stellaria media* (L.) Cyr. Planta 9: 776–791.

772. Mayr, B. 1969. Ontogenetische Studien an Myrtales-Blüten. Bot. Jahrb. 89: 210–271.

773. McCollum, G.D. 1966. Occurrence of petaloid stamens in wild carrot from Sweden. Econ. Bot. 20: 361–367.

774. McLean, D.M. 1947. Stamen morphology in the flowers of musk melon. J. Agric. Res. 74: 49–54. [Not seen.]

775. Meeuse, A.D.J. 1963. Stachyospory, phyllospory and morphogenesis. Adv. Front. Pl. Sci. 7: 115–156.

776. Meeuse, A.D.J. 1965. Angiosperms – past and present. Phylogenetic botany and interpretative floral morphology of the flowering plants. Adv. Front. Pl. Sci. 11: 1–228.

777. Meeuse, A.D.J. 1966. Fundamentals of Phytomorphology. Ronald Press, New York. 231 pp.

778. Meeuse, A.D.J. 1966. The homology concept in phytomorphology – some moot points. Acta Bot. Neerl. 15: 451–476.

779. Meeuse, A.D.J. 1972. Facts and fiction in floral morphology with special reference to the Polycarpicae. 1. A general survey. Acta Bot. Neerl. 21: 113–127. 2. Interpretation of the floral morphology of various taxonomic groups. Ibid.: 235–252. 3. Consequences and various additional aspects of the anthocorm theory. Ibid.: 351–365.

780. Meeuse, A.D.J. 1972. Angiosperm phylogeny, floral morphology and pollination ecology. Acta Biotheor. 21: 145–166.

781. Meeuse, A.D.J. 1972. Sixty-five years of theories of the multiaxial flower. Acta Biotheor. 21: 167–202.

782. Meeuse, A.D.J. 1972 [1975]. Taxonomic affinities between Piperales and Polycarpicae and their implications in interpretative floral morphology. pp. 3–27 in: Advances in Plant Morphology (edited by Y.S. Murty, B.M. Johri, H.Y.M. Ram & T.M. Varghese). Sarita Prakashan, Meerut.

783. Meeuse, A.D.J. 1973. Some fundamental principles in interpretative floral morphology. Vistas Pl. Sci. 1: 1–78.

784. Meeuse, A.D.J. 1974. Phaneranthy, aphananthy, and floral morphology: some special aspects of the early evolution of the angiosperms. Acta Bot. Indica 2: 107–119.

785. Meeuse, A.D.J. 1980 [1982]. What is polyandry? Phytomorphology 30: 388–396.

786. Meeuse, A.D.J. 1987. Comparative floral morphology of the Rosaceae and of the Rosidae in general. Phytomorphology 37: 103–112.

787. Mei, J., C. Liang & B. He. 1992. Comparison on the stamen vascular bundle between CMS lines and its maintainers in rice. Acta Bot. Austro Sin., No. 8: 200–205 + 2 plates. [Chin.; Eng. summ.]

788. Meiri, L. & R. Dulberger. 1986. Stamen filament structure in the Asteraceae: the anther collar. New Phytol. 104: 693–701.

789. Melchior, H. 1925. Die phylogenetische Entwicklung der Violaceen und die natürlichen Verwandtschaftsverhältnisse ihrer Gattungen. Rep. Spec. Nov. Reg. Veg., Beih. 36: 83–125 + 3 plates.

790. Melville, R. 1960. Contributions to the flora of Australia: VI. The pollination mechanism of *Isotoma axillaris* Lindl. and the generic status of *Isotoma* Lindl. Kew Bull. 14: 277–279.

791. Melville, R. 1963. A new theory of the angiosperm flower: II. The androecium. Kew Bull. 17: 1–63.

792. Melville, R. 1969. Studies in floral structure and evolution. I. The Magnoliales. Kew Bull. 23: 133–180.

793. Melville, R. 1983. The affinity of *Paeonia* and a second genus of Paeoniaceae. Kew Bull. 38: 87–105.

794. Menezes, N.L. de. 1980. Evolution in Velloziaceae, with special reference to androecial characters. pp. 177–138 in: Petaloid Monocotyledons (edited by C.D. Brickell, D.F. Cutler & M. Gregory). Academic Press, London.

795. Menezes, N.L. de. 1988. Evolution of the anther in the family Velloziaceae. Bol. Bot. Univ. S. Paulo 10: 33–41.

796. Mepham, R.H. & G.R. Lane. 1969. Formation and development of the tapetal periplasmodium in *Tradescantia bracteata*. Protoplasma 68: 175–192.

797. Merxmüller, H. & P. Leins. 1967. Die Verwandtschaftsbeziehungen der Kreuzblütler und Mohngewächse. Bot. Jahrb. 86: 113–129. [Eng. summ. 127]

798. Merxmüller, H. & P. Leins. 1971. Zur Entwicklungsgeschichte männlicher Begonienblüten. Flora 160: 333–339.

799. Meure, B., D. Strack & R. Wiermann. 1984. The systematic distribution of ferulic acid-sucrose esters in anthers of the Liliaceae. Pl. Med. 50: 376–380.

800. Meyer, V.G. 1966. Flower abnormalities. Bot. Rev. 32: 165–218.

801. Meyerowitz, E.M., J.L. Bowman, L.L. Brockman, G.N. Drews, T. Jack, L.E. Sieburth & D. Weigel. 1991. A genetic and molecular model for flower development in *Arabidopsis thaliana*. Development, Suppl. 1: 157–167.

802. Meyerowitz, E.M., D.R. Smyth & J.L. Bowman. 1989. Abnormal flowers and pattern formation in floral development. Development 106: 209–217.

803. Michaelis, P. 1924. Blütenmorphologische Untersuchungen an den Euphorbiaceen, unter besonderer Berücksichtigung der Phylogenie der Angiospermenblüte. Bot. Abh. 3. G. Fischer, Jena. pp. 1–150 + 41 plates.

804. Michener, C.D. 1962. An interesting method of pollen collecting by bees from flowers with tubular anthers. Revta. Biol. Trop. 10: 167–175.

805. Miki-Hirosige, H. & S. Nakumara. 1982 [1983]. Incorporation of label from myoinositol-2^{3}H by young anther of *Lilium longiflorum*. Phytomorphology 32: 85–94.

806. Milby, T.H. 1976. Studies in the floral anatomy of *Polygala* (Polygalaceae). Amer. J. Bot. 63: 1319–1326.

807. Milby, T.H. 1980. Studies in the floral anatomy of *Claytonia* (Portulacaceae). Amer. J. Bot. 67: 1046–1050.

808. Miller, W.L. 1929. Staminate flower of *Echinocystis lobata*. Bot. Gaz. 88: 262–284 + Plates XIII–XVI.

809. Millet, B. 1977. Observation, en microscopie électronique à balayage, de la surface du filet staminal de quelques espèces de Berberidacées. Comparaison avec d'autres organes sensibles. C.R. Seanc. Soc. Biol. Besançon 171: 580–584. [French; Eng. summ.]

810. Milyaeva, E.L. & N.V. Tsinger. 1968. Starch in developing anthers of *Citrus sinensis*: a cytochemical and electron microscope study. Fiziol. Rast. 15: 303–307. [Not seen.]

811. Misset, M.T. & J.P. Gourret. 1984. Accumulation of smooth cisternae in the tapetal cells of *Ulex europaeus* L. (Papilionoideae). J. Cell Sci. 72: 65–74.

812. Młodzianowski, F. & K. Idzikowska. 1978. The ultrastructure of anther wall and pollen of *Hordeum vulgare* at the microspore stage. Acta Soc. Bot. Pol. 47: 219–224.

813. Möbius, M. 1923. Über die Färbung der Antheren und des Pollens. Ber. Dtsch. Bot. Ges. 41: 12–16.

814. Mohrbutter, C. 1936. Embryologische Studien an Loganiaceen. Planta 26: 64–80.

815. Moissl, E. 1941. Vergleichende embryologische Studien über die Familie der Caprifoliaceae. Öst. Bot. Z. 90: 153–212.

816. Moncur, M.W. 1988. Floral development of tropical and subtropical fruit and nut species. An atlas of scanning electron micrographs. Natural Resources ser. 8. CSIRO, Melbourne. [Not seen.]

817. Money, L.L., I.W. Bailey & B.G.L. Swamy. 1950. The morphology and relationships of the Monimiaceae. J. Arnold Arbor. 31: 372–404.

818. Monteiro Scanavacca, W.R. 1975. Vascularização e natureza de estruturas do androceu em Lecythidaceae. Bol. Bot., Univ. S. Paulo 3: 61–73. [Eng. summ.]

819. Moore, J.A. 1936. Floral anatomy and phylogeny in the Rutaceae. New Phytol. 35: 318–322.

820. Moore, J.A. 1936. The vascular anatomy of the flower in the papilionaceous Leguminosae. I, II. Amer. J. Bot. 23: 279–290; 349–355.

821. Mori, S.A., G.T. Prance & A.B. Bolten. 1978. Additional notes on the floral biology of neotropical Lecythidaceae. Brittonia 30: 113–130.

822. Mosebach, G. 1932. Über die Schleuderbewegung der explodierenden Staubgefässe und Staminodien bei einigen Urticaceen. Planta 16: 70–115.

823. Moseley, M.F. 1958. Morphological studies of the Nymphaeaceae. I. The nature of the stamens. Phytomorphology 8: 1–29.

824. Moussel, B., C. Moussel & J.-C. Audran. 1992. La stérilité mâle nucléo-cytoplasmique chez la féverole (Vicia faba L.). IX. Fonctionnement du tapis et formations membranaires chez les génotypes fertiles. Incidence du dysfonctionnement du tapis sur l'ontogénèse de l'exine des microspores stériles (cytoplasme 447). Grana 31: 25–48. [Eng. summ.]

825. Mu, X.-J., F.-H. Wang & W.-L. Wang. 1988. Development and histochemical observations of tapetum and peritapetal membrane in anther of Pulsatilla chinensis. Acta Bot. Sin. 30: 6–13 + 2 plates. [Chin.; Eng. summ.]

826. Müller, F. 1870. Umwandlung von Staubgefässen in Stempel bei Begonia. Bot. Ztg. 28: 149–153 + Plate II.

827. Müller, K. 1908. Beiträge zur Systematik der Aizoaceen. Bot. Jahrb. 57, Beibl. 97: 54–94.

828. Müller-Doblies, U. 1969. Über die Blütenstände und Blüten sowie zur Embryologie von Sparganium. Bot. Jahrb. 89: 359–450.

829. Müller-Doblies, D. 1970. Über die Verwandtschaft von Typha und Sparganium im Infloreszenz- und Blütenbau. Bot. Jahrb. 89: 451–562. [Eng. summ. 529–531.]

830. Murbeck, S. 1912. Untersuchungen über den Blütenbau der Papaveraceen. K. Sven. Vetenkapsakad. Handl. 50 (1): 1–168.

831. Murbeck, S. 1918. Über staminale Pseudapetalie und deren Bedeutung für die Frage nach der Herkunft der Blütenkrone. Lunds Univ. Årsskr. (NF) Afd. 2, 14 (25): 1–59.

832. Murbeck, S. 1941. Untersuchungen über das Androeceum der Rosaceen. Lunds Univ. Årsskr. (NF) Afd. 2, 37 (7): 1–56.

833. Murgia, M., M. Charzynska, M. Rougier & M. Cresti. 1991. Secretory tapetum of Brassica oleracea L.: polarity and ultrastructural features. Sex. Pl. Reprod. 4: 28–35.

834. Murty, Y.S. 1954. Studies in the order Parietales. IV. Vascular anatomy of the flower of Tamaricaceae. J. Indian Bot. Soc. 33: 226–238.

835. Nábělek, F. 1909. Über die systematische Bedeutung des feineren Baues der Antherenwand. Sber. Akad. Wiss. Wien, Math.-Nat. Kl. I, 115: 1427–1490 + 4 plates.

836. Nair, N.C. 1972 [1975]. Floral morphology and embryology of Myristica malabarica Lamk. with a discussion on certain aspects of the systematics of Myristicaceae. pp. 264–277 in: Advances in Plant Morphology (edited by Y.S. Murty, B.M. Johri, H.Y.M. Ram & T.M. Varghese). Sarita Prakashan, Meerut.

837. Nair, N.C. & V. Abraham. 1962. Floral morphology of a few species of Euphorbiaceae. Proc. Indian Acad. Sci., B. 56: 1–12. [Not seen.]

838. Nair, N.C. & K.S. Nathawat. 1958. Vascular anatomy of the flower of some species of Zygophyllaceae. J. Indian Bot. Soc. 37: 172–180.

839. Nakanishi, T. 1982. Morphological and ultraviolet absorption differences between fertile and sterile anthers of Japanese apricot cultivars in relation to their pollination stimuli. Scientia Hort. 18: 57–63.

840. Nakashima, H., H.T. Horner & R.G. Palmer. 1984. Histological features of anthers from normal and ms3 mutant soybean. Crop Sci. 24: 735–739. [Not seen.]

841. Nakashima, H., C. Tsuda, K. Murata & T. Narikawa. 1980. Histological features and inheritance of male sterile adzuki bean (Vigna angularis). Jpn. J. Breed. 30: 241–245. [Eng.] [Not seen.]

842. Nakazawa, K. 1956. The vascular course of Piperales. I. Chloranthaceae. Jpn. J. Bot. 15: 199–207.

843. Namikawa, I. 1919. Über das Öffnen der Antheren

bei einigen Solanaceen. Bot. Mag., Tokyo 33: 62–69.

844. Nanda, K. & S.C. Gupta. 1971. Endothecium in *Pyrostegia venusta*. Curr. Sci. 40: 470–471. [Not seen.]

845. Nanda, K. & S.C. Gupta. 1975 [1976]. Syngenesious anthers of *Helianthus annuus* – a histochemical study. Bot. Notiser 128: 450–454.

846. Nanda, K. & S.C. Gupta. 1977 [1978]. Development of tapetal periplasmodium in *Rhoeo spathacea*. Phytomorphology 27: 308–314.

847. Nanda, K. & S.C. Gupta. 1978. Studies in the Bignoniaceae II. Ontogeny of dimorphic anther tapetum in *Tecoma*. Amer. J. Bot. 65: 400–405.

848. Nanda, K. & S.C. Gupta. 1981. Histochemical localization of total carbohydrates of insoluble polysaccharides in *Helianthus* anthers. J. Indian Bot. Soc. 60: 257–260.

849. Nanda, K. & S.C. Gupta. 1983. Histochemical studies in the Bignoniaceae. II. Total carbohydrates of insoluble polysaccharides in *Tecoma* anthers. Beitr. Biol. Pfl. 58: 243–251.

850. Narayana, L.L. 1958, 1959. Floral anatomy of Meliaceae. I. J. Indian Bot. Soc. 37: 365–374 (1958). II. Ibid. 38: 288–295 (1959).

851. Narayana, L.L. 1960. Studies in Burseraceae I. J. Indian Bot. Soc. 39: 204–209. II. Ibid.: 402–409.

852. Narayana, L.L. & D. Rao. 1978. Systematic position of Humiriaceae, Linaceae and Erythroxylaceae in the light of their comparative floral morphology and embryology – a discussion. J. Indian Bot. Soc. 57: 258–266.

853. Narayana, L.L. & M. Sayeeduddin. 1958. Floral anatomy of Simarubaceae. I. J. Indian Bot. Soc. 37: 517–522.

854. Narayana, P.S. 1986. Reproductive behaviour of *VZM 2B* – a male sterile *Sorghum* line. J. Indian Bot. Soc. 65: 131–135.

855. Nast, C.G. 1944. The comparative morphology of the Winteraceae. VI. Vascular anatomy of the flowering shoot. J. Arnold Arbor. 25: 454–466.

856. Nast, C.G. & I.W. Bailey. 1945. Morphology and relationships of *Trochodendron* and *Tetracentron*. II. Inflorescence, flower, and fruit. J. Arnold Arbor. 26: 267–276.

857. Nast, C.G. & I.W. Bailey. 1946. Morphology of *Euptelea* and comparison with *Trochodendron*. J. Arnold Arbor. 27: 186–192.

858. Nave, E.B. & V.K. Sawhney. 1986. Enzymatic changes in post-meiotic anther development in *Petunia hybrida*. I. Anther ontogeny and isozyme analyses. J. Pl. Physiol. 125: 451–465.

859. Negodi, G. 1933. Dispositivi anatomici preposti alla deiscenza dell'antera nel genere *Urtica* e *Parietaria*. Atti Soc. Nat. Matem. Modena 64: 111–116.

860. Negri, V., B. Romano & F. Ferranti. 1989. Male sterility in birdsfoot trefoil (*Lotus corniculatus* L.). Sex. Pl. Reprod. 2: 150–153.

861. Nelson, E. 1954. Gesetzmässigkeiten der Gestaltwandlung im Blütenbereich: ihre Bedeutung für das Problem der Evolution. E. Nelson Chernex, Montreux. 302 pp. + 14 plates.

862. Nelson, E. 1965. Zur organophyletischen Natur des Orchideenlabellums. Bot. Jahrb. 84: 175–214. [Eng. summ. 209–210.]

863. Nelson, E. 1967. Das Orchideenlabellum ein Homologon des einfachen medianen Petalums der Apostasiaceen oder ein zusammengesetztes Organ? Bot. Jahrb. 87: 22–35. [Eng. summ. 33–34.]

864. Nemirovich-Danchenko, E.N. 1980. The development of the androecium and the nature of polyandry in some Papaveraceae. Bot. Zh. SSSR 65: 1088–1100. [Russ.; Eng. summ.]

865. Nessler, C.L. & P.G. Mahlberg. 1976. Laticifers in stamens of *Papaver somniferum* L. Planta 129: 83–85.

866. Nester, J.E. & J.A.D. Zeevaart. 1988. Flower development in normal tomato and a gibberellin-deficient (ga-2) mutant. Amer. J. Bot. 75: 45–55.

867. Neubauer, H.F. 1959. Über das Staminodium von *Kigelia aethiopica* Decne. Öst. Bot. Z. 106: 546–550.

868. Neubauer, H.F. 1987. Bemerkungen über *Aleurites* J.R. & G. Forst.: Blüte und Blütenstand. Beitr. Biol. Pfl. 62: 57–67.

869. Neumayer, H. 1924. Die Geschichte der Blüte. Versuch einer zusammenfassenden Beantwortung der Frage nach der Vergangenheit der generativen Region bei den Anthophyten. Abh. Zool.-Bot. Ges. Wien, Bot. 14 (1): 1–112.

870. Nevling, L.I. & C.J. Niezgoda. 1978. On the genus *Schleinitzia* (Leguminosae-Mimosoideae). Adansonia, sér. 2, 18: 345–363.

871. Nicolas, G. 1919. Remarques sur l'androcée des Crucifères à propos de fleurs anormales d'*Isatis djurdjurae* Coss. et Dur. Bull. Soc. Hist. Nat. Afr. N. 10: 111–114.

872. Nietsch, H. 1941. Zur systematischen Stellung von *Cyanastrum*. Öst. Bot. Z. 90: 31–52.

873. Nikiticheva, Z.I. 1968. Anther development and microsporogenesis in some representatives of Scrophulariaceae and Orobanchaceae. Bot. Zh. SSSR 53: 1704–1715. [Russ.; Eng. summ.]

874. Nishiyama, I. 1981. Male sterility caused by cooling

treatment at the young microspore stage in rice plants. XX. Optical microscopical observations of unfixed, intact anthers. Jpn. J. Crop Sci. 50: 495–501. [Eng.]

875. Noel, A.R.A. 1983. The endothecium – a neglected criterion in taxonomy and phylogeny? Bothalia 14: 833–838.

876. Noher de Halac, I., I.A. Cismondi & C. Harte. 1990. Pollen ontogenesis in *Oenothera*: a comparison of genotypically normal anthers with the male-sterile mutant *sterilis*. Sex. Pl. Reprod. 3: 41–53.

877. Noher de Halac, I., G. Fama & I.A. Cismondi. 1992. Changes in lipids and polysaccharides during pollen ontogeny in *Oenothera* anthers. Sex. Pl. Reprod. 5: 110–116.

878. Nordenstam, B. 1978. Taxonomic studies in the tribe Senecioneae (Compositae). Opera Bot. Lund 44: 1–83.

879. Nordenstam, B. & G. El Ghazaly. 1977. Floral micromorphology and pollen ultrastructure in some Centaureinae (Compositae) mainly from Egypt. Publ. Cairo Univ. Herb., Nos. 7–8: 143–155.

880. Novák, F.J. 1971. Cytoplasmic male sterility in sweet pepper (*Capsicum annuum*). II. Tapetal development in male-sterile anther. Z. Pflanzenzucht. 65: 221–232. [Not seen.]

881. Novák, F.J. 1972. Tapetal development in the anthers of *Allium sativum* L. and *Allium longicuspis* Regel. Experientia 28: 1380–1381.

882. Nozeran, R. 1953. Sur quelques fleurs mâles d'Euphorbiacées. Rec. Trav. Lab. Bot. Géol. Zool., Montpellier, sér. Bot. 6: 99–114. [Not seen.]

883. Nozeran, R. 1955. Contribution à l'étude de quelques structures florales (essai de morphologie florale comparée). Ann. Sci. Nat., Bot., sér. 11, 16: 1–224.

884. Oehler, E. 1927. Entwicklungsgeschichtlich-zytologische Untersuchungen an einigen saprophytischen Gentianaceen. Planta 3: 641–733.

885. Oganesyan, M.G. 1975. Tapetal layer of tomato anthers. Biol. Zh. Arm. 28(9): 103–107. [Russ.] [Not seen.]

886. Ogorodnikova, V.F. 1976. The ultrastructure of tapetal films in the anthers of some cultivated plants. Trudy Prikl. Bot. Genet. Selek. 58(1): 124–129. [Russ.; Eng. summ.]

887. Ogorodnikova, V.F. 1980. Structural and functional peculiarities of the tapetal cells of the anthers in *Triticum aestivum* L. Trudy Prikl. Bot. Genet. Selek. 67(3): 150–156 + 5 plates. [Russ.; Eng. summ.]

888. Ogorodnikova, V.F. 1983. Submicroscopic changes in tapetum of rye anthers (*Secale cereale* L.) in ontogenesis. Trudy Prikl. Bot. Genet. Selek. 74: 27–36. [Russ.; Eng. summ.]

889. Ogorodnikova, V.F. 1986. The genesis and ultrastructure of the sporopollenin wall of tapetal cells in grasses. Bot. Zh. SSSR 71: 1366–1371 + 2 plates. [Russ. only.]

890. Ogorodnikova, V.F. 1987. Characteristic features of tapetum structure of lucerne anthers (*Medicago sativa* L.). Nauch.-Tekh. Byull. Inst. Rast. N.I. Vavilova 170: 22–26. [Russ.; Eng. summ.]

891. Ogorodnikova, V.F. 1990. The dynamics of cell organelles ultrastructure in tapetum of *Triticum aestivum* (Poaceae). Bot. Zh. SSSR 75: 186–192 + 4 plates. [Russ. only]

892. Olsen-Gisel, H. 1983. Development in stamens of *Viola odorata*. Dissert. Bot. 70. J. Cramer, Vaduz. 191 pp.

893. Oryol, L.I. & M.A. Zhakova. 1976. Differentiation of anther's walls and dimorphism of tapetum of tomato, *Lycopersicon esculentum* Mill. (Solanaceae). Bot. Zh. SSSR 61: 1720–1729. [Russ.]

894. Oryol, L.I. & M.A. Zhakova. 1977. The mechanism of anther dehiscence of tomato, *Lycopersicon esculentum* Mill. (Solanaceae). Bot. Zh. SSSR 62: 1720–1730. [Russ.; Eng. summ.]

895. Osche, G. 1983. Optische Signale in der Coevolution von Pflanze und Tier. Ber. Dtsch. Bot. Ges. 96: 1–27.

896. Overman, M.A. & H.E. Warmke. 1972. Cytoplasmic male sterility in *Sorghum*. II. Tapetal behavior in fertile and sterile anthers. J. Hered. 63: 226–234.

897. Ozenda, P. 1949. Recherches sur les dicotylédones apocarpiques. Contribution à l'étude des angiospermes dites primitives. Publ. Lab. Biol. II, Ecole Norm. Super. Masson & Cie, Paris. 183 pp.

898. Ozenda, P. 1952. Remarques sur quelques interprétations de l'étamine. Phytomorphology 2: 225–231.

900. Pacini, E. 1990. Tapetum and microspore functions. pp. 213–237 in: Microspores. Evolution and Ontogeny (edited by S. Blackmore & R.B. Knox). Academic Press, London.

901. Pacini, E., L.M. Bellani & R. Lozzi. 1986 [1987]. Pollen, tapetum and anther development in two cultivars of sweet cherry (*Prunus avium*). Phytomorphology 36: 197–210.

902. Pacini, E. & G. Casadoro. 1981. Tapetum plastids of *Olea europaea* L. Protoplasma 106: 289–296.

903. Pacini, E. & M. Cresti. 1978. Ultrastructure characteristics of tapetum & microspore mother cells in

Lycopersicum peruvianum during meiotic prophase. Bull. Soc. Bot. Fr. 125: 121–128.

904. Pacini, E. & G.G. Franchi. 1991. Diversification and evolution of tapetum. pp. 301–316 in: Pollen and Spores – Patterns of Diversification (edited by S. Blackmore & S.H. Barnes). Syst. Assoc., Clarendon Press, Oxford.

905. Pacini, E. & G.G. Franchi. 1993. Role of the tapetum in pollen and spore dispersal. Pl. Syst. Evol., Suppl. 7: 1–12.

906. Pacini, E., G.G. Franchi & M. Hesse. 1985. The tapetum: its form, function, and possible phylogeny in Embryophyta. Pl. Syst. Evol. 149: 155–185.

907. Pacini, E. & B.E. Juniper. 1979. The ultrastructure of pollen-grain development in the olive (*Olea europaea*) 1. Proteins in the pore. 2. Secretion by tapetal cells. New Phytol. 83: 157–163; 165–174.

908. Pacini, E. & B.E. Juniper. 1983. The ultrastructure of the formation and development of the amoeboid tapetum in *Arum italicum* Miller. Protoplasma 117: 116–129.

909. Pacini, E. & C.J. Keijzer. 1989. Ontogeny of intruding non-periplasmodial tapetum in the wild chicory, *Cichorium intybus* (Compositae). Pl. Syst. Evol. 167: 149–164.

910. Pacini, E., P.E. Taylor, M.B. Singh & R.B. Knox. 1992. Development of plastids in pollen and tapetum of rye-grass, *Lolium perenne* L. Ann. Bot. 70: 179–188.

911. Paclt, J. 1948. Sur la métamorphose des étamines chez le *Catalpa ovata* × *C. bignonioides* (Bignoniaceae). Ber. Schweiz. Bot. Ges. 58: 381–382.

912. Paetow, W. 1931. Embryologische Untersuchungen an Taccaceen, Meliaceen und Dilleniaceen. Planta 14: 441–470.

913. Pal, N. 1952. A contribution to the life-histories of *Stellaria media* Linn. and *Polycarpon loeflingiae* Benth. & Hook. Proc. Natn. Inst. Sci., India, B, 18: 363–378.

914. Paliwal, R.L. 1956. Morphological and embryological studies in some Santalaceae. Agra Univ. J. Res. Sci. 5: 193–284.

915. Palser, B.F. 1951, 1954, 1958, 1961–62, 1963. Studies of floral morphology in the Ericales. I. Organography and vascular anatomy in the Andromedeae. Bot. Gaz. 112: 447–485 (1951). III. Organography and vascular anatomy in several species of the Arbuteae. Phytomorphology 4: 335–354 (1954). IV. Observations on three members of the Gaultherieae. Trans. Illinois State Acad. Sci. 51: 24–34 (1958). V. Organography

and vascular anatomy in several United States species of the Vacciniaceae. Bot. Gaz. 123: 79–111 (1961–62). VI. The Diapensiaceae. Ibid. 124: 200–219 (1963).

916. Pan, K.-Y., J.-H. Li, A.-M. Lu & J. Wen. 1993. The embryology of *Tetracentron sinense* Oliver and its systematic significance. Cathaya 5: 49–58.

917. Panchaksharappa, M.G. & C.K. Rudramuniyappa. 1972 [1975]. Role of insoluble polysaccharides in the anther development of some members of Gramineae – millets. pp. 315–324 in: Advances in Plant Morphology (edited by Y.S. Murty, B.M. Johri, H.Y.M. Ram & T.M. Varghese). Sarita Prakashan, Meerut.

918. Panchaksharappa, M.G. & C.K. Rudramuniyappa. 1974. Localization of nucleic acids and insoluble polysaccharides in the anther of *Zea mays* L. A histochemical study. Cytologia 39: 153–160.

919. Panchaksharappa, M.G. & C.K. Rudramuniyappa. 1975. Localization of polysaccharides in the anthers of *Sorghum vulgare* Pers. and *Paspalum scrobiculatum* L. J. Karnak Univ., Sci. 20: 124–129.

920. Panchaksharappa, M.G., C.K. Rudramuniyappa & R.R. Hegde. 1985. A review on recent advances in histochemistry of anther development. J. Pl. Sci. Res. 1: 60–83.

921. Panchaksharappa, M.G. & J. Syamasendar. 1974. A cytochemical study of anther development in *Iphigenia pallida* Bak. Cytologia 39: 133–138.

922. Pande, P.C. & V. Singh. 1981. A contribution to the embryology of the Iridaceae. J. Indian Bot. Soc. 60: 160–167.

923. Parameswaran, N. 1962. Floral morphology and embryology in some taxa of the Canellaceae. Proc. Indian Acad. Sci., B, 55: 167–182.

924. Parkin, J. 1923. The strobilus theory of angiospermous descent. Proc. Linn. Soc., Lond. 135: 51–66.

925. Parkin, J. 1951. The protrusion of the connective beyond the anther and its bearing on the evolution of the stamen. Phytomorphology 1: 1–18.

926. Pass, A. 1940. Das Auftreten verholzter Zellen in Blüten und Blütenknospen. Öst. Bot. Z. 89: 119–164, 169–210.

927. Paterson, B.R. 1960–61. Studies of floral morphology in the Epacridaceae. Bot. Gaz. 122: 259–279.

928. Pauzé, F. & R. Sattler. 1978. L'androcée centripète d'*Ochna atropurpurea*. Can. J. Bot. 56: 2500–2511.

929. Pedersen, K.R., E.M. Friis, P.R. Crane & A.N. Drinnan. 1994. Reproductive structures of an extinct platanoid from the Early Cretaceous (latest Albian) of

eastern North America. Rev. Palaeobot. Palynol. 80: 291–303.

930. Perdue, T.D., C.A. Loukides & P.A. Bedinger. 1992. The formation of cytoplasmic channels between tapetal cells in *Zea mays*. Protoplasma 171: 75–79.

931. Perez De La Vega, M. & J.R. Lacadena. 1979. Cytohistological studies on anther and pollen development in alloplasmic rye. Cytologia 44: 295–304.

932. Periasamy, K. & M.K. Kandasamy. 1981. Development of the anther of *Annona squamosa* L. Ann. Bot. (NS) 48: 885–893.

933. Periasamy, K. & C. Sampoornam. 1983 [1984]. Studies on the hypanthial tube, androecium and pollination in *Arachis hypogaea* L. Beitr. Biol. Pfl. 58: 403–411.

934. Periasamy, K. & R. Sivaramakrishnan. 1979 [1980]. Anther development in *Aegiceras corniculatum* Blanco. pp. 134–139 in: Histochemistry, Developmental and Structural Anatomy of Angiosperms: a Symposium (edited by K. Periasamy). P & B Publications, Tiruchirapalli.

935. Periasamy, K. & B.G.L. Swamy. 1960. Studies in the Annonaceae. I. Microsporogenesis in *Cananga odorata* and *Miliusa wightiana*. Phytomorphology 9: 251–263.

936. Periasamy, K. & B.G.L. Swamy. 1966. Morphology of the anther tapetum of angiosperms. Curr. Sci. 35: 427–303.

937. Periasamy, K. & S. Thangavel. 1988. Anther development in *Xylopia nigricans*. Proc. Indian Acad. Sci., Pl. Sci. 98: 251–255.

938. Pesacreta, T.C., V.I. Sullivan & K.H. Hasenstein. 1993. The connective base of *Cirsium horridulum* (Asteraceae): description and comparison with the viscoelastic filament. Amer. J. Bot. 80: 411–418.

939. Pesacreta, T.C., V.I. Sullivan, K.H. Hasenstein & J.M. Durand. 1991. Thigmonasticity of thistle staminal filaments. I. Involvement of a contractile cuticle. Protoplasma 163: 174–180.

940. Petenatti, E.M. & L.A. Del Vitto. 1991. Morfologia y ontogenia de las anteras de *Solanum elaeagnifolium* (Solanaceae) y especies afines y su posible relacion con los visitantes. Kurtziana 21: 195–204. [Eng. summ.]

941. Peters, I. & S. Jain. 1987. Genetics of grain amaranths. III. Gene-cytoplasmic male sterility. J. Hered. 78: 251–256.

942. Pfeffer, W. 1872. Zur Blütenentwicklung der Primulaceen und Ampelideen. Jahrb. Wiss. Bot. 8: 194–215 + Plates XIX–XXII.

943. Pisek, A. 1924. Antherenentwicklung und meiotische

Teilung bei der Wacholdermistel. Sber. Akad. Wiss. Wien, Math.–Nat. Kl. I, 133: 1–15. [Not seen.]

944. Pizzolato, T.D. 1983. A three-dimensional reconstruction of the vascular system to the lodicules, androecium, and gynoecium of a fertile floret of *Panicum dichotomiflorum* (Gramineae). Amer. J. Bot. 70: 1173–1187.

945. Plantefol, L. 1948 [1949]. L'ontogénie de la fleur. Ann. Sci. Nat., Bot., sér. 11, 9: 35–186.

946. Poddubnaja, W. 1927. Spermatogenesis bei einigen Compositen. Planta 4: 284–298.

947. Poggendorff, W. 1932. Flowering, pollination, and natural crossing in rice. Agr. Gaz. N.S.W. 43: 898–904.

948. Pohl, F. 1931. Über sich öffnende Kristallräume in den Antheren von *Deherainia smaragdina* Jahrb. Wiss. Bot. 75: 481–493.

949. Polak, J.M. 1900. Untersuchungen über die Staminodien der Scrophulariaceen. Öst. Bot. Z. 50: 33–41, 87–90, 123–132, 164–167 + Plates II, III.

950. Politis, J. 1957. Über die Tanninoplasten oder Gerbstoffbildner der Rosaceen. Protoplasma 48: 261–268.

951. Polowick, P.L., R. Bolaria & V.K. Sawhney. 1990. Stamen ontogeny in the temperature-sensitive 'stamenless-2' mutant of tomato (*Lycopersicon esculentum* L.). New Phytol. 115: 625–631.

952. Polowick, P.L. & V.K. Sawhney. 1987. A scanning electron microscopic study on the influence of temperature on the expression of cytoplasmic male sterility in *Brassica napus*. Can. J. Bot. 65: 807–814.

953. Polowick, P.L. & V.K. Sawhney. 1988. High temperature induced male and female sterility in canola (*Brassica napus* L.). Ann. Bot. 62: 83–86.

954. Polowick, P.L. & V.K. Sawhney. 1991. Microsporogenesis in a normal line and in the *ogu* cytoplasmic male sterility line of *Brassica napus*: II. The influence of intermediate and low temperatures. Sex. Pl. Reprod. 4: 22–27.

955. Porsch, O. 1913. Die Abstammung der Monokotylen und die Blütennektarien. Ber. Dtsch. Bot. Ges. 31: 580–590.

956. Posluszny, U. & Charlton, W.A. 1993. Evolution of the helobial flower. Aquat. Bot. 44: 303–324.

957. Posluszny, U. & R. Sattler. 1973. Floral development of *Potamogeton densus*. Can. J. Bot. 51: 647–656.

958. Posluszny, U. & R. Sattler. 1974. Floral development of *Potamogeton richardsonii*. Amer. J. Bot. 61: 209–216.

959. Posluszny, U. & R. Sattler. 1974. Floral development

of *Ruppia maritima* var. *maritima*. Can. J. Bot. 52: 1607–1612.

960. Prakash, N., A.L. Lim & F.B. Sampson. 1992. Anther and ovule development in *Tasmannia* (Winteraceae). Aust. J. Bot. 40: 877–885.

961. Prance, G.T. 1976. The pollination and androphore structure of some Amazonian Lecythidaceae. Biotropica 8: 235–241.

962. Prasad, K. 1977. Histochemistry of anther and ovule in *Farsetia hamiltonii* (Royle) and *Eruca sativa* Mill. (Cruciferae). J. Indian Bot. Soc. 56: 90–99.

963. Preil, W. 1974. Über die Verweiblichung männlicher Blüten bei *Begonia semperflorens*. Untersuchungen zur Vererbung und Physiologie von zu Narben umgebildeten Antheren und freiliegenden Samenanlagen. Z. Pflanzenzücht. 72: 132–151.

964. Pullaiah, T. 1983. Studies in the embryology of Senecioneae (Compositae). Pl. Syst. Evol. 142: 61–70.

965. Pułło, E. & A. Slusarkiewicz. 1975. Development of ovules on the stamens in flowers of *Solanum tuberosum* variety Flisak. Acta Soc. Bot. Pol. 44: 519–527.

966. Puri, V. 1941. The life-history of *Moringa oleifera* Lamk. J. Indian Bot. Soc. 20: 263–284.

967. Puri, V. 1947, 1948. Studies in floral anatomy. IV. Vascular anatomy of the flower of certain species of the Passifloraceae. Amer. J. Bot. 34: 562–573. V. On the structure and nature of the corona in certain species of the Passifloraceae. J. Indian Bot. Soc. 27: 130–149 (1948).

968. Puri, V. 1951. The role of floral anatomy in the solution of morphological problems. Bot. Rev. 17: 471–553.

969. Puri, V. 1970 [1972]. Anther sacs and pollen grains: some aspects of their structure and function. J. Palynol. 6: 1–17.

970. Py, G. 1932. Recherches cytologiques sur l'assise nourricière des microspores et les microspores des plantes vasculaires. Rev. Gén. Bot. 44: 316–413, 450–462, 484–512 + Plates XVII–XIX.

971. Quiros, C.F., A. Rugama, Y.Y. Dong & T.J. Orton. 1986. Cytological and genetical studies of a male sterile celery. Euphytica 35: 867–875.

972. Raadts, E. 1979. Rasterelektronenmikroskopische und anatomische Untersuchungen an Konnektivdrüsen von *Kalanchoe* (Crassulaceae). Willdenowia 9: 169–175.

973. Raghavan, T.S. 1939. Studies in the Capparidaceae. II. Floral anatomy and some structural features of the Capparidaceous flower. J. Linn. Soc. (Bot.) 52: 239–257.

974. Raghavan, V. 1988. Anther and pollen development in rice (*Oryza sativa*). Amer. J. Bot. 75: 183–196.

975. Rainio, A.J. 1929. Über die Intersexualität bei der Gattung *Papaver*. Ann. Soc. Zool.-Bot. Fenn. Vanamo 9: 258–285.

976. Rajan, S. 1954. Vascularization of the three-stamened-three-carpelled flowers of *Nyctanthes arbortristis*. J. Madras Univ., B, 24: 237–243.

977. Raju, M.V.S. 1961. Morphology and anatomy of the Saururaceae. I. Floral anatomy and embryology. Ann. Missouri Bot. Gard. 48: 107–124.

978. Ram, H.Y.M. & I.V.R. Rao. 1984. Physiology of flower bud growth and opening. Proc. Indian Acad. Sci., Pl. Sci. 93: 253–274.

979. Ramana, R.V. & P.S.P. Rao. 1983. Tapetal dimorphism in two species of *Premna* L. Curr. Sci. 52: 1059–1061.

980. Rangarajan, R. & B.G.L. Swamy. 1979 [1980]. Studies on the procambium in certain organs of monocotyledons. I. Anther filament of *Gloriosa superba* L. pp. 208–219 in: Histochemistry, Developmental and Structural Anatomy of Angiosperms: a Symposium. (edited by K. Periasamy). P & B Publications, Tiruchirapalli.

981. Rao, A.N. 1961. Fibrous thickenings in the anther epidermis of *Wormia burbidgei* Hook. Curr. Sci. 30: 426.

982. Rao, A.N. 1962. Floral anatomy and gametogenesis in *Hopea racophloea* Dyer. J. Indian Bot. Soc. 41: 557–562.

983. Rao, A.N. & H. Singh. 1964. Stamens and carpels within the ovary of *Durio zibethinus* Murr. Gard. Bull., Singapore 20: 289–294.

984. Rao, B.H. & P.S.P. Rao. 1977. Anther in *Capparis decidua* (Forsk.) Pax. Curr. Sci. 46: 683–684. [Not seen.]

985. Rao, B.H. & P.S.P. Rao. 1984. Sporogenesis and genesis of gametophytes in *Cordia sebestena* L. Acta Bot. Indica 12: 69–76.

986. Rao, C.V. 1949. Floral anatomy of some Sterculiaceae with special reference to the position of stamens. J. Indian Bot. Soc. 28: 237–245.

987. Rao, C.V. 1952. Floral anatomy of some Malvales and its bearing on the affinities of families included in the order. J. Indian Bot. Soc. 31: 171–203.

988. Rao, C.V. 1960, 1964. Studies in the Proteaceae. I. Tribe Persoonieae. Proc. Natn. Inst. Sci. India, B, 26: 300–336 (1960). IV. Tribes Banksieae, Musgraveae and Embothrieae. Ibid. 30: 197–244 (1964).

989. Rao, C.V. & T. Ramalakshmi. 1968 [1969]. Floral

anatomy of Euphorbiaceae. I. Some non-cyathium taxa. J. Indian Bot. Soc. 47: 278–300.

990. Rao, H.S. 1954. Pollination mechanism in *Acacia catechu*. J. Indian Bot. Soc. 33: 93–97.

991. Rao, K.V.R. 1940. Gametogenesis and embryogeny in five species of the Convolvulaceae. J. Indian Bot. Soc. 19: 53–69.

992. Rao, M.K. & K.U. Devi. 1983. Variation in expression of genic male sterility in pearl millet. J. Hered. 74: 34–38.

993. Rao, P.S.P. & B.H. Rao. 1975. Anther development in *Cleome tenella* Linn. Curr. Sci. 44: 438–440. [Not seen.]

994. Rao, S.R.S. 1987. Structure, distribution and classification of plant trichomes in relation to taxonomy: Sterculiaceae. Feddes Repert. 98: 127–135.

995. Rao, S.R.S. 1991. Structure and distribution of plant trichomes in relation to taxonomy: *Hibiscus* L. Feddes Repert. 102: 335–344.

996. Rao, V.S. 1953, 1954, 1955. The floral anatomy of some Bicarpellatae. 1. Acanthaceae. J. Univ. Bombay 21 (5): 1–34 (1953). II. Bignoniaceae. Ibid. 22: 55–70 (1954). III. Pedaliaceae. Ibid. 23 (5): 18–26 (1955).

997. Rao, V.S. 1961. Floral Anatomy. Scholar's Library, New York. 25 pp. [Bibliography.]

998. Rao, V.S. 1967–8. The stamens of *Burmannia*. J. Univ. Bombay 36: 25–26.

999. Rao, V.S. & A. Ganguli. 1963. The floral anatomy of some Asclepiadaceae. Proc. Indian Acad. Sci., B, 57: 15–44.

1000. Rao, V.S. & A. Ganguli. 1963. Studies in the floral anatomy of the Apocynaceae. J. Indian Bot. Soc. 42: 419–435.

1001. Rao, V.S. & K. Gupte. 1961. The floral anatomy of some Scitamineae. Part IV. J. Univ. Bombay 29: 134–150.

1002. Rao, V.S., H. Karnik & K. Gupte. 1954. The floral anatomy of some Scitamineae. Part I. J. Indian Bot. Soc. 33: 118–147.

1003. Rao, V.S. & R.M. Pai. 1959, 1960. The floral anatomy of some Scitamineae. II. J. Univ. Bombay 28 (3): 82–114 (1959). III. Ibid. (5): 1–19 (1960).

1004. Rao, V.S., K. Sirdeshmukh & M.G. Sardar. 1958. The floral anatomy of the Leguminosae. J. Univ. Bombay (NS) 26: 65–138.

1005. Rauh, W. & H. Reznik. 1951. Histogenetische Untersuchungen an Blüten- und Infloreszenzachsen. I. Die Histogenese becherförmiger Blüten- und Infloreszenzachsen sowie der Blütenachsen einiger Rosoideen.

Sber. Heidelberger Akad. Wiss., Math.-Nat. Kl. 3: 1–71. [Not seen.]

1006. Rauh, W., H.-F. Schölch & H. Straka. 1965. Weitere Untersuchungen an Didiereaceen. 1 & 2. Infloreszenz-, blütenmorphologische und embryologische Untersuchungen mit Ausblick auf die systematische Stellung der Didiereaceen. Sber. Heidelberger Akad. Wiss., Math.-Nat. Kl 3: 5–218, 221–434.

1007. Razi, B.A. & K. Subramanyam. 1953. Embryology of the Dipsacaceae. I. Glands, the male and female gametophytes. Proc. Indian Acad. Sci., B, 36: 249–257.

1008. Reece, P.C. 1939. The floral anatomy of the avocado. Amer. J. Bot. 26: 429–433.

1009. Rendle, A.B. 1903. The origin of the perianth in seed-plants. New Phytol. 2: 66–72.

1010. Renuka, C. & K. Swarupanandan. 1986. Morphology of the flower in *Thottea siliquosa* and the existence of staminodes in Aristolochiaceae. Blumea 31: 313–318.

1011. Reynolds, J. & J. Tampion. 1983. Double Flowers: a Scientific Study. Pembridge Press, London. 183 pp.

1012. Reznickova, S.A. & H.G. Dickinson. 1982. Ultrastructural aspects of storage lipid mobilization in the tapetum of *Lilium hybrida* var. *enchantment*. Planta 155: 400–408.

1013. Reznickova, S.A. & M.T.M. Willemse. 1980. Formation of pollen in the anther of *Lilium*. II. The function of the surrounding tissues in the formation of pollen and pollen wall. Acta Bot. Neerl. 29: 141–156.

1014. Reznickova, S.A. & M.T.M. Willemse. 1981. The function of the tapetal tissue during microsporogenesis in *Lilium*. Acta Soc. Bot. Pol. 50 (1–2): 83–88.

1015. Reznickova, S.A. & M.T.M. Willemse. 1981. Electron microscopic and histochemical study of developing lily (*Lilium hybridum* cultivar *Enchantment*) anther in connection with the metabolism of reserve nutrients. Fiziol. Rast. 28: 1181–1189. [Russ.; Eng. summ.]

1016. Reznickova, S.A. & M.T.M. Willemse. 1981 [1982]. Electron-microscopic and histochemical investigation of tissues of the developing lily anther in connection with metabolism of reserve nutrient substances. Sov. Pl. Physiol. 28: 856–864. [Not seen.]

1017. Richharia, R.H. 1934. The number of microsporangia in each stamen of Asclepiadaceae. Curr. Sci. 2: 340–342.

1018. Richter, S. 1929. Über den Öffnungsmechanismus der Antheren bei einigen Vertretern der Angiospermen. Planta 8: 154–184.

1019. Rickson, F.R. 1979. Ultrastructural development of the beetle food tissue of *Calycanthus* flowers. Amer. J. Bot. 66: 80–86.

1020. Riss, M.M. 1918. Die Antherenhaare von *Cyclanthera pedata* (Schrad.) und einiger anderer Cucurbitaceen. Flora 111–112: 541–559.

1021. Risse, K. 1929. Beiträge zur Zytologie der Dipsacaceen. Bot. Arch. 23: 266–288.

1022. Risueño, M.C., G. Giménez-Martín, J.F. López-Sáez & M.I.R. García. 1969. Origin and development of sporopollenin bodies. Protoplasma 67: 361–374.

1023. Ritzerow, H. 1907. Über Bau und Befruchtung kleistogamer Blüten. Flora 98: 163–212.

1024. Rivière, S. 1951. Structure anatomique de la base de l'anthère introrse de lis (*Lilium candidum*-Liliacées). Rev. Gén. Bot. 58: 370–388 + Plates VI–IX.

1025. Rivière, S. 1952. Vérification 'in situ' de la vascularisation de l'anthère de lis au niveau de l'insertion du connectif. Bull. Soc. Bot. Fr. 99: 72–75.

1026. Rivière, S. 1952. Le cambium intra-fasciculaire de l'anthère de lis (*Lilium candidum*). Rev. Gén. Bot. 59: 209–231.

1027. Rivière, S. 1954. Quelques observations cytologiques relatives à l'anthère de *Lilium candidum*. Rev. Gén. Bot. 61: 197–228.

1028. Robertson, B.L. 1984. Tapetal cell changes and sporoderm development in *Rhigozum trichotomum* (Burch.). Ann. Bot. 53: 803–810.

1029. Robertson, R.E. & S.C. Tucker. 1979. Floral ontogeny of *Illicium floridanum*, with emphasis on stamen and carpel development. Amer. J. Bot. 66: 605–617.

1030. Robinson, H. 1977 [1978]. An analysis of the characters and relationships of the tribes Eupatorieae and Vernonieae (Asteraceae). Syst. Bot. 2: 199–208.

1031. Robinson, H. 1985. Observations on fusion and evolutionary variability in the angiosperm flower. Syst. Bot. 10: 105–109.

1032. Robinson, H. & R.M. King. 1977. Eupatorieae – systematic review. pp. 437–85 in: The Biology and Chemistry of the Compositae (edited by V.H. Heywood, J.B. Harborne & B.L. Turner), Vol. 1. Academic Press, London.

1033. Robinson Beers, K. & T.D. Pizzolato. 1987. Development of the vascular system in the fertile floret of *Anthoxanthum odoratum* L. (Gramineae). II. Sieve-element plexus, stamen traces, and the xylem discontinuity. Bot. Gaz. 148: 209–220.

1034. Robson, N.K.B. 1972. Evolutionary recall in *Hypericum* (Guttiferae)? Trans. Bot. Soc. Edinb. 41: 365–383.

1035. Roelofsen, P.A. & A.L. Houwink. 1951. Cell wall structure of staminal hairs of *Tradescantia virginica* and its relation with growth. Protoplasma 40: 1–22.

1036. Roeser, K.R. 1973. Die Staubblätter der Gladiole. Mikrokosmos 62: 268–270.

1037. Rohrhofer, J. 1931. Morphologische Studien an den Staminodien der Bignoniaceae. Öst. Bot. Z. 80: 1–30 + Plates I–VIII.

1038. Rohweder, O. 1963. Anatomische und histogenetische Untersuchungen an Laubsprossen und Blüten der Commelinaceen. Bot. Jahrb. 82: 1–99.

1039. Rohweder, O. 1967, 1970. Centrospermen-Studien. 3. Blütenentwicklung und Blütenbau bei Silenoideen (Caryophyllaceae). Bot. Jahrb. 86: 130–185 [Eng. summ. 178–179] (1967). 4. Morphologie und Anatomie der Blüten, Früchte und Samen bei Alsinoideen und Paronychioideen s. lat. (Caryophyllaceae). Ibid. 90: 201–271 [Eng. summ. 266] (1970).

1040. Rohweder, O. 1972. Das Andröcium der Malvales und der 'Konservatismus' des Leitgewebes. Bot. Jahrb. 92: 155–167.

1041. Rohweder, O. & K. Huber. 1974. Centrospermen-Studien. 7. Beobachtungen und Anmerkungen zur Morphologie und Entwicklungsgeschichte einiger Nyctaginaceen. Bot. Jahrb. 94: 327–359.

1042. Rohwer, J. & K. Kubitzki. 1985. Entwicklungslinien im *Ocotea*-Komplex (Lauraceae). Bot. Jahrb. 107: 129–135.

1043. Roland-Heydacker, F. 1979. Aspects ultrastructuraux de l'ontogénie du pollen et du tapis chez *Mahonia aquifolium* Nutt. Berberidaceae. Pollen Spores 21: 259–278.

1044. Ronse Decraene, L.P. 1988. Two types of ringwall formation in the development of complex polyandry. Bull. Soc. Roy. Bot. Belg. 121: 122–124.

1045. Ronse Decraene, L.P. 1989. Floral development of *Cochlospermum tinctorium* and *Bixa orellana* with special emphasis on the androecium. Amer. J. Bot. 76: 1344–1359.

1046. Ronse Decraene, L.P. 1990. Morphological studies in Tamaricales I: Floral ontogeny and anatomy of *Reaumuria vermiculata* L. Beitr. Biol. Pfl. 65: 181–203.

1047. Ronse Decraene, L.P., D. Clinckemaillie & E. Smets. 1993. Stamen-petal complexes in Magnoliatae. Bull. Jard. Bot. Natn. Belg. 62: 97–112.

1048. Ronse Decraene, L.P. & E. Smets. 1987. The distribution and the systematic relevance of the androecial characters oligomery and polymery in the Magnoli-ophytina. Nord. J. Bot. 7: 239–253.

1049. Ronse Decraene, L.P. & E. Smets. 1990. The systematic relationship between Begoniaceae and Papaveraceae: a comparative study of their floral development. Bull. Jard. Bot. Natn. Belg. 60: 229–273.

1050. Ronse Decraene, L.P. & E. Smets. 1990. The floral development of *Popowia whitei* (Annonaceae). Nord. J. Bot. 10: 411–420.

1051. Ronse Decraene, L.P. & E. Smets. 1991. Androecium and floral nectaries of *Harungana madagascariensis* (Clusiaceae). Pl. Syst. Evol. 178: 179–194.

1052. Ronse Decraene, L.P. & E.F. Smets. 1991. Morphological studies in Zygophyllaceae. I. The floral development and vascular anatomy of *Nitraria retusa*. Amer. J. Bot. 78: 1438–1448.

1053. Ronse Decraene, L.P. & E. Smets. 1991. The impact of receptacular growth on polyandry in the Myrtales. Bot. J. Linn. Soc. 105: 257–269.

1054. Ronse Decraene, L.P. & E. Smets. 1991. The floral ontogeny of some members of the Phytolaccaceae (subfamily Rivinoideae) with a discussion of the evolution of the androecium in the Rivinoideae. Biol. Jaarb. Dodonaea 59: 77–99.

1055. Ronse Decraene, L.P. & E.F. Smets. 1992. An updated interpretation of the androecium of the Fumariaceae. Can. J. Bot. 70: 1765–1776.

1056. Ronse Decraene, L.P. & E.F. Smets. 1992. Complex polyandry in the Magnoliatae: definition, distribution and systematic value. Nord. J. Bot. 12: 621–649.

1057. Ronse Decraene, L.P. & E.F. Smets. 1993. Dédoublement revisited: towards a renewed interpretation of the androecium of the Magnoliophytina. Bot. J. Linn. Soc. 113: 103–124.

1058. Ronse Decraene, L.P. & E.F. Smets. 1993 [1994]. The distribution and systematic relevance of the androecial character polymery. Bot. J. Linn. Soc. 113: 285–350.

1059. Ross, R. 1982. Initiation of stamens, carpels, and receptacle in the Cactaceae. Amer. J. Bot. 69: 369–379.

1060. Roth, I. 1959. Histogenese und morphologische Deutung der Kronblätter von *Primula*. Bot. Jahrb. 79: 1–16.

1061. Rousseau, D. 1927. Contribution à l'anatomie comparete des Piperacées. Mem. Acad. Roy. Belg., Cl. Sci. 9: 3–45. [Not seen.]

1062. Rowley, J.R. 1963. Ubisch body development in *Poa annuua*. Grana Palynol. 4: 25–36.

1063. Rowley, J.R. 1987. Plasmodesmata-like processes of tapetal cells. Cellule 74: 227–242.

1064. Rowley, J.R. 1993. Cycles of hyperactivity in tapetal cells. Pl. Syst. Evol., Suppl. 7: 23–38.

1065. Rowley, J.R., N.I. Gabarayeva & B. Walles. 1992. Cyclic invasion of tapetal cells into loculi during microspore development in *Nymphaea colorata* (Nymphaeaceae). Amer. J. Bot. 79: 801–808.

1066. Rubsamen-Weustenfeld, T., V. Mukielka & U. Hamann. 1994. Zur Embryologie, Morphologie und systematischen Stellung von *Geosiris aphylla* Baillon (Monocotyledoneae-Geosiridaceae/Iridaceae). Mit einigen embryologischen Daten zur Samenanlage von *Isophysis tasmanica* (Hook.) T. Moore (Iridaceae). Bot. Jahrb. 115: 475–545.

1067. Rudramuniyappa, C.K. 1985. A histochemical study of developing sporogenous tissue and periplasmodial tapetum in the anther of *Parthenium* (Compositae). Cytologia 50: 891–898.

1068. Rudramuniyappa, C.K. & B.G. Annigeri. 1984. A histochemical study of meiocytes, microspores, pollen and the tapetum in *Kalanchoe*. Nord. J. Bot. 4: 661–667.

1069. Rudramuniyappa, C.K. & B.G. Annigeri. 1985. Histochemical observations on the sporogenous tissue and tapetum in the anther of *Euphorbia*. Cytologia 50: 39–48.

1070. Rudramuniyappa, C.K. & P.B. Mahajan. 1991 [1992]. Histochemical and fluorescence microscopic study of anther development in *Spathodea campanulata* Beauv. Phytomorphology 41: 175–188.

1071. Rudramuniyappa, C.K. & M.G. Panchaksharappa. 1980. Pollen development in *Triticum durum* Desf.: a histochemical study. J. S. Afr. Bot. 46: 33–43.

1072. Rudramuniyappa, C.K. & M.G. Panchaksharappa. 1982 [1983]. Histochemistry of anther in *Pennisetum* and *Setaria*. Beitr. Biol. Pfl. 57: 193–203.

1073. Safwat, F.M. 1962. The floral morphology of *Secamone* and the evolution of the pollinating apparatus in Asclepiadaceae. Ann. Missouri Bot. Gard. 49: 95–129.

1074. Said, C. 1960. Recherches anatomiques sur l'androcée du *Commelina chamissonis* (Klotzch). Bull. Soc. Hist. Nat. Afr. N. 51: 159–174.

1075. Saini, H.S., M. Sedgley & D. Aspinall. 1984. Developmental anatomy in wheat of male sterility induced by heat stress, water deficit or abscisic acid. Aust. J. Pl. Physiol. 11: 243–253.

1076. Salisbury, E.J. 1919. Variation in *Eranthis hyemalis*,

Ficaria verna, and other members of the Ranunculaceae, with special reference to trimery and the origin of the perianth. Ann. Bot. 33: 47–79.

1077. Salisbury, E.J. 1920. Variation in *Anemone apennina* L. and *Clematis vitalba* L., with special reference to trimery and abortion. Ann. Bot. 34: 107–116.

1078. Salisbury, E.J. 1926. Floral construction in the Helobiales. Ann. Bot. 40: 419–445.

1079. Salisbury, E. 1974. The variations in the reproductive organs of *Stellaria media* (*sensu stricto*) and allied species with special regard to their relative frequency and prevalent modes of pollination. Proc. Roy. Soc., Lond. B, 185: 331–342.

1080. Sampson, F.B. 1963. The floral morphology of *Pseudowintera*, the New Zealand member of the vesselless Winteraceae. Phytomorphology 13: 403–423.

1081. Sampson, F.B. 1969. Studies on the Monimiaceae. I. Floral morphology and gametophyte development of *Hedycarya arborea* J.R. et G. Forst. (subfamily Monimioideae). Aust. J. Bot. 17: 403–424. II. Floral morphology of *Laurelia novae-zelandiae*. A. Cunn. (subfamily Atherospermoideae.). N.Z. J. Bot. 7: 214–240.

1082. Sampson, F.B. 1987. Stamen venation in the Winteraceae. Blumea 32: 79–89.

1083. Samuelsson, G. 1913. Studien über die Entwicklungsgeschichte der Blüten einiger Bicornes-Typen. Ein Beitrag zur Kenntnis der systematischen Stellung der Diapensiaceen und Empetraceen. Svensk Bot. Tidskr. 7: 97–188.

1084. Sands, M.J.S. 1973. New aspects of the floral vascular anatomy in some members of the Order Rhoeadales sensu Hutch. Kew Bull. 28: 211–256.

1085. Santos, A., I. Abreu & R. Salema. 1979. Elaborate system of RER and degenerescence of tapetum during pollen development in some dicotyledons. J. Submicr. Cytol. 11: 99–107.

1086. Sass, J.E. 1944. The initiation and development of foliar and floral organs in the tulip. Iowa St. Coll. J. Sci. 18: 447–456. [Not seen.]

1087. Sastri, R.L.N. 1958. Floral morphology and embryology of some Dilleniaceae. Bot. Not. Lund 111: 495–511.

1088. Sastri, R.L.N. 1969. Floral morphology, embryology, and relationships of the Berberidaceae. Aust. J. Bot. 17: 69–79.

1089. Satina, S. & A.F. Blakeslee. 1941. Periclinal chimeras in *Datura stramonium* in relation to development of leaf and flower. Amer. J. Bot. 28: 862–871.

1090. Sato, T., M.K. Thorsness, M.K. Kandasamy, T. Nishio, M. Hirai, J.B. Nasrallah & M.E. Nasrallah. 1991. Activity of an S locus gene promoter in pistils and anthers of transgenic *Brassica*. Pl. Cell 3: 867–876.

1091. Sato, Y. & M. Kato. 1990. Anther tapetum of *Aletris luteoviridis* (Liliaceae) and *Abelia grandiflora* (Caprifoliaceae). Sci. Rep. Yokohama Natn. Univ., Sect. II, Biol. Geol., 37: 13–25. [Eng.] [Not seen.]

1092. Sattler, R. 1962. Zur frühen Infloreszenz- und Blütenentwicklung der Primulales sensu lato mit besonderer Berücksichtigung der Stamen-Petalum-Entwicklung. Bot. Jahrb. 81: 358–396.

1093. Sattler, R. 1965. Perianth development of *Potamogeton richardsonii*. Amer. J. Bot. 52: 35–41.

1094. Sattler, R. 1972 [1975]. Centrifugal primordial inception in floral development. pp. 170–178 in: Advances in Plant Morphology (edited by Y.S. Murty, B.M. Johri, H.Y.M. Ram & T.M. Varghese). Sarita Prakashan, Meerut.

1095. Sattler, R. 1973. Organogenesis of Flowers. A Photographic Text-Atlas. University of Toronto Press, Toronto. 207 pp.

1096. Sattler, R. 1988. Homeosis in plants. Amer. J. Bot. 75: 1606–1617.

1097. Sattler, R. & L. Perlin. 1982. Floral development of *Bougainvillea spectabilis* Willd., *Boerhaavia diffusa* L. and *Mirabilis jalapa* L. (Nyctaginaceae). Bot. J. Linn. Soc. 84: 161–182.

1098. Sattler, R. & V. Singh. 1973. Floral development of *Hydrocleis nymphoides*. Can. J. Bot. 51: 2455–2458.

1099. Sattler, R. & V. Singh. 1977. Floral organogenesis of *Limnocharis flava*. Can. J. Bot. 55: 1076–1086.

1100. Sattler, R. & V. Singh. 1978. Floral organogenesis of *Echinodorus amazonicus* Rataj and floral construction of the Alismatales. Bot. J. Linn. Soc. 77: 141–156.

1101. Sauer, H. 1933. Blüte und Frucht der Oxalidaceen, Linaceen, Geraniaceen, Tropaeolaceen und Balsaminaceen. Vergleichend-entwicklungsgeschichtliche Untersuchungen. Planta 19: 417–481.

1102. Saunders, E.R. 1932. On some recent contributions and criticisms dealing with morphology in angiosperms. New Phytol. 31: 174–219.

1103. Saunders, E.R. 1934. Comments on 'Floral anatomy and its morphological interpretation.' New Phytol. 33: 127–170.

1104. Saunders, E.R. 1936. On certain features of floral construction and arrangement in the Malvaceae. Ann. Bot. 50: 247–282.

1105. Saunders, E.R. 1936. The vascular ground-plan as a guide to the floral ground-plan: illustrated from Cistaceae. New Phytol. 35: 47–67.

1106. Saunders, E.R. 1936. On rhythmic development and radial organisation in the flower. J. Linn. Soc., Bot. 50: 291–322.

1107. Saunders, E.R. 1937–9. Floral Morphology. 2 vols. W. Heffer, Cambridge.

1108. Saunders, E.R. 1940. Further observations on the morphology and anatomy of the flower in *Salvia*. Ann. Bot. (NS) 4: 629–633.

1109. Saunders, E.R. 1941. The significance of certain morphological variations of common occurrence in flowers of *Primula*. New Phytol. 40: 64–85.

1110. Savich, E.I. 1968. The formation of archesporium and the origin of tapetum in Helobiae. Bot. J. USSR 53: 514–523. [Russ.; Eng. summ.]

1111. Sawada, M. 1971. Floral vascularization of *Paeonia japonica* with some consideration on systematic position of the Paeoniaceae. Bot. Mag., Tokyo 84: 51–60.

1112. Sawhney, V.K. 1981. Abnormalities in pepper (*Capsicum annuum*) flowers induced by gibberellic acid. Can. J. Bot. 59: 8–16.

1113. Sawhney, V.K. & R.I. Greyson. 1973. Morphogenesis of the stamenless-2 mutant in tomato. I. Comparative description of the flowers and ontogeny of stamens in the normal and mutant plants. Amer. J. Bot. 60: 514–523. II. Modifications of sex organs in the mutant and normal flowers by plant hormones. Can. J. Bot. 51: 2473–2479.

1114. Sawhney, V.K. & R.I. Greyson. 1979. Interpretations of determination and canalisation of stamen development in a tomato mutant. Can. J. Bot. 57: 2471–2477.

1115. Sawhney, V.K. & E.B. Nave. 1986. Enzymatic changes in post-meiotic anther development in *Petunia hybrida*. II. Histochemical localization of esterase, peroxidase, malate- and alcohol dehydrogenase. J. Pl. Physiol. 125: 467–473.

1116. Sawhney, V.K. & P.L. Polowick. 1986. Temperature-induced modifications in the surface features of stamens of a tomato mutant: an SEM study. Protoplasma 131: 75–81.

1117. Schachner, J. 1924. Beiträge zur Kenntnis der Blüten- und Samenentwicklung der Scitamineen. Flora 117: 16–40.

1118. Schaeppi, H. 1939. Vergleichend-morphologische Untersuchungen an der Staubblättern der Monocotyledonen. Nova Acta Leopoldina (N.F.) 6: 389–447.

1119. Schaeppi, H. 1976. Über die männlichen Blüten einiger Menispermaceen. Beitr. Biol. Pfl. 52: 207–215.

1120. Schaeppi, H. & F. Steindl. 1942. Blütenmorphologische und embryologische Untersuchungen an Loranthoideen. Vierteljahrsschr. Naturf. Ges. Zürich 87: 301–372.

1121. Schaeppi, H. & F. Steindl. 1945. Blütenmorphologische und embryologische Untersuchungen an Viscoideen. Vierteljahrsschr. Naturforsch. Ges. Zürich 90: 1–46.

1122. Schaffner, J.H. 1897. The development of the stamens and carpels of *Typha latifolia*. Bot. Gaz. 24: 93–102 + Plates IV–VI.

1123. Schick, B. 1982. Untersuchungen über die Biotechnik der Apocynaceenblüte II. Bau und Funktion des Bestäubungsapparates. Flora 172: 347–371.

1124. Schips, M. 1913. Zur Öffnungsmechanik der Antheren. Beih. Bot. Zbl. 31(1): 119–208. Also: Diss., Freiburg.

1125. Schmid, R. 1972. Floral bundle fusion and vascular conservatism. Taxon 21: 429–446.

1126. Schmid, R. 1976. Filament histology and anther dehiscence. Bot. J. Linn. Soc. 73: 303–315.

1127. Schmid, R. 1977. Edith R. Saunders and floral anatomy: bibliography and index to families she studied. Bot. J. Linn. Soc. 74: 179–187.

1128. Schmid, R. 1978. Reproductive anatomy of *Actinidia chinensis* (Actinidiaceae). Bot. Jahrb. 100: 149–195.

1129. Schmid, R. 1980. Comparative anatomy and morphology of *Psiloxylon* and *Heteropyxis*, the subfamilial and tribal classification of Myrtaceae. Taxon 29: 559–595.

1130. Schmid, R. & P.H. Alpert. 1977. A test of Burck's hypothesis relating anther dehiscence to nectar secretion. New Phytol. 78: 487–498.

1131. Schmidt, E. 1928. Untersuchungen über Berberidaceen. Beih. Bot. Zbl. 45(2): 329–396.

1132. Schnarf, K. 1923. Über das Verhalten des Antherentapetums einiger Pflanzen. Öst. Bot. Z. 72: 242–245.

1133. Schneider, E.L. 1976. The floral anatomy of *Victoria* Schomb. (Nymphaeaceae). Bot. J. Linn. Soc. 72: 115–148.

1134. Schneider, E.L. & J.D. Buchanan. 1980. Morphological studies of the Nymphaeaceae. XI. The floral biology of *Nelumbo pentapetala*. Amer. J. Bot. 67: 182–193.

1135. Schneider, J.M. 1908. Der Öffnungsmechanismus der *Tulipa* Anthere. Ber. Dtsch. Bot. Ges. 26: 394–398.

1136. Schneider, J.M. 1909. Zur ersten und zweiten Hauptfrage der Antherenmechanik. Ber. Dtsch. Bot. Ges. 27: 196–201.

1137. Schneider, J.M. 1911. Über das Öffnen des Nahtgewebes der Antheren. Ber. Dtsch. Bot. Ges. 29: 406–416.

1138. Schnepf, E., F. Witzig & R. Schill. 1979. Über Bildung und Feinstruktur des Translators der Pollinarien von *Asclepias curassavica* und *Gomphocarpus fruticosus* (Asclepiadaceae). Trop. Subtrop. Pfl. Welt 25: 1–39.

1139. Schoch-Bodmer, H. 1939. Beiträge zur Kenntnis des Streckungswachstums der Gramineen-Filamente. Planta 30: 168–204.

1140. Schöffel, K. 1932. Untersuchungen über den Blütenbau der Ranunculaceen. Planta 17: 315–371.

1141. Schoute, J.C. 1932. On pleiomery and meiomery in the flower. Rec. Trav. Bot. Neerl. 29: 164–226.

1142. Schoute, J.C. 1936. On the contort aestivation of the androecium of *Sidalcea*. Rec. Trav. Bot. Neerl. 33: 645–648.

1143. Schrodt, J. 1901. Zur Oeffnungsmechanik der Staubbeutel. Ber. Dtsch. Bot. Ges. 19: 483–488.

1144. Schultz, E.A., F.B. Pickett & G.W. Haughn. 1991. The *FLO10* gene product regulates the expression domain of homeotic genes *AP3* and *P1* in *Arabidopsis* flowers. Plant Cell 3: 1221–1237.

1145. Schultze Motel, W. 1959. Entwicklungsgeschichtliche und vergleichend-morphologische Untersuchungen im Blütenbereich der Cyperaceae. Bot. Jahrb. 78: 129–170.

1146. Schwarze, C. 1914. Vergleichende entwicklungsgeschichtliche und histologische Untersuchungen reduzierter Staubblätter. Jahrb. Wiss. Bot. 54: 189–242 + Plates I–IV. Also: Diss., Tubingen.

1147. Schweitzer, H.-J. 1977. Die räto-jurassischen Floren des Iran und Afghanistans. 4. Die rätische Zwitterblüte *Irania hermaphroditica* nov. spec. und ihre Bedeutung für die Phylogenie der Angiospermen. Palaeontographica, B, 161: 98–145.

1148. Scott, R.W. 1985. Microcharacters as generic markers in the Eupatorieae. Taxon 34: 26–30.

1149. Seth, N. & M.R. Vijayaraghavan. 1991. Sequential pathways of anther dehisence in *Sesbania speciosa*. Cytologia 56: 621–626.

1150. Shah, C.K., S.N. Pandey & P.N. Bhatt. 1974. Cytophotometric study of ascorbic acid in the developing anthers of *Aponogeton natans* L. Indian J. Pl. Physiol. 17: 16–22.

1151. Shamrov, I.I. 1981. Some peculiar features of the development of the anther in *Ceratophyllum demersum* and *C. pentacanthum* (Ceratophyllaceae). Bot. Zh. SSSR 66: 1464–1473. [Russ. only.]

1152. Shamrov, I.I. 1983. The structure of the anther and some peculiar features of the microsporogenesis and pollen grain development in the representatives of the genus *Ceratophyllum* (Ceratophyllaceae). Bot. Zh. SSSR 68: 1662–1667. [Russ. only.]

1153. Shamrov, I.I. 1986. Anther development in *Gentiana lutea* (Gentianaceae). Bot. Zh. SSSR 71: 733–739. [Russ.; Eng. summ.]

1154. Sharma, B.B. & D.D. Awasthi. 1975. On the androecium of *Canna indica* L. Geophytology 5(1): 30–32. [Not seen.]

1155. Sharma, H.P. 1954, 1963. Studies in the order Centrospermales. I. Vascular anatomy of the flower of certain species of the Portulacaceae. J. Indian Bot. Soc. 33: 98–111 (1954). II. Vascular anatomy of the flower of certain species of the Molluginaceae. Ibid. 42: 19–32 (1963).

1156. Sharma, P.N., C. Chatterjee, C.P. Sharma & S.C. Agarwala. 1987. Zinc deficiency and anther development in maize. Pl. Cell Physiol. 28: 11–18.

1157. Sharman, B.C. 1960. Developmental anatomy of the stamen and carpel primordia in *Anthoxanthum odoratum*. Bot. Gaz. 121: 192–198.

1158. Shealy, H.E. & J.M. Herr. 1973. Carpelloid stamens in *Rubus trivialis* Michx. Bot. Gaz. 134: 77–87.

1159. Sheel, K. & N.N. Bhandari. 1990 [1991]. Ontogenetic and histochemical studies on the anther development of *Carica papaya* L. Phytomorphology 40: 85–94.

1160. Shimizu, T., S. Takao & A. Takao. 1990. Trisporangiate anthers found in the genus *Impatiens* (Balsaminaceae). Bot. Mag., Tokyo 103: 335–337.

1161. Siddiqi, M.R. & T.K. Wilson. 1976. Floral anatomy of the genus *Knema* (Myristicaceae). Biologia, Lahore 22: 127–141.

1162. Siddiqui, S.A. & F.A. Khan. 1988. Ontogeny and dehiscence of anther in Solanaceae. Bull. Soc. Bot. Fr. 135, Lett. Bot. (2): 101–109.

1163. Simonenko, V.K. 1982. Anther and microspore development in fertile and cytoplasmic male sterile lines of the sunflower (*Helianthus petiolaris*). Tsitol. Genet. 16(5): 34–41. [Russ.; Eng. summ.] [Not seen.]

1164. Simpson, B.B., J.L. Neff & D.S. Seigler. 1983. Floral biology and floral rewards of *Lysimachia* (Primulaceae). Amer. Midl. Nat. 110: 249–256.

1165. Singh, D. & S.K. Mahna. 1977. Development of

anther in normal and induced *Physalis ixocarpa* Brot. J. Indian Bot. Soc. 56: 281–286.

1166. Singh, I.S., J.C. Vallee & H.L. Dulieu. 1979. Pollen fertility and free amino acid content in anthers of *Petunia hybrida*. Pfl. Zucht 82(2): 116–132.

1167. Singh, V. 1965, 1966. Morphological and anatomical studies in Helobiae. II. Vascular anatomy of the flower of Potamogetonaceae. Bot. Gaz. 126: 137–144. III. Vascular anatomy of the node and flower of Najadaceae. Proc. Indian Acad. Sci. B, 61: 98–108 (1965). VI. Vascular anatomy of the flower of Alismataceae. Proc. Natn. Acad. Sci. India, B, 36: 329–344 (1966). [Not seen.]

1168. Singh, V. & D.K. Jain. 1975. Floral development of *Justicia gendarussa* (Acanthaceae). Bot. J. Linn. Soc. 70: 243–253.

1169. Singh, V. & R. Sattler. 1973. Nonspiral androecium and gynoecium of *Sagittaria latifolia*. Can. J. Bot. 51: 1093–1095.

1170. Singh, V. & R. Sattler. 1974. Floral development of *Butomus umbellatus*. Can. J. Bot. 52: 223–230.

1171. Singh, V. & R. Sattler. 1977. Development of the inflorescence and flower of *Sagittaria cuneata*. Can. J. Bot. 55: 1087–1105.

1172. Singh, V. & R. Sattler. 1977. Floral development of *Aponogeton natans* and *A. undulatus*. Can. J. Bot. 55: 1106–1120.

1173. Skarby, A. 1986. *Normapolles* anthers from the Upper Cretaceous of southern Sweden. Rev. Palaeobot. Palynol. 46: 235–256.

1174. Skipworth, J.P. 1970 [1971]. Development of floral vasculature in the Magnoliaceae. Phytomorphology 20: 228–236.

1175. Skipworth, J.P. & W.R. Philipson. 1966. The cortical vascular system and the interpretation of the *Magnolia* flower. Phytomorphology 16: 463–469.

1176. Skubatz, H., P.S. Williamson, E.L. Schneider & B.J.D. Meeuse. 1990. Cyanide-insensitive respiration in thermogenic flowers of *Victoria* and *Nelumbo*. J. Exp. Bot. 41: 1335–1339.

1177. Small, J. 1917. On the floral anatomy of some Compositae. I, II. J. Linn. Soc., Bot. 43: 517–525, 995–1010.

1178. Small, J. 1917. The origin and development of the Compositae. New Phytol. 16: 159–177; 17: 13–40, 69–94, 114–142, 200–230.

1179. Smets, E. 1986. Localization and systematic importance of the floral nectaries in the Magnoliatae (Dicotyledons). Bull. Jard. Bot. Natn. Belg. 56: 51–76.

1180. Smith, C.M. 1929. Development of *Dionaea muscipula*: 1. Flower and seed. Bot. Gaz. 87: 507–530 + Plates XX–XXIV.

1181. Smith, F.H. & E.C. Smith. 1942. Floral Anatomy of the Santalaceae and Some Related Forms. Oregon State Mon., Studies in Botany, No. 5, 93 pp.

1182. Smith, G.H. 1926, 1928. Vascular anatomy of Ranalian flowers. I. Ranunculaceae. Bot. Gaz. 82: 1–29 (1926). II. Ranunculaceae (continued), Menispermaceae, Calycanthaceae, Annonaceae. Ibid. 85: 152–177 (1928).

1183. Sobick, U. 1983. Blütenentwicklungsgeschichtliche Untersuchungen an Resedaceen unter besonderer Berücksichtigung von Androeceum und Gynoeceum. Bot. Jahrb. 104: 203–248.

1184. Soejarto, D.D. 1969. Aspects of reproduction in *Saurauia*. J. Arnold Arbor. 50: 180–196.

1185. Sokhi, J. & R.N. Kapil. 1984 [1985]. Morphogenetic changes induced by Trioza in flowers of *Terminalia arjuna*. I. Androecium. Phytomorphology 34: 117–128.

1186. Šopova, M. 1983. The nature and presence of crystals in the anthers of *Capsicum annuum* and *Solanum lycopersicum*. God. Zb. Med. Fak. Skopje 36: 59–63. [Serbo-Croat; Eng. summ.]

1187. Souvré, A. & L. Albertini. 1982. Étude des modifications de l'ultrastructure du tapis plasmodial du *Rhoeo discolor* Hance au cours du développement de l'anthère, en relation avec les données cytochimiques et autoradiographiques. Rev. Cytol. Biol. Vég., Bot. 5: 151–169.

1188. Sporne, K.R. 1958. Some aspects of floral vascular systems. Proc. Linn. Soc., Lond. 169: 75–84.

1189. Sporne, K.R. 1973. A note on the evolutionary status of tapetal types in dicotyledons. New Phytol. 72: 1173–1174.

1190. Sporne, K.R. 1977. Girdling vascular bundles in dicotyledonous flowers. Gard. Bull. Singapore 29: 165–173.

1191. Sprague, T.A. 1927. The morphology and taxonomic position of the Adoxaceae. J. Linn. Soc., Bot. 47: 471–487.

1192. Sridhar & D. Singh. 1986. Development of anther and male gametophyte in Cucurbitaceae. J. Indian Bot. Soc. 65: 487–493.

1193. Staedtler, G. 1923. Über Reduktionserscheinungen im Bau der Antherenwand von Angiospermen-Blüten. Flora (Jena) 116: 85–108 + Plates II–III.

1194. Stanley, R.G. & E.G. Kirby. 1973. Shedding of pollen and seeds. pp. 295–340 in: Shedding of Plant

Parts (edited by T.T. Kozlowski,). Academic Press, New York & London.

1195. Stebbins, G.L. 1973. Morphogenesis, vascularization and phylogeny in angiosperms. Breviora, No. 418: 1–19.

1196. Steer, M.W. 1977. Differentiation of the tapetum in *Avena*. 1. The cell surface. J. Cell. Sci. 25: 125–138. II. The endoplasmic reticulum and Golgi apparatus. Ibid. 28: 71–86.

1197. Steeves, T.A., M.W. Steeves & A.R. Olson. 1991. Flower development in *Amelanchier alnifolia* (Maloideae). Can. J. Bot. 69: 844–857.

1198. Steffen, K. & W. Landmann. 1958. Entwicklungsgeschichtliche und cytologische Untersuchungen am Balkentapetum von *Gentiana cruciata* L. und *Impatiens glandulifera* Royle. Planta 50: 423–460.

1199. Steinbrinck, C. 1909. Zu der Mitteilung von J.M. Schneider über den Öffnungsmechanismus der Tulpenanthere. Ber. Dtsch. Bot. Ges. 27: 2–10.

1200. Steinbrinck, C. 1909. Über den ersten Öffnungsvorgang bei Antheren. Ber. Dtsch. Bot. Ges. 27: 300–312.

1201. Steinbrinck, C. 1915. Über den Nachweis von Kohäsionsfalten in geschrumpelten Antherengeweben. Ber. Dtsch. Bot. Ges. 33: 66–72.

1202. Steinecke, H. 1993. Embryologische, morphologische und systematische Untersuchungen ausgewählter Annonaceae. Dissert. Bot. no. 205: 1–237. J. Cramer, Berlin.

1203. Stejskal-Streit, V. 1939, 1940. Vergleichende Untersuchungen gehemmter Staubblätter. I. Öst. Bot. Z. 88: 269–300 (1939). II. Ibid. 89: 1–56 (1940).

1204. Stelly, D.M. & R.G. Palmer. 1982. Variable development in anthers of partially male-sterile soybeans. J. Heredity 73: 101–108.

1205. Sterk, A.A. 1970. Reduction of the androecium in *Spergularia marina* (Caryophyllaceae). Acta Bot. Neerl. 19: 488–494.

1206. Sterling, C. 1977. Comparative morphology in the carpel of the Liliaceae: tepallary and staminal vascularization in the Wurmbaeoideae. Bot. J. Linn. Soc. 74: 63–69.

1207. Stevens, V.A.M. & B.G. Murray. 1981. Studies on heteromorphic self-incompatibility systems: the cytochemistry and ultrastructure of the tapetum of *Primula obconica*. J. Cell Sci. 50: 419–431.

1208. Stone, B.C. 1968 [1969]. Morphological studies in Pandanaceae. I. Staminodia and pistillodia of *Pandanus* and their hypothetical significance. Phytomorphology 18: 498–509.

1209. Stone, B.C. 1986. The genus *Pandanus* (Pandanaceae) on Christmas Island, Indian Ocean. Gard. Bull., Singapore 39: 193–202.

1210. Stone, B.C. & K.L. Huynh. 1982 [1983]. The identity, affinities, and staminate floral structure of *Pandanus pendulinus* Marteli (Pandanaceae). Gard. Bull., Singapore 35: 197–207.

1211. Stoudt, H.N. 1941. The floral morphology of some of the Capparidaceae. Amer. J. Bot. 28: 664–675.

1212. Straw, R.M. 1956. Adaptive morphology of the *Penstemon* flower. Phytomorphology 6: 112–119.

1213. Stringer, S. & J.G. Conran. 1991. Stamen and seed cuticle morphology in some *Arthropodium* and *Dichopogon* species (Anthericaceae). Aust. J. Bot. 39: 129–135.

1214. Stroebl, F. 1925. Die Obdiplostemonie in den Blüten. Bot. Arch. 9: 210–224.

1215. Struckmeyer, B.E., T. Heikkinen & K.C. Berger. 1961. Developmental anatomy of tassel and ear shoots of corn grown with different levels of boron. Bot. Gaz. 123: 111–116.

1216. Struckmeyer, B.E. & P. Simon. 1986. Anatomy of fertile and male-sterile carrot flowers from different genetic sources. J. Amer. Soc. Hort. Sci. 111: 965–968.

1217. Stützel, T. 1985. Die Bedeutung monothecatbisporangiater Antheren als systematisches Merkmal zur Gliederung der Eriocaulaceen. Bot. Jahrb. 105: 433–438. [Eng. summ.]

1218. Suarez-Cervera, M. & J.A. Seoane-Camba. 1986. Ontogénèse des grains de pollen de *Lavandula dentata* L. et évolution des cellules tapétales. Pollen Spores 28: 5–28.

1219. Subramanyam, K. 1949. On the nectary in the stamen of *Memecylon heyneanum* Benth. Curr. Sci. 18: 415.

1220. Subramanyam, K. & H.S. Narayana. 1972 [1975]. Some aspects of the floral morphology and embryology of *Flagellaria indica* Linn. pp. 211–217 in: Advances in Plant Morphology (edited by Y.S. Murty, B.M. Johri, H.Y.M. Ram & T.M. Varghese). Sarita Prakashan, Meerut.

1221. Sugaya, S. 1962. Transformation interchangeable between the members of the stamen and floral nectary of *Salix chaenomeloides* Kimura. Ecol. Rev. (Japan) 15: 231–234. [Eng.]

1222. Sun, F. & R.A. Stockey. 1992. A new species of *Palaeocarpinus* (Betulaceae) based on infructescences, fruits, and associated staminate inflorescences and leaves from the Paleocene of Alberta, Canada. Int. J. Pl. Sci. 153: 136–146.

1223. Sun, M. & F.R. Ganders. 1987. Microsporogenesis in male-sterile and hermaphroditic plants of nine gynodioecious taxa of Hawaiian *Bidens* (Asteraceae). Amer. J. Bot. 74: 209–217.

1224. Sundberg, M.D. 1982. Petal-stamen initiation in the genus *Cyclamen* (Primulaceae). Amer. J. Bot. 69: 1707–1709.

1225. Sundberg, S. 1985. Micromorphological characters as generic markers in the Astereae. Taxon 34: 31–37.

1226. Surányi, D. 1974. Correlation between gynoecium and androecium in Prunoideae species. Acta. Bot. Acad. Sci. Hung. 20: 379–388.

1227. Swamy, B.G.L. 1948. Vascular anatomy of orchid flowers. Bot. Mus. Leafl. Harvard Univ. 13: 61–95.

1228. Swamy, B.G.L. 1949. Further contributions to the morphology of the Degeneriaceae. J. Arnold Arbor. 30: 10–38.

1229. Swamy, B.G.L. 1953. The morphology and relationships of the Chloranthaceae. J. Arnold Arbor. 34: 375–408.

1230. Swamy, B.G.L. 1953. On the floral structure of *Scyphostegia*. Proc. Natn. Inst. Sci. India 19: 127–142.

1231. Swamy, B.G.L. & I.W. Bailey. 1949. The morphology and relationships of *Cercidiphyllum*. J. Arnold Arbor. 30: 187–210.

1232. Symon, D.E. 1979. Sex forms in *Solanum* (Solanaceae) and the role of pollen collecting insects. pp. 385–397 in: The Biology and Taxonomy of the Solanaceae (edited by J.G. Hawkes, R.N. Lester & A.D. Skelding). Linn. Soc. Symp. ser. No. 7. Academic Press, London.

1233. Szabó, M., S. Gulyás & J. Frank. 1984. Comparative anatomy of the androecium of male sterile and fertile sunflowers (*Helianthus*). Acta Bot. Hung. 30: 67–73.

1234. Takahashi, M. & K. Sohma. 1979. Pollen wall formation and tapetum in *Disporum smilacinum* A. Gray (Liliaceae). Sci. Rep. Tôhoku Univ., ser. IV, 37: 273–281.

1235. Tamura, M. 1972. Morphology and phyletic relationship of the Glaucidiaceae. Bot. Mag., Tokyo 85: 29–41.

1236. Tamura, M. 1985. Stamens and carpels in the primitive angiosperms. Proc. Jpn. Soc. Pl. Taxon. 5: 49–55. [Jpn.; Eng. summ.]

1237. Tamura, M.N. 1991. Biosystematic studies on the genus *Polygonatum* (Liliaceae). II. Morphology of staminal filaments of species indigenous to Japan and its adjacent regions. Acta Phytotax. Geobot. 42: 1–18. [Eng.]

1238. Tamura, M.N. 1993. Biosystematic studies on the genus *Polygonatum* (Liliaceae) III. Morphology of staminal filaments and karyology of eleven Eurasian species. Bot. Jahrb. 115: 1–26.

1239. Tatintseva, S.S. 1972. Development of stamen in *Sorghum caffrorum* Jakuschev. Bot. Zh. SSSR 57: 916–921. [Russ. only.]

1240. Taylor, D.W. 1988. Eocene floral evidence of Lauraceae: corroboration of the North American megafossil record. Amer. J. Bot. 75: 948–957.

1241. Teichman, I. von, P.J. Robbertse & C.F. Van der Merwe. 1982. Contributions to the floral morphology and embryology of *Pavetta gardeniifolia* A. Rich. 3. Microsporogenesis and pollen structure. S. Afr. J. Bot. 1: 28–30.

1242. Tepfer, S.S. 1953. Floral anatomy and ontogeny in *Aquilegia formosa* var. *truncata* and *Ranunculus repens*. Univ. Calif. Publ. Bot. 25: 513–648.

1243. Terabayashi, S. 1977, 1978, 1981, 1983. Studies in morphology and systematics of Berberidaceae. I. Floral anatomy of *Ranzania japonica*. Acta Phytotax. Geobot. 28: 45–57 [Eng.] (1977). II. Floral anatomy of *Mahonia japonica* (Thunb.) D. and *Berberis thunbergii* DC. Ibid. 29: 106–118 [Eng.] (1978). IV. Floral anatomy of *Plagiorhegma dubia* Maxim., *Jeffersonia diphylla* (L.) Pers. and *Achlys triphylla* (Smith) DC. ssp. *japonica* (Maxim.) Kitam. Bot. Mag., Tokyo 94: 141–157 (1981). V. Floral anatomy of *Caulophyllum* Michx., *Leontice* L., *Gymnospermium* Spach and *Bongardia* Mey. Mem. Fac. Sci. Kyoto Univ., ser. Biol. 8: 197–217. VI. Floral anatomy of *Diphylleia* Michx., *Podophyllum* L. and *Dysosma* Woodson. Acta Phytotax. Geobot. 34: 27–47 [Eng.]. VII. Floral anatomy of *Nandina domestica* Thunb. J. Phytogeog. Taxon. 31: 16–21 [Eng.] (1983).

1244. Terabayashi, S. 1985. The comparative floral anatomy and systematics of the Berberidaceae. I. Morphology. Mem. Fac. Sci. Kyoto Univ., ser. Biol. 10: 73–90.

1245. Terziiski, D. 1976, 1977. Electron microscopic studies on the generative organs of leguminous plants. I. Ultrastructure of *Vicia sativa* L. and *V. dumetorum* L. anther epidermis cells. Fitologiya, Sofiya 5: 21–33 [Bulg.; Eng. summ.] (1976). II. Cell ultrastructure of the anther fibrous layer in *Vicia sativa* L. and *V. dumetorum* L. Ibid. 6: 47–58 [Bulg.; Eng. summ.] (1977).

1246. Teschner, C. 1931. Beiträge zur Entwicklungsgeschichte und Anatomie der Blüte von *Erica* und ihrer systematischen Bedeutung. Diss., Berlin. [Not seen.]

1247. Testillano, P.S., P. Gonzalez–Melendi, B. Fadon, A. Sanchez–Pina, A. Olmedilla & M.C. Risueño. 1993.

Immunolocalization of nuclear antigens and ultra-structural cytochemistry on tapetal cells of *Scilla peruviana* and *Capsicum annuum*. Pl. Syst. Evol. Suppl. 7: 75–90.

1248. Theis, R. & G. Röbbelen. 1990. Anther and microspore development in different male sterile lines of oilseed rape (*Brassica napus* L.). Angew. Bot. 64: 419–434.

1249. Thiele, E.-M. 1988. Bau und Funktion des Antheren-Griffel-Komplexes der Compositen. Diss. Bot. 117. J. Cramer, Berlin & Stuttgart. 169 pp.

1250. Thompson, H.J. & W.R. Ernst. 1967. Floral biology and systematics of *Eucnide* (Loasaceae). J. Arnold Arbor. 48: 56–88.

1251. Thompson, J.M. 1924, 1925, 1929, 1931, 1933, 1934. Studies in advancing sterility. I. The Amherstieae. Hartley Bot. Lab. Publ. no. 1, University of Liverpool, 54 pp. (1924). II. The Cassieae. Ibid., No. 2. 44 pp. (1925). IV. The legume. Ibid., No. 6. 47 pp. (1929). V. The theory of the leguminous strobilus. Ibid., No. 7. 79 pp. (1931). VI. The theory of scitaminean flowering. Ibid., No. 11. pp. 1–111. (1933). VII. The state of flowering known as angiospermy. Ibid., No. 12. pp. 1–47. (1934).

1252. Thompson, J.M. 1927. A study in advancing gigantism with staminal sterility with special reference to the Lecythideae. Hartley Bot. Lab. Publ. no. 4, University of Liverpool. 44 pp.

1253. Thompson, J.M. 1936. On the floral morphology of *Elettaria cardamomum* Maton. Hartley Bot. Lab. Publ. no. 14, University of Liverpool. pp. 1–23.

1254. Thomson, B.F. 1942. The floral morphology of the Caryophyllaceae. Amer. J. Bot. 29: 333–349.

1255. Tiagi, Y.D. 1955. Studies in floral morphology. II. Vascular anatomy of the flower of certain species of the Cactaceae. J. Indian Bot. Soc. 34: 408–428.

1256. Tiagi, Y.D. 1961. Studies in floral morphology VII. The development of the anther and ovule of *Mammillaria carnea* Zucc. with a discussion on the systematic position of the family Cactaceae. Madhya Bharati (J. Univ. Saugar) II B, 10 (10), 12–24. [Not seen.]

1257. Tiagi, Y.D. 1963. Studies in floral morphology. VII & VIII. A further study of the vascular anatomy of the flower of certain species of the Cactaceae. J. Indian Bot. Soc. 42: 545–558, 559–573.

1258. Tiagi, Y.D. 1963. Vascular anatomy of the flower of certain species of the Calycanthaceae. Proc. Indian Acad. Sci. 58B: 224–234.

1259. Tiagi, Y.D. & S. Kshetrapal. 1972 [1975]. Studies on the floral anatomy, evolution of the gynoecium and

relationships of the family Loganiaceae. pp. 408–416 in Advances in: Plant Morphology (edited by Y.S. Murty, B.M. Johri, H.Y.M. Ram & T.M. Varghese). Sarita Prakashan, Meerut.

1260. Tillson, A.H. 1940. The floral anatomy of the Kalanchoideae. Amer. J. Bot. 27: 595–600.

1261. Tillson, A.H. & R. Bamford. 1938. The floral anatomy of the Aurantioideae. Amer. J. Bot. 25: 780–793.

1262. Tischler, G. 1915. Die Periplasmodiumbildung in den Antheren der Commelinaceen und Ausblicke auf das Verhalten der Tapetenzellen bei den übrigen Monokotylen. Jahrb. Wiss. Bot. 55: 53–90 + Plate I.

1263. Tiwari, S.C. & B.E.S. Gunning. 1986. Cytoskeleton, cell surface and the development of invasive plasmodial tapetum in *Tradescantia virginiana* L. Protoplasma 133: 89–99.

1264. Tiwari, S.C. & B.E.S. Gunning. 1986. An ultrastructural, cytochemical and immunofluorescence study of postmeiotic development of plasmodial tapetum in *Tradescantia virginiana* L. and its relevance to the pathway of sporopollenin secretion. Protoplasma 133: 100–114.

1265. Tiwari, S.C. & B.E.S. Gunning. 1986. Colchicine inhibits plasmodium formation and disrupts pathways of sporopollenin secretion in the anther tapetum of *Tradescantia virginiana* L. Protoplasma 133: 115–128.

1266. Tiwari, S.C. & B.E.S. Gunning. 1986. Development and cell surface of a non-syncytial invasive tapetum in *Canna*: ultrastructural, freeze-substitution, cytochemical and immunofluorescence study. Protoplasma 134: 1–16.

1267. Tiwari, S.C. & B.E.S. Gunning. 1986. Development of tapetum and microspores in *Canna* L.: an example of an invasive but non-syncytial tapetum. Ann. Bot. 57: 557–563.

1268. Tobe, H. 1980. Morphological studies on the genus *Clematis* Linn. VI. Vascular anatomy of the androecial and gynoecial regions of the floral receptacle. Bot. Mag., Tokyo 93: 125–133.

1269. Tobe, H. & B.E. Hammel. 1993. Floral morphology, embryology, and seed anatomy of *Ruptiliocarpon caracolito* (Lepidobotryaceae). Novon 3: 423–428.

1270. Tobe, H. & P.H. Raven. 1983. An embryological analysis of Myrtales: its definition and characteristics. Ann. Missouri Bot. Gard. 70: 71–94.

1271. Tobe, H. & P.H. Raven. 1984. The embryology and relationships of Oliniaceae. Pl. Syst. Evol. 146: 105–116.

1272. Tobe, H. & P.H. Raven. 1986. Evolution of polysporangiate anthers in Onagraceae. Amer. J. Bot. 73: 475–488.

1273. Tohda, H. 1965. Morphological investigations on the staminodia of the female plants in *Moehringia lateriflora* Fenz. 1. On the development of tapetal cells and pollen mother cells. Sci. Rep. Tôhoku Univ., IV, Biol. 31: 83–92. [Eng.]

1274. Tokumasu, S. 1976. A comparative study in anther development on male-fertile and male-sterile plants of *Pelargonium crispum*. L'Her. ex Ait. Euphytica 25: 151–159.

1275. Tong, K. 1930. Studien über die Familie der Hamamelidaceae, mit besonderer Berücksichtigung der Systematik und Entwicklungsgeschichte von *Corylopsis*. Bull. Dept. Biol. No. 2, Sun Yatsen Univ. 72 + 14 pp.

1276. Tongiorgi, E. 1935. Il tipo delle dicotiledoni et il tipo delle monocotiledoni nella formazione del vero periplasmodio nella famiglia delle Lauraceae. Nuovo G. Bot. Ital. 42: 387–397.

1277. Trapp, A. 1954. Staubblattbildung und Bestäubungsmechanismus von *Incarvillea variabilis* Batalin. Öst. Bot. Z. 101: 208–219.

1278. Trapp, A. 1956. Entwicklungsgeschichtliche Untersuchungen über die Antherengestaltung sympetaler Blüten. Beitr. Biol. Pfl. 32: 279–312.

1279. Trapp, A. 1956. Zur Morphologie und Entwicklungsgeschichte der Staubblätter sympetaler Blüten. Bot. Studien 5: 1–93.

1280. Troll, W. 1922. Über Staubblatt- und Griffelbewegungen und ihre teleologische Bedeutung. Flora (NF) 15: 191–250 + Plates IV–VI.

1281. Troll, W. 1927. Zur Frage nach der Herkunft der Blumenblätter. Flora 122: 57–75.

1282. Troll, W. 1928. Organisation und Gestalt im Bereich der Blüte. J. Springer, Berlin. 413 pp.

1283. Troll, W. 1928. Über Antherenbau, Pollen und Pollination von *Galanthus* L. Flora 123: 321–343.

1284. Troll, W. 1929. *Roscoea purpurea* Sm., eine Zingiberacee mit Hebelmechanismus in den Blüten. Mit Bemerkungen über die Entfaltungsbewegungen der fertilen Staubblätter von *Salvia*. Planta 7: 1–28.

1285. Trull, M.C., B.L. Holaway, W.E. Friedman & R.L. Malmberg. 1991. Developmentally regulated antigen associated with calcium crystals in tobacco anthers. Planta 186: 13–16.

1286. Trull, M.C., B.L. Holaway & R.L. Malmberg. 1992. Development of stigmatoid anthers in a tobacco mutant: implications for regulation of stigma differentiation. Can. J. Bot. 70: 2339–2346.

1287. Tschirch, A. 1904. Sind die Antheren der Kompositen verwachsen oder verklebt? Flora 93: 51–55 + Plate II.

1288. Tucker, S.C. 1972 [1975]. The role of ontogenetic evidence in floral morphology. pp. 359–369 in: Advances in Plant Morphology (edited by Y.S. Murty, B.M. Johri, H.Y.M. Ram & T.M. Varghese). Sarita Prakashan, Meerut.

1289. Tucker, S.C. 1975. Floral development in *Saururus cernuus* (Saururaceae): 1. Floral initiation and stamen development. Amer. J. Bot. 62: 993–1007.

1290. Tucker, S.C. 1976. Intrusive growth of secretory oil cells in *Saururus cernuus*. Bot. Gaz. 137: 341–347.

1291. Tucker, S.C. 1980, 1982. Inflorescence and flower development in the Piperaceae. I. *Peperomia*. Amer. J. Bot. 67: 686–702 (1980). III. Floral ontogeny of *Piper*. Ibid. 69: 1389–1404 (1982).

1292. Tucker, S.C. 1981. Inflorescence and floral development in *Houttuynia cordata* (Saururaceae). Amer. J. Bot. 68: 1017–1032.

1293. Tucker, S.C. 1984. Unidirectional organ initiation in leguminous flowers. Amer. J. Bot. 71: 1139–1148.

1294. Tucker, S.C. 1984. Origin of symmetry in flowers. pp. 351–395 in: Contemporary Problems in Plant Anatomy (edited by R.A. White & W.C. Dickison). Academic Press, Orlando.

1295. Tucker, S.C. 1987. Floral initiation and development in legumes. pp. 183–239 in: Advances in Legume Systematics, Part 3 (edited by C.H. Stirton). Royal Botanic Gardens, Kew.

1296. Tucker, S.C. 1988. Loss versus suppression of floral organs. pp. 69–82 in: Aspects of Floral Development (edited by P. Leins, S.C. Tucker & P.K. Endress). J. Cramer, Berlin.

1297. Tucker, S.C. 1989. Evolutionary implications of floral ontogeny in legumes. pp. 59–75 in: Advances in Legume Biology (edited by C.H. Stirton & J.L. Zarucchi). Mon. Syst. Bot. Missouri Bot. Gard., No. 29.

1298. Tuyama, T. 1980. On the genera *Camelliastrum* and *Theopsis* (Theaceae). J. Jpn. Bot. 55: 215–222. [Jpn.; full Eng. summ.]

1299. Ubisch, G. von. 1927. Zur Entwicklungsgeschichte der Antheren. Planta 3: 490–495.

1300. Uchino, A. 1979. Combination of metamorphosed flowers among different shoots on a single corm of *Trillium smallii*. Kumamoto J. Sci. Biol. 14 (2): 17–26. [Not seen.]

1301. Ueda, K. 1986. Vascular systems in the Magnoliaceae. Bot. Mag., Tokyo 99: 333–349.

1302. Uexküll-Gyllenband, M. 1901. Phylogenie der Blüt-
enformen und der Geschlechtverteilung bei den
Composen. Inaug. Diss. Univ. Zürich. E. Nägele,
Stuttgart. 80 pp. + 2 plates.

1303. Uhl, N.W. 1976. Developmental studies in *Ptycho-
sperma* (Palmae). II. The staminate and pistillate
flowers. Amer. J. Bot. 63: 97–109.

1304. Uhl, N.W. 1988. Floral organogenesis in palms. pp.
25–44 in: Aspects of Floral Development (edited by
P., Leins, S.C. Tucker & P.K. Endress). J. Cramer,
Berlin.

1305. Uhl, N.W. & J. Dransfield. 1984. Development of
the inflorescence, androecium, and gynoecium with
reference to palms. pp. 397–449 in: Contemporary
Problems in Plant Anatomy (edited by R.A. White &
W.C. Dickison). Academic Press, Orlando.

1306. Uhl, N.W. & H.E. Moore. 1977. Centrifugal stamen
initiation in phytelephantoid palms. Amer. J. Bot. 64:
1152–1161.

1307. Uhl, N.W. & H.E. Moore. 1980. Androecial development in six polyandrous genera representing five
major groups of palms. Ann. Bot. 45: 57–75.

1308. Untawale, A.G. & R.K. Bhasin. 1973. On endothecial
thickenings in some monocotyledonous families.
Curr. Sci. 42: 398–400.

1309. Ursin, V.M., J. Yamaguchi & S. McCormick. 1989.
Gametophytic and sporophytic expression of anther-
specific genes in developing tomato anthers. Plant
Cell 1: 727–736.

1310. Vallade, J., D. Maizonnier & A. Cornu. 1987. La
morphogenèse florale chez le pétunia. I. Analyse d'un
mutant à corolle staminée. Can. J. Bot. 65: 761–764.

1311. Van der Pijl, L. 1952. The stamens of *Ricinus*. Phytomorphology 2: 130–132.

1312. Van der Werff, H. & P.K. Endress. 1991. *Gamanthera*
(Lauraceae), a new genus from Costa Rica. Ann. Missouri Bot. Gard. 78: 401–408.

1313. Van Heel, W.A. 1962. Miscellaneous teratological
notes together with some general considerations. I.
Hyacinthus orientalis L.; II. *Hibiscus rosa-sinensis* L.;
III. *Pisum sativum* L. Proc. K. Ned. Akad. Wet., C,
65: 392–406.

1314. Van Heel, W.A. 1966. Morphology of the androecium
in Malvales. Blumea 13: 177–394.

1315. Van Heel, W.A. 1969. The synangial nature of pollen
sacs on the strength of 'congenital fusion' and 'conservatism of the vascular bundle system', with special
reference to some Malvales, I, II. Proc. K. Ned. Akad.
Wet., C, 72: 172–206.

1316. Van Heel, W.A. 1977. The pattern of vascular

1317. Van Heel, W.A. 1987. Androecium development in
Actinidia chinensis and *A. melanandra* (Actinidiaceae).
Bot. Jahrb. 109: 17–23.

1318. Van Heel, W.A. 1987. Note on the morphology of
the male inflorescences in *Cercidiphyllum*
(Cercidiphyllaceae). Blumea 32: 303–309.

1319. Van Heusden, E.C.H. 1992. Flowers of Annonaceae:
morphology, classification, and evolution. Blumea,
suppl. 7: 1–218.

1320. Van Rensburg, H.J., P.J. Robbertse & J.G.C. Small.
1985. Morphology of the anther, microsporogenesis
and pollen structure of *Momordica balsamina*. S. Afr.
J. Bot. 51: 125–132.

1321. Van Tieghem, P. 1895. Observations sur la structure
et la déhiscence des anthères des Loranthacées, suiv-
ies de remarques sur la structure et la déhiscence de
l'anthère en général. Bull. Soc. Bot. Fr. 42: 363–368.

1322. Van Tieghem, P. 1903. Structure de l'étamine chez
les Scrofulariacées. Ann. Sci. Nat., Bot., sér. 8, 17:
363–371.

1323. Van Tieghem, P. 1907. Sur les anthères symétrique-
ment hétérogènes. Ann. Sci. Nat., Bot., sér. 9, 5: 364–
370.

1324. Vanvinckenroye, P., E. Cresens, L.-P. Ronse Decraene & E. Smets. 1993. A comparative floral developmental study in *Pisonia*, *Bougainvillea* and *Mirabilis*
(Nyctaginaceae) with special emphasis on the gynoecium and floral nectaries. Bull. Jard. Bot. Nat. Belg.
62: 69–96.

1325. Van Went, J.L. 1981. Some cytological and ultrastructural aspects of male sterility in *Impatiens*. Acta
Soc. Bot. Pol. 50: 249–252.

1326. Varghese, T.M. 1972 [1973]. Structure of the anther
in some Scrophulariaceae. J. Palynol. 8: 84–88.

1327. Vasil, I.K. 1967. Physiology and cytology of anther
development. Biol. Rev. 42: 327–373.

1328. Vaughan, J.G. 1955. The morphology and growth of
the vegetative and reproductive apices of *Arabidopsis
thaliana* (L.) Heynh., *Capsella bursa-pastoris* (L.)
Medic. and *Anagallis arvensis* L. J. Linn. Soc. Lond.
(Bot.) 55: 279–301.

1329. Vautier, S. 1949. La vascularisation florale chez les
Polygonacées. Candollea 12: 219–343.

1330. Vázquez-Santana, S., J. Márquez-Guzmán, M.
Engleman & A. Martinez-Mena. 1992 [1993]. Devel-
opment of ovule and anther of *Cuscuta tinctoria*
(Cuscutaceae). Phytomorphology 42: 195–202.

1331. Veldkamp, J.F., N.A.P. Franken, M.C. Roos & M.P.

bundles in the stamens of *Nymphaea lotus* L. and its
bearing on stamen morphology. Blumea 23: 345–348.

Nayar. 1978. A revision of *Diplectria* (Melastomataceae). Blumea 24: 405–430.

1332. Velenovsky, J. 1910. Vergleichende Morphologie der Pflanzen. Vol. III. F. Řivnáč, Prague.

1333. Venkatesh, C.S. 1952. The anther and pollen grains of *Zannichellia palustris*. Curr. Sci. 21: 225–226. [Not seen.]

1334. Venkatesh, C.S. 1955. The structure and dehiscence of the anther in *Memecylon* and *Mouriria*. Phytomorphology 5: 435–440.

1335. Venkatesh, C.S. 1955 [1956]. Structure and dehiscence of the anther in *Exacum pedunculatum* L. Lloydia 18: 143–148.

1336. Venkatesh, C.S. 1956. Structure and dehiscence of the anther in *Najas*. Bot. Notiser 109: 75–82.

1337. Venkatesh, C.S. 1956. The curious anther of *Bixa* – its structure and dehiscence. Amer. Midl. Nat. 55: 473–476.

1338. Venkatesh, C.S. 1956. The special mode of dehiscence of anthers of *Polygala* and its significance in autogamy. Bull. Torrey Bot. Cl. 83: 19–26.

1339. Venkatesh, C.S. 1956, 1957 [1958]. The form, structure and special modes of dehiscence in anthers of *Cassia*. I. Subgenus *Fistula*. Phytomorphology 6: 168–176. II. Subgenus *Lasiorhegma*. Ibid.: 272–277 (1956). III. Subgenus *Senna*. Ibid. 7: 253–273 (1957 [1958]).

1340. Venkatesh, C.S. & K. Subramanyam. 1984. Structure and poricidal dehiscence of the anther in *Sonerila wallichii* Benn. (Melastomataceae). Indian Bot. Reporter, Prof. K.B. Deshpande commem. vol.: 43–6. [Current studies in botany, ed. R.M. Pai.]

1341. Venkateswarlu, J. & Rao, P.S.P. 1972. Embryological studies in some Combretaceae. Bot. Notiser 125: 161–179.

1342. Vennigerholz, F. & B. Walles. 1987. Cytochemical studies of pectin digestion in epidermis with specific cell separation. Protoplasma 140: 110–117.

1343. Venturelli, M. 1983. Estudos embriológicos em Loranthaceae: gênero *Tripodanthus*. Kurtziana 16: 71–90. [Eng. summ.]

1344. Verbeke, J.A. 1992. Fusion events during floral morphogenesis. Ann. Rev. Pl. Physiol. Pl. Mol. Biol. 43: 583–598.

1345. Vermeulen, P. 1953. The vanished stamens. Amer. Orchid Soc. Bull. 22: 650–655.

1346. Vermeulen, P. 1955. The rostellum of the Ophrydeae. Amer. Orchid Soc. Bull. 24: 239–245.

1347. Vermeulen, P. 1959. The different structure of the rostellum in Ophrydeae and Neottieae. Acta Bot. Neerl. 8: 338–355.

1348. Vijayaraghavan, M.R. & U. Dhar. 1975. *Kadsura heteroclita* – microsporangium and pollen. J. Arnold Arbor. 56: 176–182.

1349. Vijayaraghavan, M.R., S. Kumra & V. Sujata. 1987. Ontogeny and histochemistry of anther wall in *Saxifraga ciliata* Lindl. Proc. Indian Acad. Sci., Pl. Sci. 97: 301–307.

1350. Vijayaraghavan, M.R. & K.N. Marwah. 1969. Studies in the family Ranunculaceae – microsporangium, microsporogenesis and Ubisch granules in *Nigella damascena* Phyton (Horn, Austria) 13: 203–209.

1351. Vijayaraghavan, M.R. & S. Ratnaparkhi. 1973. Dual origin and dimorphism of the anther tapetum in *Alectra thomsoni* Hook. Ann. Bot. 37: 355–359.

1352. Vijayaraghavan, M.R. & S. Ratnaparkhi. 1979. Histological dynamics of anther tapetum in *Heuchera micrantha*. Proc. Indian Acad. Sci., B, 88: 309–316. [Not seen.]

1353. Villari, R. 1990. Embryological observations on *Tipuana tipu* (Benth.) O. Kuntze (Dalbergieae, Papilionaceae). G. Bot. Ital. 124: 293–300.

1354. Vincent, P.L.D. & F.M. Getliffe. 1988. The endothecium in *Senecio* (Asteraceae). Bot. J. Linn. Soc. 97: 63–71.

1355. Vink, W. 1970, 1977. The Winteraceae of the Old World. I. *Pseudowintera* and *Drimys* – morphology and taxonomy. Blumea 18: 225–354 (1970). II. *Zygogynum* – morphology and taxonomy. Ibid. 23: 219–250 (1977).

1356. Vishenskaya, T.D. 1980. Polymerous androecium and its development in the flower of *Thea sinensis* L. (Theaceae). Bot. Zh. SSSR 65: 39–50 + 2 plates. [Russ.; Eng. summ.]

1357. Vishenskaya, T.D. 1980. The development of the polymerous androecium in *Stuartia pseudocamellia* (Theaceae). Bot. Zh. SSSR 65: 948–957 + 1 plate. [Russ.; Eng. summ.]

1358. Vishnjakova, M.A. [Vishnyakova, M.A.] & V.K. Lebsky. 1986. Morphology of anther tapetum sporopollenin walls in some angiosperms. Bot. Zh. SSSR 71: 754–759 + 2 plates. [Russ. only.]

1359. Vogel, S. 1978. Evolutionary shifts from reward to deception in pollen flowers. pp. 89–96 in: The Pollination of Flowers by Insects (edited by A.J. Richards). Linn. Soc. Symp. ser. No. 6. Academic Press, London.

1360. Vogel, S. 1981. Die Klebstoffhaare an den Antheren von *Cyclanthera pedata* (Cucurbitaceae). Pl. Syst. Evol. 137: 291–316.

1361. Vogel, S. 1986. Ölblumen und ölsammelnde Bienen.

2. *Lysimachia* und *Macropis*. Trop. Subtrop. Pflwelt. 54: 1–168. [Eng. summ. 160–162.]

1362. Vogel, S. & A. Cocucci. 1988. Pollen threads in *Impatiens*: their nature and function. Beitr. Biol. Pfl. 63: 271–287.

1363. Volgin, S.A. 1982. Morphological interpretation of the cactus flower: 2. Receptacle, perianth, androecium. Nauch. Dokl. Vyssh. Shk., Biol. Nauki (5), 75–80. [Russ. only].

1364. Volgin, S.A. 1988. Vergleichende Morphologie und Gefässbündelanatomie der Blüte bei den Rivinoideae (Phytolaccaceae). Flora 181: 325–337.

1365. Volkens, G. 1899. Ueber die Bestäubung einiger Loranthaceen und Proteaceen. Ein Beitrag zur Ornithophilie. pp. 251–270 + Plate X in: Botanische Untersuchungen: S. Schwendener zum 10 Februar 1899 dargebracht. Borntraeger, Berlin.

1366. Wagner, B.L. 1983. Genesis of the vacuolar apparatus responsible for druse formation in *Capsicum annuum* L. (Solanaceae) anthers. Scanning Electron Microsc. 2: 905–912.

1367. Walter, H. 1906. Die Diagramme der Phytolaccaceen. Bot. Jahrb. 37, Beibl. 85: 1–57.

1368. Wang, C.-S., L.L. Walling, K.J. Eckard & E.M. Lord. 1992. Patterns of protein accumulation in developing anthers of *Lilium longiflorum* correlate with histological events. Amer. J. Bot. 79: 118–127.

1369. Wang, Y., L. Tu, Y. Zhu & L. Guan. 1992. The ultrastructural aspect of the changes of the tapetum in *Caragana stenophylla* Pojark. Acta Sci. Nat. Univ. Intramongolicae 23: 245–248 + Plates 3–5. [Chin.; Eng. summ.]

1370. Wang, Z. & Y. Zhang. 1991. A discussion on the formation and evolution of flower type of tree peony and herb peony by observing flower bud differentiation of herb peony. Acta Hort. Sin. 18 (2): 163–168. [Chin.] [Not seen.]

1371. Warmke, H.E. & S.-L.J. Lee. 1977. Mitochondrial degeneration in Texas cytoplasmic male-sterile corn anthers. J. Hered. 68: 213–222.

1372. Wassmer, A. 1955. Vergleichend-morphologische Untersuchungen an der Blüten der Crassulaceen. Inst. Allg. Bot., Univ. Zürich, ser. A, No. 7: 1–112. [Not seen.]

1373. Webb, C.J. & Lloyd, D.G. 1986. The avoidance of interference between the presentation of pollen and stigmas in angiosperms. II. Herkogamy. N.Z. J. Bot. 24: 163–178.

1374. Weber, A. 1976. Beiträge zur Morphologie und Systematik der Klugieae und Loxonieae (Gesneriaceae).

II. Morphologie, Anatomie und Ontogenese der Blüte von *Monophyllaea* R. Br. Bot. Jahrb. 95: 435–454.

1375. Weber, A. 1976. Wuchsform, Infloreszenz- und Blütenmorphologie von *Epithema* (Gesneriaceae). Pl. Syst. Evol. 126: 287–322.

1376. Weber, G.F.T. 1928. Vergleichend-morphologische Untersuchungen über die Oleaceenblüte. Planta 6: 591–658.

1377. Weberling, F. 1981. Morphologie der Blüten und der Blütenstände. Verlag Eugen Ulmer, Stuttgart. 391 pp.

1378. Weberling, F. 1989. Morphology of Flowers and Inflorescences. [Transl. by R.J. Pankhurst.] Cambridge University Press, Cambridge. 405 pp.

1379. Weberling, F. & M. Hildenbrand. 1982 [1983]. Zur Tapetumentwicklung bei *Triosteum* L., *Leycesteria* Wall. und *Kolkwitzia* Graebn. (Caprifoliaceae). Beitr. Biol. Pfl. 57: 481–486. [Eng. summ.].

1380. Weberling, F. & M. Hildenbrand. 1986. Weitere Untersuchungen zur Tapetumentwicklung der Caprifoliaceae. Beitr. Biol. Pfl. 61: 3–20.

1381. Webster, B.D., S.P. Lynch & C.L. Tucker. 1979. A morphological study of the development of reproductive structures of *Phaseolus lunatus* L. J. Amer. Soc. Hort. Sci. 104: 240–243.

1382. Wee, Y.C. & A.N. Rao. 1980. Anthesis and variations in floral structure of *Parkia javanica*. Malay. Forester 43: 493–499.

1383. Wernham, H.F. 1911, 1912. Floral evolution; with particular reference to the sympetalous dicotyledons. New Phytol. 10: 73–83, 109–120, 145–159, 217–226, 293–305 (1911); Ibid. 11: 145–166, 217–255, 290–305, 373–397 (1912).

1384. Westwood, M.N. & J.S. Challice. 1978. Morphology and surface topography of pollen and anthers of *Pyrus* species. J. Amer. Soc. Hort. Sci. 103: 28–37.

1385. Wetter, M.A. 1983. Micromorphological characters and generic delimitation of some New World Senecioneae (Asteraceae). Brittonia 35: 1–22.

1386. Wettstein, R. von. 1893. Ueber das Androeceum von *Philadelphus*. Ber. Dtsch. Bot. Ges. 11: 480–484 + Plate XXIV.

1387. Wetzstein, H.Y. & D. Sparks. 1984. The morphology of staminate flower differentiation in pecan. J. Amer. Soc. Hort. Sci. 109: 245–252.

1388. Whallon, J.H., W. Tai & G.R. Hooper. 1982. Cytoplasmic plate complexes: unusual subcellular structures in tapetal cells of *Hordeum*. Micron 13: 479–480.

1389. Whatley, J.M. 1982. Fine structure of the endothecium and developing xylem in *Phaseolus vulgaris*. New Phytol. 91: 561–570.

1390. Wilde, W.J.J.O. de. 1974. The genera of tribe Passifloreae (Passifloraceae), with special reference to flower morphology. Blumea 22: 37–50.

1391. Wilder, G.J. 1987. Contributions to taxonomy and morphology of *Schultesiophytum chorianthum* Harl. and *Dicranopygium mirabile* Harl. (Cyclanthaceae). Opera Bot. 92: 277–291.

1392. Wilkinson, A.M. 1944. Floral anatomy of some species of *Cornus*. Bull. Torrey Bot. Cl. 71: 276–301.

1393. Wilkinson, A.M. 1948. Floral anatomy and morphology of some species of the tribe Lonicereae of the Caprifoliaceae. Amer. J. Bot. 35: 261–271.

1394. Wilkinson, A.M. 1948. Floral anatomy and morphology of some species of the tribes Linnaeeae and Sambuceae of the Caprifoliaceae. Amer. J. Bot. 35: 365–371.

1395. Wilkinson, A.M. 1949. Floral anatomy and morphology of *Triosteum* and of the Caprifoliaceae in general. Amer. J. Bot. 36: 481–489.

1396. Willemse, M.T.M. 1993. Calcium and calmodulin distribution in the tapetum of *Gasteria verrucosa* during anther development. Pl. Syst. Evol. Suppl. 7: 107–116.

1397. Williams, G. & Adam, P. 1993. Ballistic pollen release in Australian members of the Moraceae. Biotropica 25: 478–480.

1398. Wilson, C.L. 1937. The phylogeny of the stamen. Amer. J. Bot. 24: 686–699.

1399. Wilson, C.L. 1942. The telome theory and the origin of the stamen. Amer. J. Bot. 29: 759–764.

1400. Wilson, C.L. 1950. Vasculation of the stamen in the Melastomaceae, with some phyletic implications. Amer. J. Bot. 37: 431–444.

1401. Wilson, C.L. 1965; 1973 [1974]. The floral anatomy of the Dilleniaceae. I. *Hibbertia* Andr. Phytomorphology 15: 248–274 (1965). II. Genera other than *Hibbertia*. Ibid. 23: 25–42 (1973 [1974]).

1402. Wilson, C.L. 1976. Floral anatomy of *Idiospermum australiense* (Idiospermaceae). Amer. J. Bot. 63: 987–996.

1403. Wilson, C.L. 1982. Vestigial structures and the flower. Amer. J. Bot. 69: 1356–1365.

1404. Wilson, C.L. & Just, T. 1939. The morphology of the flower. Bot. Rev. 5: 97–131.

1405. Wilson, T.K. 1966. The comparative morphology of the Canellaceae. IV. Floral morphology and conclusions. Amer. J. Bot. 53: 336–343.

1406. Wilson, T.K. & L.M. Maculans. 1967. The mor-

phology of the Myristicaceae. I. Flowers of *Myristica fragrans* and *M. malabarica*. Amer. J. Bot. 54: 214–220.

1407. Wojciechowska, W. 1981. 'Carpelloid stamens' in *Lotus* sp. Acta Soc. Bot. Pol. 50: 405–408.

1408. Wolff, G.P. 1924. Zur vergleichenden Entwicklungsgeschichte und biologischen Bedeutung der Blütennektarien. Bot. Archiv 8: 305–344.

1409. Wolter, M., C. Seuffert & R. Schill. 1988. The ontogeny of pollinia and elastoviscin in the anther of *Doritis pulcherrima* (Orchidaceae). Nord. J. Bot. 8: 77–88.

1410. Woodson, R.E. 1935. The floral anatomy and probable affinities of the genus *Grisebachiella*. Bull. Torrey Bot. Club 62: 471–478.

1411. Woon, C. & H. Keng. 1979. Observations on stamens of the Dipterocarpaceae. Gard. Bull., Singapore 32: 1–55.

1412. Worsdell, W.C. 1915–16. The Principles of Plant-Teratology, 2 vols. Ray Society, London.

1413. Wóycicki, Z. 1924. Recherches sur la déhiscence des anthères et le rôle du stomium. Rev. Gén. Bot. 36: 196–212, 253–268.

1414. Wóycicki, Z. 1932, 1933, 1935. Quelques détails du développement des anthères et du pollen chez certains réprésentants du genre *Gentiana*. I. *Gentiana asclepiadea* L. Acta Soc. Bot. Pol. 9: 7–30 (1932). II. *Gentiana fetisowi* Rgl. et Winkler. Ibid. 10: 1–24 (1933). III. *Gentiana lutea* L. Ibid. 12: 207–226 (1935). [Pol.; French or Germ. summ.]

1415. Wunderlich, R. 1950. Die Agavaceae Hutchinsons im Lichte ihrer Embryologie, ihres Gynözeum-, Staubblatt- und Blattbaues. Öst. Bot. Z. 97: 437–502.

1416. Wunderlich, R. 1954. Über das Antherentapetum mit besonderer Berücksichtigung seiner Kernzahl. Öst. Bot. Z. 101: 1–63.

1417. Wunderlich, R. 1971[1972]. Die systematische Stellung von *Theligonum*. Öst. Bot. Z. 119: 329–394.

1418. Wyatt, J.E. 1984. An indehiscent anther mutant in the common bean. J. Amer. Soc. Hort. Sci. 109: 484–487.

1419. Xi, X.-Y. 1991. A comparative study of anther and pollen development in male-fertile and male-sterile green onion (*Allium fistulosum* L.). Acta Bot. Sin. 33: 770–775 + 3 plates. [Chin.; Eng. summ.]

1420. Xu, H.-Q., Q.-Y. Huang, K. Shen & Z.-G. Shen. 1993. Anatomical studies on the effects of boron on the development of stamen and pistil of Rape (*Brassica napus*). Acta Bot. Sin. 35: 453–457 + 2 plates. [Chin.; Eng. summ.]

1421. Yakobson, L.Ya. 1968. Development of the anthers in *Rhododendron*. Izv. Akad. Nauk Latv. SSR (12): 101–109. [Eng. summ.] [Not seen.]

1422. Yamazaki, T. 1991. Morphological structure of the anther of *Tsusiophyllum tanakae* Maxim. J. Jpn. Bot. 66: 35–38. [Jpn.; Eng. summ.]

1423. Yates, I.E. & D. Sparks. 1992. External morphological characteristics for histogenesis in pecan anthers. J. Amer. Soc. Hort. Sci. 117: 181–189.

1424. Yeo, P.F. 1993. Secondary pollen presentation. Form, function and evolution. Pl. Syst. Evol., Suppl. 6: 268 pp.

1425. Yonemori, K., A. Sugiura, K. Tanaka & K. Kameda. 1993. Floral ontogeny and sex determination in monoecious-type Persimmons. J. Amer. Soc. Hort. Sci. 118: 293–297.

1426. Yu, F. & T. Fu. 1990. Cytomorphological research on anther development of several male-sterile lines in *Brassica napus* L. J. Wuhan Bot. Res. 8: 209–216 + 2 plates. [Chin.; Eng. summ.]

1427. Yurukova-Grancharova, P.D. & T.D. Daskalova. 1992. A cytoembryological study of *Salvia officinalis* L. (Lamiaceae Lindl.). I. Histological structure of the anthers and microsporogenesis. Fitologija (Sofia) 43: 36–43. [Eng.]

1428. Zandonella, P. 1964. Structure des faisceaux staminaux chez les Centrospermales. C.R. Acad. Sci., Paris 259: 3335–3336.

1429. Zandonella, P. 1967. Les nectaires des Alsinoideae: *Stellaria* et *Cerastium* sensu lato. C.R. Acad. Sci. Paris, D, 264: 2466–2469.

1430. Zenkteler, M. 1962. Microsporogenesis and tapetal development in normal and male sterile carrots (*Daucus carota*). Amer. J. Bot. 49: 341–348.

1431. Zhang, M., Y. Wang & H. Ding. 1991. Anther development of *Angelica sinensis*. Acta Bot. Bor.-Occ. Sin. 11: 200–205 + 2 plates. [Chin.; Eng. summ.]

1432. Zhang, S.-Z. & X. Zhu. 1990. The development of the organoid of endothecium of *Triticale* anther. Acta Bot. Bor.-Occ. Sin. 10(3): 166–169 + 1 plate. [Chin.; Eng. summ.]

1433. Ziegler, A. 1925. Beiträge zur Kenntnis des Androeceums und der Samenentwicklung einiger Melastomaceen. Bot. Arch. 9: 398–467.

1434. Zimmermann, A. 1922. Die Klebstoffhaare der Antheren. pp. 110–124 in Die Cucurbitaceen, Vol. 2. G. Fischer, Jena. [Not seen.]

1435. Zimmermann, W. 1957. Phylogenie der Blüte. Phyton (Horn, Austria) 7: 162–182.

1436. Zimmermann, W. 1965. Die Telomtheorie. Fortschr. Evolutionsforch. 1: 1–235. G. Fischer Verlag, Stuttgart.

1437. Zohary, M. & B. Baum. 1965. On the androecium of *Tamarix* flower and its evolutionary trends. Israel J. Bot. 14: 101–111.

SUBJECT INDEX

Appendages

16, 194, 294, 311, 317, 323, 354, 382, 397, 420, 451, 534, 596, 624, 627, 658, 661–5, 689, 691, 739, 767, 789, 892, 915, 959, 972, 1000, 1010, 1019, 1030, 1032, 1081, 1093, 1123, 1138, 1148, 1176, 1213, 1229, 1249, 1258, 1277, 1280, 1314, 1331, 1391, 1400, 1411, 1433.

Crystals

6, 35, 75, 81, 114, 139, 140, 259, 523, 524, 534, 535, 579, 607, 759, 760, 846, 894, 948, 1126, 1128, 1186, 1203, 1285, 1335, 1348, 1366.

Dehiscence

7, 8, 32, 39, 74, 83, 85, 114, 126, 135, 156, 187, 210, 211, 222, 233, 250, 253, 288, 309, 310, 316, 319, 323, 325, 326, 329, 341, 354, 376, 382, 392, 398, 472, 474–6, 481, 523, 529, 533, 598, 611, 620, 654, 675, 767, 768, 795, 835, 843, 859, 892, 894, 922, 931, 1018, 1083, 1124, 1126, 1130, 1135–7, 1143, 1149, 1162, 1191, 1194, 1199–201, 1283, 1287, 1321, 1323, 1334–40, 1342, 1343, 1359, 1362, 1391, 1397, 1408, 1413, 1418, 1433.

Development

9, 14, 34, 36, 44, 45, 58, 61, 63, 64, 66, 69, 70, 73, 75–7, 79, 80, 86, 89, 92–5, 102, 110, 111, 118, 119, 123, 124, 130–2, 142, 143, 146, 149, 184, 185, 198, 204, 205, 207, 213, 221, 233, 240, 243, 247, 273, 277, 284, 285, 298–302, 310, 314, 315, 318, 321, 330–5, 342, 343, 346, 356, 362, 363, 367, 377, 390, 395, 396, 399, 407, 408, 430, 448, 449, 455, 460, 465, 466, 480, 486, 503, 507, 508, 512, 529, 530, 538, 540–2, 545, 554, 568, 578–91, 593, 597–604, 606, 613, 621, 626–8, 633, 641, 642, 656–8, 661–3, 665, 666, 671, 677–9, 682, 698–702, 705–15, 726, 727, 729, 732, 734, 735, 736a, 739, 743, 757, 769, 772, 797, 798, 801, 802, 807, 808, 810, 816, 828–30, 847, 864, 873, 876, 877, 881, 884, 892, 893, 897, 914, 917, 920, 928, 932–35, 937, 942, 943, 951, 952, 957–60,

974, 980, 1005, 1006, 1011, 1029, 1033, 1038, 1039, 1044–47, 1049–60, 1080, 1086, 1089, 1092–5, 1097–100, 1110, 1121, 1122, 1125, 1133, 1144–6, 1153, 1157, 1168–72, 1180, 1183, 1192, 1195, 1197, 1198, 1203, 1207, 1214, 1215, 1224, 1228, 1239, 1242, 1246, 1251–3, 1256, 1263, 1274, 1278, 1279, 1286, 1288, 1289, 1291–7, 1303–7, 1310, 1314, 1315, 1317, 1318, 1324, 1327, 1328, 1330, 1344, 1350, 1353, 1355–7, 1362, 1370, 1374–6, 1381, 1387, 1414, 1421, 1425, 1431, 1433.

Endothecial thickenings

12, 19, 38, 96, 97, 107, 175, 178, 242, 269, 291, 341, 372, 373, 375, 376, 398, 453, 490, 595, 596, 620, 624, 649, 747, 835, 844, 875, 878, 879, 981, 1018, 1023, 1080, 1193, 1225, 1249, 1308, 1326, 1354, 1385, 1389.

Evolution and phylogeny

17, 28, 54, 59, 115, 124, 136, 137, 145, 152, 169, 196, 197, 214, 225, 260, 286, 312–4, 316, 318–21, 323, 328, 329, 354, 377, 380, 381, 383, 415, 533, 534, 558, 619, 670, 697, 704, 710, 716, 747, 775–7, 779–85, 791, 792, 794, 795, 883, 895, 897, 898, 900, 904, 906, 924, 925, 945, 956, 987, 1028, 1031, 1045, 1048, 1054, 1056, 1058, 1147, 1179, 1189, 1195, 1208, 1251, 1272, 1297, 1302, 1319, 1383, 1398, 1400, 1401, 1403, 1411, 1424, 1435, 1436.

General

7, 11, 13, 27, 28, 30, 54, 59, 66, 67, 70, 76, 84, 87, 88, 99, 124–6, 135, 136, 145, 147, 168, 174, 196, 197, 204, 205, 213, 214, 220, 224, 225, 234, 265, 270, 279, 280, 290, 311–3, 319, 320, 328, 331, 342, 346, 379, 380, 383, 394, 398, 402, 404–7, 415, 423, 424, 434, 441–3, 445, 472, 475, 494, 497, 499, 501, 510, 546, 557–9, 570, 576, 606, 611, 617, 640, 648, 649, 657, 670, 674, 683, 688, 697, 703, 704, 716, 725, 736a, 741, 748, 755, 762, 764, 775–86, 791, 800, 802, 813, 816, 831, 861, 869, 883, 895, 898, 900, 904–6, 920, 924–6, 936, 945, 955, 956, 968, 969, 978, 997, 1009, 1011, 1023, 1028, 1031, 1044, 1048, 1056–8, 1094–6, 1102, 1103, 1106, 1107, 1118, 1124–7, 1129, 1130, 1136, 1137, 1141, 1143, 1173, 1179, 1188–90, 1193, 1194, 1201, 1236, 1251, 1279, 1281, 1282, 1288, 1294, 1321, 1327, 1332, 1344, 1358, 1359, 1377, 1378, 1383, 1398, 1399, 1403, 1404, 1408, 1412, 1416, 1424, 1428, 1435, 1436.

Growth movements

133, 157, 340, 358, 360, 361, 404, 463, 492, 510, 552, 634, 672, 809, 822, 939, 1279.

Histochemistry

10, 41, 42, 71, 90–2, 94, 170–3, 201, 202, 216, 240, 245, 268, 355, 366, 386, 410, 454, 456, 484, 485, 523, 524, 611, 616, 620, 622, 640, 799, 805, 810, 813, 825, 845, 848, 849, 858, 877, 917–21, 939, 950, 962, 970, 1015, 1016, 1067–72, 1090, 1115, 1150, 1159, 1166, 1187, 1207, 1247, 1264, 1266, 1342, 1349, 1368.

Homologies

13, 26, 28–30, 36, 67–9, 79, 84, 131, 136, 144, 160, 214, 225, 228, 241, 276, 285, 290, 294, 297, 363, 364, 402, 424, 425, 432–7, 439–45, 488, 508, 557–9, 591, 606, 621, 626, 657, 661, 670, 681, 683, 684, 686, 692, 694, 696, 697, 706, 716, 736a, 764, 765, 776–9, 782, 784, 791, 792, 862, 863, 869, 897, 924, 925, 956, 975, 1002, 1009, 1039, 1057, 1060, 1092, 1096, 1117, 1118, 1158, 1182, 1188, 1236, 1242, 1251, 1254, 1281, 1282, 1314, 1345–7, 1398, 1399, 1412, 1436.

Mutants and sterility

103, 104, 117, 134, 176, 177, 184, 188, 189, 204, 205, 208, 221, 240, 242, 278, 336, 418, 421, 428, 446, 468, 474, 491, 497, 521, 546, 560, 563, 565, 572, 612, 622, 623, 637, 652, 653, 656, 674, 732, 744, 750, 787, 801, 802, 824, 839–41, 854, 860, 866, 874, 876, 880, 896, 920, 931, 941, 951–4, 971, 992, 1075, 1113, 1114, 1116, 1144, 1146, 1163, 1165, 1203, 1204, 1216, 1223, 1233, 1248, 1274, 1286, 1310, 1325, 1396, 1418, 1419, 1426, 1430.

Palaeobotany

214, 218–20, 266, 272, 311, 312, 316, 320, 324, 378–83, 775, 929, 945, 1147, 1173, 1222, 1240, 1399.

Pollination

32, 33, 37, 87, 122, 126–8, 138, 145, 152, 157, 158, 222, 228, 265, 270, 271, 288, 308, 309, 318, 320, 328, 332, 349, 358, 389, 400, 404, 414, 415, 419, 464, 499, 509, 525, 539, 579, 590, 634, 658, 660, 667, 672, 717, 725, 780, 790, 804, 821, 892, 895, 905, 933, 947, 961, 990, 1073, 1079, 1123, 1134, 1138, 1164, 1212, 1232, 1249, 1250, 1277, 1284, 1359, 1360, 1361, 1365, 1373, 1424, 1433, 1434.

Secretory tissue

21, 23, 26, 35, 71, 82, 119, 120, 124, 127, 170, 229–31, 239, 266, 296, 298, 304, 307, 308, 314, 322, 332, 354, 359, 364, 388, 419, 517, 592, 598, 608, 624, 647, 689, 715, 739, 764–

6, 789, 803, 820, 865, 870, 892, 955, 972, 990, 1002, 1006, 1030, 1039, 1041, 1051, 1164, 1179, 1219, 1221, 1254, 1269, 1290, 1295, 1360, 1361, 1408, 1429.

Scanning electron microscopy (SEM)

1, 2, 34, 43, 46, 73, 110, 114, 117, 119, 127, 134, 138, 153, 154, 156, 183–5, 189, 214, 216–8, 221, 226, 239, 243, 247, 262, 264, 267, 272, 273, 287, 288, 302, 306–8, 310, 311, 313–5, 317, 321, 322, 324, 325, 327–9, 331–6, 344, 354, 362, 367, 378, 381, 382, 387, 396, 397, 399, 421, 458, 480, 513, 522–4, 530, 531, 533–6, 538, 539, 541, 545, 563, 578, 586, 603, 611, 620, 631, 656, 661–5, 705, 706, 708, 709, 712, 713, 722, 734, 739, 788, 809, 816, 839, 866, 870, 929, 951–3, 971, 972, 1019, 1029, 1045–7, 1049–53, 1055–8, 1116, 1134, 1144, 1149, 1164, 1183, 1197, 1210, 1212, 1217, 1224, 1234, 1237, 1238, 1240, 1286, 1289–97, 1304–7, 1312, 1317, 1318, 1324, 1362, 1366, 1381, 1384, 1385, 1387, 1420, 1423, 1425.

Staminode

16, 17, 33, 51, 52, 82, 109, 110, 116, 120, 122, 138, 143, 169, 171, 172, 229, 231, 268, 272, 294, 298, 301, 303, 304, 306–8, 311, 317, 320, 387, 400, 401, 417, 419, 425, 458, 534, 535, 538, 540, 599, 600, 601a, 602, 626, 627, 660, 661, 803, 817, 822, 867, 911, 915, 949, 987, 1002, 1006, 1010, 1034, 1037, 1050, 1051, 1081, 1098, 1103, 1117, 1146, 1182, 1203, 1208, 1212, 1228, 1242, 1252, 1273, 1281, 1295, 1331, 1339, 1401.

Tapetum

1–3, 6, 10, 11, 22, 41, 42, 53, 81, 88, 90, 91, 94, 103, 104, 115, 125, 139–42, 147, 148, 150, 155, 161, 168, 174, 181, 182, 195, 198, 200–2, 206, 208, 216, 224, 235, 254, 274, 277, 279–81, 293, 355, 357, 365, 384, 386, 410–2, 418, 421, 426, 427, 453–5, 457, 465, 467, 468, 482, 483, 493–7, 500–2, 504–6, 515, 518, 521, 526, 529, 550, 555, 561, 563, 576, 593, 611, 614, 615, 639, 640, 644, 646, 652, 653, 668, 674, 717, 719, 724, 728, 737, 744, 755–62, 796, 805, 811, 814, 824, 825, 833, 846, 847, 854, 880, 881, 885–91, 893, 900–10, 916, 920, 930, 934–6, 946, 954, 969–71, 979, 1012–4, 1021, 1022, 1027, 1043, 1062–5, 1067–9, 1085, 1091, 1110, 1132, 1138, 1187, 1189, 1196, 1207, 1218, 1234, 1247, 1248, 1262, 1263, 1265–7, 1274, 1276, 1299, 1325, 1327, 1350–2, 1358, 1369, 1371, 1379, 1380, 1388, 1396, 1409, 1414, 1416, 1420, 1430.

Transmission electron microscopy (TEM)

1–3, 5, 6, 10, 41, 42, 53, 57, 71, 88, 93, 102, 104, 114, 150, 155, 161, 181, 182, 186, 187, 189, 193, 195, 198, 202, 206, 208, 216, 217, 254, 281, 293, 361, 365, 366, 386, 409–12, 418, 421, 428, 468, 471, 483, 494, 495, 500–2, 504–6, 515, 516, 521, 523, 526, 611–5, 645, 646, 671, 673, 724, 728, 737, 752, 796, 805, 810–2, 824, 833, 865, 887–91, 896, 901–3, 907–10, 930, 938, 939, 954, 1012–6, 1022, 1035, 1043, 1062–5, 1085, 1091, 1138, 1151, 1176, 1187, 1196, 1207, 1218, 1245, 1247, 1263–6, 1325, 1342, 1360, 1361, 1366, 1369, 1371, 1388, 1389, 1409, 1432.

Teratology

26, 117, 163, 275, 276, 363, 394, 413, 433, 434, 442, 445, 459, 592, 637, 643, 657, 683, 693–5, 800, 826, 871, 911, 965, 975, 983, 1112, 1158, 1300, 1310, 1313, 1386, 1407, 1412.

Vasculature

4, 16, 23–7, 29–31, 35, 36, 44, 45, 47, 49–52, 60, 72, 82, 98, 105, 108–10, 113, 121, 122, 144, 145, 150a, 151, 164–6, 180, 191, 194, 210, 212, 228, 232, 236–8, 245, 252, 256–9, 261–4, 289, 298, 302–4, 314, 345, 364, 367, 369, 370, 374, 393, 430, 433, 462, 477, 487, 488, 498, 511, 519, 520, 532, 535, 542, 547, 548, 553, 556, 564, 574, 575, 590, 598–601a, 605–608, 610, 621, 635, 651, 655, 676, 678–680, 687, 688, 690, 693, 722, 723, 736a, 787, 791–3, 806, 808, 818–20, 823, 831, 834, 838, 842, 850, 851, 853, 855, 856, 897, 914, 915, 927, 944, 967, 968, 973, 976, 977, 980, 982, 986–9, 996, 999–1004, 1008, 1024–6, 1033, 1034, 1037–40, 1045, 1046, 1051, 1052, 1073, 1081, 1082, 1084, 1088, 1101, 1103–6, 1108, 1109, 1111, 1118, 1120, 1126, 1128, 1155, 1156, 1161, 1167, 1174, 1175, 1177, 1182, 1188, 1190, 1195, 1206, 1211, 1220, 1227–31, 1235, 1242–4, 1252–5, 1257–61, 1268, 1288, 1301, 1306, 1307, 1312, 1314–6, 1329, 1356, 1357, 1364, 1392–5, 1398, 1400–3, 1405, 1406, 1410, 1411, 1415, 1428.

SYSTEMATIC INDEX

Dicotyledons

General

85, 87, 99, 115, 124, 125, 136, 145, 196, 197, 214, 234, 260, 312, 316, 318, 319, 323, 328, 329, 331, 346, 359, 378, 380–383, 394, 405, 406, 415, 475, 528, 533, 570, 609, 617, 620, 640, 648, 649, 657, 703, 710, 779, 782, 791, 816, 831, 861, 900, 904, 906, 945, 1005, 1048, 1056–1058, 1095, 1107, 1126, 1127, 1129, 1147, 1179, 1195, 1236, 1270, 1279, 1282, 1288, 1358, 1383, 1412, 1416, 1424.

Acanthaceae

70, 321, 359, 576, 657, 835, 926, 996, 1023, 1103, 1168, 1193, 1277–1279.

Aceraceae
76, 329, 490.

Actinidiaceae
123, 213, 256, 1128, 1184, 1317.

Adoxaceae
1191.

Aizoaceae
60, 377, 460, 517, 540, 827, 1188, 1428.

Alangiaceae
106, 345, 520.

Alseuosmiaceae
329.

Amaranthaceae
143, 294, 377, 573, 941, 1428.

Amborellaceae
311, 323, 325, 817.

Anacardiaceae
76, 329.

Annonaceae
245, 246, 309, 311, 323, 325, 328, 414, 508, 708, 717, 897, 932, 935, 937, 1050, 1056, 1058, 1182, 1202, 1319.

Apiaceae see Umbelliferae

Apocynaceae
15, 65, 69, 111, 216, 217, 244, 293, 354, 470, 544, 657, 1000, 1123, 1424.

Aquifoliaceae
76, 328.

Araliaceae
431, 754.

Aristolochiaceae
267, 323, 325, 532, 709, 743, 1010, 1058.

Asclepiadaceae
42, 65, 244, 318, 328, 639, 657, 658, 669, 742, 970, 999, 1017, 1073, 1123, 1138, 1410.

Asteraceae see Compositae

Austrobaileyaceae
52, 303, 306, 307, 311, 320, 323, 325.

Balanopaceae
329, 514.

Balanophoraceae
350, 351, 1193.

Balsaminaceae
717, 1101, 1160, 1198, 1325, 1362.

Basellaceae
143, 371, 666, 1428.

Bataceae
282, 568.

Begoniaceae
112, 393, 683, 798, 826, 963, 1049, 1058, 1190.

Berberidaceae
66, 119, 130, 243, 315, 325, 359, 366, 463, 508, 584, 809, 835, 1043, 1047, 1058, 1088, 1131, 1243, 1244, 1323.

Betulaceae
4, 298, 489, 490, 514, 533, 548, 1222.

Bignoniaceae
321, 454–456, 652, 653, 688, 717, 844, 847, 849, 867, 911, 996, 1028, 1037, 1070, 1146, 1277, 1279.

Bixaceae
213, 610, 1045, 1337, 1398.

Bombacaceae
236–238, 268, 286, 401, 512, 983, 987, 1040, 1314, 1315, 1398.

Boraginaceae
420, 758, 985, 1283.

Brassicaceae see Cruciferae

Brunelliaceae
260.

Bruniaceae
329.

Burseraceae
329, 851.

Buxaceae
329.

Byblidaceae
329.

Cabombaceae
62, 323, 508, 547.

Cactaceae
359, 463, 713, 1059, 1255–1257, 1363.

Callitrichaceae
1193.

Calycanthaceae
229, 265, 323, 325, 334, 419, 508, 1019, 1182, 1258, 1402.

Calyceraceae
332, 576, 1424.

Campanulaceae
43, 66, 152, 539, 576, 590, 591, 667, 790, 1023, 1190, 1424.

Canellaceae
323, 923, 1405.

Cannabaceae
72, 239, 494, 1193.

Capparaceae
289, 321, 412, 606, 636, 712, 797, 973, 984, 993, 1057, 1211.

Caprifoliaceae
359, 520, 576, 815, 1091, 1379, 1380, 1393–1395.

Caricaceae
1159, 1190.

Caryocaraceae
262.

Caryophyllaceae
29, 41, 143, 231, 285, 377, 478, 494, 642, 657, 683, 731, 734, 735, 765, 766, 771, 875, 913, 1039, 1047, 1058, 1079, 1106, 1146, 1200, 1205, 1214, 1254, 1273, 1428, 1429.

Casuarinaceae
367, 533, 548, 791, 1193.

Cecropiaceae
83.

Celastraceae
76.

Cephalotaceae
329, 1106.

Ceratophyllaceae
62, 323, 325, 1151, 1152.

Cercidiphyllaceae
310, 316, 533, 791, 792, 1231, 1318.

Chenopodiaceae
175, 377, 409, 467, 468, 515, 516, 563, 565, 668, 744, 886, 1058, 1428.

Chloranthaceae
218, 311, 314, 323, 381, 716, 738, 842, 1229.

Chrysobalanaceae
329, 433, 832.

Circaeasteraceae
368, 580.

Cistaceae
463, 512, 553, 634, 1023, 1105.

Clethraceae
39, 607, 699, 1083.

Clusiaceae *see* Guttiferae

Cneoraceae
329.

Cobaeaceae
67, 70, 576, 606, 688, 1280.

Cochlospermaceae
610, 1045, 1398.

Combretaceae
1185, 1270, 1341.

Compositae
18, 19, 21, 92, 130, 233, 235, 242, 269, 291, 292, 295, 328,
496, 521, 566, 576, 589, 595, 596, 624, 625, 635, 667, 678,
718, 788, 845, 848, 878, 879, 909, 938, 939, 946, 964, 970,
1011, 1030, 1032, 1067, 1132, 1148, 1163, 1177, 1178, 1223,
1225, 1233, 1249, 1262, 1287, 1302, 1354, 1385, 1424.

Connaraceae
162.

Convolvulaceae
357, 433, 611, 717, 875, 991, 1330.

Coriariaceae
76, 329, 1106.

Cornaceae
519, 520, 754, 1392.

Corylaceae
4, 489, 490, 514, 533, 548, 688.

Crassulaceae
70, 155, 285, 329, 359, 433, 444, 972, 1068, 1106, 1214,
1260, 1372.

Crossosomataceae
329, 1398.

Cruciferae
26, 77, 117, 159, 163, 171–173, 179, 204, 221, 240, 254, 278,
289, 321, 413, 418, 438, 451, 459, 518, 561, 576, 637, 656,
685, 750, 791, 797, 801, 833, 871, 875, 952–954, 962, 1057,
1058, 1090, 1102, 1144, 1248, 1280, 1328, 1420, 1426.

Cucurbitaceae
57, 65, 98, 100, 101, 164–166, 175, 178, 198, 248, 385, 486,
541, 542, 576, 587, 630, 657, 711, 774, 808, 835, 1020, 1085,
1190, 1192, 1320, 1360, 1434.

Cunoniaceae
82, 257, 258, 260, 329.

Cyrillaceae
212.

Daphniphyllaceae
533, 581.

Datiscaceae
232, 707.

Davidsoniaceae
329.

Degeneriaceae
51, 311, 323, 325, 686, 792, 1228.

Diapensiaceae
1083.

Didiereaceae
1006.

Diegodendraceae
17.

Dilleniaceae
213, 255, 793, 897, 912, 981, 1087, 1188, 1398, 1401.

Dipsacaceae
148, 576, 585, 677, 875, 1007, 1021, 1200.

Dipterocarpaceae
982, 1411.

Droseraceae
717, 1180.

Ebenaceae
1425.

Elaeagnaceae
329, 583, 638, 1270.

Elaeocarpaceae
987, 1314.

Elatinaceae
1106, 1270.

Empetraceae
1083.

Epacridaceae
39, 328, 835, 927, 1083, 1424.

Ericaceae
39, 63, 130, 194, 209–211, 359, 461, 657, 695, 699, 767, 769,
835, 915, 1018, 1083, 1106, 1190, 1214, 1246, 1279, 1421,
1422.

Erythroxylaceae
852.

Escalloniaceae
82, 329, 520.

Eucommiaceae
283, 533.

Eucryphiaceae
259.

Euphorbiaceae
76, 95, 359, 391, 501, 546, 567, 571, 621, 657, 727, 791, 803, 837, 868, 875, 882, 970, 989, 1057, 1069, 1181, 1193, 1279, 1311.

Eupomatiaceae
33, 265, 301, 307, 308, 310, 311, 320, 323, 325, 501, 508.

Eupteleaceae
299, 316, 533, 791, 857.

Fabaceae *see* Leguminosae

Fagaceae
490, 514, 533, 548, 791, 1064.

Flacourtiaceae
328, 330, 1190, 1398.

Garryaceae
520.

Gentianaceae
70, 155, 448, 631, 632, 717, 723, 728, 884, 970, 1132, 1153, 1190, 1198, 1280, 1335, 1414.

Geraniaceae
155, 285, 474, 576, 1047, 1057, 1101, 1200, 1214, 1274.

Gesneriaceae
66, 67, 70, 321, 359, 527, 657, 671, 688, 835, 1103, 1146, 1279, 1374, 1375.

Glaucidiaceae
1235.

Globulariaceae
1279.

Gomortegaceae
120, 323.

Goodeniaceae
150a, 152, 667, 1424.

Greyiaceae
329.

Griseliniaceae
520.

Grossulariaceae
70, 82, 285, 329, 395, 560, 633, 717.

Grubbiaceae
328, 1181.

Guttiferae
70, 512, 608, 698, 710, 1034, 1047, 1051, 1106, 1188, 1398.

Gyrostemonaceae
284.

Haloragaceae
1047, 1270.

Hamamelidaceae
108–110, 298, 300, 302, 316, 317, 324, 328, 380, 533, 1058, 1275, 1323.

Helwingiaceae
520.

Hernandiaceae
323, 328, 647.

Himantandraceae
50, 301, 307, 311, 323, 325, 792.

Hippocastanaceae
76, 329, 576, 683, 1018.

Humiriaceae
852.

Hydnoraceae
203, 1193.

Hydrangeaceae
68, 82, 160, 276, 285, 329, 395, 553, 633, 683, 686, 1200, 1386.

Hydrophyllaceae
46.

Icacinaceae
328.

Idiospermaceae
323, 1402.

Illiciaceae
323, 325, 334, 482, 508, 791, 792, 897, 1029, 1058.

Irvingiaceae
285.

Juglandaceae
380, 480, 490, 514, 533, 548, 733, 749, 791, 1387, 1423.

Labiatae
66, 70, 130, 277, 446, 447, 457, 511, 657, 743, 1103, 1108, 1218, 1278, 1279, 1284, 1427.

Lactoridaceae
323.

Lamiaceae *see* Labiatae

Lardizabalaceae
325, 508, 1058.

Lauraceae
130, 156, 230, 272, 311, 323, 328, 598, 791, 835, 1008, 1042, 1058, 1193, 1240, 1276, 1279, 1312.

Lecythidaceae
512, 702, 818, 821, 961, 1252, 1270.

Leguminosae
9, 29, 48, 102, 134, 139–141, 170, 219, 242, 247, 252, 266, 274, 328, 329, 396, 400, 403, 421, 476, 501, 512, 641, 657, 667, 672, 675, 688, 717, 737, 743, 768, 791, 804, 811, 820, 824, 840, 841, 860, 870, 890, 926, 933, 990, 1004, 1018, 1023, 1085, 1149, 1190, 1204, 1245, 1251, 1293–1297, 1339, 1353, 1369, 1381, 1382, 1389, 1399, 1407, 1418, 1424.

Leitneriaceae
514.

Lentibulariaceae
1278, 1279.

Limnanthaceae
285.

Linaceae
69, 576, 685, 688, 852, 1101, 1214, 1269.

Loasaceae
122, 512, 530, 715, 1190, 1250, 1424.

Loganiaceae
576, 814, 1259.

Loranthaceae
12, 96, 97, 207, 328, 358, 650, 717, 943, 1120, 1121, 1321, 1343, 1365.

Lythraceae
705, 772, 1023, 1053, 1106, 1141, 1270.

Magnoliaceae
144, 145, 180, 200, 245, 320, 323, 325, 333, 334, 482, 525, 686, 688, 791, 792, 835, 897, 1174, 1175, 1200, 1301.

Malpighiaceae
1023.

Malvaceae
12, 286, 359, 463, 512, 576, 791, 987, 995, 1023, 1040, 1104, 1142, 1313–1315, 1398.

Medusagynaceae
261.

Melanophyllaceae
520.

Melastomataceae
127, 551, 657, 691, 768, 1106, 1190, 1219, 1270, 1280, 1331, 1334, 1340, 1400, 1433.

Meliaceae
76, 328, 329, 667, 850, 1424.

Melianthaceae
329, 912.

Menispermaceae
325, 508, 1058, 1119, 1182, 1279.

Molluginaceae
377, 517, 1047, 1155.

Monimiaceae
304, 305, 322, 323, 325, 326, 328, 791, 817, 1081, 1279.

Montiniaceae
329.

Moraceae
72, 73, 389, 533, 654, 791, 1193, 1397.

Moringaceae
966.

Myoporaceae
67.

Myricaceae
380, 514, 548, 736a.

Myristicaceae
34, 35, 323, 836, 1161, 1406, 1424.

Myrothamnaceae
316, 533.

Myrsinaceae
328, 934, 1047, 1092.

Myrtaceae
56, 68, 71, 131, 153, 154, 273, 397, 512, 569, 683, 700, 772, 791, 926, 1048, 1053, 1056, 1129, 1270, 1424.

Nelumbonaceae
61, 62, 323, 452, 1134, 1176.

Neuradaceae
329, 832.

Nyctaginaceae
143, 377, 575, 875, 1041, 1097, 1324, 1428.

Nymphaeaceae
61, 62, 200, 323, 325, 488, 508, 547, 606, 686, 688, 823, 1058, 1064, 1065, 1133, 1200, 1262, 1316.

Ochnaceae
17, 768, 791, 875, 928, 1398.

Olacaceae
162, 1181.

Oleaceae
159, 576, 902, 907, 1376.

Oliniaceae
1270, 1271.

Onagraceae
47, 75, 113, 132, 276, 328, 347, 359, 576, 611, 717, 772, 875–877, 1047, 1063, 1190, 1214, 1270, 1272, 1280.

Oxalidaceae
16, 149, 1101, 1193, 1214, 1299.

Paeoniaceae
68, 121, 507, 508, 710, 752, 793, 835, 897, 1111, 1188, 1370.

Papaveraceae
26, 27, 70, 86, 199, 289, 325, 339, 349, 440, 450, 509, 512, 592, 616, 657, 682, 743, 791, 797, 830, 835, 864, 865, 975, 1049, 1055, 1058, 1084, 1102, 1235, 1424.

Parnassiaceae
23, 24, 82, 329, 633, 1280.

Passifloraceae
70, 576, 967, 1190, 1280, 1390.

Pedaliaceae
175, 178, 996.

Penaeaceae
1270.

Pentaphragmataceae
501.

Penthoraceae
329.

Physenaceae
264.

Phytolaccaceae
143, 377, 1047, 1054, 1057, 1058, 1364, 1367, 1428.

Piperaceae
323, 875, 1061, 1193, 1291.

Pittosporaceae
329.

Plantaginaceae
490, 1018.

Platanaceae
316, 380, 533, 929.

Plumbaginaceae
377, 388, 717, 1047.

Polemoniaceae
477, 576, 706, 729, 1023.

Polygalaceae
206, 564, 667, 806, 875, 1023, 1338, 1424.

Polygonaceae
64, 70, 296, 390, 429, 562, 676, 1057, 1058, 1141, 1329.

Portulacaceae
143, 371, 377, 463, 552, 807, 1047, 1058, 1155, 1428.

Primulaceae
27, 719, 835, 942, 1047, 1060, 1092, 1103, 1109, 1164, 1207, 1224, 1328, 1361.

Proteaceae
329, 462, 605, 667, 743, 926, 988, 1193, 1365, 1424.

Rafflesiaceae
323, 328, 338, 1424.

Ranunculaceae
54, 70, 90, 91, 121, 142, 169, 175, 178, 187, 200, 241, 276, 281, 325, 359, 386, 427, 464, 508, 532, 555, 686, 717, 743, 793, 825, 835, 970, 1011, 1018, 1056, 1058, 1076, 1077, 1106, 1140, 1182, 1188, 1200, 1203, 1235, 1242, 1251, 1268, 1280, 1299.

Resedaceae
492, 1183.

Rhamnaceae
76, 79, 1047.

Rhizophoraceae
328, 578, 579, 717, 1047, 1270, 1424.

Rosaceae
5, 66, 68, 193, 329, 359, 436, 458, 479, 490, 512, 532, 545, 549, 553, 588, 683, 786, 791, 832, 839, 901, 950, 1011, 1018, 1056, 1057, 1158, 1197, 1226, 1279, 1384.

Rubiaceae
22, 129, 190, 328, 476, 576, 667, 1023, 1241, 1424.

Rutaceae
70, 76, 285, 329, 359, 701, 810, 819, 1214, 1261, 1280.

Salicaceae
364, 442, 490, 514, 791, 1221.

Santalaceae
37, 352, 914, 1181, 1193.

Sapindaceae
76, 329, 684.

Sapotaceae
1103.

Saururaceae
323, 325, 722, 977, 1289, 1290, 1292, 1294.

Saxifragaceae
2, 3, 12, 70, 82, 285, 329, 359, 395, 505, 506, 633, 685, 1106, 1214, 1349, 1352.

Schisandraceae
323, 325, 482, 556, 792, 897, 1348.

Scrophulariaceae
27, 32, 66, 67, 70, 107, 167, 204, 321, 359, 439, 503, 531, 586, 593, 597, 657, 688, 728, 736, 739, 743, 762, 873, 949, 970, 1018, 1023, 1103, 1146, 1203, 1212, 1278, 1279, 1322, 1326, 1351.

Scyphostegiaceae
1230.

Scytopetalaceae
286.

Simaroubaceae
76, 285, 329, 685, 853.

Solanaceae
6, 14, 20, 31, 55, 70, 93, 94, 103, 104, 114, 128, 175–179,

226, 227, 242, 251, 287, 355, 481, 484, 485, 491, 522–524, 529, 606, 611, 643, 743, 756, 757, 762, 804, 835, 843, 858, 866, 880, 885, 886, 893, 894, 903, 940, 951, 965, 1011, 1085, 1089, 1112–1116, 1146, 1162, 1165, 1166, 1186, 1232, 1247, 1285, 1286, 1309, 1310, 1366.

Stachyuraceae
1106.

Staphyleaceae
76, 329.

Stegnospermataceae
377.

Sterculiaceae
70, 145, 286, 328, 359, 512, 746, 875, 986, 987, 994, 1047, 1057, 1314, 1315, 1399, 1424.

Strasburgeriaceae
17.

Stylidiaceae
151, 152, 360, 361, 1190.

Styracaceae
263.

Surianaceae
329.

Tamaricaceae
834, 1046, 1056, 1437.

Tetracentraceae
310, 316, 533, 792, 856, 916.

Theaceae
619, 710, 1047, 1298, 1356, 1357.

Theligonaceae
1417.

Theophrastaceae
948, 1047, 1092, 1103.

Thymelaeaceae
14, 131, 384, 487, 680, 1270.

Tiliaceae
133, 286, 463, 500, 502, 504, 512, 576, 987, 1190, 1203, 1314.

Tovariaceae
362.

Trapaceae
690, 1270.

Tremandraceae
192, 768.

Trimeniaceae
311, 323, 325, 327, 817.

Trochodendraceae
310, 316, 533, 618, 792, 856.

Tropaeolaceae
275, 276, 553, 606, 743, 1101, 1190.

Turneraceae
740.

Ulmaceae
72, 191, 222, 533, 576, 717, 1193.

Umbelliferae
1, 38, 70, 161, 179, 335, 336, 426, 466, 576, 743, 773, 971, 1216, 1430, 1431.

Urticaceae
72, 791, 822, 859, 1193.

Valerianaceae
40, 576, 582, 875, 1190.

Verbenaceae
976, 979, 1103, 1132, 1279, 1280.

Violaceae
36, 328, 353, 657, 689, 789, 875, 892, 1023, 1193.

Viscaceae
12, 1121.

Vitaceae
76, 399, 942, 1047.

Vochysiaceae
1424.

Winteraceae
49, 59, 311, 323, 325, 334, 508, 532, 792, 855, 960, 1056, 1080, 1082, 1355.

Zygophyllaceae
76, 285, 329, 685, 838, 1052, 1057.

Monocotyledons
General
85, 87, 99, 115, 124, 125, 136, 196, 197, 234, 328, 331, 398, 405, 406, 443, 475, 570, 617, 640, 648, 649, 748, 831, 861, 900, 904, 906, 945, 955, 956, 1058, 1095, 1107, 1118, 1126, 1127, 1262, 1282, 1288, 1308, 1412, 1416, 1424.

Agavaceae
430, 449, 543, 693, 1118, 1415.

Alismataceae
200, 231, 599, 714, 875, 956, 1064, 1078, 1100, 1110, 1118, 1167, 1169, 1171, 1193, 1262.

Alliaceae
175, 410, 881, 886, 970, 1018, 1022, 1118, 1419.

Alstroemeriaceae
1018.

Amaryllidaceae
29, 276, 398, 435, 611, 692, 1011, 1018, 1118, 1280, 1283, 1413.

Anarthriaceae
748.

Aponogetonaceae
200, 751, 956, 1118, 1150, 1172, 1262.

Araceae
200, 348, 372–375, 422, 550, 576, 681, 683, 761, 908, 926, 1018.

Arecaceae *see* Palmae

Bromeliaceae
70, 200, 343, 344, 504, 1118, 1262.

Burmanniaceae
998.

Butomaceae
70, 200, 600, 601, 714, 956–959, 1078, 1098–1100, 1110, 1118, 1170, 1262.

Cannaceae
606, 626, 628, 629, 660, 1117, 1154, 1251, 1266, 1267, 1424.

Centrolepidaceae
469, 748.

Commelinaceae
10, 200, 349, 387, 398, 683, 759, 760, 763, 770, 796, 846, 1035, 1038, 1074, 1106, 1118, 1187, 1262–1265.

Costaceae
627–629, 686, 1002, 1251.

Cyanastraceae
398, 872.

Cyclanthaceae
473, 1391.

Cyperaceae
58, 78, 105, 146, 150, 200, 745, 748, 1145.

Dioscoreaceae
1118, 1262.

Ecdeiocoleaceae
748.

Eriocaulaceae
748, 1217.

Flagellariaceae
748, 1220.

Gramineae
8, 25, 44, 45, 53, 58, 63, 80, 81, 89, 175, 183–186, 188, 189, 195, 200, 208, 223, 242, 250, 253, 356, 363, 408, 428, 471, 483, 513, 546, 572, 622, 623, 655, 673, 726, 730, 732, 787, 812, 854, 874, 886–889, 891, 896, 910, 917–919, 930, 931, 944, 947, 974, 992, 1018, 1023, 1033, 1062, 1071, 1072, 1075, 1118, 1139, 1156, 1157, 1196, 1215, 1239, 1371, 1388, 1432.

Heliconiaceae
629, 1251.

Hydrocharitaceae
200, 231, 340, 341, 417, 554, 601a, 602, 604, 717, 751, 956, 1078, 1118, 1167, 1193, 1336, 1424.

Hypoxidaceae
1118.

Iridaceae
70, 200, 453, 576, 721, 747, 799, 922, 1036, 1066, 1118.

Joinvilleaceae
748.

Juncaceae
58, 118, 200, 1023, 1118.

Juncaginaceae
200, 751, 956, 1110, 1118.

Lemnaceae
717, 1193.

Lilaeaceae
29, 751.

Liliaceae s.1.
58, 70, 74, 181, 182, 201, 202, 249, 254, 288, 297, 337, 369, 370, 398, 411, 416, 432, 433, 437, 449, 464, 493, 495, 498, 574, 576, 606, 611–615, 644, 679, 686–688, 694, 696, 706, 717, 721, 724, 726, 762, 799, 805, 875, 886, 921, 956, 980, 1012–1016, 1018, 1024–1027, 1086, 1091, 1106, 1118, 1132, 1135, 1199, 1200, 1206, 1213, 1234, 1237, 1238, 1247, 1262, 1280, 1283, 1299, 1368, 1396, 1413, 1424.

Limnocharitaceae
956, 1099, 1100.

Lowiaceae
629.

Marantaceae
626, 628, 629, 660, 1117, 1251, 1424.

Mayacaceae
398, 1106.

Musaceae
629, 1118, 1251.

Orchidaceae
137, 138, 157, 158, 228, 271, 365, 376, 526, 594, 651, 661–665, 862, 863, 1028, 1227, 1345–1347, 1409.

Palmae
116, 392, 577, 1303, 1305–1307.

Pandanaceae
534–538, 1208–1210, 1304.

Philesiaceae
398.

Poaceae *see* Gramineae

Pontederiaceae
398, 1023.

Posidoniaceae
751.

Potamogetonaceae
200, 603, 751, 956, 1093, 1110, 1167, 1262.

Rapateaceae
398, 1106.

Restionaceae
748.

Ruscaceae
1118.

Scheuchzeriaceae
751, 956, 1110.

Smilacaceae
328, 1118.

Strelitziaceae
629, 645, 646, 1251, 1342.

Taccaceae
200, 465, 912.

Tecophilaeaceae
398.

Trilliaceae
69, 720, 753, 1118, 1300.

Typhaceae
200, 828, 829, 1118, 1122, 1262.

Velloziaceae
200, 721, 794, 795.

Xanthorrhoeaceae
398.

Xyridaceae
748, 1424.

Zannichelliaceae
751, 1167, 1193, 1333.

Zingiberaceae
215, 425, 628, 629, 686, 1001–1003, 1117, 1118, 1251, 1253, 1284.

Zosteraceae
751, 956, 1193.

ADDENDUM: Important references seen after preparation of manuscript

Crane, P.R., E.M. Friis & K.R. Pedersen. 1994. Palaeobotanical evidence on the early radiation of magnoliid angiosperms. Pl. Syst. Evol. Suppl. 8: 51–72.

Endress, P.K. 1994. Diversity and Evolutionary Biology of Tropical Flowers. Cambridge University Press, Cambridge, UK. 511 pp.

Endress, P.K. & E.M. Friis (eds.). 1994. Early evolution of flowers. Pl. Syst. Evol. Suppl. 8. Springer Verlag, Wien. 229 pp.

Erbar, C. & P. Leins. 1994. Flowers in Magnoliidae and the origin of flowers in other subclasses of the angiosperms. I. The relationships between flowers of Magnoliidae and Alismatidae. Pl. Syst. Evol. Suppl. 8: 193–208.

Greyson, R.I. 1994. The Development of Flowers. Oxford University Press, New York. 314 pp.

Howell, G.J., A.T. Slater & R.B. Knox. 1993. Secondary pollen presentation in angiosperms and its biological significance. Aust. J. Bot. 41: 417–438.

Ladd, P.G. 1994. Pollen presenters in the flowering plants – form and function. Bot. J. Linn. Soc. 115: 165–195.

Leins, P. & C. Erbar. 1994. Flowers in Magnoliidae and the origin of flowers in other subclasses of the angiosperms. II. The relationships between flowers of Magnoliidae, Dilleniidae, and Caryophyllidae. Pl. Syst. Evol. Suppl. 8: 208–218.

Ronse Decraene, L.P. & E.F. Smets. 1994. Merosity in flowers: definition, origin, and taxonomic significance. Pl. Syst. Evol. 191: 83–104.

Sullivan, V.I., T.C. Pesacreta, J. Durand & K.H. Hasenstein. 1994. A survey of autofluorescent patterns in the staminal connective base epidermis in 60 species of Asteraceae. Amer. J. Bot. 81: 1119–1127.

Tsou, C.-H. 1994. The embryology, reproductive morphology, and systematics of Lecythidaceae. Mem. N.Y. Bot. Gard. 71: 1–110.

Index

abciss, 31–32, 35, 62, 92, 107, 193, 238
aberrant, 5
abiotic, 49
abortion, 129
absorbent, 201
absorption, 196
Acacia, 140, 196, 200–211, 214, 245, 247
Acanthaceae, 94, 103, 155, 198
acarpellate, 114
acetic, 162, 164, 236, 258, 262
acetolysis, 267
aceto–orcein, 265
Achatocarpaceae, 154
acicular, 159, 173, 177
Acnistus, 164
Aconitum, 213
Acoraceae, 105
Acorus, 13, 83, 86, 103, 105, 149
Acrocarpus, 240
acropetal, 247
Acrotriche, 193
Actaea, 213
Actinidia, 197
Actinocalyx, 39, 47
Adansonia, 16
Adema, 15
Adenanthereae, 238
adnation, 15, 19, 37, 86
aerodynamic, 13
Aerosol-OT, 258, 266
Aeschynanthus, 95, 96
Aeschynomene, 236, 238, 244
Afrothismia, 86
Afzelia, 237, 238
Agapanthus, 156

agar, 267
Agavaceae, 153, 156
Airyantha, 237
Aizoaceae, 154
Akebia, 62–63, 80
Albian, 27–28, 30
Albizzia, 237–238, 241
alcian blue, 236
Alismataceae, 105, 147, 153, 157
Alisma, 83, 105, 157
Aloe, 15, 156
Alstroemeriaceae, 153, 157
Amara, 206
Amaranthaceae, 154
Amaryllidaceae, 95–96, 100, 103, 153, 156, 176, 196
Amblystigma, 229
Amborella, 74, 78–80
Amegilla, 207
Amherstia, 95–97, 249
Amherstieae, 238, 240, 245, 249
amino acids, 197
amoeboid, 255
Amorpha, 237
Amphicoma, 194, 213
amphicribral, 12
Anarthriaceae, 137, 154, 158
anastomosis, 138–145, 149–150, 151–158
anatropous, 29
Androceras, 100
Androcymbium, 157
androphore, 66–67, 75, 203, 205, 213–214
Anemone, 194
anemophily, 13, 16, 49
Anemopsis, 74, 206, 213

Angelica, 148
Anigozanthus, 97, 107, 146, 157
aniline blue, 264
animal, 2, 13–14, 16, 47, 69, 95–97, 112, 131, 159, 181, 192, 197–198, 202–203, 206, 215, 264,
animal-pollinated, 14, 192, 197–198, 200, 204, 215
Annonaceae, 65, 95, 199
annuli, 138–139, 141, 143, 150, 153–155, 157–158
Anomochloa, 103
anther guide, 205
anther sac, 221
Anthericaceae, 8, 100, 106, 153, 156, 175
Anthericum, 156, 175
antheridium, 2,
anthesis, 6, 86, 93, 106, 162, 166, 193, 237
Antholyza, 96, 97
Anthonotha, 249
anthophilous, 29, 45, 51
anthophoridae, 207–208, 212
anthophyte, 65–66, 69–74, 76, 87
Anthurium, 105, 174
antibody, 171, 178
antigen, 178
Antirrhinum, 114
ant, 181, 183, 214
anucleate, 197
aperture, 3, 27, 32, 43, 51, 256
apetalous, 37
Apiaceae, 136–137, 148, 155
Apiales, 37
apices, 29, 77, 82, 86, 106–107, 142, 195, 202–203, 205, 208, 211

apicule, 8
Apidae, 196, 212
Apios, 237
Apis, 12, 207, 210, 212
Apocynaceae, 8, 17–19, 155, 222
Apodolirion, 142, 156
Apoidea, 197, 200, 212
apomorphic, 157, 222
apomorphy, 157, 179, 222–223, 231
Aponogetonaceae, 105, 153, 157
Apostasia, 19, 153, 214
Aptian, 27, 28
aquatic, 7, 13, 72, 74, 179
Aquilegia, 202, 213
Arabidopsis, 120
arabinogalactan, 128
Araceae, 18, 83, 86, 95, 103, 105,
    113, 137, 145, 147, 149, 153, 157,
    174, 177, 180, 196, 214
Arales, 86, 153, 157
Araliaceae, 37, 155
Archaeanthus, 112
Archamamelis, 47
archesporial, 9, 10, 119–122, 124,
    126–129, 131, 255
Archidendron, 237
Arcoa, 240, 249
arcuate, 61, 62, 74, 79, 86
Arecaceae, 154, 158, 175
Arecales, 158
Areciflorae, 154, 158
Arenaria, 120, 127–128
Arethusa, 212
Argynnis, 213
Ariflorae, 153, 157
Aristea, 139
Aristolochiaceae, 62, 74, 86, 104, 154
Aristolochia, 74–75
Artemesia, 13, 18
Arthropodium, 106, 195, 202, 211
arthropods, 14, 175, 181
Arum, 174
Asarum, 62–63, 75, 86
Ascarina, 27, 53, 104
ascidiate, 114
Asclepiadaceae, 8, 18–19, 106–107,
    196–197, 200, 204, 221–224,
    231–233
Asclepias, 106, 200, 229
Asparagaceae, 152, 156

Asparagales, 103, 152, 156, 195, 202
Asphodelaceae, 153, 156, 175, 272
Aspidistra, 105
Astephaninae, 229
Asteraceae, 1, 8, 9, 13, 17–19, 95,
    103, 106–107, 136–138, 148, 150,
    156, 193, 260
Asteridae, 1, 15, 18, 30, 47, 49, 155,
    171, 173
Asterogyne, 181
Asteropollis, 27
Atalaya, 139
atavistic, 112
atectate, 27, 32
Ateleia, 237, 239, 243, 247
Atherosperma, 80–81
Atherurus, 174
attractant, 14, 28, 30, 45, 49, 51, 69,
    193–194, 202–203, 206, 215
attraction, 14, 51, 159, 166,
attractive, 28, 45, 51–53, 106, 194–
    195, 249
Augouardia, 240, 249
auricles, 17
auriculate, 32
Austrobaileya, 4, 62, 78, 79, 104, 111,
    113
Austrobaileyaceae, 78
autapomorphy, 113, 145
autofluorescent, 264
automimicry, 211
autumnal, 214
awns, 43, 52, 201, 204, 208, 213
axes, 27, 65, 67, 74, 86
axial, 65
axile, 51, 124
axillary, 3, 114
axis, 4, 5, 8, 61–65, 67, 73, 78, 83,
    86–87, 114, 138, 140, 143, 177
Azolla, 130

Bactris, 199
Baeckea, 202
Balsaminaceae, 95, 159, 174
bamboos, 103
Bambusoideae, 150
Bankzellen, 141
Baphia, 237, 240
Baptisia, 237
Barberetta, 157

barium, 177
Barleria, 8, 16
Barnhartia, 141
Barremian, 45
baseplate, 137–141, 143–158
basic fuchsin, 137, 260, 267
basifixed, 8, 16, 37, 82–83, 85, 106,
    145, 236–237, 239
Basellaceae, 154
bat, 15, 16, 95, 181, 183, 197, 200,
    203
Bauhinia, 95, 145, 237, 238, 239, 240,
    241, 243, 249
BB-4½, 262
beetle, 13–14, 29–30, 45, 49, 51, 53,
    74, 76–79, 87, 107, 111, 175, 181,
    194, 197–200, 202, 204–205, 207–
    208, 212–215
bee, 12, 14, 16, 18, 43, 45, 85, 103,
    162, 166, 173, 174–176, 180–194,
    196–198, 200, 202–203, 205–208,
    210–215, 233, 249, 263
Begoniaceae, 155
Begonia, 178, 211
behavior, 14, 15, 29, 49, 93, 131,
    202, 212
Behnia, 156
Bennettistemon, 67, 69
Bennettitales, 25, 27, 58, 65, 66–67,
    69–71, 76, 87
benzenoid, 196
benzyl benzoate, 262
Berberidaceae, 82
Berberis, 82
Bertiera, 107
bicarpellate, 37, 39, 43
bicolor, 156
Bignoniaceae, 155
bilabiate, 202, 205, 213
bilateral, 13, 43, 202, 213–214
bills (bird), 193, 203, 214
binucleate, 124
biochemistry, 196–197, 201, 215, 255,
    264
biophysical, 93
biotic, 69, 181
biphasic, 122
birds, 14–15, 95, 181, 183, 184, 193,
    198, 214
birefringent, 162, 164, 262

bird-pollination, 183, 214, 238
bisexual, 25, 66, 192, 197–198, 211
bisporangiate, 13, 27, 65, 66–67, 79, 113
bitegmic, 29, 58
Bixa, 6, 15
Bletiinae, 214
Blighia, 139
blueberry, 178, 180
Boerlagiodendron, 214
Boganiidae, 198
Bombacaceae, 16
Bombus, 82, 198, 212–213
boron, 267
Botrantha, 104
Bougainvillea, 16
Bougardia, 147
brachiate stamens, 82
Brachistus, 163
Brachycylix, 249
Brachystegia, 249
Brasenia, 74, 75, 76, 87
breeding system, 127, 129
Brenierea, 240
bristles, 201, 221, 231, 233
Bromeliaceae, 153, 157, 174, 177
Bromeliales, 153, 157
Bromeliiflorae, 153, 157
Browallia, 165
Brownea, 237, 245
Browneopsis, 245
Brunfelsia, 166
Bruniaceae, 155
Brunoniaceae, 193
Bryonia, 194
bryophytes, 2, 10
Bubbia, 82, 86
Buddlejaceae, 155
bugs, 13
Bulbinella, 175
Bulbine, 156, 175, 195, 202
Burkea, 238
Burmannia, 156
Burmanniaceae, 86, 153, 156
Burmanniales, 153, 156
Burnatia, 147, 157
Burseraceae, 155
Bussea, 237
butanol, 236, 258
butanol-paraffin, 258

Butomaceae, 103, 105
Butomus, 103, 105
butterfly, 14–15, 95, 183, 213, 221
Buxales, 29
buzz-pollination, 14, 16, 18, 45, 95, 99–100, 159, 174, 180–184, 193, 196, 198, 202, 206–208, 211–214, 249, 266

Caberla's solution, 260
Cabomba, 74, 75, 104
Cabombaceae, 104
Cactaceae, 154, 205
Caesalpiniaceae, 95–98, 137, 145, 150, 153, 155, 176, 236–240, 245, 247–249
Caesalpinia, 98, 237
Caladenia, 196, 212
Calathea, 158, 193
calcareous, 178
Calceolaria, 202
calcification, 162
calciphilic, 178
calciphobic, 178
calcium, 8, 14, 159, 162, 164, 166, 167, 171, 174, 176–185
calcium chloride, 260, 261
calcium oxalate dihydrate, 177
calcium oxalate dihydride, 180
Caldesia, 105
Calla, 83, 86, 174
Calliandra, 204, 237–238, 242, 245
Callitrichaceae, 143, 147, 149, 155
Callitriche, 13, 95, 149
callose, 138
Calluna, 173
callus, 266
Calopogon, 212
Calpurnea, 237
Caltha, 81, 85
Calycanthaceae, 27, 30, 32, 35, 37, 78, 199
Calycanthus, 14, 32, 35, 78–80, 199, 205
Calyceraceae, 8, 18
calyptrate, 32, 35
Camelia, 18
Campanian, 28, 37, 45, 47
Campanulaceae, 107, 155, 193
Campanulales, 155, 193

Campanula, 18
Campsiandra, 245
canaliculate, 10, 11, 16, 17
canalization, 14
Canavalia, 140
Canellaceae, 60
Cannaceae, 14, 154, 158
Canna, 158
cantharophily, 76, 77
Capparaceae, 95, 155
Capparales, 39, 45
Capparis, 196
Capsicum, 162–164, 168, 171, 177, 182
capsules, 182, 263
carbohydrate, 204, 263
carboniferous, 13
carotenoids, 265
carpellate, 239, 243
carpel, 1–3, 31–32, 35, 37, 41, 43, 51, 58, 111–114, 199, 204, 210, 221, 236–237, 239, 243–247
Caryophyllaceae, 154, 201
Caryophyllidae, 145, 154
Cassia, 140, 145, 153, 176, 237–238, 240–241, 248–249
Cassieae, 198, 206, 236–237, 240, 245, 248–249
Castanospermum, 237–239
Catereitidae, 198
catkins, 13
caudicle, 204, 221–223
Caulophyllum, 82
Cavendishia, 173
Caytoniales, 25
Caytonia, 69–72
cDNA, 265
Cecropia, 8, 16
Celastraceae, 155
cellulose, 95, 138, 258–261, 267
cell and organismal theories, 1, 178–179, 196, 201, 204, 211, 237, 242, 255, 263–265
cell-to-cell, 238
Cenomanian, 28–30, 37, 49
Centaurea, 106
Centranthus, 17
centrifixed, 103
centrifugal development, 5
centripetal development, 5

Centrolepidaceae, 137, 154, 158
Centrosema, 247
Centrospermae, 7
Ceratiola, 173
Ceratonia, 238, 240
Ceratophyllaceae, 104
Ceratophyllum, 1, 8, 13, 72, 74, 95, 103–104
Cerberioideae, 19
Cercideae, 240, 249
Cercidiphyllum, 85
Cercidium, 237
Cercis, 237, 240
Cesalpiniaceae, 197, 206
Cestrum, 16, 166, 184
Cevallia, 63
Chamaecrista, 237–238, 240, 247, 249–250
Chamaedaphne, 201
charcoalification, 30, 45
chasmogamous, 127
chemistry, 19, 214, 266
Chenopodiaceae, 154
Chenopodium, 264
chewing (pollen, tissues), 15, 49, 181
chimeras, 112, 121
Chimonanthes, 79
chloral hydrate, 260–261, 265
Chloranthaceae, 13, 27–29, 32, 47, 49–52, 78, 82, 85, 104
Chloranthistemon, 31–32, 47
chloranthoid, 27–28, 32, 49, 51
Chloranthus, 27, 52, 82, 85, 104, 112
chlorazol black E, 259, 262
chlorine, 177
Chlorophytum, 147, 156, 175
chloroplast, 171, 173
chloroplast gene *rbc*L, 1
chromic acid, 258, 264
chromosome, 2, 113
Chrysobalanaceae, 137, 155
Chrysomelidae, 198
Chrysophylla, 237, 240
Cibirhiza, 222
Cimicifuga, 213
Cinnamomum, 199
Cirsium, 264
Cistaceae, 205
citronellyl, 196
citrus-rose-like, 213

clade, 65, 71, 72, 74, 76, 78, 80, 85, 87, 103, 145, 147, 150–151, 171–180, 224
cladistic, 1, 20, 25–26, 39, 58, 71, 113, 136, 154, 222, 272
cladogram, 58, 69–74, 76–77, 223
Clavatipollenites, 27, 29, 49
cleistogamous, 7, 12, 93, 127, 131
Clematis, 147
Cleome, 16, 95
Clethraceae, 176
Clethra, 176
climate, 30, 77, 179, 193, 267
Clitoria, 238, 244
clove oil, 259, 261–262
Clusiaceae, 39, 43, 45
Clusia, 39, 45
Cnestis, 147
coagulant formula, 258, 263
Cochliostema, 100
cocoon, 49
Coix, 145, 153
Colchicaceae, 95, 153, 157
Coleoptera, 29, 30, 35, 49, 87, 52, 198, 199, 211
collembolan, 181
collenchyma, 261
Colletidae, 207
Collomia, 120, 127
Colophospermum, 245
color, 14, 161, 194, 195, 206, 211, 215, 221, 256, 265
colporate, 51
columellate, 32
Colvillea, 237
Combretaceae, 47, 155
Commelina, 142, 158, 197
Commelinaceae, 99–101, 154, 158, 196–197, 205–206, 211
Commelinales, 100, 154, 158
Commeliniflorae, 154, 158
Conanthera, 100
conduplicate, 67
confocal microscopy, 264
conifers, 2, 4, 13
Connaraceae, 137, 155
connation, 8, 19, 66, 67, 86, 100
connective, 7–12, 15–16, 19, 25, 28–29, 32, 35, 37, 39–40, 43, 45, 47, 49, 51, 58, 59, 61–62, 65, 82–83,

85–87, 92, 103, 106–107, 139, 146, 151, 177, 196, 200–201, 204, 206, 221–222, 231, 236–239, 241, 242, 249, 255
Conospermum, 193
Convallariaceae, 105
Convallaria, 105
Convolulus, 171
coontail, 1
copper, 259
copulation, 14
corollary, 232
corolla, 3, 5, 12, 16–17, 39, 47, 63, 86, 122, 173, 182, 193, 201, 213, 221, 265
Coronaria, 194
corona, 221–222, 231, 233
corpusculum, 203, 221
corpus, 119
cortex, 7, 9, 12, 112
Corymborkis, 146
cosexual, 25, 47, 53
Costaceae, 94, 154, 158
Costus, 94
Couperites, 29
Couroupita, 205, 206, 214
CRAF fixative, 258, 263
cranberry, 178, 180
Crassulaceae, 103, 155
cresyl violet acetate, 260
Cretaceous, 2, 13, 26–30, 37, 45, 47, 49, 51, 53, 66
Crinum, 95, 156, 196
Cronquistiflora, 31–32, 35, 45, 51
cross-pollination, 12, 206, 222
Crotalaria, 237
cryosectioning, 265
Cryptosepalum, 240, 249
crystalline, 177, 182
crystal, 10, 159–168, 171, 173–183, 255, 261, 266–267
Cucurbitaceae, 181, 204
Cucurlionidae, 45
cupule, 31–32, 35, 37, 111, 114
curculionid, 35, 49, 76, 77
Cuscuta, 10, 17
cuticle, 261, 263
Cyanastraceae, 100, 101
Cyanastrum, 100, 101
Cyanella, 100, 101, 156

Cyanotis, 142, 158
Cycadeoidea, 25, 67
cycads, 13
Cyclamen, 8, 16
Cyclocephala, 199
Cymodoceaceae, 13
Cynanchinae, 222
Cynanchum, 222, 224, 227, 229–231, 233
Cynometra, 237
Cyperaceae, 16, 137, 147, 154, 158
Cyperales, 154, 158
Cyperus, 146
Cyphomandra, 9, 164, 166, 171, 182, 196, 198, 205, 215
Cypripediaceae, 153, 204
Cyrtanthus, 156
cytokinesis, 255
cytological, 121, 197, 258, 262
cytoplasm, 121, 197–198

Dactylorhiza, 264
Dalea, 238
DAPI, 266
Darwin, C., 12
Dascillus, 199
dasheen, 177
Datura, 119, 168
deception, 76, 195–196, 206, 249
declinate, 238
dédoublement, 5
Degeneria, 4, 60, 62, 65, 77, 104, 111–112
Degeneriaceae, 65, 104
Deherania, 161, 173, 183, 204, 205–209, 211–212, 214–215, 236–237, 240, 247, 249, 255, 266–267
dehiscence, 7, 11, 17–18, 26, 28, 29, 32, 35, 37, 39, 43, 47, 58, 60, 62–64, 66, 68–69, 76, 79, 80, 82–83, 85–87, 92–93, 95, 100, 102–103, 137, 145, 150, 160–164, 166–168, 174, 176, 179–182, 192–193, 195, 198, 200–202, 204, 205–209, 211–212, 236–237, 240, 247, 249, 255, 266–267
dehydrogenase, 265
Deinbollia, 139
Delonix, 95, 97–98
Delphinium, 213

Delpino, F., 12, 206
Demosthenesia, 93–94
Deprea, 165
dermatitis, 263
desiccating, 3, 93
Desmanthus, 240, 245
Desmodium, 237
Detarieae, 238, 240, 249
Detarium, 237–238, 241
Detrusandra, 32, 35, 45
diadelphous, 8, 238–239, 244
Dialictus, 194
Dialium, 240, 243
diamidino, 266
diandrous, 214
Dianella, 100, 102, 195, 202, 205, 211
dianthoid, 143
dibutyl phthalate, 262
Dicentra, 80, 82
Dichidanthera, 141
dichogamy, 192, 211
Dichopogon, 195, 202, 211
Dichorisandra, 99–101, 158
Dichrostachys, 140, 237–238, 249
dicliny, 29, 211
Dicorynia, 240
Dicoryphinae, 103
Didelotia, 240, 249
Didieriaceae, 154
digital imaging, 256
dihydroflavonol, 265
Dilleniaceae, 65, 70–71, 100, 197, 206–207
Dilleniidae, 30, 39, 45, 47, 49, 70, 155
dimethylsulfoxide, 265
Dimorphandra, 240, 249
dimorphic, 77, 127, 197, 203, 206, 247
dinosaurs, 47
Dioclea, 237
dioecy, 5, 13, 197
Dioscoreaceae, 62, 86, 105, 152, 156
Dioscorea, 105, 156
diplostemony, 5
Dipodium, 205
Dipsacaceae, 156
Dipsacales, 156
Diptera, 29, 49, 51–52, 183, 194, 198, 211

dispersal, 3, 8, 13, 19, 26–30, 47, 49, 53, 69, 174, 193, 232, 267
Distemonanthus, 249
disulcate, 32, 37
DNA, 1, 113, 171, 173, 265
Dodecatheon, 206
dorsifixed, 8, 18, 29, 37, 41, 47, 49, 82, 86, 106, 236, 239
dorsum, 264
Doryanthaceae, 103
Doryanthes, 103
Doryphora, 104
Dracaena, 139, 156
Dracaenaceae, 152, 156
Dracophyllum, 171
Drimiopsis, 156
Drimys, 64, 65, 82, 199
Droseraceae, 155
druse, 159, 162, 168, 173, 175, 176, 177, 179, 180, 181, 182, 183
Duftdrüsen, 183
Dumasia, 140
Dunalia, 11
Dyckia, 157
dynamics, 13, 180
dynastid scarab beetles, 74

Ecdeiocolea, 141, 146, 149
Ecdeiocoleaceae, 137, 147, 149, 154, 158
Echeandia, 8, 100, 202, 205, 211
ecological, 30, 92, 145, 192–193, 202, 211–212, 214, 266,
EDTA, 262
Edwardsia, 238
egg, 2, 3
Eichhornia, 147, 153, 157
ektexine, 32, 43, 47
Elaeis, 175
electron microscopy, 262, 263
electrophoresis, 266
elephant, 213
eliaphor, 203
Elleschodes, 45
Elythranthera, 212
embryogeny, 262
embryology, 261, 255, 264
embryos, 261
Empodium, 146, 156
enantiostyly, 207, 249

endoaperture, 43, 47
endodermal, 261
endogenously, 196, 204
endophyte, 178
endoplasmic reticulum, 177
endosperm, 3, 11, 58
endothecium, 5, 7–11, 13, 16–19, 43,
    74, 78, 95, 100, 103, 121, 136–152,
    156, 161, 165–168, 171, 236, 255,
    264, 272
Englerodendron, 245, 249
Enkianthus, 6, 16, 171, 173
entomophilous, 13, 85
environmentals, 179
Epacridaceae, 100, 173, 184, 193
Epacris, 171
Ephedra, 1, 2, 13, 25, 66, 67, 69, 76,
    113
Ephedripites, 69
Epiblema, 212
Epidendroideae, 145, 150, 197, 212
epidermal, 7, 9–10, 32, 39, 43, 47,
    51, 95, 98, 100, 120, 124, 126, 136,
    159–160, 196, 211, 237, 242
epifluorescence, 266
epigeic, 181
epigynous, 8
Epilobium, 6, 175
Epimedium, 82
epiphytic, 145
epitope, 178
Equisetum, 2
Erica, 171, 173, 179, 201, 213
Ericaceae, 9, 16, 19, 43, 92–95, 100,
    159–160, 168, 171, 173–174, 175,
    177–178, 180, 183, 193, 201, 206,
    212
Ericales, 12, 16, 30, 39, 43, 45, 47,
    49, 51, 71, 173, 175, 185
ericoid, 178, 201
Eriocaulaceae, 137, 154, 158
Eriocaulon, 158
Eriochilus, 212
Erythrina, 95, 140, 237–238, 244
Erythronium, 103
Erythroxylaceae, 155
Eschweilera, 214
Esheandia, 175
esterases, 265
ethanol, 236, 258–259, 261, 263, 266

ethereal, 35, 45
Eucalyptus, 202
Eucnide, 61, 63
Eucomis, 156
eudicots, 72, 82–83, 85–87
eugenol, 196
euglossine, 45, 196–197, 215
Euphorbiaceae, 155, 214
Eupomatia, 35, 45, 77, 104, 199
Eupomatiaceae, 32, 104, 199
Eurypetalum, 237
Eurystemon, 100
eusocial, 212
Eustrephus, 100, 102, 212–214, 224,
    234, 247, 249, 256, 266–267
Exacum, 180
exinase, 181
exine, 28, 32, 35, 43, 47, 51, 69, 181,
    197, 204, 267
Exodeconus, 163
exogenously, 196, 204
exomorphic, 236
Exoneura, 207, 212
exoskeleton, 204
Exospermum, 82, 104
exothecium, 9, 136
exserted, 203
extrorse, 62, 103, 145
extra-floral rewards, 203, 214
exudate, 35, 51, 69, 77, 200
exude, 9, 69
Eysenhardtia, 237

FAA, 161, 236, 258, 261, 263
Fabaceae, 95, 137, 145, 147, 150–151,
    155
Fabales, 143, 147, 150
Fairchildiana, 238, 244
faschia, 18
Fast Green-FCF, 259
fatty acid, 263
fauna, 29, 53
ferns, 2
fertilization, 2–3, 12, 13, 113
fibrous-thickened, 9, 168
fission, 130
fitness, 192
Flagellariaceae, 137, 158
flavanone, 265
flavone 3-hydroxylase, 265

flavonoids, 265
Fleishmannia, 17
flies, 13–15, 29, 45, 53, 95, 173, 183,
    197–198, 202, 205–207, 211–215,
    221
flight, 181
floated, 259
Floscopa, 158
fluorescence, 264, 265
fluorochrome, 266
fluorogenic, 264
flying pollinators, visitors, 16, 19, 182
fly, *see* flies
fly-pollination, 95
Fockea, 222
Fockeae, 222–224, 231–232
Folotsia, 229, 231–233
foragers, 196, 202, 204, 206, 211–212,
    215
foraging, 197–198, 200, 202–207,
    211–213, 221
formaldehyde, 258, 263
formaldehyde-glutaraldehyde, 263
formalin, 236, 258, 259
Fortunearia, 62, 64
fossil, 2, 4, 13–16, 25–30, 32, 35, 37,
    39, 43, 45, 47, 49, 51–53, 66, 69–
    70, 74, 87, 111–113, 267
Fothergilla, 63
Four-and-a-half fluid, 261
foveolate, 28, 32, 47
FPA, 258, 261, 263
fractionation, 82
fragrances, 193
freezing, 258, 263, 265
freeze-substitution, 263
Freycinetia, 92, 214
fritillary, 213
fruit, 3, 27, 76, 32, 43
Fuchsia, 15
Fuertesia, 63
Fumarioideae, 82
fungus, 178, 182–183

Gagea, 103
Galactia, 140
Galanthus, 100, 176
Galbulimima, 4, 62, 77
Galega, 238, 247
gametogenesis, 137

gametophytes, 129, 266
Garcinia, 39
Gasteria, 107
Gastrodia, 211–212
Gaultheria, 171, 201
Geitonoplesium, 100, 102
Genescreen, 265
genetic, 130–131, 178, 180, 255, 264, 267
gene, 1, 5, 114, 118, 128, 130–131, 265
geniculate, 45
genotype, 206
Gentianaceae, 155
gentianales, 155
gene-tracing, 11
Geoblasteae, 214
Geosiridaceae, 138
Geraniaceae, 37, 155
Geraniales, 37, 155
Geranium, 16
germinal, 197
germinate, 197, 199
germination, 3, 223, 233, 267
Gesneriaceae, 17–18, 95–96, 100, 155
gesneriads, 196
Gilbertiodendron, 237–238, 249
Gillettiodendron, 237
Ginkgo, 2, 10, 13, 178
Gladiolus, 15
gland, 39, 43, 49, 51, 144, 173, 196–197, 200–201, 205, 214–215, 236, 238, 242
gleaning, 14, 16–17, 182
Gleditsia, 237, 240
Globba, 94
Gloriosa, 15, 95, 156
Glossodia, 196, 212
Glossopteris, 69–72
glycerol, 137, 161, 182, 258–260, 262, 266
glycolic acid, 180
glycol methacrylate, 265
Gnaphalium, 17
gnats, 182
Gnetalean, 67–69
Gnetales, 25, 58, 65–66, 69–71, 75, 87, 113–114
Gnetum, 1, 10, 13, 25, 66–67, 76, 113
Gochnatia, 17

Goethe, 3, 5, 111
gold–palladium, 236
Gomortegaceae, 78, 80–81, 104
Gomortega, 104
Gonolobeae, 222–224, 232–233
gonophyll, 4
Goodeniaceae, 193
Goodeniales, 193
Goydera, 229
Graboswkia, 162
Gramineae, 9, 16
granular pouches, 159, 164, 171, 176, 201
graphite, 236
grasses, 47, 49
Griffzellen, 141
Guatteria, 199
guide rails, 107, 221, 223–224, 228–233
gum arabic, 260
Gunneraceae, 155
GUS, 265
Gymnocladus, 237
gymnosperms, 2–3, 4, 9–11, 13, 114, 192
gynoecium, 5, 19, 74–75, 94, 103, 112–114, 120, 123, 196, 199, 203–204, 208, 221, 261
gynostegium, 8, 221–224, 227–229, 231, 233
gynostemium, 8, 94, 106, 203–204
gynostrobilus, 1
Gyrocarpus, 78

Haemanthus, 156
Haematoxylon, 237–239
Haemodoraceae, 97, 100, 107, 147, 153, 157
Haemodorales, 153, 157
Haemodorum, 100
Haitingeria, 67–69
Halictidae, 207, 212
halictid, 207–208, 210
Haloragaceae, 155
Hamamelidaceae, 35, 37, 47, 51, 63, 70, 82, 85, 93, 103
hamamelidacean, 40
hamamelid, 10, 27–30, 35, 37, 47, 49, 51, 53, 62, 85, 86
haploid, 1, 2, 3, 266

haplomorphic, 13
Harleyodendron, 237
hawk moth, 16, 95, 183
Hedychium, 158
Hedyosmum, 27, 53
Hedysarum, 238
Helmholtzia, 102
Hemalum, 262
Hemerocallidaceae, 153, 156
Hemerocallis, 15, 156
hemipterans, 69
Henslow, G., 205
herbarium, 137, 161, 176, 222, 255, 262, 266
herbivory, 179, 181–183
herb, 111
hermaphrodite, 246
Hernandiaceae, 78, 80–81
Heteranthera, 100
heteranthery, 99, 100, 197–198, 215, 249
heteroblastic, 130–131
heterochrony, 19, 118–119, 127–132
heterostameny, 246, 248
Heterostemon, 249
heterostyly, 5, 207
Hibbertia, 206–210
Hibiscus, 15–16, 18
Himantandraceae, 32
Himantandra, 4, 111–112
histochemistry, 260, 256, 264
histoclear, 258–259, 261
histogenesis, 114, 126, 127
histomount, 259
histone, 265
Homalictus, 207
homeotic, 118, 131
homology, 1–3, 4, 9–10, 25, 37, 60, 65, 111–113, 114, 118, 139, 150, 177, 249, 272
homoplastic, 70, 80
homoplasy, 9, 70, 80
homozygotic, 266
honey, 12, 162, 174–176, 180–182, 205, 207, 212, 233
Hopkinsia, 147, 153, 154
Hortonia, 81, 104
host-restricted, 178
Houstoniana, 237–238, 242, 245
Houttuynia, 61, 74

hovering, 14–15, 16, 202–203
Humboldtia, 249
hummingbirds, 173, 181, 204
Huttonaea, 139, 144
Hyacinthaceae, 153, 156, 272
Hyacinthus, 175
Hybanthus, 201, 204, 208, 213
hybridization, 265, 266
Hydnoraceae, 95
Hydrangea, 92
Hydrocharitaceae, 153, 157
hydrochloric acid, 162, 164
Hydrocleys, 105
hydrophily, 13
hydrophobic, 200, 263
hygroscopic, 7, 168, 260
hylacoid, 143
Hylaeus, 207
Hylodendron, 240
Hymenocallis, 96
Hymenoptera, 29, 37, 39, 45, 49, 51,
    53, 183
Hymenostegia, 238
hypanthium, 15, 37, 51, 71
hypanthium-bearing, 51
Hypericaceae, 65, 70–71
hypermorphosis, 130
Hyphaene, 158
hypodermal, 9, 10, 11, 119, 121, 159,
    166–168, 171, 176
hypotrophied, 153
Hypoxidaceae, 153, 156
Hypoxis, 156

Icacinaceae, 92, 155
idioblasts, 176–177
Illiciales, 78, 83, 85
Illicium, 114
Immobilon, 265
immunological, 265
Impatiens, 95, 161, 168, 174, 182
inaperturate, 197
incompatibility, 264
Indigofera, 238, 241
indusium, 193
Inga, 238
Ingeae, 245
insect, 12–19, 25–26, 28–29, 39, 45,
    49, 51–53, 70, 159, 174, 180–183,
    193, 196–197, 200–201, 204–207,

211, 214–215, 221–222, 231, 233–
    234, 266
instar, 181
integument, 262
intercalary, 238, 267
interphase, 121, 125
interstaminal, 233
introrse, 43, 62, 95, 103, 145, 201,
    237
Intsia, 238
Inuleae, 17
Iochroma, 183, 184
iodine, 262, 267
Iridaceae, 8, 16, 95–97, 137–138, 143,
    147, 150–151, 153, 157, 202
Iris, 256
Isoetes, 2
isozyme, 265
Ixioideae, 143, 150

Jaquinia, 161
Jeffersonia, 82
Joinvilleaceae, 137, 154, 158
Juncaceae, 154, 158
Juncaginaceae, 105, 153, 157
Jurassic, 66–67, 69

Kadsura, 82, 104
kalancooid, 143
Kalappia, 249
Kalmia, 12, 167, 168, 171, 173–175,
    179, 193, 201, 213
Karimbolea, 224
Keppetipola, 104
Keraudrenia, 106
KI-4 1/2, 262
kiwifruit, 177
Klattia, 150
Kniphofia, 156
Koehler illumination, 256
KOH, 267
Koompassia, 240
Krameriaceae, 137, 155
Krameria, 138–139, 149, 202

labellum, 194, 211–212
labia, 199
Labichea, 240

Lacandonia, 5
lactic acid, 137, 261
lactophenol, 266–267
Lactoridaceae, 74, 104
Lactoris, 104
Lamiaceae, 94, 103, 155, 202
laminar, 4–6, 8, 12, 14–15, 19, 28–
    29, 35, 58, 60–62, 65, 67, 69–70,
    73–79, 87, 111, 113, 231
Lamium, 127–128
Lanaria, 157
Lantana, 16
Lardizabalaceae, 80, 82
larvae, 181, 196–197, 200
Lasioglossum, 198, 212
Lathyrus, 238
latrorse, 28, 49, 62, 95
Lauraceae, 29, 35, 37, 39–60, 63, 70,
    78–81, 92–93, 104, 196–197, 199,
    201
Laurales, 30, 72, 74, 78, 80–81, 85,
    111
Laureliopsis, 80
Laurus, 199
lauryl sulfate, 262
leaf, 9, 26, 28–29, 47, 111–114, 119,
    214
leaves, 2–4, 26–29, 37, 47, 111, 113–
    114, 118, 176, 178, 237, 259
leaf-traps, 113
Lecythidaceae, 197, 203, 205, 206,
    213
Lecythis, 202, 205, 206, 214
Ledebouria, 156
Ledum, 201
Leersia, 139
legumes, 181, 193, 200, 202, 205,
    236–238, 246–247
Leguminanthus, 67, 69
Leguminosae, 13, 18, 176, 249
Lemnaceae, 159, 174, 177, 178
Lemna, 174, 180
Lepianthes, 93–94
Lepidoptera, 52, 78, 181, 183, 198
Leptostemonum, 164, 167, 171, 184
Lethomasites, 27
Leucaena, 237
Leucanthum, 228, 229
Leucomphalos, 141
Leucostegane, 249

lignification, 30, 45, 138, 199, 258, 260, 264
Liliaceae, 16, 103, 153, 157, 159, 163, 175, 177, 181, 194, 201, 272
lilioid, 87, 99, 100, 103, 138, 152–153, 156–157, 175, 196, 205, 212
Lilium, 7, 10, 15, 93, 121–122, 125–126, 141, 144, 157, 264–265
Limnocharitaceae, 105
Limnophyton, 141, 144, 147, 157
limonene, 258–259
Linaria, 202
Lindera, 60, 79
lingulate, 73–74, 78, 87
lipid, 10, 95, 161, 177, 183, 197, 200, 204, 263
liquid-fixed, 258
Litsea, 104
Loasaceae, 63, 99
Lobeliaceae, 8
Lobelia, 18
Lonchocarpus, 140, 237
Loranthaceae, 193, 213
loriid, 193, 197
Loropetalum, 85
Lotus, 238
Ludwigia, 168, 175, 182
Lupinus, 237
Luzuriagaceae, 100, 102
Lycianthes, 18, 167, 171
Lycium, 16
Lycopersicon, 8–9, 18, 100, 163–165, 171, 184, 263, 265
Lycopodium, 2
Lythraceae, 155

Maastrichtian, 47
Macropiper, 104
Macrosperma, 243, 247
Macrosphyra, 17
macrosystematic, 95
Magnolia, 4, 10, 75, 107, 113, 198–199
Magnoliaceae, 28, 62, 65, 74, 85, 112, 199
Magnoliales, 1, 4, 27–30, 32, 35, 39, 45, 49, 51, 62, 70, 72–74, 76–79, 85–87, 92, 103, 106–107, 111, 113, 147, 154, 198

Mahonia, 8, 82
Maingaya, 85
malate dehydrogenase, 265
Malvaceae, 8, 65, 70, 155
malvalean, 45, 65, 71
mammals, 16, 181
mandibles, 199
manganese, 177
Maniltoa, 245
Marantaceae, 154, 158
Maranthaceae, 14
Mariceae, 150
Markea, 183
Marsdenieae, 222, 224, 229, 231–233
marsupials, 193
Martiodendron, 240
Masaridae, 181
massulae, 8
maxillae, 199
Mayacaceae, 100–101
Mayaca, 7, 9, 100–101
Medicago, 238
medifixed, 8, 10, 15–18
megachilid, 200
megagametophyte, 3
megasporangium, 1
megasporocyte, 262
megasporophyll, 112
meiosis, 2, 119, 121–122, 124, 126–127, 129, 162, 261
meiotic, 124, 161
Melampyrum, 202
Melangyna, 207
Melanthiaceae, 103, 105
Melanthiales, 103
Melastomataceae, 100, 160, 176, 197–198, 200–201
Meliaceae, 92
Melianthaceae, 95
Melianthus, 95
Melicytus, 201, 208–209, 213
Meliponinae, 45
Mellittieae, 150, 145
Melstomataceae, 12
Memecylon, 201
Mendoravia, 245
Mentha, 202
Mentzelia, 99
meristematic, 9, 112, 114, 119–121, 126–128, 130–131, 249, 262, 264

Mesozoic, 58, 267
metachromatic, 260–261
metamorphoses, 3, 111
Metastelma, 222, 229
methyl salicylate, 259, 262
methyl violet, 161
microdissections, 260
microgametophyte, 3, 264, 255
Microloma, 231
micromorphology, 28, 39, 43, 47
micropterigid, 52–53, 198
Micropterigidae, 29
micropylar, 68–69
microsporangial, 2, 25, 60, 62, 65, 67, 118–119, 121, 128, 131
microsporangiophores, 113
microsporangium, 2–4, 6, 9, 25, 27–28, 32, 58, 60–67, 69–70, 74, 76, 78, 80, 82, 85–86, 92, 114, 119–123, 127, 129, 131, 249, 255
microspores, 3, 255, 266
microspore, 3, 122, 129, 255, 266
microsporocytes, 121, 122, 124, 129, 130
microsporocyte, 119, 121, 122, 124, 129, 130
microsporogenesis, 128, 272
microsporophyll, 2, 4, 25, 58, 60, 66–75, 87
microstrobilus, 2, 67, 68
microsynangia, 2
mimectic, 206, 211
mimicry, 14, 180, 211
Mimosaceae, 106–107, 137, 145, 147, 150, 155, 205, 238
Mimosa, 238, 242, 245
mimosoid, 19, 205, 214, 236–238, 240, 245, 247, 249
mitochondria, 128, 177
mitosis, 122, 129–130, 131
mitostigma, 221
mitotic, 121–122, 124–127, 129–130
Moldenhauera, 237, 247
Molluginaceae, 154
monadelphy, 8, 238–239, 244
monads, 43, 197
monandrous, 214
Monimiaceae, 78, 80–82, 85, 104
Monimia, 81
monocarpellate, 112–113

Monochoria, 100, 102
monoclonal, 178
monoecious, 5
monophagy, 181
monophyletic, 25, 27, 51, 66, 69, 78, 111, 143, 150, 233
monosulcate, 28–29, 32
monosymmetric, 95, 100, 103
Moquinia, 17
Moraea, 141
Morrenia, 231–232
mosaicism, 27–29, 39, 265
mosquitoes, 181, 183
moth, 14, 16, 29, 53, 95, 183, 198, 221
mountant, 259–260
Mouriri, 9, 14
mouthparts, 17, 37, 49, 198, 202
mRNAs, 265
mucilage, 203–204
multicarpellate, 114
multicellular, 51, 176
multistaminate, 112, 198, 205, 247
multiwhorled, 207
multi-sporangiate, 7
mutant, 118, 130
Mutiseae, 17
mutualism, 25
mycorrhizal, 178
Myricaceae, 60
Myristicaceae, 8, 67, 199
Myristica, 199
Myroxylon, 236, 238–239
Myrtaceae, 155, 197, 200, 202, 205
myrtalean, 7, 37, 43

N-butanol, 258
Najadales, 153, 157
Narthecium, 105
necrosis, 11, 179
nectaries, 9, 15, 30, 37, 43, 45, 47, 49, 51–52, 184, 194, 200–203, 205–206, 208, 213–214
nectar, 16–19, 37, 39, 43, 45, 51, 82, 95–97, 175–176, 180, 182–184, 193–194, 196–202, 204–207, 211–215, 221
nectar guide, 206, 212
nectiferous, 37, 49, 51
Nelumbo, 199

Nelumbonaceae, 199
Nemestrinidae, 198
neoteny, 128
Neptunia, 238, 242, 245–246, 248–249
Netzfasern, 141
neutral red, 196
Nicandra, 166–167, 171
Nierembergia, 8, 166, 168, 202
night blooming, 196
nitidulid, 35, 49, 78, 198
nitrocellulose, 265
Nivenia, 150, 202
Noahdendron, 63
nocturnal flowers, 196
Nomarski, 262
nonaptation, 182, 184
noodle-squeezer, 18
Normapolles, 47
Nothoscordum, 156
nuclear, 122, 177, 197, 266
nucleic, 263, 265–266
nuclei, 2–3, 11, 121, 255, 265–266
Nuphar, 61–62, 74, 76, 79, 85
Nyctaginaceae, 154
Nymphaeaceae, 4, 74–75, 104, 111, 154, 199
Nymphaealean, 63, 72–75, 78, 85–86, 111, 154, 198
Nymphaea, 5, 74, 76, 104
Nytran, 265

obdiplostemony, 5
Ochna, 147, 176
Ochnaceae, 155, 176
Ocotea, 104
Oddoniodendron, 237
odor, 197, 199, 215
oedemerid, 181, 199
Öffnungsgewebe, 160
oilpalm, 175
oils, 9, 14, 35, 45, 184, 196, 198, 202, 215, 258–259, 261–262, 265, 267
oil-collecting, 166
olfactory, 14, 177, 194
oligandry, 198, 205–206
oligolectic, 200
Onagraceae, 95, 175
Oncidiinae, 150
Oncidium, 106
onion, 112

Onobrychis, 238
Ononis, 237
Onoserioides, 17
Onoseris, 17
ontogeny, 79, 111–114, 118–119, 122, 126–131, 160, 200–201, 212, 215, 222, 233, 263, 265–266
OP (oxalate package), 10, 159, 161–169, 170–171, 173–185
Ophiopogon, 105
optical, 106, 256–257, 262–263, 267
optics, 137, 261, 262, 266
Orchidaceae, 18–19, 94–95, 106–107, 137, 143, 145, 148, 150, 153, 197, 204, 212, 214
orchid, 8, 19, 92, 95, 101, 112, 138, 148, 150, 153, 157, 182, 196–197, 204, 211–212, 214, 264
Orchis, 194
organogenesis, 131
organography, 255
Ornithogalum, 156, 175
Orobanche, 211
Oryza, 265
osmium tetroxide, 263, 264
osmophores, 9, 10, 183, 196, 203
os, 209
OT (opening tissue), 160, 163, 176, 180
Ottelia, 142, 157
ovary, 8, 15, 19, 37, 39, 41, 43, 45, 49, 51, 112, 194, 208, 261–262
ovate, 32, 37
oxalate, 8, 10, 14, 159–160, 162–167, 169–171, 174–185
oxalate package, *see* OP
Oxalidaceae, 155, 201
Oxalis, 204, 211
Oxypetalum, 221, 231

Paleoenkianthus, 43, 52
paleoherb, 72–76, 83, 85–87
Paleozoic, 267
palm, 47, 159, 199
Paloveopsis, 249
palynoflora, 27
palynology, 28, 255, 260, 267
palynomorph, 27, 267
Pancratium, 95
Pandanaceae, 92

Papaver, 181, 194, 198, 200, 214
Papaveraceae, 80–82, 154, 194
Papaverales, 87, 154
Paphia, 213
Papilionoideae, 18, 193, 201–202, 236–239, 243–244, 247, 249
Parachidendron, 242
Paradisia, 175
paraffin embedding, 236, 257–259
Paramacrolobium, 249
paraplast, 258
parenchyma, 9, 10, 12, 159, 164, 176
Parietales, 65
Parkia, 240
Parkieae, 150, 238
Parkinsonia, 237
parrots, 193, 197
passerines, 193, 214
Passiflora, 15, 204
Passifloraceae, 95, 155
Patersonia, 141, 144, 147, 150, 153
pathogens, 179
pathway, 180, 183, 214, 266
PAUP, 222
peat, 267
pea plant, 128
Pedaliaceae, 155
Pellegriniodendron, 240, 249
Peltophorum, 237–238
Pentaclethra, 140, 238, 241, 249
Pentatropis, 231, 232
Pentoxylon, 65–66, 69–70
Peperomia, 104
Perarchidendron, 237
perfume, 106
perfume-collecting, 196
perianth, 8, 14–15, 19, 29, 51, 67, 75, 81, 83, 87, 106, 125, 192–196, 202–207, 211, 213–215
periplocoideae, 222, 224, 231–232
permanganate, 262
peroxidases, 265
Persea, 199
Perseanthus, 35
Perseeae, 35
petalodia, 246, 249
petaloid, 14, 32, 205
Petalostemon, 238
Petalostylis, 237, 240, 249

Petunia, 11, 171, 265
Peumus, 81, 85
Phaseolus, 138
phenolics, 260
phenol, 261
phenotypic, 127–128, 130
phenylindole, 266
Philesiaceae, 152, 156
Philydraceae, 102
phloroglucinol, 260
Phlox, 181
Phormiaceae, 100, 102, 196
photographic techniques, 202, 256, 264, 266
photomicrography, 256, 264, 260
Phyllodoce, 173
phyllomes, 4
phyllotaxy, 67–68
Physalis, 8, 16, 171
physaloid, 143
Phytolaccaceae, 154
Pilea, 178
Pinellia, 174
Pinus, 8
Piperaceae, 74, 93, 94, 104, 154
piperalean, 27, 72, 74, 85, 154
Piperales, 27, 72, 75, 85, 154
Piper, 85
Piroconites, 69, 75
pistilode, 32, 35, 209
pistil, 18–19, 37, 41, 43, 49, 112–113, 221 265
piston, 18
Pisum, 237–238
Pittosporaceae, 100
plant breeding, 118, 266
Plantaginaceae, 155
Plantaginales, 155
Plantago, 13, 166, 177–181, 183, 192, 200, 233, 255, 257–258, 260–261, 263–267, 272
plasma, 128
plastochron, 122, 126, 128
Platananthus, 47
platanoid, 28, 47, 51
Platylepis, 144
Pleistocene, 267
pleomorphic, 13
plesiomorphic, 4, 28, 58, 64, 65, 69, 71, 75–77, 85, 153, 173, 223

Pleurandra, 207
Pleurothyrium, 196, 199
Plumbaginaceae, 154
Plumerioideae, 19
Poaceae, 137, 145, 147, 149, 153–154, 158
Poales, 95, 154, 158
Podophyllum, 82
Pogonia, 212
polarised light, 11, 150–151, 154, 161–163, 166, 168, 176, 261, 266
pollenkitt, 47, 196, 204,
pollen-collecting, 100
pollen-eating, 213, 214
pollen-feeding, 181
pollen-foraging, 197
pollen-seeking, 14
pollinarium, 19, 107, 204, 205, 221–223, 231, 233
pollination, 12–18, 25–26, 28–30, 35, 39, 43, 45, 49, 51–62, 69–70, 74, 79, 82–83, 85–87, 95, 99, 103, 111, 127, 166, 173, 180–182, 184, 192–193, 202–208, 210, 221, 233, 249, 255, 256, 266, 272
pollination-by-deceit, 206, 212
pollinator, 12–16, 18–19, 28–29, 37, 39, 45, 49, 51, 53, 69, 77, 85–87, 95, 106–159, 173–174, 179–180, 182–183, 192–194, 196–198, 200, 202–205, 207–208, 211–212, 214–215 249, 266
pollinium, 8, 19, 197, 221–224, 231–233, 266
polyads, 19, 45, 197, 214
polyandry, 198, 203, 205–206, 213, 247
Polygalaceae, 137, 155
Polygonaceae, 154
polymers, 258
polypropylene, 266
Polystemonanthus, 245
polyvinyl lactophenol, 266, 267
Pontederiaceae, 100, 102, 147, 153, 157
Pontederia, 157
Pontederiales, 153, 157
porate, 26, 63, 86, 100, 193, 198, 203, 205, 212–213, 240, 247–248
Porcelia, 95

Portulacaceae, 154
Posidoniaceae, 105
Posidonia, 105
post-anthesis, 161
post-meiotic, 129
Potamogeton, 105, 157
Potamogetonaceae, 105, 153, 157
potassium, 177
potassium iodide, 262
potato, 181
Potentilla, 194
Potomac, 28
PP-4½, 262
PPBB-4½, 262
predation, 3, 77
primordium, 5, 68–69, 71, 79, 113,
    119–120, 123–124, 126–129, 131,
    240, 243, 245, 247–249
prionium, 158
probing, 85, 193, 194, 204, 206, 211
proboscis, 18, 202, 204, 221, 233
prolate, 29
propionic acid, 258
propylene, 263
Prosopis, 106, 200
protandry, 192
Protasparagus, 156
Protea, 214
Proteaceae, 15, 19, 193, 213
protein, 79, 128, 200, 263, 266
protoangiosperm, 198
protoderm, 119
protogyny, 192, 211
protostigmas, 193
pseudannular, 138, 143, 146, 153–155
pseudanthery, 196, 203, 211–212,
    214–215
pseudanthia, 214
Pseudarthria, 140
pseudocopulatory, 212, 214
pseudofruits, 214
pseudomacrolobium, 245
pseudomonadelphy, 238–239, 244
pseudopollen, 211
pseudostipe, 228
pseudo-raphides, 177
Psilotum, 2
Psychopsis, 94
Pterocaulon, 17
Pyrrorhiza, 100

queen bees, 213
Quercus, 8, 16
quinone-imine dyes, 260
Quisqualis, 147

Rafflesia, 92
Rafflesiaceae, 95, 180
Ranunculaceae, 81, 83, 85, 154, 194,
    205, 213
ranunculiid, 35, 82–83, 85, 87, 147,
Ranunculus, 6, 8, 10, 13, 15, 19, 194,
    213
Rapateaceae, 100, 137, 154, 158
Raphidenpollen, 159, 174
raphides, 159, 161, 168, 174–177,
    180–183
*rbc*L, 1, 72, 170–171, 175, 180, 183
Reineckea, 105
reproductive, 1, 4, 25–29, 45, 47, 67,
    69, 85, 87, 112, 122, 127, 130, 192,
    213, 255, 264, 265, 267
resins, 39, 45, 262, 264
resin, 39, 45, 214, 257, 259, 262–
    265
resin-embedded, 260
Resorptionsgewebe, 159
resorption tissue, 159, 171, 176, 179
Restionaceae, 137, 147, 153–154,
    158
Restionales, 95
rewards (pollinator), 9, 14–17, 25, 28,
    39, 45, 49, 51, 77, 87, 183–184,
    193, 195–200, 202–207, 214–215,
    249
Rhamnales, 29
Rhexia, 176
Rhizophoraceae, 155
Rhododendron, 18, 95, 173
Rhodohypoxis, 156
Rhoeas, 194
Rhynchanthera, 100
Richardia, 174
RNA, 265
robbers, 181
Robinia, 238
rodents, 214
Rohdea, 105
Rosaceae, 137, 155
Rosales, 37, 155
Rosa, 6, 15–16, 196

Roscoea, 94, 103
Rosea, 140
rosids, 15, 28–30, 37, 41, 45, 47, 49,
    51, 82, 85, 103, 155
Rothmannia, 93
rRNA, 72
Rubiaceae, 13, 93, 107, 156, 193
Rubiales, 156
Rubricaulis, 104
Ruppia, 157
Ruscus, 67
Rutaceae, 15, 19, 92

Safranin-O, 161, 236, 259, 267
Sagittaria, 61, 63, 83, 105, 147, 157
Salix, 8, 16
Salpiglossis, 164
Salvia, 6, 8, 16, 94, 103, 202
Samanea, 243
Sandersonia, 146, 147, 157
Sanguinaria, 61, 81
Sanguinea, 158
Sanmiguelia, 112
Sansevieria, 156
Santonian, 28, 37, 45, 47
Sapindaceae, 15, 19, 28, 85, 137, 139,
    147, 155
Sapindopsis, 28
Saraca, 240, 243, 249
Sarcandra, 52, 85, 104
Sarcostemma, 27, 229, 231
Sargentodoxa, 82
Sarracenia, 113
Saurauia, 197
Sauromatum, 144
Saururaceae, 63, 74, 104
Saururus, 104
sawflies, 212
Saxifragaceae, 103
saxifragalean, 41, 47
saxifragalean-type, 37
Scadoxus, 156
Scania, 30, 32, 45, 47
scarab, 74, 194, 199, 213
scarabaeids, 77
Scarabinaeidae, 76
scent, 9, 10, 14, 17, 76–77, 106, 177,
    183, 194, 196, 198, 205–206, 211,
    213, 215, 266
scent-gathering, 182

scent-making, 177, 183
Scheuchzeriaceae, 105
Scheuchzeria, 105
Schiekia, 100
Schisandraceae, 82, 104
Schismocarpus, 99
Schizanthus, 10, 11, 16–17, 161, 164–165, 167, 171, 177, 182–183
Schizostephanus, 229, 231
Schwenckia, 164, 166, 178
Scilla, 156
Scilloideae, 175
sclereids, 176
Sclerolobium, 237
Scoliopus, 175
scopal, 208
Scrophulariaceae, 17–18, 155, 213
Scrophulariales, 155
seagrass, 13
Secamoneae, 222, 231–232
Secamonoideae, 222
secrete, 196, 198, 200–201, 214
secretion, 43, 51, 68, 69, 183, 199–202, 204–205
secretory, 107, 196, 206, 208
sectioning, 161, 257–260, 262, 264–265
Sedum, 15
seed, 1–3, 10–12, 19, 25, 32, 35, 37, 47, 51, 58, 65, 69, 71, 76, 82, 123, 181–183, 192
seedlings, 3
self-sterile, 173
SEM (scanning electron microscope), 68–69, 113, 123, 209, 236, 249, 262–266
senecioid, 148
Senecioneae, 151
Senecio, 141, 148, 151
Senna, 237, 246, 248–249
Sepiaria, 98
septal arms, 7, 10, 160, 162, 166–167, 171
septum, 7–11, 67, 92, 159–160, 162–168, 171, 174, 177–178, 180–181, 183
serum adhesive, 259
serum (bovine, horse, swine), 259
short-styled flower, 211
short-tongued bees, 198

Siparuna, 80, 81, 104
siphonogamy, 3
Skytanthus, 8
slugs, 181
Smargadina, 161, 173
smears (cytological), 260, 261, 265
Smilacaceae, 152, 156
Smilax, 156
sodium hydroxide, 262
sodium hypochlorite, 162, 164
sodium salicylate, 259
soils, 178, 179
Solanaceae, 8, 16–17, 19, 99–100, 106, 147, 153, 155, 159–163, 165–166, 168–171, 173, 175–183, 185, 197, 249
solanaceous, 160, 162, 164, 178, 196
Solanales, 145
Solanum, 17–18, 99–100, 162, 164–168, 171, 181, 184, 197–198, 206
Solanum-type, 99–102, 106, 198, 202, 206, 211
Solidago, 18
Sollya, 100
sonication (see buzz-pollination)
Sophora, 237–239, 241
Sophoreae, 237–239, 247
Soraria, 208
spadices, 83, 86, 113, 174
Spanomera, 29
Sparganiaceae, 176
Sparganium, 176
Sparmannia, 176, 182
spathe, 113
Spathicarpa, 95
Spathiphyllum, 174
Spathodoid, 143,
sperm, 1–3, 12
spermaceti, 257
spermatophytes, 3
sphecid wasp, 45
spherical, 28, 32, 35, 43
spikelets, 122
spike, 122
spines, 35
Spiranthoideae, 150
Spirodela, 174, 177, 180
sporangial, 10, 15, 65, 131, 237
sporangiophores, 65
sporangium, 4–7, 32, 37, 39, 47, 58,

60, 64–68, 70, 72, 87, 92, 93, 111, 114, 236–238, 240
sporogenous, 7–9, 11, 93, 119–122, 124–126, 129, 130, 262
sporophyll, 60, 65, 68, 69
sporophyte, 2, 266
sporopollenin, 8, 14, 35, 95, 265, 267
Sprengelia, 173, 184
spur, 6, 16, 19, 43, 51, 201, 203, 207–208, 213
Spurr's resin, 263
squash preparations, 261, 264, 266
staining (tissues), 10, 124, 161, 162, 166, 177, 200, 236, 257, 259–262, 265, 267
stamen-scars, 112
Stapelieae, 200, 222–224, 231–232
starch, 198, 262, 267
Stelis, 182–183
Stemonaceae, 62
Stemona, 86
Stemonocoleus, 240
Stenanthium, 175
Stenodrepanum, 237
Stenomeris, 86
Stephanocolpites, 27
Sterculiaceae, 106
Sternbergia, 100
stigma, 1, 3, 13, 18–19, 35, 37, 39, 41, 43, 45, 47, 49, 51, 69, 80, 82, 86, 112, 183, 193, 196, 200, 203–204, 206–208, 210–211, 215, 264
stigmatic, 32, 35, 51, 183, 192, 196, 199, 204–205, 221, 231
Stockwell's bleach, 260, 262
stomium, 7, 10–11, 58, 60–64, 74–75, 78–83, 85–87, 92, 95, 98, 139, 145, 151, 159, 160, 162–164, 166–168, 171, 176, 179, 250, 255, 265
Storkiella, 240, 245
Strelitzia, 92–95, 98, 158
Strelitziaceae, 94–95, 98, 154, 158
Stromanthe, 141, 158
strontium, 177
Strumariinae, 103
Stuhlmannia, 237
style, 5, 8, 13, 18, 32, 43, 47, 51, 52, 107, 112, 118, 167, 193, 202, 205, 207–208, 215, 221, 233
Stylidiaceae, 204

stylodia, 208, 211
styloids, 177
stylopodium, 37, 41
Stypandra, 195, 202, 211
subhypodermal, 10, 166–167, 171, 173
substomial, 265
sucrose, 267
Sudan, 177
sugar, 200, 211
sulfur, 177
sulfuric acid, 162, 164
sunflowers, 112
Swartzia 243, 246–247
Swartziae, 247
Swertia, 194
Sympetalae, 15, 19
synangia, 25, 65–72, 87
synanthery, 99
syncrescent, 75
synfloresences, 214
syngamy, 2
syrphid, 18, 45, 181, 183, 198, 200, 205–207, 211–214

T-butanol, 258
tabular conformation, 7, 10–11
Tacca, 63–64, 86, 105, 156
Taccaceae, 105, 152, 156
Talauma, 199, 204
Tamarindus, 237–239, 249
Tambourissa, 81–82
tanniniferous, 260, 262
tapetum, 8, 10–12, 93, 121, 124, 162–163, 168, 179, 181, 255, 265, 272
Taraxacum, 17
taro, 177
Tasmannia, 82
Tecophilaeaceae, 100–101, 147, 152–153, 156
tectum, 32, 39
telome, 4, 8, 12, 65, 71
teratology, 112
Ternstroemia, 206
terpenes, 196
terpenoids, 196
tertiary, 47, 49, 75, 236, 258
Tessmannia, 237
Tetracentron, 29, 85
tetrads (pollen), 27, 43, 222–223, 255

Tetrapterocarpon, 240
tetrasporangiate, 25, 27–29, 39, 47, 60, 62–63, 80, 124, 237
Tetrastylis, 95
Thalictrum, 81, 85
thealean, 39, 43, 45, 47, 49, 51
Theales, 39, 49
theca, 5–11, 16, 28, 35, 37, 39, 43, 45, 47, 58–65, 70, 74–76, 78–83, 85–87, 92–93, 95, 100, 103, 106, 114, 151, 160, 163, 193, 198, 203–204, 207, 237
thecal arcs, 7, 10
Thelymitra, 196, 212
Thelymitreae, 212, 214
Theophrastaceae, 159, 173, 177
thieves (of pollen), 211
thionin, 260, 261
thoracic, 198, 207, 212–213
threads, viscin 2–3, 43, 51–52, 95, 98, 168
thrips, 14–15, 29, 52, 173, 175, 181, 198
Thryptomene, 200
Thunbergia, 94, 103, 155
thunbergioid, 143
Thymelaeaceae, 155
Tibouchina, 176
Tigrideae, 150
Tilia, 175–176, 182
Tiliaceae, 175–177, 197
tissue restoring, 257–258, 261
tobacco, 112, 122–124, 126, 129, 265
Tofieldia, 105
toluene, 259, 260
Toluidine blue-O, 124, 260, 265
toxicity, chemical, 180, 183, 258–259
Tradescantia, 158
Trautvetteria, 85
trap-flower, 62, 86
Tremandraceae, 100
Trennungsgewebe, 159–160
Triassic, 66–67, 69, 112
Tricalysia, 107
trichomes, 9, 43, 51, 195–196, 199, 204, 212, 237
Tricoryne, 202
Trifolium, 238
Triglochin, 105, 147, 157
Trilliaceae, 105

Trimeniaceae, 74, 78
Tripogandra, 100
Trisulca, 174
Triticum, 122
Trochodendraceae, 199
Trochodendrales, 29
Trochodendron, 85, 199
Tropaeolaeum, 8, 16
Tubiflorae, 7, 160
Tubocapsicum, 164
Tulbaghia, 142, 156
Tulipa, 103, 194, 265
tunica, 9, 10, 119
tuning fork (vibrating), 161, 174, 180, 206, 266
Turonian, 26–27, 30, 32, 35, 37, 39, 45, 51
Tylophora, 229, 232–233
Typha, 66, 67, 157, 161, 176
Typhaceae, 153, 157, 176
Typhales, 153, 157

U-shaped thickenings, 10, 138–145, 147–158
ultraviolet light (see UV)
Umbellularia, 79, 81
UV (ultraviolet light), 14, 162, 182, 194, 264

Vaccinium, 171, 173, 178, 180, 201, 213
Valerianaceae, 17, 156
Valerianella, 17
Vancouveria, 82
Vassobia, 177
vectors, 12–14, 49, 193, 197, 202, 204, 206, 214
Vellozia, 92, 141, 146, 157
Velloziaceae, 92, 100, 102, 153, 157
Velloziales, 153, 157
Veratrum, 103, 105
Verbenaceae, 155
vertebrates, 193
vescicular-arbuscular mycorrhizae, 178
vibratile (see buzz-pollination)
Viburnum, 15, 16
Vicia, 238
Vinca, 194
Viola, 8, 127, 200–201, 204, 208, 213

Violaceae, 200–201
violet, 161, 194, 215, 260–261, 264
Virginia, 27–28
Viscaceae, 92, 95, 180
viscidium, 203, 204
viscin, 43, 51, 95, 98, 168
viscoelastic filaments, 264
Viscum, 60
Volucella, 18, 198

Wachendorfia, 147, 157
Wagatea, 237
Walleria, 100, 140, 146–147, 153, 156
wasps, 45, 181, 214
water-pollinated, 95
wavelengths, 264
weddelite, 177
weevils, 45, 175
Weltrichia, 67
Welwitschia, 1, 13, 25, 67, 68–69, 76, 113
whewellite, 177

Williamsonia, 67
Williamsoniella, 25, 67
Winteraceae, 27–28, 52, 82, 86, 104, 154, 199
wind-pollinated, 13–14, 18, 47, 49, 53, 69
Withania, 168
Witheringia, 164, 177
Witsenia, 150
Wurmbea, 157

Xanthoceras, 28, 85
Xanthosoma, 95
Xerophyta, 92, 100, 102, 157
Xiphidium, 100
xylem, 12, 261
xylene, 259, 261
Xylocopa, 198
Xymalos, 81
Xyridaceae, 100, 137, 154, 158
Xyris, 7, 9, 158

yeast, 130, 131
Yucca, 156

Zamia, 10
Zamiaceae, 181
Zantedeschia, 174
Zapoteca, 245
Zea, 142, 145, 153
Zebrina, 158, 178
Zenia, 249
Zenobia, 173, 201, 213
Zhakova, 159
Zingiberaceae, 14, 47, 94, 103, 154, 158
Zingiberales, 154, 158
Zirkle's TBA schedule, 258
zoophily, 192–193, 212
Zygadenus, 105, 194
Zygogynum, 18, 82
Zygophyllaceae, 155
zygote, 2